FUNDAMENTALS OF FLUID POWER CONTROL

This exciting new reference text is concerned with fluid power control. It is an ideal reference for the practicing engineer and a textbook for advanced courses in fluid power control. In applications in which large forces and/or torques are required, often with a fast response time, oil-hydraulic control systems are essential. They excel in environmentally difficult applications because the drive part can be designed with no electrical components, and they almost always have a more competitive power–weight ratio than electrically actuated systems. Fluid power systems have the capability to control several parameters, such as pressure, speed, and position, to a high degree of accuracy at high power levels. In practice, there are many exciting challenges facing the fluid power engineer, who now must have a broad skill set.

John Watton entered industry in 1960 working on the design of heat exchangers. He then studied Mechanical Engineering at Cardiff University, obtaining his BSc degree followed by his PhD degree. In 1969, he returned to industry as a Senior Systems Engineer working on the electrohydraulic control of guided pipe-laying machines. Following a period at Huddersfield University, he returned to Cardiff University in 1979 and was appointed Professor of Fluid Power in 1996, receiving his DSc degree in the same year. He was awarded the Institution of Mechanical Engineers Bramah Medal in 1999 and a special award from the Japan Fluid Power Society in 2005, both for outstanding research contributions to fluid power.

Professor Watton has been continually active as a researcher and consultant with industry in the past 40 years. He has worked on components and systems design, manufacturing plant monitoring, and the design of new mobile machines, and he has acted as an Expert Witness on a variety of fluid power issues. He is a Chartered Engineer, a Fellow of the Institution of Mechanical Engineers, and was elected a Fellow of the Royal Academy of Engineering in 2007.

Fundamentals of Fluid Power Control

John Watton, DSc FREng
Cardiff University, School of Engineering

CAMBRIDGE
UNIVERSITY PRESS

CAMBRIDGE UNIVERSITY PRESS
Cambridge, New York, Melbourne, Madrid, Cape Town, Singapore,
São Paulo, Delhi, Dubai, Tokyo

Cambridge University Press
32 Avenue of the Americas, New York, NY 10013-2473, USA

www.cambridge.org
Information on this title: www.cambridge.org/9780521762502

First published 2009

Printed in the United States of America

A catalog record for this publication is available from the British Library.

Library of Congress Cataloging in Publication data

Watton, J., 1944–
Fundamentals of fluid power control / John Watton.
    p.   cm.
Includes bibliographical references and index.
ISBN 978-0-521-76250-2 (hardback)
1. Fluid power technology.   2. Hydraulic control.   3. Component analysis.   I. Title.
TJ843.W383   2009
629.8′042 – dc22   2008054781

ISBN   978-0-521-76250-2 Hardback

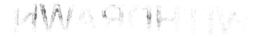

Contents

Preface

This book is aimed at undergraduate students as a second-year and beyond entry stage to fluid power. There is much material that will also appeal to technicians regarding the background to fluid power and the operation of components and systems. Fluid power is often considered a specialist subject but should not be so given that the same would not be said for electrical power. In fact, there are many applications for which fluid power control is the only possibility because of force/torque/power/environmental demands. In the past 20 years, a number of groups around the world have made significant steps forward in both the understanding and the application of theory and control, complementing the R&D activity undertaken within the manufacturing industry. Details of just one organization involving many participating fluid power centers around the world are available at www.fluid.power.net. I embarked on this book ostensibly as a replacement for my first book, *Fluid Power Systems – Modelling, Simulation, Analog and Microcomputer Control*, published by Prentice-Hall in 1989 and now out of print. However, the result is a much different book and perhaps not surprising, given the developments in fluid power in the past 20 years. Following many constructive comments by undergraduate students, friends in industry, and academic friends who still use my first book for teaching, it was clear that a new book was needed. It was felt that a new book should integrate far more fundamental background theory with its application to real components and systems, but without the book becoming research orientated; this is the intention. Validation of theory has been significantly aided by advances in computer modeling of fluid mechanics and system dynamic issues, together with advances in sensors and instrumentation for experimental validation of component and systems performance. These aspects are introduced where appropriate.

Chapter 1 introduces fluid power, indicating its need, circuit symbols, various standard circuits, and associated components. Practical examples of fluid power control are given with the intention of conveying the power-level breadth and application breadth of the subject, varying from precision micrometer position control to primary processing of materials and products. Some common circuit components are presented with their operating concepts, and a further reading list includes textbooks and related industrial literature.

Chapter 2 introduces fluid physical properties for different applications that now must seriously begin to consider the use of less mineral-oil content as both supply and environmental issues begin to dominate many new applications. Fluid bulk modulus issues are presented in some detail, particularly for flexible-hose applications for which its reduction can be dramatic. Fluid cleanliness is also introduced, as is the importance of understanding the effects of cavitation conditions on material erosion. Electrorheological and magnetorheological fluids are now emerging in fluid power applications following many years of awareness, and this is presented for a student racing car suspension real-time control application. A further reading list is included.

Chapter 3 is the first substantial chapter; it discusses the steady-state characteristics of circuit components. It begins with essentials of fluid flow theory and moves on to applications involving restrictors, control gaps, and leakage gaps used in components. Unique solutions are presented where appropriate, with practical data and supporting computation fluid dynamics simulations introduced for the first time. A section on flow-reaction forces is essential and considered in some detail. Developments in servovalves are also briefly discussed and their characteristics analyzed. Positive-displacement pumps and motors are discussed with respect to generic losses and supported by measurements, particularly with respect to efficiency. A section on servovalve behavior is included, together with other control valves and accumulators commonly used in circuits. Finally, the concept of design of experiments is introduced to aid experimental testing to determine performance characteristics. Many worked examples are also included, together with a further reading list.

Chapter 4 is concerned with the steady-state performance of drive systems; it discusses the interconnection of valves, servovalves, pumps, and motors in a variety of configurations. The relatively unknown theory of power transfer units for aircraft applications is discussed and compared with practice in a qualitative sense. This chapter covers graphical and explicit design approaches to understanding steady-state behavior. Several worked examples are also included as well as a further reading list.

Chapter 5, the second substantial chapter, is concerned with system dynamics – that is, time-varying behavior. The philosophy of this chapter is to derive the basic mass flow and force–torque continuity equations, integrate them into typical components and circuits, and then consider solutions to determine the dynamic response of common components and circuits. Linear differential equations are considered, together with frequency response and transfer function concepts. The concept of linearizing equations is introduced to aid analysis when components have nonlinear pressure–flow characteristics such as servovalves. Transmission-line effects are covered in some detail with practical validation. State-space analysis is introduced as a basis for control-theory developments in the next chapter. Finally, an overview of data-based modeling is considered as a means of growing importance when considering the determination of a dynamic model with some knowledge of its probable form. Various methods are introduced, such as the group method of data handling, artificial neural networks, and time-series modeling, with practical validation. Many additional worked examples are also included, together with a further reading list.

Chapter 6 is concerned with controlling fluid power systems and therefore calls on the work of previous chapters. The third substantial chapter, it brings together basic background theory for closed-loop stability, digital control, closed-loop response improvement, and feedback control implementation. The concepts are applied to typical circuits, including the effect of long lines. State feedback is developed for both analog and digital feedback control and extended to include state estimation for state control and linear quadratic control. Again, many examples and additional worked examples are included. On–off switching of valves is then considered as an alternative to conventional control techniques because this is gaining popularity, particularly for high-water-content fluid applications. This part of the chapter is dominated by the practical aspect, but real application results are shown. Finally, an introduction to fuzzy-logic and neural network control is added to whet the appetite for these relatively new approaches for hydraulic systems control. Developing these aspects further is beyond the scope of this book, although some practical results are shown to allow the reader to obtain a feel for the approaches used. Again, a further reading list is included.

Chapter 7 is the final substantial chapter; it consists of just five of the many advanced studies undertaken by me, colleagues, and undergraduate students who have worked with me on a range of applications. The idea here is to develop existing concepts presented in the previous chapters, not to present a collection of research papers but to show a continuing thread of what usually happens in practice. Hence, many aspects of each study are not included but may be taken further from the references given. The first study is concerned with extending hydrostatic pump slipper theory to the case in which the slipper has a groove, rotation, and tilt, the last giving rise to hydrodynamic effects. The second study is concerned with modeling and real control of a forging press cylinder, including both proportional and switched valve systems. The third study is concerned with the modeling and control of a real vehicle wheel active suspension and includes model identification, control by computer simulation, and practical computer control. The fourth study is concerned with the performance of a commercially used car power-steering unit and, in particular, the crucial performance of the power-steering valve. The fifth study is concerned with progress toward intelligent monitoring of pump cylinder pressures using onboard electronics. These five studies embrace theory and practice with practical data to show the effectiveness and limitations of the approaches taken.

John Watton
jwatton@fluidpowerconsultants.com
Llandaff, Cardiff, July 2008

1   Introduction, Applications, and Concepts

1.1 The Need for Fluid Power

In applications for which large forces, torques, or both are required, often with a fast response time, it is inevitable that oil-hydraulic control systems will be called on. They may be used in environmentally difficult applications because the drive part can be designed with no electrical components, and often they are the only feasible means of obtaining the forces required, particularly for linear actuation. A particularly important feature is that they almost always have a more competitive power–weight ratio when compared with electrically actuated systems, and they are the inherent choice for mobile machines and plants. Fluid power systems also have the capability of being able to control several parameters, such as pressure, speed, and position, to a high degree of accuracy and at high power levels. The latest developments are now achieving position control to an accuracy expressed in micrometers and with high-water-content fluids. In practice, there are many exciting challenges facing the fluid power engineer, who now must preferably have skills in several of the following topics:

- Materials selection, water-based fluids, higher working pressures
- Fluid mechanics and thermodynamics studies
- Wear and lubrication
- The use of alternative fluids, given the environmental aspects of mineral oil, together with the extremely important issue of future supplies of mineral oil
- Energy efficiency
- Vibration and noise analysis
- Condition monitoring and fault diagnosis
- Component design, steady-state and dynamic
- Circuit design, steady-state and dynamic
- Machine design and its integrated hydraulics
- Sensor technologies
- Electrical–electromagnetic design
- Computer control techniques
- Signal processing and associated algorithms
- Modern control theory and artificial intelligence

Hydraulic control applications cover a vast range of industries and power levels:

- Ore and mineral extraction, mining, and transportation
- Materials primary processing, steel mills, forging presses
- Product forming and shaping from metal and plastic stock
- Wood processing, paper production
- General production-line machines, injection molding
- General testing machines, test beds, four-poster rigs for vehicle testing
- Bridges, canal-barrage locks
- Transport, road vehicles, rail, shipping, aircraft
- Military vehicles, aerospace
- Mobile machines for construction
- Public services, road cleaning, health, maintenance, elevators
- Leisure, theme parks, wave generators, animation, theater stage control

Figure 1.1 shows a photograph of a hot steel strip finishing mill that forms the final stage of a series of operations involving hydraulic control systems and transforming iron ore to high-quality steel strips. The strip is then either passed on to customers – for example, for vehicle body pressing – or for further processing by means of cold rolling, tinning, or both. Work roll bending (WRB), automatic gauge control (AGC), and work roll shift (WRS) operations are dominated by hydraulic control on different stands. Each of the WRB cylinders and the AGC capsules is controlled by a servovalve–actuator unit, and most of the control systems are reproduced on all the mill stands.

Also shown in Fig. 1.1 is part of a condition monitoring and fault diagnostic system developed by members of the author's research group. Data acquisition is undertaken using National Instruments hardware and Labview software, and an *expert systems* approach significantly aids the fault diagnostic task. The fault diagnostic system developed automatically analyzes the performance of 28 servo-valve systems, indicating when their condition is such that they need to be changed or repaired, thus avoiding mill downtime. The effect of this is to improve cost effectiveness, increase production, improve safety, and ensure customer supply on time.

Figure 1.2 shows a high-torque low-speed motor drive from just two of the many large-scale applications undertaken by Hagglunds, noted for its specialism in this area, among many others. These two applications indicate the need for hydraulic drives in bulk-materials handling and in the chemical-processing industry, and with a level of control sophistication. Other applications for this type of drive include pulp and paper, mining, rubber, recycling, sugar, conveyors, merchant, dryers, and evaporators, and many more.

Figure 1.3 illustrates some other fluid power applications – for example, rock drilling, underground tunneling, component or materials testing with cylinder drives used to create linear motion.

The mobile machine market relies on fluid power for cleaning, loading, lifting, excavating, quarrying, and so on, and with an impressive array of machines, many with multifunctional capabilities and advanced control technologies. For example, Fig. 1.4 shows a machine for cleaning city center buildings, repair work with machines similar to those used in fruit-picking and horticultural areas.

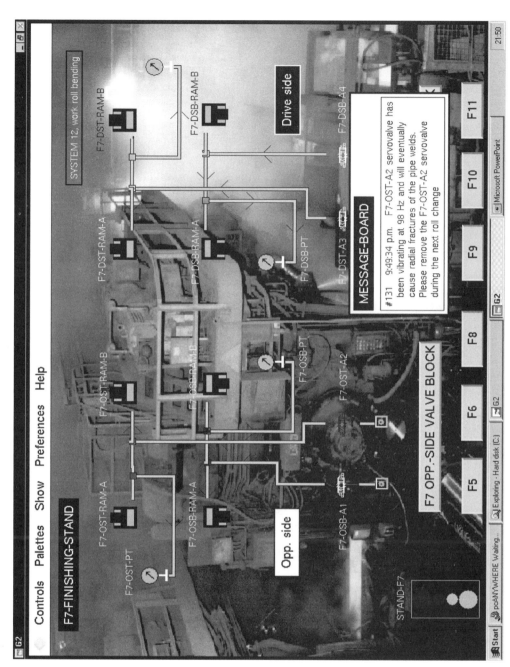

Figure 1.1. Part of a hot steel strip rolling mill (Gale and Watton, 1999).

(a) A bulk-handling application

(b) A chemical-processing application

Figure 1.2. Bulk-material handling and chemical-processing use of Hagglunds high-torque low-speed motor drives (www.hagglunds.com).

Figure 1.5 shows machines for quite different market requirements, one a three-wheeled machine for third-world operations, the other two from a major manufacturer and used for excavation and construction applications.

A feature of the low-cost three-wheeled loader is that both the loading boom and the two front wheel drives are hydraulically operated using load-sensing proportional control-valve technology. The two drive wheels are independently controlled, allowing the machine to turn a tight circle, almost about its own axis, by virtue of the free-wheeling pivoted wheel at the rear.

Considering the specifications of the JCB Ltd. range of mobile machines reveals the innovations in vehicle suspension, power transmission, and fluid power control necessary to ensure continual improvements in machine efficiency, performance, safety, and reliability (www.jcb.com).

Rock-drilling and tunneling applications

Materials–components testing machines

Figure 1.3. Some further examples of hydraulically controlled machines.

The leisure and entertainment industries are increasingly calling on hydraulic control systems such as the three-axis motion ride, a simplification of vehicle testing systems and flight simulators, and the modern interpretation of the fairground Ferris wheel, shown in Fig. 1.6.

The London Eye has four separate drive units, two on each side of the rim, each with four drive wheels operating in pairs that grip beams fixed along each side of the rim's outer frame. In normal operation, all 16 wheels will run in unison, but the system has been designed with sufficient capacity to allow individual pairs of wheels to be retracted, should a problem occur, with no effect on the running speed. The Eye can be run normally with only 12 wheels in operation and can be safely evacuated with as few as 8, though turning at a slightly lower speed. The running

Figure 1.4. A multiaxis mobile machine being used for city center building cleaning.

(a) A low-cost loader co-designed by the author, HR Wright, and Compact Loaders Ltd., the UK manufacturer

(b) A JCB UK Ltd. Fastrac tractor (www.jcb.com)

(c) A JCB UK Ltd. tracked excavator (www.jcb.com)

Figure 1.5. Further examples of mobile machines.

(a) Motion control simulator (b) The London Eye

Figure 1.6. Examples of fluid power for entertainment and leisure purposes.

beam has a high-grip coating, and each pair of wheels is fitted with sensors that increase the drive pressure automatically should any slippage be detected. It is a fairly standard system and is considered very reliable. A high level of redundancy has been built in that should guarantee near-permanent operation. There are two separate hydraulic supply lines, for example, and each drive unit can be isolated and run independently. Should all hydraulic pressure be lost, mechanical brakes have been installed within the hub of each wheel; safety for the passengers and the operating staff is paramount (www.londoneye.com).

Figure 1.7 shows just one of many applications of cylinder drives in the general navigation–maritime–marine area. It illustrates bridge-lifting and the integral lock-gate parallel actuation by means of computer control.

Aerospace also relies on fluid power, not only for testing systems but also for flight controls, as shown in Fig. 1.8. Moog Inc. is a worldwide designer, manufacturer, and integrator of precision-control components and systems. In general, electrohydraulic servovalves are used for primary flight controls, such as aileron, elevator, and rudder actuation. Secondary flight controls include spoilers and air-brake actuation. High-lift devices such as leading- and trailing-edge slats use power supplies with hydraulic motor rotary actuation. In addition, hydraulic auxiliary power units and hydraulic motor control of emergency generators illustrate the crucial importance of fluid power control in aircraft. The advantageous power–weight ratio, relatively benign failure modes, and the pedigree of flight reliability experience may explain why a change to purely electrical power control is many years away, as far as the author can deduce.

(a) One of three Bascule bridges

(b) Control of a pair of lock gates

Figure 1.7. The bridge-lifting system and lock-gate control at the Cardiff Bay Barrage, UK.

1.2 Circuits and Symbols

It is clear from just the few examples shown that fluid power systems can vary significantly in both circuit complexity and operating strategy. However, some basic functional requirements common to all systems are as follows:

- A hydraulic power source – pumps
- A means of distributing the power – steel pipes and flexible hose

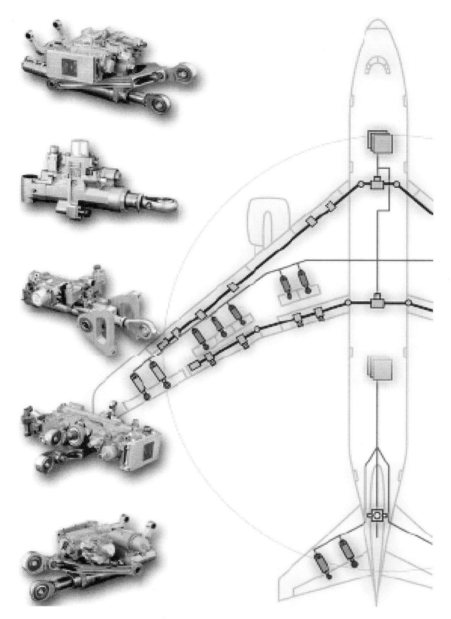

Aircraft wing surface position control

Figure 1.8. Applications in the aerospace industry. Supplied by Moog Inc. (www.moog.com).

- A means of controlling the fluid power – pressure and flow control valves
- A means to provide load actuation – cylinders and motors

Consider a simple circuit, for example, Fig. 1.9, which shows a cylinder and a motor drive circuit illustrating basic system components. The circuit requires a tank with its fluid, a pump, a pressure-relief valve (PRV), a directional control valve, and a cylinder to provide the force to move the load.

Pump (1) draws oil from tank (2), and the pump output line will contain high-pressure filter (3) to prevent dangerous particles from passing into the system and causing damage. PRV (4) is required to set the working pressure and also to

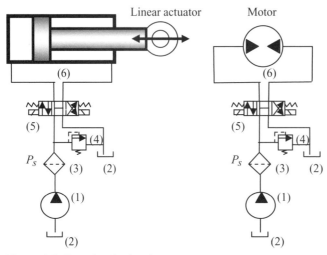

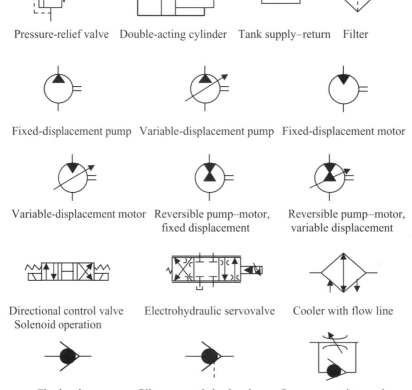

Figure 1.9. Two simple circuits.

Pressure-relief valve   Double-acting cylinder   Tank supply–return   Filter

Fixed-displacement pump   Variable-displacement pump   Fixed-displacement motor

Variable-displacement motor   Reversible pump–motor,   Reversible pump–motor,
                                fixed displacement   variable displacement

Directional control valve   Electrohydraulic servovalve   Cooler with flow line
Solenoid operation

Check valve   Pilot-operated check valve   One-way restrictor valve

Figure 1.10. Some common fluid power component symbols.

protect the system from catastrophic failure should the pump output flow not be required by the load. Operation of directional valve (5), usually by means of an electrical signal, allows fluid to flow in either of two directions, as indicated on the valve symbol. Actuator (6) will move in the appropriate direction, depending on the input signal selected. Notice the use of standard, internationally recognized symbols; see ISO 1219 and ISO 9461. Just a few of the common symbols are shown in Fig. 1.10.

1.3 Pumps and Motors

The starting point for a hydraulic system is its power supply, which is provided by a *positive-displacement pump*. For static applications, this is usually driven by an electric ac induction motor at fixed speed; for mobile applications, it is connected to the power takeoff point at the diesel engine. There are many types of positive-displacement pumps that operate on one of three principles – rotary gear, rotary vane, or piston displacement – and *many are reversible in principle to also act as a motor*. Figure 1.11 shows some common types of gear, vane, and piston-pumping principles of operation.

Gear-type machines tend to be at the lower-cost and lower-power end of the market, with vane types at the medium power level and piston types at the high

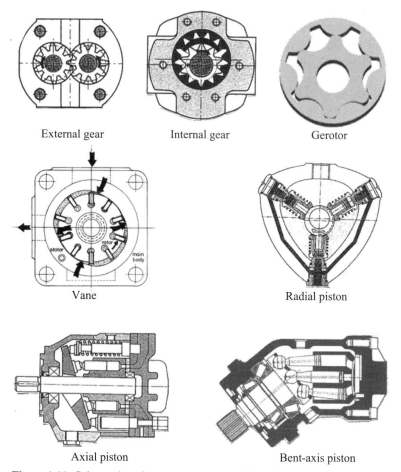

External gear   Internal gear   Gerotor

Vane   Radial piston

Axial piston   Bent-axis piston

Figure 1.11. Schematics of some common positive-displacement machines.

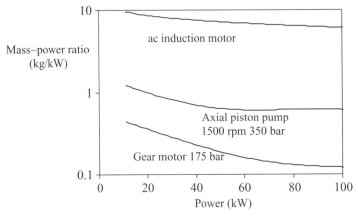

Figure 1.12. Mass–power ratios for electrical and hydraulic drives.

power level, and with pressure now being demanded at the 1000-bar level. Whatever the type selected for the application, one significant feature, as mentioned earlier, is that they almost always have a more competitive weight–power ratio when compared with electrically actuated systems. Figure 1.12 shows a graph of typical mass–power ratios for an electrical motor and an axial piston pump, taken from just two manufacturers' literature. Also shown is a gear motor for comparison as an actuator.

Other hydraulic pumps and motors tend to fall within the range of the pump and motor shown in Fig. 1.12, and dc motors tend to have a better mass–power ratio than ac motors. However, it is clear that hydraulic machines are superior to electrical machines by a factor of typically greater than 10. Figure 1.13 shows a 7.5-kW and a 22-kW hydraulic power supply unit, indicating this size and mass difference between its electric motor and axial piston pump. The 7.5-kW unit shows a small gear pump connected to the drive motor to act as a make-up pump to the circuit. Notice that the larger pump is connected to the electric motor drive by a bell housing, the preferred

Figure 1.13. A 7.5-kW and a 22-kW power supply unit with an ac induction motor drive and a pressure-compensated pump.

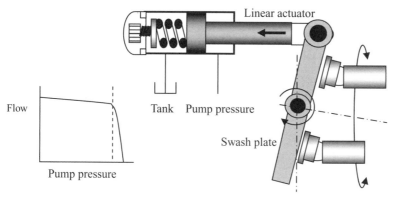

Figure 1.14. Flow off-loading of an axial piston swash-plate pump.

method. It also has rubber mountings between the assembled unit and its locating frame to minimize vibration transmission.

The pump swash-plate angle of an axial piston machine is set by a hydrome-chanical mechanism, usually by means of a spring-loaded actuator, when the load pressure exceeds a set maximum value. The swash plate is brought back toward a zero angle to significantly reduce the flow rate when it is not required, and the pump is referred to as a pressure-compensated, or off-loading, pump. This mechanism is shown schematically in Fig. 1.14.

Figure 1.15 shows a variable swash-plate axial piston pump with an electrohy-draulic spool valve system used to position the swash-plate actuator by means of position feedback; therefore, it has the ability to move the swash plate to any angle within its range. This pump is therefore a true variable-displacement pump.

Figure 1.16 shows a basic control device consisting of a spool valve, force-generating device, and actuator to control the swash plate of an axial piston pump to achieve variable-flow control. The force may be generated by a proportional solenoid or by an electrohydraulic servovalve.

Perhaps *one obvious disadvantage of fluid power is the noise level*, particularly from pumps, although there is a continual drive from manufacturers to reduce noise levels. Some typical measurements for axial piston pumps having a common design are shown in Fig 1.17.

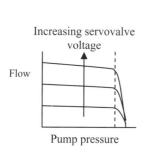

Figure 1.15. Axial piston pump with position control of the swash plate.

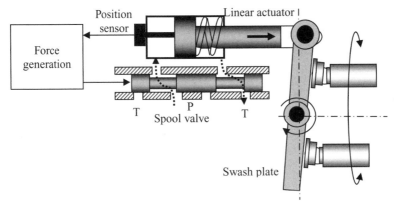

Figure 1.16. Spool-valve–actuator swash-plate control of a piston pump.

It is important to understand how noise level is measured and the way different noise sources are added. The sound-pressure level L is defined for airborne noise (ABN) as follows:

$$L = 10 \log \left(\frac{P}{P_{\text{ref}}} \right) \text{ dB}, \qquad (1.1)$$

where P is the sound-pressure intensity, measured by the sound-level meter by means of a calibrated integral microphone. The reference pressure P_{ref} used for ABN analysis corresponds to the minimum audible sound at 1 kHz and an absolute pressure of 20 µPa, and is designated the 0-dB sound-pressure level. The human ear is able to detect a sound-pressure level up to 200 Pa, and this factor of 10^7 gives rise to a decibel range of 140 dB, the threshold of pain.

Sound-level meters have integrated filters that shape the frequency spectrum of the noise measured before the mean noise level is calculated. This filter is designed to represent the frequency response of the human ear and is known as an A weighted filter. Hence, a measurement obtained from a sound-level meter in this manner is referred to as a dB(A) measurement. *Hearing protection is required, particularly for daily exposure levels above 85 dB(A).* Therefore, if different sound sources are present, the total sound-pressure intensity must first be determined.

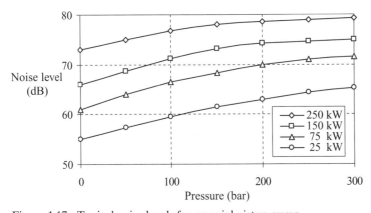

Figure 1.17. Typical noise levels for an axial piston pump.

Example 1.1

A sound-level meter recorded an average background noise level in a machine room of $L_b = 75\,\text{dB}$ with the pump switched off. The pump was then switched on, and the total noise level reading increased to $L_t = 80\,\text{dB}$. Determine the pump noise level L_p:

$$\text{total sound power,} \quad P_t = P_p + P_b, \tag{1.2}$$

$$\frac{P_t}{P_\text{ref}} = \frac{P_p}{P_\text{ref}} + \frac{P_b}{P_\text{ref}},$$

$$10^{L_t/10} = 10^{L_p/10} + 10^{L_b/10}. \tag{1.3}$$

Inserting the measured values gives:

$$10^8 = 10^{L_p/10} + 10^{7.5},$$

$$10^{L_p/10} = 0.684 \times 10^8,$$

$$L_p/10 = 7.835 \rightarrow L_p = 78.4\,\text{dB}. \tag{1.4}$$

Example 1.2

The conditions apply as given in Example 1.1, but a second identical pump is to be added to the hydraulic power supply system in the room. Determine the expected new overall noise level:

$$\frac{P_t}{P_\text{ref}} = \frac{P_p}{P_\text{ref}} + \frac{P_p}{P_\text{ref}} + \frac{P_b}{P_\text{ref}},$$

$$10^{L_t/10} = 2 \times 10^{L_p/10} + 10^{L_b/10}, \tag{1.5}$$

$$10^{L_t/10} = 2 \times 10^{7.84} + 10^{7.5} = 1.684 \times 10^8,$$

$$L_t = 82.3\,\text{dB}.$$

Therefore, the addition of a second pump, having a noise level of 78.4 dB, has increased the room noise level from 80 to 82.3 dB.

The *measured performance characteristics of a pump–motor* are often supplied by the manufacturer in graphical form and are therefore typical of that type and size of machine. One way of presenting the performance characteristic is as shown in Fig. 1.18, which represents an axial piston machine acting as either a pump or a motor.

It can be seen that the effect of real machine losses means that optimum efficiency is located around a preferred pressure setting and speed. This optimum condition can be designed to some extent by the manufacturer for a preferred operating condition. The optimum efficiency condition may also change with machine size, and it is therefore important to select the correct pump size to match the required system performance and also to ensure that the best efficiency is achieved. Note that the conditions for optimum efficiency are not the same when the machine role is reversed. It certainly occurs at a lower pressure when acting as a motor.

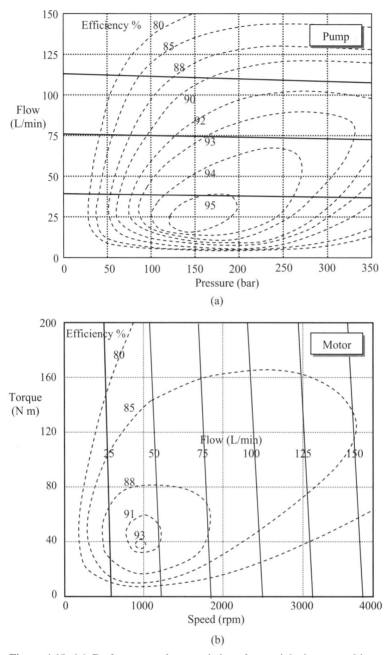

Figure 1.18. (a) Performance characteristics of an axial piston machine acting as a pump. (b) Performance characteristics of an axial piston machine acting as a motor.

1.4 Cylinders

Figure 1.19 shows some features of a cylinder, with many designs available from low-cost self-build to high-cost low-friction servoactuators for precision control applications. Figure 1.19 illustrates a single-rod actuator. Double-rod actuators are also used, particularly for accurate position control applications in which advantages of

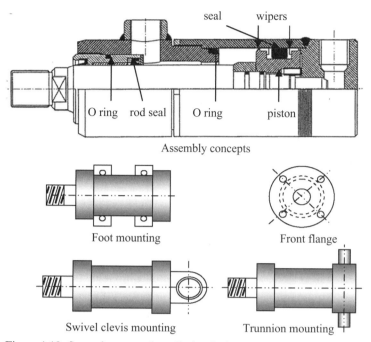

Figure 1.19. Some features of a cylinder design.

similar gains when both extending and retracting become desirable. Piston rods are usually produced in standard sizes and are precision grounded to typically 0.03 mm with a hard chrome deposit of typically 0.25-µm surface finish. Cylinders may be foot mounted or have swivel clevis–trunnion, or flange end connections. Another important feature that may be added is end cushioning, whereby fluid is forced through a variable restrictor as the piston rod moves toward the end of its stroke. This requires rod-end modification, as shown conceptually in Fig. 1.20.

1.5 Valves

The circuit diagrams shown in Fig. 1.9 illustrate a *pressure-relief valve* (*PRV*) at the pump outlet. This should preferably be a two-stage valve because of its improved controllability and stability compared with those of a single-stage PRV. Figure 1.21 shows a schematic of a two-stage PRV. Its main advantage over a single-stage PRV is that the poppet lift is controlled by a weak spring (actually plus a flow-reaction force) compared with the stiff second-stage spring that sets the main pressure. Thus, the main poppet can respond more quickly to flow relief under transient and excessive pressure-increase conditions.

Figure 1.20. Cylinder end-cushioning concept.

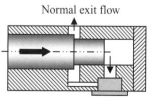

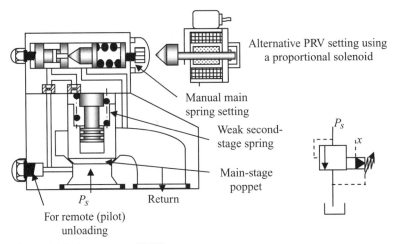

Alternative PRV setting using
a proportional solenoid

Manual main
spring setting

Weak second-
stage spring

Main-stage
poppet

P_s

Return

For remote (pilot)
unloading

Figure 1.21. A two-stage PRV.

The PRV shown in Fig. 1.22 is a manually set valve similar to that shown in Fig. 1.21 but with an additional unloading valve that is actually a directional-control valve. With no voltage applied to the solenoid, the pump flow is diverted through a valve path back to tank with negligible resistance. This type of PRV with pressure off-loading is invaluable, sometimes absolutely necessary, because it allows a pump to be switched on without load, thus avoiding start-up problems with power overloading. Once the pump is run up to speed, the solenoid valve is switched on, the pump flow path is diverted to the load, and the PRV operates in its normal mode.

The circuits in Fig. 1.9 also show a solenoid-operated *directional valve*, which requires a voltage to be applied to the solenoid at the appropriate end of the directional valve, usually 12/24 V dc or 110/240 V ac. A directional valve is therefore a three-state valve, the internal spool that governs the flow paths being moved either to the far left or to the far right by means of an actuating force provided by the

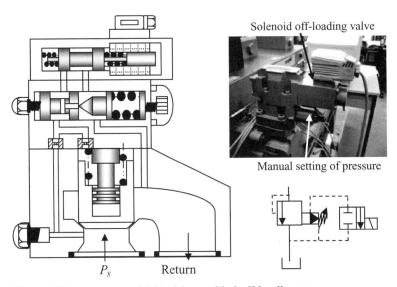

Solenoid off-loading valve

Manual setting of pressure

P_s Return

Figure 1.22. A two-stage PRV with an added off-loading stage.

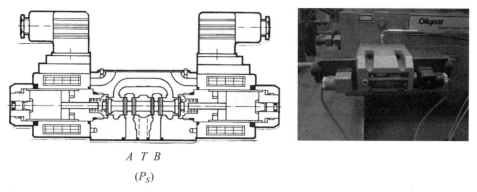

A T B

(P_S)

Figure 1.23. A directional control valve with solenoid actuation.

activated solenoid at the appropriate end. A directional valve is shown in Fig. 1.23. This valve is a five-way valve, although the two internal return lines are connected to a single port to tank. Actuation of the appropriate solenoid rapidly moves the spool either fully left or fully right, thus connecting either $P_s \to A$ with $B \to T$ or $P_s \to B$ with $A \to T$.

There are many porting and spool design combinations. Two common types of spool land design are overlapped lands and underlapped lands; Fig. 1.24 shows just the spool land configurations, with both overlap and underlap being exaggerated for the purpose of explanation. The overlapped spool gives a more positive shutoff when in the central, or neutral, position, but keep in mind that exactly matched ports, or critically lapped ports, could give a small leakage at the neutral position. The underlapped spool clearly places both ports A and B at the same pressure – in this case, supply pressure. Different porting arrangements are possible, just a few of the many possible also shown in Fig. 1.24.

In more advanced systems for which precise and fast control of flow is needed, the directional valve is replaced with a proportional valve, or *electrohydraulic servovalve*, as shown in Fig. 1.25. These valves can vary the flow rate by being able

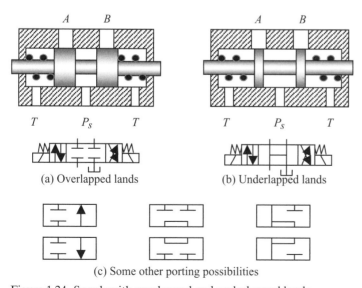

(a) Overlapped lands   (b) Underlapped lands

(c) Some other porting possibilities

Figure 1.24. Spools with overlapped and underlapped lands.

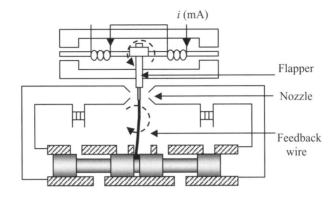

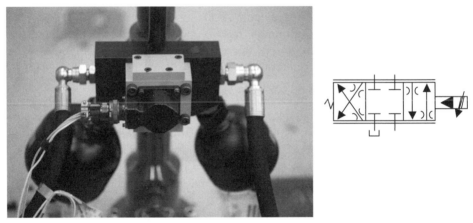

Figure 1.25. An electrohydraulic servovalve connected to a cylinder.

to move the internal spool to precise positions using an internal spool-position-generating stage, the position being proportional to input voltage. The servovalve shown is a common type known as a force-feedback type. A small current applied to the electromagnetic first stage causes the flapper to rotate a very small amount because its armature is sitting between a pair of permanent magnets. The displacement of the flapper between the pair of nozzles creates a pressure differential that moves the spool. The feedback wire provides a torque that balances the electromagnetic torque, and the spool comes to a rest position with its feedback wire virtually centered; the spool displacement is substantially proportional to the applied current at a constant-load pressure differential. As with directional valves, there is a variety of servovalves available that use different techniques to both generate the pressure differential to move the spool and to generate spool-position feedback.

Directional valves and servovalves have spools that move within a housing, or bush, but at the same time do not allow a severe leakage flow loss across them. The spools are machined to a high precision, with servovalves having the severest requirement for radial clearance, typically 2–4 µm compared with directional-valve spools with a clearance of typically 2–20 µm. This requirement, together with that for other internal components within a servovalve, makes them far more expensive than directional valves. For aerospace applications, the cost is further increased because of more rigorous quality and performance checks.

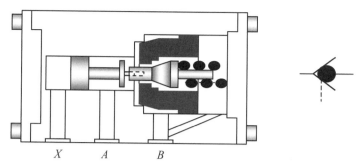

Figure 1.26. A pilot-operated check valve.

Many other control valves exist, again in many forms, such as *check valves*, flow control valves, pressure-reducing valves, and counterbalance valves. A schematic of a *pilot-operated check valve* is shown in Fig. 1.26. These valves allow free flow in one direction and blocked flow in the other direction. The pilot-operated check valves use a pilot pressure to open the valve poppet to allow flow in the normally blocked direction and can be useful, for example, for flow regeneration or sequencing operations.

With no pilot pressure applied at X, free flow occurs from A to B and with an associated small pressure drop set by the initial spring force. This cracking pressure is of the order of 3–7 bar and is generated by movement of the poppet assembly to the left. Flow is blocked from B to A because of closure of the poppet assembly that cannot move to the left. Therefore, with no pilot pressure at X, the unit behaves like a basic check valve. When a pilot pressure is applied at X, there is enforced movement of the poppet assembly to the right and flow can then occur from B to A. The pilot two-poppet assembly improves the decompression effects of the fluid under pressure.

Consider the lifting system in Fig. 1.27. Without check-valve control, a dangerous condition occurs when the load is demanded to be lowered because it will collapse rapidly, the runaway condition. When the directional-valve spool is in its

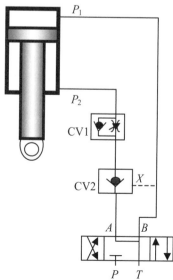

Figure 1.27. Check-valve (CV) control of a lifting–lowering circuit.

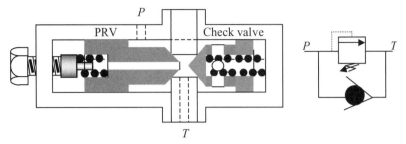

Figure 1.28. A counterbalance valve.

neutral condition as shown, then the pressure P_2 is governed by the load divided by the annulus cross-sectional area. This pressure allows the check valve to be shut off; hence, the load is held and cannot lower. Check-valve CV2 (pilot-operated) closure is aided by the lower side pressure A being set to tank pressure T by the directional valve. When lowering is demanded, the directional valve path is $P \rightarrow B$ and $A \rightarrow T$. The generated pressure P_1 is sufficient to pilot CV2, which allows flow to return by means of the variable-setting flow-control valve in CV1.

Therefore, sudden collapse is avoided, and the lowering speed is set by the flow-control valve. When lifting is required, the directional-valve path is $P \rightarrow A$ and $B \rightarrow T$. The flow-control valve in CV1 is bypassed, and the lifting speed is set by the pump flow rate. It is possible to configure four check valves such that the flow-control valve is used for both lifting and lowering.

Other ways of avoiding sudden collapse is to use either a *counterbalance valve* or an *overcenter valve*. Both are essentially a check valve combined with a PRV, the latter being pilot operated. A counterbalance valve will not open until the preset pressure is reached, which is typically 1.3 times the load pressure generated. For an overcenter valve, a pilot signal of typically 50% of the counterbalance-valve setting is required. Figure 1.28 shows a schematic of a counterbalance valve that is a combination of a check valve and a PRV. The port connection P is at the cylinder's lower point. Therefore, when the directional valve is in its neutral condition and the load is stationary, the static pressure induced at P, because of the load mass, is insufficient to open the PRV because it has been set 30% higher than the static pressure. In addition, the check valve is driven to its shutoff position.

When the directional valve is opened to lower the load, then the annulus pressure builds up and opens the PRV. When the directional valve is reversed to raise the load, then there is free flow by means of the check valve. Obviously, for lowering, the annulus pressure is set to the PRV pressure and greater than that without the counterbalance valve. Therefore, the upper, bore, pressure is also higher than that without the counterbalance valve and more power is needed. This pressure increase can be reduced if an overcenter valve is used with its pilot pressure signal taken from the upper, main bore, side of the cylinder. An overcenter valve is shown in Fig. 1.29.

A relatively low pilot signal is used to move the PRV poppet, thereby setting the line pressure to tank pressure or a low pressure, as determined by the pressure drop across the directional valve. Note that if the load attempts to run away, then the pilot pressure X will drop, thus bringing the counterbalance feature back into action. This will decelerate the load and is highly desirable. Figure 1.30 shows a mobile machine left in a suspended condition with no operator control.

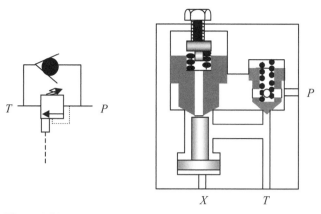

Figure 1.29. An overcenter valve.

Flow-control valves are used to set a desired flow rate that can be maintained reasonably accurately under variable load pressure conditions. This is achieved by pressure compensation, as shown in the two forms in Fig. 1.31.

The required flow rate is set by a manual adjuster that controls the opening of the metering orifice. Flow through the variable orifice produces a pressure drop $P_1 - P_2$, which acts across the spool unit. Any tendency to exceed the flow setting therefore generates an increased force that moves the spool unit to the left, closing the metering orifice and resulting in the required reduction in flow rate. These valves do not achieve the desired constant-flow-rate control until a small pressure drop across the valve is generated, a typical characteristic being shown as Fig. 1.32.

Flow-rate control can be meter-in or meter-out, as shown in Fig. 1.33. In each case, the pressures on either side of the actuator and, hence, the directional valve controlling the system will be different. Note also that meter-in flow control will not prevent runaway when the actuator is extending.

Cartridge-valve technology is based around screw-in units and has made rapid advances and can fulfill most industrial applications. A vast range of products is

Figure 1.30. A mobile machine in a suspended condition using a pilot-operated check valve.

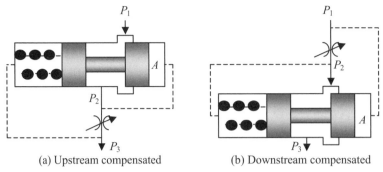

(a) Upstream compensated   (b) Downstream compensated

Figure 1.31. A pressure-compensated flow-control valve.

provided by the Sun Hydraulics Corporation (www.sunhydraulics.com). This company manufacturers hydraulic cartridge valves, load-control valves, counterbalance valves, pressure-control valves, flow-control valves, solenoid directional-control valves, sandwich valves, manifold blocks, relief valves, and proportional valves. Products are now suitable to operate at flows of up to 800 L/min with pressures up to 350 bar. Cartridge valves are essentially screw-in components assembled in either an integral manifold block or a customized manifold "valvepak." Once a customer's hydraulic circuit has been developed, it is incorporated into a single, custom manifold designed to fit into a defined location. Benefits to customers using Sun's valvepak solutions include order simplification, reduced assembly time and cost, and consolidation of the hydraulic control system. All manifolds can be manufactured in either T-6061 aluminum, 210 bar, or 65–45–12 high-strength ductile iron, 350 bar. Sun's process is different from that of most other custom manifold suppliers, in part because of the extensive use of compound angle drilling. This means that ports, mounting surfaces and holes, and cartridges can be located just where they are needed. The true benefit, though, is the generally smaller size with fewer potential leakage points and construction drillings. Figure 1.34 shows a valvepak incorporating cartridge valves, directional valves, and so on.

Note the removal of pipe-work required for conventional face-mounted valve technology and the simplicity of screw-in technology, particularly for

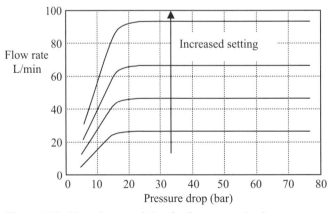

Figure 1.32. Flow characteristic of a flow-control valve.

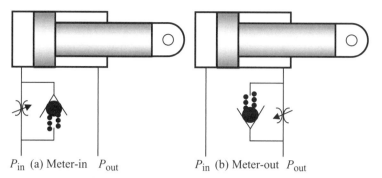

P_{in} (a) Meter-in P_{out} P_{in} (b) Meter-out P_{out}

Figure 1.33. Meter-in and meter-out flow control of an actuator.

solenoid-operated valves. Consider one application by the author concerning a check-valve–flow make-up–pressure-control bridge used in loading circuit to test a cylinder, as shown in Fig. 1.35. The force loading in each direction is almost equal and is $\approx P_2 A_2$ for a low load make-up pressure (often termed *boost pressure*) and a working test pressure. Flow continuity is provided by the suitably sized make-up pump when the test cylinder is retracting. A particular advantage of the use of cartridge technology is that the PRV can be rapidly replaced with a flow-control valve for other tests; the assembly is a compact and convenient test unit that has several uses.

1.6 Servoactuators

Servovalves generally drive cylinders and motors with just a few application exceptions. Therefore, it makes sense to mount the servovalve onto the cylinder or motor to form what is termed a *servoactuator*, as shown in Fig. 1.36.

Compact arrangements as shown in Fig. 1.36 minimize fluid volumes between the servovalve ports and the actuator ports, thus minimizing fluid compressibility effects; the highest hydraulic undamped natural frequency is obtained. If side loads are anticipated in linear drive applications, then the cylinder rod may be fitted with *hydrostatic bearings*. These are essentially narrow grooves around the rod supplied with high pressure, typically half the working pressure, to maintain the rod in its near-central position. This is shown schematically in Fig. 1.37.

Figure 1.34. Cartridge-valve technology provided by Sun Hydraulics Corporation (www.sunhydraulics.com).

(a)

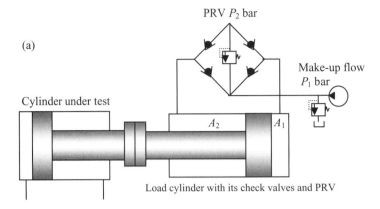

PRV P_2 bar

Make-up flow
P_1 bar

Cylinder under test

A_2 A_1

Load cylinder with its check valves and PRV

(b)

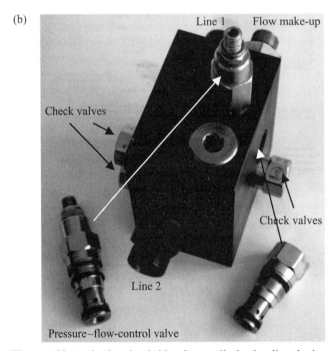

Line 1   Flow make-up

Check valves

Check valves

Line 2

Pressure–flow-control valve

Figure 1.35. A check-valve bridge for a cylinder-loading device.

With some units the friction level can reduced, but note that the use of hydrostatic bearings results in an increased power loss.

1.7 Power Packs and Ancillary Components

In practice, many applications use an integral pump–tank–PRV–filter set, known as a *power pack*, such as that shown in Fig. 1.38.

Notice in Fig. 1.38(a) that the power pack also includes an *oil cooler* – in this case, an air blast unit attached to the power pack. Water coolers are also popular and usually available at a lower cost for a similar power rating. Figure 1.38(b) has two different features, one being that the four gear pumps at the rear of the power pack are submerged in the oil to reduce noise, the other being that two cooling

(a)

Rotary motion

(b)

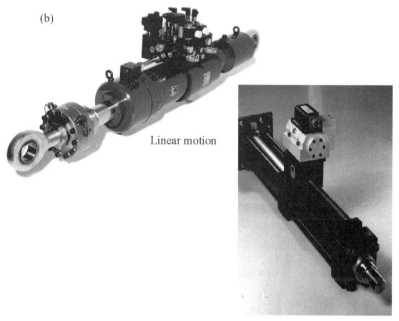

Linear motion

Figure 1.36. Servoactuator units utilizing Star Hydraulics servovalves (www.star-hydraulics. co.uk).

pumps at the front of the power pack actually recirculate the oil from the cooling unit outside the building.

The design of the tank is an important issue to ensure the correct flow of fluid though it. Figure 1.39 illustrates some of the important points to consider, bearing

Figure 1.37. Hydrostatic bearing added to a cylinder rod to counteract side loads (clearances exaggerated).

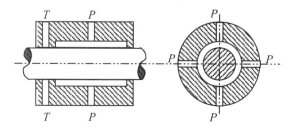

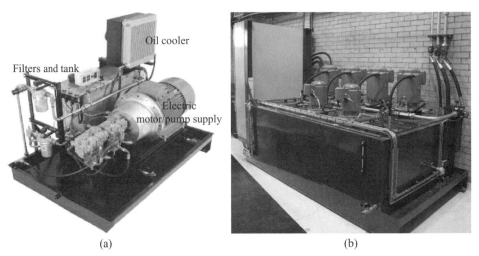

(a)   (b)

Figure 1.38. Two variations in power-pack design: (a) a commercial power pack and just one of a type provided by Hagglunds Denison (www.hagglunds.com); (b) a multipump power pack in use at Cardiff University, School of Engineering.

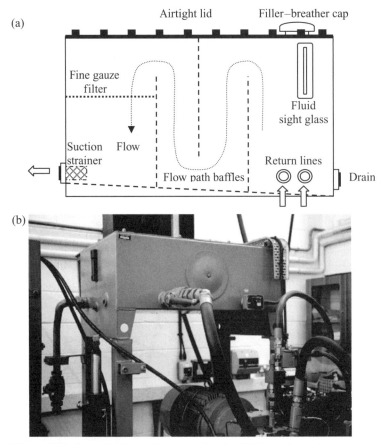

Figure 1.39. Some basic requirements for the fluid supply tank.

(a)   (b)

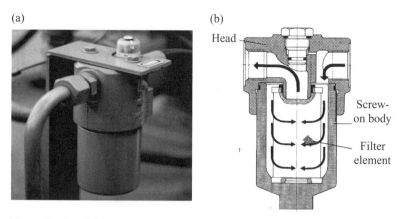

Figure 1.40. A high-pressure filter.

in mind that the tank volume should be at least 3–5 times the maximum pump flow rate. In other words, it should take 3–5 min to empty the tank if a major line failure occurs. It is also preferable to sit the tank outlet above the pump inlet to ensure a positive suction head, remembering to calculate the friction pressure drop down the pump inlet pipe.

Cavitation must be avoided by ensuring a net positive suction head (NPSH) above the vapor pressure of the fluid being used. It is essential that the *fluid used be properly filtered* to remove any unwanted particles that will inevitably be created over the longer term. It is usual practice to place a high-pressure filter, say 10 µm, in the pump outlet line with perhaps a low-pressure filter, say 30 µm, in the suction line at the tank or in the tank return line. Reducing the filter size gives an increased pressure drop, and the demand for fine filtration below 3 µm can create significant pressure-drop problems. For example, one application by the writer resulted in one manufacturer quoting a pressure drop of 25 bar for a 10-µm filter and 75 bar for a 3-µm filter. Figure 1.40 shows a typical high-pressure filter placed in a pump outlet line.

The lower part of the filter, containing the filter element, is screwed to the upper part for ease of filter replacement. Fluid filtration is extremely important to avoid component failures and resulting high-cost downtime, and it follows that regular filter checks should be made together with checks of the condition of the fluid. Particle contamination counts should be made regularly, and this may be done either off-line or on-line. The linear servovalve–actuator circuit shown in Fig. 1.25 has two *accumulators* fitted into the servovalve manifold block, one at the supply pressure port to reduce oscillation effects, the other at the tank return to set a minimum pressure. Accumulators are usually nitrogen gas precharged to match the system pressure required. They have a variety of functions:

• As an energy storage device to reduce installed power
• Maintaining pressure for a given time during system repetitive operation
• Reduction of pulsations caused by a pump or motor
• Providing vehicle suspension damping and shock absorbing
• Pressure transfer (e.g., between air and oil), known as a transfer barrier
• Accommodating thermal expansion in closed systems

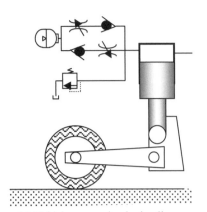

(a) Vehicle suspension hydraulics   (b) Sewage treatment plant (Fawcett Christie Ltd.)

Figure 1.41. Accumulator applications.

Two applications of accumulators are shown in Fig. 1.41, one representing a small-scale suspension application and the other representing a large-scale sewage treatment plant application. Accumulators can be bladder type, in which the bladder material depends on the application, commonly nitrile or butyl, diaphragm type, or piston type, in which the bladder is replaced with a steel piston. The latter are used where large pressure fluctuations are expected or where the loss of a gas in the former could lead to a serious failure. Figure 1.42 shows a schematic of the basic accumulator types.

Accumulators are usually precharged to typically 90% of the minimum pressure required and are sized by considering the pressure changes expected during the duty cycle of a circuit. This allows calculation of the change in volume using the gas laws, isothermal or adiabatic expansion, where appropriate. For details of accumulators and associated applications, for example, as shown in Fig. 1.41, see Fawcett Christie Hydraulics Ltd. (www.fch.co.uk).

Considering the interconnection of components, it is crucial that the pressure capability must be given to *steel pipe–flexible hose selection*. Pressure drops must be minimized while appreciating that increasing pipe diameter reduces the maximum pressure that can be tolerated. From a component vibration point of view, it is better to have short flexible hoses at the end of a steel line – for example, that couples a pump to a valve; steel pipe also radiates less noise than flexible hose. Based on a

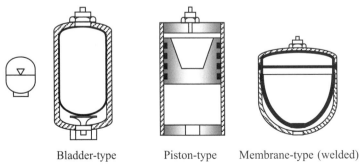

        Bladder-type   Piston-type   Membrane-type (welded)

Figure 1.42 Schematic of common accumulators.

Table 1.1. *Maximum working pressures for steel pipe*

Outside diameter (mm)	Wall thickness (mm)	Max working pressure (bar)
8	1.5	340
16	1.5	170
16	3.0	340
20	2.0	180
25	3.0	220
25	4.0	290

Table 1.2. *Maximum working pressures for two-braid hose*

Internal diameter (mm)	Max working pressure (bar)
6.4	400
9.5	330
12.7	275
19.1	215
25.4	160

safety factor of 4, Tables 1.1 and 1.2 show some maximum working pressures that may be tolerated for some typical steel pipes and flexible hose.

It should be noted that there are many flexible hose designs using different materials, often nitrile–neoprene, different numbers of braids, and different numbers of spirals. A specific hose manufacturer will supply details, but similar designs from different manufacturers may have different pressure ratings.

1.8 References and Further Reading

BOOKS AND PAPERS

Anderson WR [1988]. *Controlling Electrohydraulic Systems*, Marcel Dekker, Inc.

Banks DS and Banks DD [1988]. *Industrial Hydraulic Systems*, Prentice-Hall.

Chapple PJ [2003]. *Principles of Hydraulic Systems Design*, Coxmoor.

Darlington MJ, Culley SJ, Potter S [2001]. Knowledge and reasoning: Issues raised in automating the conceptual design of fluid power systems. *Int. J. Fluid Power* 2 (2), 75–86.

Gale K and Watton J [1999]. *A real-time Expert System for the condition monitoring of hydraulic control systems in a Hot Strip Finishing Mill*. Proceedings of the IMechE, *Journal of Systems and Control Engineering*, Vol. 213, pp. 359–374.

Goodwin AB [1976]. *Fluid Power Systems – Theory, Worked Examples and Problems*, Macmillan.

Guillon M [1969]. *Hydraulic Servo Systems*, Butterworths.

Hodges P [1996]. *Hydraulic Fluids*, Arnold.

Ivantysyn J and Ivantysynova M [2001]. *Hydrostatic Pumps and Motors*, Akademia Books International.

Johansson A and Palmberg J-O [2002]. Quieter hydraulic systems – design considerations. In *Proceedings of the Fifth JFPS International Symposium on Fluid Power*, The Japan Fluid Power Society, pp. 799–804.

Kleman A [1989]. *Interfacing Microprocessors in Hydraulic Systems*, Marcel Dekker, Inc.

Kojima E and Ichiyanagi T [2000]. Research on pulsation attenuation characteristics of silencers in practical fluid power systems. *Int. J. of Fluid Power* 1(2), 29–38.

Konami S and Nishiumi T [1999]. *Hydraulic Control Systems*, TDU Press.

Korn J [1969]. *Hydrostatic Transmission Systems*. Intertext Books.

Lambeck RP [1983]. *Hydraulic Pumps and Motors – Selection and Application for Hydraulic Power Control Systems*, Marcel Dekker.

Martin H [1995]. *The Design of Hydraulic Components and Systems*, Ellis Horwood.

McCloy D and Martin HR [1980]. *Control of Fluid Power – Analysis and Design*, Ellis Horwood.

Merritt HE [1967]. *Hydraulic Control Systems*, Wiley.

Nervegna N. *Oleodinamica E Pneumatica, Politecnico di Torino*. Fluid Power Research Laboratory Course Notes. Vol. 1, Sistemi; Vol. 2, Componenti; Vol. 3, Esercitazioni.

Norvelle FD [1995]. *Fluid Power Technology*, West Publishing.

Paszota Z [2002]. *Aspects Energetiques Des Transmissions Hydrostatiques*, Wydawnictwo Politechniki Gdanskiej.

Peters RJ [2002]. *Noise and Acoustics*, Coxmoor.

Pinches MJ and Ashby JG [1988]. *Power Hydraulics*, Prentice-Hall.

Pippenger J and Hicks T [1979]. *Industrial Hydraulics*, McGraw-Hill Kogakusha.

Reed EW and Larman IS [1985]. *Fluid Power with Microprocessor Control*, Prentice-Hall International.

Rohner P [1988] (1) *Industrial Hydraulic Control*; (2) *Industrial Hydraulic Control* – Workbook, Teacher's Edition; (3) *Industrial Hydraulic Control* – Workbook, AE Press.

Stecki JS and Garbacik AJ [2002]. *Design and Steady State Analysis of Hydraulic Control Systems*, Fluid Power Net Publications.

Thoma JU [1964]. *Hydrostatic Power Transmission*, Trade and Technical Press.

Trostmann E [1996]. *Water Hydraulics Control Technology*, Marcel Dekker, Inc.

Turner IC [1996]. *Engineering Applications of Pneumatics and Hydraulics*, Arnold.

Watton J [1989]. *Fluid Power Systems – Modelling, Simulation, Analogue and Microcomputer Control*, Prentice-Hall.

Watton J [1992]. *Condition Monitoring and Fault Diagnosis in Fluid Power Systems*, Ellis Horwood.

Watton J [2007]. *Modelling, Monitoring and Diagnostic Techniques for Fluid Power Systems*, Springer-Verlag.

Yeaple F [1995]. *Fluid Power Design Handbook*, Marcel Dekker, Inc.; available at www.fluid.power.net.

Younkin GW [2002]. *Industrial Servo Control Systems, Fundamentals and Applications*, Marcel Dekker, Inc.

COMMERCIAL DESIGN LITERATURE

Mannesmann Rexroth series on fluid power [1991], second issue:
  Vol. 1, *Basic Principles and Components of Fluid Power Technology*.
  Vol. 2, *Proportional and Servo Valve Technology*.
  Vol. 3, *Planning and Design of Hydraulic Systems*.
  Vol. 4, *Logic Element Technology*.
  Vol. 6, *Hydrostatic Drives with Control of the Secondary Circuit*.
Parker Hannifin Design Engineers Handbook [2001], Vol. 1, *Hydraulics Servo and Proportional Systems Catalog*, Moog Corporation Inc., East Aurora, New York 14052–0018.
Vickers Industrial Hydraulics Manual [1992]. Vickers Incorporated Training Center, Rochester Hills, Michigan.
Vickers Closed Loop Electrohydraulics Systems Manual [1992]. Vickers Incorporated Training Center, Rochester Hills, Michigan.

2 An Introduction to Fluid Properties

2.1 Fluid Types

The preferred working fluid for most applications is mineral oil, although in some applications there is a requirement for a water-based or synthetic fluid, mainly for reasons of fire hazards and increasingly for environmental considerations. The drive toward nonmineral oil fluids has seen a renewed attitude to pure water hydraulics together with the emergence of biodegradable and vegetable-based fluids. Fire-resistant fluids in use fall under the following classifications:

HFA 5/95 oil-in-water emulsion, typically 5% oil and 95% water
HFB 60/40 water-in-oil emulsion, typically 60% oil and 40% water
HFC 60/40 water-in-glycol emulsion, typically 60% glycol and 40% water
HFD synthetic fluid containing no water
HFE synthetic biodegradable fluid

The use of water-based fluids has implications for component material selection – for example, the use of stainless steel, plastics, and ceramics. In addition, serious consideration of fluid properties must also be given, particularly viscosity, which can be very high at low temperatures in some cases. Fluids are being continually developed, and the following information is intended to reflect the general trend and is not considered as definitive because this would require an overview of many suppliers from many countries around the world – for example, see www.shell.com.

Type HFA 5/95 oil-in-water emulsions are fire-resistant emulsions that exhibit enhanced stability, lubrication, and antiwear characteristics and have the following important aspects:

- They have much improved stability toward variations in temperature, pressure, shear, and bacterial attack.
- The performance limitations become obvious for systems operating well above 70 bar, reliability and efficiency often being sacrificed where fire resistance is of paramount importance.
- A main concern stems from low fluid viscosity, the critical effect this has on pump performance, and the relatively poor hydrodynamic lubrication properties of most conventional fluids. A loss of volumetric efficiency is to be expected because of the viscosity being similar to that of water, approximately 1 mm^2/s

33

at 40°C. Internal leakage will depend on the pump type, with axial and radial piston pumps generally offering the maximum efficiency.

- Standard vane pumps and motors are not usually recommended for use with high-water-based fluids, although modified designs are capable of operating satisfactorily at system pressures of around 70 bar.
- Difficulties associated with filterability, emulsion instability, and bacterial degradation of these conventional emulsions were not uncommon and contributed to limiting the growth of high-water-based fluids outside of the serious fire-hazard situations.

Shell Tellus mineral oils are premium-quality, solvent-refined, high-viscosity-index fluids generally acknowledged to be the "standard-setter" in the field of industrial hydraulic and fluid power lubrication and may be used for industrial hydraulic systems, mobile hydraulic fluid power transmission systems, and marine hydraulic systems. Key performance features and benefits are as follows:

- Thermal stability in modern hydraulic systems working in extreme conditions of load and temperature. Tellus oils are highly resistant to degradation and sludge formation, therefore improving system reliability and cleanliness.
- Oxidation resistant in the presence of air, water, and copper. Turbine oil stability test (TOST) results show outstanding performance for Tellus oils: low acidity, low sludge formation, and low copper loss. Therefore, they extend oil drain interval life and minimize maintenance costs.
- Hydrolytic stability that is due to good chemical stability in the presence of moisture, which ensures long oil life and reduces the risk of corrosion and rusting.
- Low friction because Tellus oils possess high-lubrication properties and excellent low-friction characteristics in hydraulic systems operating at low or high speed. Prevent stick–slip problems in critical applications, enabling very fine control of machinery.
- Excellent air release and antifoam properties that are due to the careful use of additives to ensure quick air release without excessive foaming. Quick air release minimizes cavitation and slows oxidation, maintaining system and fluid performance.
- Good water-separation properties (demulsibility). They resist the formation of water-in-oil emulsions and prevent consequent system and pump damage.
- Outstanding antiwear performance because proven antiwear additives are incorporated to be effective throughout the range of operating conditions, including low and severe-duty high-load conditions. Outstanding performance has been achieved in a range of piston and vane pump tests, including the tough Denison T6C (dry and wet versions) and the demanding Vickers 35VQ25.
- Superior filterability because Tellus oils are suitable for ultrafine filtration, an essential requirement in today's hydraulic systems. Unaffected by the usual products of contamination, such as water and calcium, which are known to cause blockage of fine filters. Customers can use finer filters, therefore achieving all the benefits of having cleaner fluids in use.

Typical characteristics for Shell Tellus mineral oils are shown in Table 2.1.

Table 2.1. *Typical physical characteristics of Shell Tellus mineral oils*

Shell Tellus oil	22	32	37	46	68
ISO oil type	HM	HM	HM	HM	HM
Kinematic viscosity					
0°C (cSt)	180	338	440	580	1040
40°C (cSt)	22	32	37	46	68
100°C (cSt) (IP 71)	4.3	5.4	5.9	6.7	8.6
Viscosity index (IP 226)	100	99	99	98	97
Density at 15°C (kg/m^3) (IP 365)	0.866	0.875	0.875	0.879	0.886
Flash-point (°C) (IP 34)	204	209	212	218	223
Pour point (°C) (IP 15)	−30	−30	−30	−30	−24

Note: IP, Institute of Petroleum; HM, Hydraulic Mineral.

HFB-Type – Shell Irus Fluid BLT

These are premium-performance water-in-oil emulsion-type, fire-resistant, hydraulic fluids containing approximately 42% water, by volume. The formulation includes a unique combination of emulsifiers to provide a homogeneous dispersion of submicron water droplets in a continuous oil phase, whereas other sophisticated additives enhance their mechanical performance and corrosion inhibition properties. They may be used in areas where there is a high fire risk – for example, in mines or steel works. Some important features are as follows:

- The optimum system temperature should be about 40°C and should never exceed 65°C (water-in-oil systems generally run about 10°C cooler than those lubricated by conventional mineral oils).
- To reduce the possibility of cavitation, inlet systems should avoid negative pressures by having adequate and unrestricted pipelines and by siting the pump to give a full head. If suction filters are necessary, extreme care must be taken to avoid undue restrictions, and twice the normal capacity for oil is typical.
- The reservoir should be sealed to prevent evaporation but fitted with a breather, and any baffles in the reservoir should be so placed as to ensure circulation and avoid stagnant pockets.
- Systems should also avoid pockets of stagnant fluid, such as ram heads and non-circulating sections, to ensure good emulsion stability over long periods.
- They will not burn readily when they contact a source of ignition because the water evaporates, forming a steam blanket that displaces oxygen from the immediate area and insulates the oil from the ignition source. After the water evaporates, the residue may burn. In the event of a fluid being sprayed onto a hot surface or if the fluid cascades onto a hot surface and runs off, the fluid will not support its own combustion.
- The fire-resistant properties of water-in-oil emulsions are greatly dependent on the ability of the fluid to hold the water content in an even dispersion over long periods. Should a slight oil separation occur, the fluid will readily re-emulsify with agitation. Significant water separation is unacceptable and steps should be taken to rectify the situation immediately.
- Water content is normally 42% by volume and should be maintained at between 35% and 45%. An accurate measurement of water content can be made by the

Table 2.2. *Typical physical characteristics of a Shell HFB fluid*

Irus BLT (White opaque fluid)		68	100
Fluid-type ISO designation		HFB	HFB
ISO viscosity grade	ISO 3448	68	100
Kinematic viscosity (mm^2/s)	ASTM D 445		
at 20°C		167	239
at 40°C		70	97
at 55°C		49	67
at 65°C		31	42
Density at 15°C (kg/m^3)	ISO 12185	934	933
Pour point (°C)	ISO 3016	−30	−27

laboratory test method IP (Institute of Petroleum) 74 (Dean and Stark method). For an approximate check of water content, the density method is sufficient, providing that the sample is representative and well shaken.

- Irus BLTs have proved their ability for use in plain bearings, moderately loaded ball bearings, and roller bearings. The life of heavily loaded ball and roller bearings tends to be reduced when water-containing fluids are used, and the manufacturer's recommendations should be followed. Distinct differences in bearing life have been found in evaluating various HFB fluids.

Typical characteristics are shown in Table 2.2.

HFD-R type – Shell Irus fluid DR is a triaryl phosphate ester fire-resistant hydraulic fluid and contains carefully selected additives to give superior oxidation and hydrolytic stability characteristics. Applications include steel and mining industries, die-casting machines, billet loaders, electric arc furnaces, forging presses, welding robots, continuous casting machines, hydraulic presses, and extrusion presses. Performance features and benefits include good fire resistance, nontoxic under European Economic Community (EEC) regulations, extended fluid change intervals, pump life similar to life with mineral hydraulic oils, fire resistance maintained during the life of the fluid, and compatible with most seal materials.

- Irus DR has excellent fire resistance. The fire resistance is inherent and is not achieved by the use of additives and therefore will not change with time. Protection is available throughout all parts of the system and the whole time the fluid is in the system. This is demonstrated in numerous standard tests designed to simulate its performance in the three most common fire-risk scenarios:

    (i) Ignitability of a spray or jet of fluid
    (ii) Spillage onto a hot surface or molten metal
    (iii) Ignitability of the fluid when soaked into an adsorbent material

- Lubrication properties of Irus DR compare favorably with those of an equivalent mineral oil of the same viscosity. This is not surprising in view of the wide use of phosphate esters as antiwear additives in oil. As a result, in many pumps, they show a similar performance (bearing life and wear properties) to that of mineral oil, although some slight derating may be necessary at very high loads. Contact with the pump manufacturer is advisable before use.

- Stability of phosphate esters is due to a natural resistance to oxidation and, for Irus DR, is further enhanced by the inclusion of an antioxidant in its formulation to give long life at normal temperatures between 60°C and 80°C and transient temperatures up to 150°C. A property of a phosphate ester fluid is that when it is contaminated with water, hydrolysis of the phosphate ester can occur, leading to the formation of strong inorganic acids. These acids can chemically attack metallic components. Normal, good housekeeping is required for minimizing water contamination. However, to provide still further protection, Irus DR is formulated with a hydrolysis stabilizer, a particular feature associated with this product.

- Viscosity–temperature properties and shear stability have a more marked change of viscosity with temperature than occurs in conventional mineral hydraulic oils. At low temperatures, it may be necessary to warm the fluid slightly prior to switching on the main pumps. A viscosity of 850 cSt is generally regarded as the reasonable maximum at which a hydraulic pump may be started, and the viscosity at which Irus fluid DR reaches this value is approximately 6°C. Because Irus fluid DR contains no thickeners or viscosity index improvers, the product is shear stable and the pump selection can be made on the basis of the listed viscosity data.

- Rusting resistance in the presence of water in the fluid has already been noted as potential concern with regard to hydrolysis. It can also cause rusting and galvanic attack on metals. Fortunately, this does not occur in the liquid phase unless free water is present and, as a result of the much higher solubility of water in phosphate esters than in mineral oils, this is rarely a problem. Rusting that is due to condensation has occasionally been found above the liquid level in mild steel tanks. This can easily be overcome by ensuring adequate ventilation.

- Contamination with mineral oils in the presence of up to 0.5% of mineral oil will not affect the properties of Irus fluid DR, but a greater degree of contamination will affect its fire-resistant properties. There is no method of reclaiming fluid affected in this way because mineral oils are incompatible with phosphate ester fluids. Contamination with mineral oils should be avoided.

- Fluid life and recyclability are possible, and it is common practice to pass the fluid through an adsorbent solid that removes the acid as it is formed. In this way, the life of Irus fluid DR (and system components) can be greatly extended. If it becomes necessary to replace the fluid, reclamation may be possible so that there is minimum impact on the environment.

- Compatibility with seals such as butyl and viton is acceptable, but further advice from suppliers of ethylene–propylene is recommended. Paints such as epoxy resin and common constructional metals are compatible. Aluminum and its alloys should be hard anodized and not used as bearing surfaces.

- Health and safety guidance from the fluid manufacturer should always be studied, and users are strongly urged to consider the environment by taking the used oil to an authorized collection point and not discharging it into drains, soil, or water.

Typical characteristics are shown in Table 2.3.

Table 2.3. *Typical physical characteristics of a Shell HFD fluid*

Irus DR		
ISO viscosity grade	ISO 3448	46
	ISO 6743/4	HFD-R
Kinematic viscosity (mm^2/s)	ASTM D 445	
0°C		1600
40°C		43
50°C		26
100°C		5.3
Viscosity index	ISO 2909	15
Density 15°C (kg/m^3)	ISO 12185	1125
Pour point (°C)	ISO 3016	−18

Shell Naturelle HFE fluids are advanced biodegradable hydraulic fluids for use in power transmission and hydraulic systems working in environmentally sensitive areas. Synthetic esters blended with specially tailored additive systems provide Shell Naturelle HF-E fluids with a superior balance of biodegradability, lubrication performance, and compatibility with the environment. Applications are in heavy-duty systems for construction and earth-moving equipment, machine tool systems, hydrostatic drive gears, general industrial control equipment and hydraulic systems, and moderately rated gearboxes for which an antiwear hydraulic oil is specified. Performance features are as follows:

- Readily biodegradable because of the high potential to be broken down rapidly and extensively by microorganisms in the environment to ultimately yield carbon dioxide and water as end products.
- Excellent viscosity–temperature characteristics with minimum changes of viscosity with variations in operating temperature, giving true "multigrade" characteristics. High shear stability ensures effective lubrication and efficient system operation.
- Excellent corrosion protection with long-term protection for common construction materials, including most metals, nonmetals, and seal materials such as viton and high nitrite. Good oxidation resistance to the formation of acidic products generated when working at high operating temperatures.
- Optimum wear protection and effective under all operating conditions, including low and severe duty situations.
- Compatibility with mineral oils. Shell Naturelle HF-E is miscible with conventional mineral-oil-based hydraulic oils in all proportions. However, to ensure that biodegradability properties are maintained, the system should be drained and flushed prior to changeover.
- Owing to the surface wetting properties of Shell Naturelle HF-E, if systems were previously operated with authorize-based hydraulic oils, deposits formed in the system during operation may be loosened and deposited in system filters. The hydraulic filters should therefore initially be checked at regular intervals.
- Caution is given because Shell Naturelle HF-E is not suitable for use in engines. During maintenance, care must be taken to use separate, clean containers for

Table 2.4. *Typical physical characteristics of Shell HFE biodegradable fluid*

Shell Naturelle	HFE15	HFE32	HFE46	HFE68
ISO viscosity grade (ISO 3348)	15	32	46	68
Color	Green	Green	Green	Green
Kinematic viscosity				
40°C (cSt)	14.1	31.6	46.1	64.9
100°C (cSt)	4.2	6.3	9.1	12.1
IP 71				
Viscosity index (IP 226)	232	156	182	187
Density at 15°C (kg/m^3) (IP 365)	0.892	0.918	0.919	0.928
Pour point (°C) (IP 15)	−54	−60	−51	−39
Flash-point Cleveland Open Cup (°C) (IP 36)	202	236	219	226

filling engine oil and Shell Naturelle HF-E. Precautions should also be taken to exclude moisture from the fluid, both during storage and in service.

- Seal and paint compatibility with all seal materials and paints normally specified for use with petroleum mineral oils. Certain plastics and industrial adhesives may be adversely affected, and advice should be sought from the respective manufacturers.
- Operating temperatures should not be allowed to exceed 90°C, and optimum fluid life will be realized if operating temperatures are maintained at approximately 55°C.
- The environment must be protected: take used oil to an authorized collection point; do not discharge into drains, soil, or water.

Typical characteristics are shown in Table 2.4.

2.2 Fluid Density

A typical variation of density for different Shell fluids previously discussed and at a temperature of 15°C is shown in Table 2.5.

Fluid density is comprehensively covered for mineral oil; it is well known that it increases with pressure and decreases with temperature. For example, Fig. 2.1 shows such a characteristic for an ISO 32 mineral oil.

Table 2.5. *Typical density characteristics of Shell fluids*

Fluid	Density (kg/m^3)
Shell Tellus ISO 32 mineral oil	875
Shell HFB 60% oil, 40% water	933
Shell HFC 60% glycol, 40% water	1084
Shell HFD phosphate ester	1125
Shell Naturelle HFE32	918

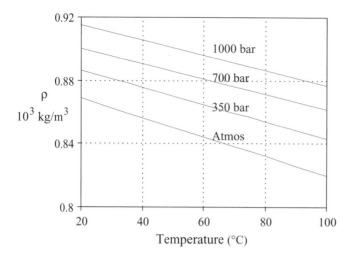

Figure 2.1. Density variation for a Shell Tellus ISO 32 mineral oil.

2.3 Fluid Viscosity

Considering fluid flow across a surface with a varying velocity profile, the fluid dynamic viscosity is defined by the Newtonian shear-stress equation at a point on the velocity profile:

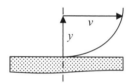

$$\tau = \mu \frac{dv}{dy}. \tag{2.1}$$

Dynamic viscosity is usually expressed in units of centiPoise, where:

$$1 \, \text{cP} = 10^{-3} \, \text{N s/m}^2. \tag{2.2}$$

Alternatively, Eq. (2.1) can be expressed as a momentum equation:

$$\tau = \frac{\mu}{\rho} \frac{d(\rho v)}{dy}. \tag{2.3}$$

The term $\mu/\rho = \nu$ is called the kinematic viscosity, which is usually expressed in units of centiStokes, where:

$$1 \, \text{cSt} = 10^{-6} \, \text{m}^2/\text{s}. \tag{2.4}$$

It therefore also follows that if the fluid density ρ is expressed in kilograms per cubic meter, and the kinematic viscosity ν is expressed in units of centiStokes, then:

$$\mu(\text{N s/m}^2) = 10^{-6} \rho \nu. \tag{2.5}$$

A comparison of kinematic viscosities for a variety of fluids is shown in Fig. 2.2(a) and for low operating pressures. A more detailed characteristic for an ISO 32

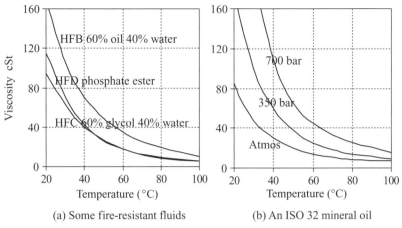

(a) Some fire-resistant fluids   (b) An ISO 32 mineral oil

Figure 2.2. Typical kinematic viscosities for a range of fluids.

mineral oil is shown in Fig. 2.2(b), illustrating the important effects of both temperature and pressure.

The data shown in Fig. 2.2 make it clear that experimental testing must specify both the pressure and temperature so that comparisons between related studies may be compared with at least a minimum of confidence. It is common that computer dynamic simulations of hydraulic systems usually assume a mean temperature in the sense that temperature will not vary significantly during the milliseconds-to-seconds of transient behavior. However, it may be necessary to model the effect of pressure on viscosity if large fluctuations in pressure are expected, although its effect may well be of secondary significance.

2.4 Bulk Modulus

Bulk modulus is a measure of the compressibility of a fluid and will be required when it is desired to calculate oil volume changes for high-pressure, large-volume systems such as forging presses or natural frequencies generally caused by the interaction of fluid compressibility and moving mass. Bulk modulus defines the compression of a fluid usually in one of two ways:

$$\text{isothermal tangent bulk modulus,} \quad \beta_T = -V\left(\frac{\partial P}{\partial V}\right)_T, \tag{2.6}$$

$$\text{isentropic tangent bulk modulus,} \quad \beta_s = -V\left(\frac{\partial P}{\partial V}\right)_s. \tag{2.7}$$

T and s refer to conditions of constant temperature and entropy, respectively. Isentropic tangent bulk modulus is usually taken from the manufacturer's data in practice, and Fig. 2.3 shows the variation for a range of fluids under perfect conditions with no pressure and dissolved air.

It is clear that mineral oil has the lowest bulk modulus, phosphate esters and water glycol fluids having a significant increase. Because the velocity of sound propagation in a fluid is given by:

$$C_0 = \sqrt{\frac{\beta_0}{\rho}}, \tag{2.8}$$

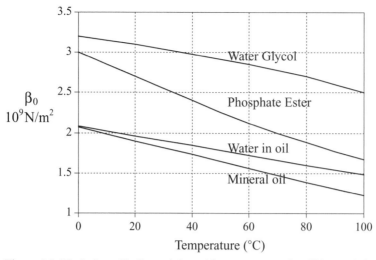

Figure 2.3. Variation of bulk modulus with temperature for different fluids.

it follows that C_0 increases for phosphate esters and water glycol fluids. This means that resonant frequencies, because of fluid compressibility–moving mass interaction, will also be increased and potentially advantageous. The effect of pressure must also be considered, and Fig. 2.4 shows such a variation in isentropic bulk modulus for a typical ISO 32 mineral oil. For hydraulic circuit calculations, it is extremely important to realize that, in practice, the fluid ideal bulk modulus characteristic shown in Fig. 2.4 is significantly reduced by:

- inherent dissolved air in solution even with the fluid as supplied by the manufacturer,
- entrained air that might, for example, be induced by cavitation or leaks when in operation,
- pipe or container elasticity effects, particularly when flexible hose is used.

In practice, it is usual to refer to the effective bulk modulus β_e, which embraces all these aspects and must be calculated for each hydraulic system being considered.

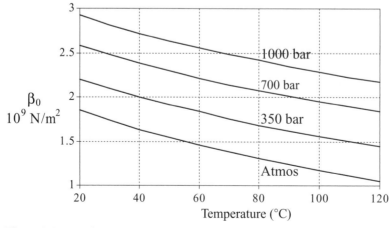

Figure 2.4. A typical ISO 32 mineral oil isentropic tangent bulk modulus characteristic.

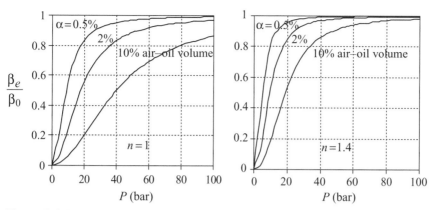

Figure 2.5. The effect of air–oil volume on oil bulk modulus. ISO 32 mineral oil, $\beta_0 = 1.6 \times 10^9$ N/m^2.

The effect of air content may be determined by an approximate analysis assuming the gas law $[PV^n = \text{constant}]$ for the air volume V within the total fluid volume. This results in an effective bulk modulus given by:

$$\frac{\beta_e}{\beta_0} = \frac{\left(\dfrac{P}{P_0}\right)^n + \alpha}{\left(\dfrac{P}{P_0}\right)^n + \dfrac{\alpha\beta_0}{nP}}. \tag{2.9}$$

The ideal fluid bulk modulus is β_0, P is the working pressure, P_0 is atmospheric pressure, α is the air–fluid volume ratio, and $n = 1.0$ for isothermal conditions or $n = 1.4$ for adiabatic conditions. A set of characteristics is shown in Fig. 2.5.

It would appear that for realistic air–oil volumes at a typical hydraulic pressure and temperature, the bulk modulus of the oil shown in Fig. 2.5 would be reduced to a value of typically $\beta_0 = 1.4 \times 10^9$ N/m^2.

Consider next the effect of different connected pipe and/or section volumes as shown in Fig. 2.6 and subject to a pressure increase P.

Adopting a finite-difference approximation bulk modulus, Eq. (2.6) gives:

$$\beta \approx -V\frac{\delta P}{\delta V} \rightarrow \delta V = -V\frac{\delta P}{\beta}. \tag{2.10}$$

If the total volume of fluid is V_t, then it follows that:

$$\delta V_t = -\delta V_1 - \delta V_2 - \delta V_3, \tag{2.11}$$

$$V_t \frac{\delta P}{\beta_e} = V_1 \frac{\delta P}{\beta_{e1}} + V_2 \frac{\delta P}{\beta_{e2}} + V_3 \frac{\delta P}{\beta_{e3}},$$

$$\frac{1}{\beta_e} = \frac{V_1}{V_t}\frac{1}{\beta_{e1}} + \frac{V_2}{V_t}\frac{1}{\beta_{e2}} + \frac{V_3}{V_t}\frac{1}{\beta_{e3}} = \sum_{i=1}^{n\,\text{sections}} \frac{V_i}{V_t}\frac{1}{\beta_{ei}}. \tag{2.12}$$

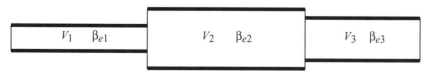

Figure 2.6. The effect of different volumes on the effective bulk modulus.

If there are no pipe–container elasticity effects, then, of course, $\beta_{ei} = \beta_0$ for each section and the summation is trivial in that $\beta_e = \beta_0$. Consider next the effect of such elasticity effects. It can be deduced from the theory of elasticity that the bulk modulus for a pipe β_c of internal diameter d_i and external diameter d_o is given by

$$\frac{1}{\beta_c} = \frac{2}{E} \left(\frac{d_o^2 + d_i^2}{d_o^2 - d_i^2} + \mu \right), \tag{2.13}$$

where E is the modulus of elasticity for the pipe–container material and μ is Poisson's ratio for the material, typically $\mu = 0.3$ and $E = 2 \times 10^{11}\,\mathrm{N/m^2}$ for steel. Note that for a material wall thickness of $t \ll d_i \approx d_o$, then Eq. (2.13) can be approximated by:

$$\frac{1}{\beta_c} \approx \frac{d}{tE}. \tag{2.14}$$

Recognizing that the effect of material elasticity is to also create an effective bulk modulus by adding the effect of the oil and the material in parallel for each section, we have:

$$\frac{1}{\beta_{ei}} = \frac{1}{\beta_{ci}} + \frac{1}{\beta_0},$$

$$\frac{1}{\beta_e} = \sum_{i=1}^{n\,\mathrm{sections}} \frac{V_i}{V_t} \frac{1}{\beta_{ei}} = \sum_{i=1}^{n\,\mathrm{sections}} \frac{V_i}{V_t} \left[\frac{1}{\beta_{ci}} + \frac{1}{\beta_0} \right],$$

$$\frac{1}{\beta_e} = \frac{1}{\beta_0} + \sum_{i=1}^{n\,\mathrm{sections}} \frac{V_i}{V_t} \frac{1}{\beta_{ci}}. \tag{2.15}$$

Worked Example 2.1

Consider the following line connected to an actuator with an assessed oil bulk modulus of $\beta_0 = 1.4 \times 10^9\,\mathrm{N/m^2}$ at the working conditions. Determine the effective bulk modulus at minimum and maximum actuator positions.

Steel line 2 m long
internal dia 13 mm
external dia 16 mm

Actuator–steel
internal dia 44 mm, thickness 10 mm
minimum length 20 mm
maximum length 120 mm

Worked Example 2.1

(i) *At minimum actuator position:*

$$\text{line volume, } V_1 = 2.65 \times 10^{-4}\,\mathrm{m^3};$$
$$\text{actuator volume, } V_2 = 0.30 \times 10^{-4}\,\mathrm{m^3};$$
$$\text{total volume, } V_t = 2.95 \times 10^{-4}\,\mathrm{m^3};$$

$$\frac{1}{\beta_{c1}} = \frac{0.052}{10^9} \quad \frac{1}{\beta_{c2}} = \frac{0.036}{10^9},$$

$$\frac{1}{\beta_e} = \frac{1}{\beta_0} + \sum_{i=1}^{n\,\text{sections}} \frac{V_i}{V_t} \frac{1}{\beta_{ci}}$$

$$\frac{1}{\beta_e} = \frac{1}{1.4 \times 10^9} + \sum_{i=1}^{n=2} \left(\frac{2.65 \times 10^{-4}}{2.95 \times 10^{-4}} \right) \frac{0.052}{10^9} + \left(\frac{0.3 \times 10^{-4}}{2.95 \times 10^{-4}} \right) \frac{0.036}{10^9}$$

$$= \frac{0.71}{10^9} + \frac{0.051}{10^9} = \frac{0.76}{10^9},$$

$$\beta_e = 1.32 \times 10^9 \, \text{N/m}^2.$$

(ii) *At maximum actuator position:*

$$\text{line volume, } V_1 = 2.65 \times 10^{-4} \, \text{m}^3;$$
$$\text{actuator volume, } V_2 = 1.82 \times 10^{-4} \, \text{m}^3;$$
$$\text{total volume, } V_t = 4.47 \times 10^{-4} \, \text{m}^3;$$

$$\frac{1}{\beta_{c1}} = \frac{0.052}{10^9} \quad \frac{1}{\beta_{c2}} = \frac{0.036}{10^9},$$

$$\frac{1}{\beta_e} = \frac{1}{\beta_0} + \sum_{i=1}^{n\,\text{sections}} \frac{V_i}{V_t} \frac{1}{\beta_{ci}},$$

$$\frac{1}{\beta_e} = \frac{1}{1.4 \times 10^9} + \sum_{i=1}^{n=2} \left(\frac{2.65 \times 10^{-4}}{4.47 \times 10^{-4}} \right) \frac{0.052}{10^9} + \left(\frac{1.82 \times 10^{-4}}{4.47 \times 10^{-4}} \right) \frac{0.036}{10^9}$$

$$= \frac{0.71}{10^9} + \frac{0.046}{10^9} = \frac{0.76}{10^9},$$

$$\beta_e = 1.32 \times 10^9 \, \text{N/m}^2.$$

It can be deduced that the effective bulk modulus does not change significantly with actuator position and the fluid contribution is reduced by just 6% because of pipe-wall and cylinder-wall elasticity effects.

Worked Example 2.2

A forging press has four main pressing cylinders 2 m in diameter, and the pump units to each cylinder deliver 4200 L/min. If the maximum rate of change of pressure was found to be 50 bar/s, determine the speed of the press given $\beta = 1.4 \times 10^9 \, \text{N/m}^2$, actuator volume $V = 1.2 \, \text{m}^3$.

Recalling Eq. (2.10), we have:

$$\beta \approx -V \frac{\delta P}{\delta V} \rightarrow \delta V = -V \frac{\delta P}{\beta}. \tag{2.16}$$

If pressure changes with time, a compressibility flow rate is generated:

$$Q_c = \lim_{\delta t \to 0} \frac{\delta V}{\delta t} = \frac{dV}{dt} = -\frac{V}{\beta}\frac{dP}{dt}. \tag{2.17}$$

The flow rate into the actuator required to service this compressibility flow rate effect is given by:

$$Q_c = \frac{V}{\beta}\frac{dP}{dt}. \tag{2.18}$$

Using the system data gives:

$$Q_c = \frac{V}{\beta}\frac{dP}{dt} = \left(\frac{1.2}{1.4 \times 10^9}\right)\left(\frac{50 \times 10^5}{1}\right) = 42.86 \times 10^{-4}\,\text{m}^3/\text{s} = 257\,\text{L/min}.$$

It can be seen that compressibility flow rates can be very significant for systems such as this and cannot be neglected when sizing the supply pumps.

Subtracting the compressibility flow rate from the pump flow rate gives the net flow rate available to move the press, giving:

$$Q_{\text{pump}} = Q_c + AU,$$
$$(4200 - 257)\frac{10^{-3}}{60} = 3.14 \times U \to U = 21\,\text{mm/s}.$$

The effect of fluid compressibility is to reduce the press speed by 6.1% compared with an incompressible fluid assumption.

The effective bulk modulus of a fluid is significantly different if the steel pipe is replaced with a flexible hose, and this now needs to be considered. Manufacturers' data are not readily available, although the effect has been studied and can actually be determined experimentally by using fast-acting flow meters at either end of the test hose. Note that a very long hose will introduce transmission line effects, making the data analysis much more complex. Consider, therefore, the simple arrangement shown in Fig. 2.7.

The fast-acting flow meters used were the poppet/LVDT (LVDT is a linear variable-differential transformer) displacement type (parker.com) with a response characteristic capable of capturing frequency components comfortably up to 250 Hz. Details are shown in Fig. 2.8.

A mean pressure is set and transient pressure changes are created by a servovalve at the supply side. A flow restrictor is used at the end of the line, and conditions are set such that the transient variation of pressure is typically ±10 bar from the mean. This means that the effect of pressure on bulk modulus during the test may be neglected. From the flow continuity equation with no moving boundary

Figure 2.7. Determining effective bulk modulus by dynamic testing.

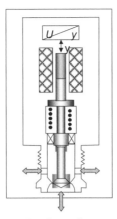

Figure 2.8. Details of a fast-acting flow meter for dynamic measurements.

and recalling Eq. (2.18):

$$Q_i - Q_o = Q_c = \frac{V}{\beta}\frac{dP}{dt},$$
$$\int_0^t (Q_i - Q_o)dt = \frac{V}{\beta}(P_2 - P_1),$$

(2.19)

where P_1 and P_2 are the pressure beginning and end values of the test, the difference being set to 20 bar. Two examples of the approach will be presented:

- A rigid-steel accumulator-type pressure test vessel with a volume of 4.92 L to determine the oil bulk modulus
- A flexible hose of a nitrile–two-wire mesh–neoprene design. The hose is 2.23 m long with an internal diameter of 20.75 mm, giving a volume of 7.54×10^{-4} m^3, with an additional volume of 3.04×10^{-4} m^3 that is due to the flow meters and fittings

A computer-based data-acquisition system was used to acquire the data and then filter the noise and pump-ripple effects by use of a 14th-order Butterworth filter on all pressure and flow-rate signals. A cutoff frequency of 20 Hz was set to be just below the flow-ripple frequency measured at 24 Hz. Typical measured and filtered transient pressure and flow-rate data are shown in Fig. 2.9 for the rigid-container test.

Integrating the flow-rate data and using Eq. (2.19) then allows determination of the bulk modulus for a range of mean pressures. Considering the rigid-container test means that the direct oil bulk modulus β_0 may be determined because wall elasticity effects are negligible. The effective bulk modulus β_e is first determined with the flexible-hose test section. The hose bulk modulus may then be estimated from Eq. (2.15), assuming an oil bulk modulus determined from the rigid-container measurement:

$$\frac{1}{\beta_{\text{hose}}} = \frac{1}{\beta_e} - \frac{1}{\beta_0}.$$

(2.20)

The results are shown in Fig. 2.10 for a range of mean pressures and one working temperature of 50°C, and the test repeatability is better than ±5%. Clearly, the hose bulk modulus is of the order of 25% that of the mineral oil and can be dominant in a circuit if the hose volume is significant. It is usually good practice to have a short

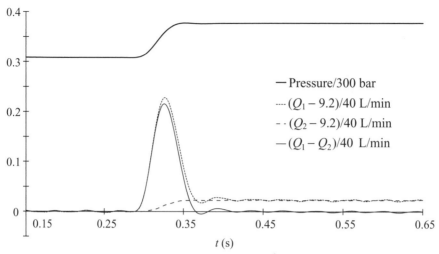

Figure 2.9. Transient test results for a rigid pressure vessel (Watton and Xue, 1994).

hose to minimize vibration transmission from components such as pumps. It is also good practice for the majority of the line length to be constructed from rigid steel pipe to minimize noise radiation. A short length of hose at either end of a steel line should therefore be checked for the possible effect on effective bulk modulus.

Worked Example 2.3

A fluid transmission line consists of three components, hose/steel pipe/hose, as shown here:

(1) Hose　　　　　　　　　　(2) Steel line　　　　　　　　　　(3) Hose

Worked Example 2.3

Hoses (1) and (3) are the same as described earlier, $\beta_{hose} = 0.5 \times 10^9$ N/m^2, volume 0.5×10^{-4} m^3.

Steel line (2) and fittings, oil bulk modulus, $\beta_0 = 1.6 \times 10^9$ N/m^2, volume 25×10^{-4} m^3.

Wall elasticity effects may be neglected:

$$\frac{1}{\beta_e} = \frac{1}{\beta_0} + \frac{V_1}{V_t}\frac{1}{\beta_{c1}} + \frac{V_2}{V_t}\frac{1}{\beta_{c2}} + \frac{V_3}{V_t}\frac{1}{\beta_{c3}},$$

$$\frac{1}{\beta_e} = \frac{1}{1.6 \times 10^9} + \frac{0.5 \times 10^{-4}}{26 \times 10^{-4}}\frac{1}{0.5 \times 10^9} + \frac{25 \times 10^{-4}}{26 \times 10^{-4}}\frac{1}{\infty} + \frac{0.5 \times 10^{-4}}{26 \times 10^{-4}}\frac{1}{0.5 \times 10^9},$$

$$\frac{1}{\beta_e} = \frac{0.625}{10^9} + \frac{0.077}{10^9} = \frac{0.702}{10^9} \rightarrow \beta_e = 1.42 \text{ GN/m}^2.$$

Therefore, the use of short isolating flexible hose at either end, each having a volume of just 2% of the steel pipe volume, has reduced the oil bulk modulus by 11.3%.

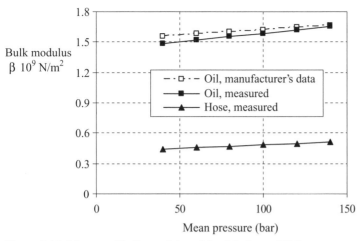

Figure 2.10. Measured bulk modulus of flexible hose, 50°C.

2.5 Fluid Cleanliness

Fluid cleanliness is crucial for hydraulic circuits, and every effort should be made to ensure effective filtration and a means of regularly checking the fluid condition. The importance of this becomes apparent when typical clearances of hydraulic flow paths are considered, as shown in Fig. 2.11.

Filters must be capable of removing particles having dimensions varying from 0.05 to 100 μm. In reality, a lower limit is usually 2–5 μm because of both filter pressure-drop limitations and a view held by some that lower values remove additives, resulting in a poorer lubrication characteristic. Some common sources of contamination are:

- Inherent contamination within new components
- Inherent particles in new–replenishing oil
- Particle ingress; for example, through a tank filler–breather

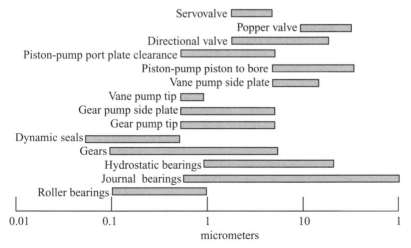

Figure 2.11. Typical clearances for hydraulic flow paths (data supplied by Star Hydraulics, UK).

- Particle ingress through cylinder rod external seals
- Particles introduced during maintenance

The process of contamination wear arises from a number of sources:

- Abrasive wear that is due to particles moving between adjacent moving surfaces
- Erosive wear that is due to particles moving within a high-velocity fluid
- Erosive wear that is due to cavitation, air bubbles being generated within a low-pressure region, and carried to a high-pressure region, causing bubble implosion followed by shock waves that can cause rapid material removal
- Water–oil separation for water-based fluids, causing loss of lubricating properties
- Adhesive wear that is due to metal-to-metal contact; lubrication films are destroyed when fine clearances are demanded to reduce flow losses
- Corrosive wear that is due to water or chemicals reacting with metallic components, particularly for water-based fluids

Considering a mineral oil in its barrel condition, measurements have shown the number of particles/milliliter of oil greater than 2 μm can be as high as 33,000. This can rise further to 80,000 for contaminated oil, falling to 10,000 for good filtration and even down to 80 for clearance-protected filtration but at a higher cost. To aid particle classification, the ISO 4406 code has evolved in conjunction with automatic particle counters that can be used in-line or off-line with a number of oil-sampling techniques.

To quantify a cleanliness requirement in practice, it is then a matter of specifying the required codes at particle sizes of 4 μm, 6 μm, and 14 μm. This is usually referred to as 4/6/14(C), indicating a certified count. For applications using microscopes for particle counting or counters pre-1999, two sizes of 5 μm and 15 μm are acceptable, and the classification is usually referred to as −/5/15. Table 2.6 shows the code. The appropriate 4 μm, 6 μm, and 14 μm ISO 4406 points are:

Code 17	4 μm	640–1300 particles/ml
Code 15	6 μm	160–320 particles/ml
Code 12	14 μm	20–40 particles/ml

In addition to particle-size analysis, it can be equally important to determine the nature of the particles. This requires more advanced equipment, such as scanning electron microscopes, image analysis, debris detectors, and ferrographic analysis. These facilities are usually made available through specialist oil-analysis companies that undertake planned sampling and analysis followed by a comprehensive report presentation. The type of particles can give an indication of where the wear is occurring and may lead to a quicker resolution of the problem, often with an effective reduction in cost and, of course, the consequences of failure.

2.6 Fluid Vapor Pressure and Cavitation

Cavitation isan extremely important phenomenon to avoid in fluid power systems design because of its often catastrophic damaging effect. Cavitation arises in perhaps the following main areas:

(i) At pump inlets where a combination of high viscosity (e.g., under low temperature conditions) and/or poor inlet pipe diameter design because of its small diameter, low head, and long length.

Table 2.6 *ISO 4406 standard for particle count classification*

Number of particles/milliliter oil			Range code
More than	Up to and including		
80,000→	160,000		24
40,000→	80,000		23
20,000→	40,000		22
10,000→	20,000		21
5000→	10,000		20
2500→	5000		19
1300→	2500		18
640→	**1300**	**4 µm**	**17**
320→	640		16
160→	**320**	**6 µm**	**15**
80→	160		14
40→	80		13
20→	**40**	**14 µm**	**12**
10→	20		11
5→	10		10
2.5→	5		9
1.3→	2.5		8
0.64→	1.3		7
0.32→	0.64		6

(ii) A lack of efficient filtration at the pump particularly when high-water-based fluids are used, in which larger size filters are used compared with mineral oil applications. These should be located in the high-pressure line and in the return line to the reservoir after the load valve, or as recommended. Generally, the volumetric capacity of filters should be such that they are able to pass two–three times the output of the pump at the operating viscosity.

(iii) The lack of a positive static fluid head at the pump inlet, and a value of at least 0.5 m, obtained from a reservoir located above the level of the pump, is usually adequate to provide a sufficient suction port pressure.

(iv) As a result of high-velocity jets caused by either badly designed flow paths within components and/or small control areas such as orifices or spool ports in control valves.

A graph of vapor-pressure variation with temperatures for water and a typical mineral oil is shown as Fig. 2.12.

It can be seen that vapor-pressure effects are more problematic for water, the value for oil being typically 0.06 N/m^2 at 40°C. Over usual working temperatures, mineral oil has a vapor pressure typically less than 1.0 N/m^2. This means that a pump operating with mineral oil would require a negative suction head equivalent to about 1 bar to create a cavitation problem. This is possible because of inlet pipe friction losses if the pipe is incorrectly sized; this problem will be pursued later. A similar effect is, of course, achieved if a high fluid viscosity occurs, such as at a low temperature, and particularly for some water-based fluids.

Cavitation can rapidly remove metallic material and has been known to destroy a component in a matter of minutes; for example, see the effects of cavitation on a PRV, shown in Fig. 2.13.

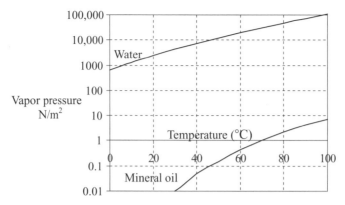

Figure 2.12. Vapor pressure for water and a typical mineral oil.

More detailed information on fluid maintenance and system condition monitoring may be found in Watton (2007).

It might be expected that cavitation erosion rate of wear is dependent on the material type, its surface hardness characteristic, and the fluid being used. A comprehensive study by Urata (1998, 2002) looked at this issue, using a comprehensive range of material specimens and fluids. A commercially available vibratory horn is used at a frequency of 19.5 kHz and 30-μm amplitude, the test specimen being attached to the end of the horn and placed within the working fluid, as shown in Fig. 2.14.

(a)

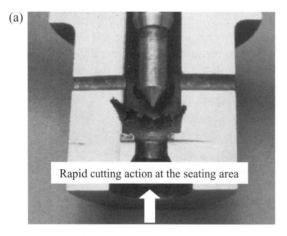

Rapid cutting action at the seating area

Figure 2.13. Severe damage to a PRV body that is due to cavitation with a water-based fluid.

(b)

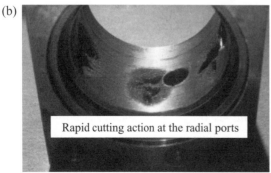

Rapid cutting action at the radial ports

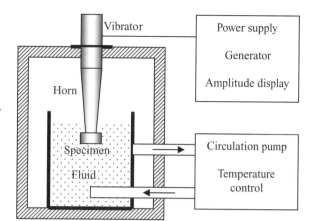

Figure 2.14. Test setup for creating cavitation erosion.

A cavitation cloud of bubbles is formed at the lower surface of the specimen, and it is the continual implosion of these bubbles that creates the erosion. Some highly selected results from the large amount of data presented are shown in Fig. 2.15 for metallic materials and for different erosion times.

It can be seen from the results that the type of material and its surface hardness can significantly change the mass loss that is due to cavitation. It does seem from the data that Silicolloy-A2 had the higher resistance to erosion than other metals, including a titanium alloy. Also, low-temperature thermal-sprayed ceramics were easily peeled off by the cavitation attack and were much poorer than common metals. It was also found that erosion is less when the fluid is a mineral oil or oil-in-water emulsion, compared with pure water, as might be expected. It is also interesting to note that for the 39 different fluids used, water produced the maximum erosion.

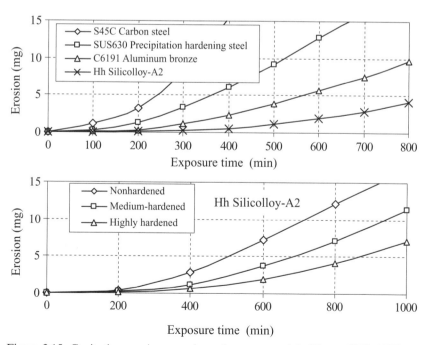

Figure 2.15. Cavitation erosion mass loss of some materials (Urata, 1998, 2002).

2.7 Electrorheological (ER) Fluids and Magnetorheological (MR) Fluids

ER and MR fluids have been known since the 1940s and are now eventually evolving, albeit slowly, as possible alternative candidates for some aspects of fluid power. These fluids exhibit a drastic change in their rheological behavior when exposed to a magnetic or electric potential across the fluid volume and because of magnetic particles in suspension within the fluid. The net result may be considered as producing a massive increase in viscosity; hence, the potential for torque transmission and suspension controlled damping and/or active control. This transformation in the fluid characteristic is due to the initially weak random dipole distributions of magnetic particles becoming realigned in a preferred direction, creating a much stronger dipole moment. ER fluids require high voltages and low current, whereas MR fluids require high currents at low voltage; therefore, MR fluids tend to be preferred. It does seem that applications show promise at the microactuator level and, therefore, MR actuators are not a threat to or even a competitor with traditional fluid power actuators. A very useful overview of the subject may be found in Agrawal et al. (2001) and Elliott (2007).

Typical MR fluids contain 20%–40% by volume iron powder suspended in a carrier medium. The maximum yield stress and the off-state viscosity are both functions of the volume fraction, so a compromise is typically required. Reduced carbonyl iron (CI) powder is the most popular choice of iron powder for MR fluids; the particles have a spherical shape that makes them robust and durable, and the magnetic properties are enhanced by the typically high level of chemical purity (99.5% pure). The carrier fluid chosen is dependent on the application, but common fluids include petroleum-based oils, silicone oils, mineral oils, and synthetic hydrocarbon oils. The off-state viscosity of the fluid is largely dependent on the carrier fluid and typically ranges from 0.01 to 1.0 N s/m^2 at a temperature of 40°C. Maintaining a well-dispersed mixture is critical to the consistent performance of MR devices. The suspension of the iron particles in the host solution is achieved through the use of additives that inhibit settling and agglomeration (Elliott, 2007).

The behavior of an MR fluid is usually characterized by the Bingham shear-stress model and related to the yield stress τ_y as follows:

$$\tau = G\gamma, \quad \tau < \tau_y, \text{ no flow,}$$

$$\tau = \tau_y + \eta\dot{\gamma}, \quad \tau \geq \tau_y, \text{ flow,} \qquad (2.21)$$

where γ is the strain, $\dot{\gamma}$ is the shear rate, and G (N/m^2) is the complex modulus. The magnetic characteristics of a MR fluid show a nonlinear behavior; for example, Fig. 2.16 shows the relationship between the yield stress τ_y and the magnetic-field strength H together with the relationship between the flux density B and the magnetic-field strength H.

To be able to utilize the properties of an MR fluid, it must be integrated within a correctly designed electromagnetic circuit to provide the appropriate magnetic flux density – for example, if used for a vehicle damper. Hence, design knowledge of electrical and electromagnetic circuits is required. A cobalt–iron alloy such as Permandur 49 may be used for the magnetic core because it has a very high maximum flux density and has the advantage that it allows the minimum cross-section to be used. This results in a more compact and lighter design.

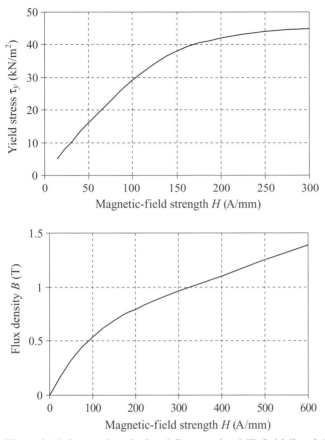

Figure 2.16. Properties of a Lord Corporation MR fluid (Lord, 2002).

If an MR fluid is to be used within boundaries that move, such as a vehicle damper, then seal design must also be considered. MR fluids have a poor compatibility with some common seal materials, and care has to be taken with seal selection; for example, the use of special materials (www.gtweed.com).

The result of a detailed design study (Elliott, 2007) produced a new MR damper design that has been manufactured, and four units fitted to the suspension of a small racing car designed and built by mechanical engineering undergraduate students at Cardiff University. The MR damper is shown in Fig. 2.17 and combines an electromagnetic circuit and MR fluid integrated within a gas spring unit.

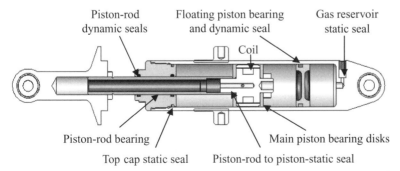

Figure 2.17. MR damper unit indicating sealing requirements (Elliott, 2007).

To determine the damper force–velocity relationship, a swept sine wave test was used over a range of current inputs. The position input signal consecutively cycled the damper with 10-mm amplitude from 1 to 8 rad/s (0.159 to 1.273 Hz) in 1-rad/s steps. This provided a detailed view of the damper performance in the 0–80 mm/s range. The test was repeated from 0 to 2.0 A applied coil currents in steps of 0.2 A. Figure 2.18 shows the results for an isolated single damper sweep with a peak velocity of +/− 70 mm/s. The relationship among the force, velocity, stroke, and coil current can be seen.

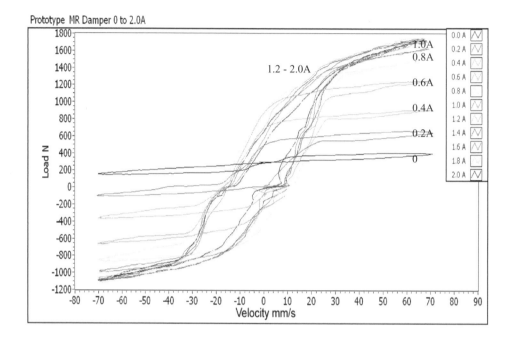

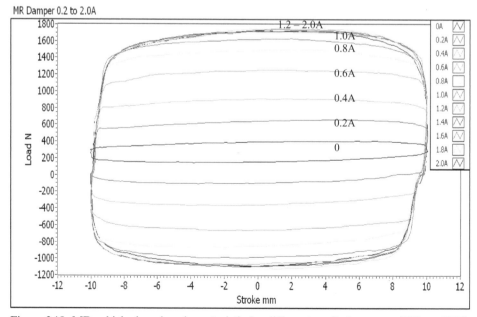

Figure 2.18. MR vehicle damping characteristic for different applied currents (Elliott, 2007).

Figure 2.19. Cardiff Racing CR03 race vehicle (Elliott, 2007).

The velocity damping was small in comparison with the damping force produced by the MR–coil current effect, and it is also evident that the velocity damping gradient is not greatly affected by the coil current. It is thought that the slope of the force–velocity plot at low speeds (less than 20 mm/s) is due to a bypass bleed of

(a) Rear MR dampers and position sensors

(b) Front MR dampers and position sensors

Figure 2.20. Installation of the MR dampers on the Cardiff CR03 race vehicle.

fluid around the edge of the main piston. Notice also the hysteresis effect that is always present with this type of MR fluid damping. The damper force properties in compression are the same as those in rebound, with the obvious exception of the offset load that is due to the gas precharge used to prevent cavitation (Elliott, 2007).

This MR damper now has a variable-force characteristic that may be adjusted while the vehicle is in motion using vehicle motion sensors and an onboard computer. This system has been successfully applied, and details may be found in Elliott (2007). The car and MR dampers are shown in Figs. 2.19 and 2.20.

2.8 References and Further Reading

Agrawal A, Kulkarni P, Vieira SL, Naganathan N [2001]. An overview of magneto and electrorheological fluids and their applications in fluid power systems. *Int. J. Fluid Power* 2(2), 5–36.

Bowns DE and Worton-Griffiths J [1973]. The effect of air in the fluid on the operating characteristics of a hydrostatic transmission. In *Proceedings of the 3rd BHRA International Fluid Power Symposium.*

Conrad F [2005]. Trend in design of water hydraulics. In *Proceedings of the 6th JFPS International Symposium on Fluid Power*, 420–431.

Dalibert A [1971]. Fire-resistant hydraulic fluids. In *Proceedings of the 2nd BHRA Fluid Power Symposium*, Paper F3.

Day MJ, Way NR, Thompson K [1987]. The use of particle counting techniques in the condition monitoring of fluid power systems. In *Condition Monitoring '87*, edited by MH Jones, Pineridge Press Ltd., 322–339.

Elliott R [2007]. Development of a magnetorheological active damping system to improve the yaw response of a racing vehicle. Ph.D. thesis, Cardiff University, School of Engineering.

Environmental Impact of Fluid Power Systems [1999]. Organized by the Institution of Mechanical Engineers. London, collection of 11 papers.

Evans JS and Hunt TM [2004]. *Oil Analysis*, Coxmoor.

Gao H, Fu X, Yang H, Tsukiji T [2002]. Numerical and experimental investigation of cavitating flow in oil hydraulic ball valve. In *Proceedings of the 5th JFPS International Symposium on Fluid Power*, 923–928.

Gohler O-C [2006]. Approach to the simulation of ageing of environmentally compatible fluids in hydraulic systems. *Int. J. Fluid Power* 7(2), 19–28.

Gohler O-C, Murrenhoff H, Schmidt M [2004]. Ageing simulation of biodegradable fluids by means of neural networks. In *Proceedings of the Power Transmission and Motion Control Workshop, PTMC 2004*. Professional Engineering Publications Ltd., 71–83.

Hayward ATJ [1961]. Aeration in hydraulic systems: its assessment and control. In *Proceedings of the Oil in Hydraulics Conference*, Institute of Mechanical Engineering, pp. 216–224.

Hodges P [1996]. *Hydraulic Fluids*, Arnold.

Hunt TM and Tilley DG [1984]. Techniques for the assessment of contamination in hydraulic oils. In *Contamination Control in Hydraulic Systems*, Institute of Mechanical Engineering, 57–63.

IMechE [1984]. *Contamination Control in Hydraulic Systems* [1984]. Institution of Mechanical Engineering, 65–77.

Iudicello F and Baseley S [1999]. Fluid-borne noise characteristics of hydraulic and electrohydraulic pumps. In *Power Transmission and Motion Control*, Professional Engineering Publications Ltd., 313–323.

Jinghong Y, Zhaoneng C, Yuanzhang L [1994]. The variation of oil effective bulk modulus in hydraulic systems. *Trans ASME, J. Dyn. Syst. Meas. Control* 116, 146–150.

Johnston DN and Edge KA [1991]. In-situ measurement of the wave speed and bulk modulus in hydraulic systems. *Proc. Inst. Mech. Eng. Part I*, 205(I3), 191–197.

Kalin M, Majdič F, Vižintin J, Perždirnik J, Velkarrh I [2008]. Analysis of the long-term performance of an axial piston pump using diamond-like carbon-coated piston shoes and biodegradable oil. *ASME J. Tribol.* 130, 011013/1–8.

Kelly ES [1973]. Fire-resistant fluids – factors in heating equipment and circuit design. In *Proceedings of the 3rd BHRA Fluid Power Symposium.*

Knight GC [1977]. Water hydraulics: Application of water-based fluids in hydraulic systems. *Tribol. Int.*, 105–107.

Lewis RT [1987]. Analysis of ferrous wear debris. In *Condition Monitoring '87*, edited by MH Jones, Pineridge Press Ltd., 360–370.

Li ZY, Yu ZY, He XF, Yang SD [1999]. The development and perspective of water hydraulics. In *Proceedings of the 4th JHPS International Symposium on Fluid Power*, 335–344.

Lichtarowicz A [1979]. Cavitating jet apparatus for cavitation erosion test. ASTM STP 664, 530–549.

Lord [2002]. MR fluid product bulletins. Available at www.lord.com.

Maxwell JF [1979]. Water-based hydraulic fluids: Coping with the steel industry – A manufacturer's point of view. *Iron Steel Eng.* 56(8), 57–60.

McCullagh P [1984]. Ferrography and particle analysis in hydraulic power systems. In *Contamination Control in Hydraulic Systems*, Institution of Mechanical Engineers, pp. 65–77.

Molyet KE, Ciocanel C, Yamamoto H, Naganathan NG [2006]. Design and performance of a MR torque transfer device. *Int. J. Fluid Power* 7(3), 21–28.

Nakano M, Yamashita K, Kawakami Y, Okamura H [2005]. Dynamic shear flow of electr-rheological fluids between two rotating plates. In *Proceedings of the 6th* JFPS International Symposium on Fluid Power, 612–617.

Nikkila P and Vilenius M [2003]. The simulation of cleanliness level in hydraulics. In *Proceedings of the 1st International Conference in Fluid Power Technology, Methods for Solving Practical Problems in Design and Control*, Fluid Power Net Publications, 233–244.

Oshima S, Leiro T, Linjama M, Koskinen KT, Vilenius M [2001]. Effect of cavitation in water hydraulic poppet valves. *Int. J. Fluid Power* 2(3), 5–14.

Price AL, Roylance BJ, Zie LX [1987]. The PQ – A method for the rapid quantification of wear debris. In *Condition Monitoring '87*, edited by MH Jones, Pineridge Press Ltd., 391–405.

Radhakrishnan M [2003]. *Hydraulic Fluids*, ASME Press.

Raw I [1987]. Particle size analyser based on filter blockage. In *Condition Monitoring '87*, edited by MH Jones, Pineridge Press Ltd., 875–894.

Riipinen H, Varjus S, Soini S, Puhakka JA, Koskinen KT, Vilenius M [2002]. Effects of microbial growth and particles on filtration in water hydraulic systems. In *Proceedings of the 5th JFPS International Symposium on Fluid Power*, 173–176.

Rinkinen J and Kiiso T [1993]. Using portable particle counter in oil system contamination control. In *Proceedings of the 3rd Scandinavian International Conference on Fluid Power*, Vol. 1, 309–328.

Ritchie T and Thomson J [1971]. An emulsifying hydraulic fluid for submarine systems. In *Proceedings of the 2nd BHRA Fluid Power Symposium.*

Roylance BJ and Hunt TM [1999]. *Wear Debris Analysis.* Coxmoor.

Sandt J, Rinkinen J, Laukka J [1997]. Particle and water on-line monitoring for hydraulic system diagnosis. In *Proceedings of the 5th Scandinavian Conference on Fluid Power*, 257–268.

Shutto S and Toscano J [2005]. Magnetorheological (MR) fluid and its application. In *Proceedings of the 6th JFPS International Symposium on Fluid Power*, pp. 590–594.

Silva G [1990]. Wear generation in hydraulic pumps. In *Proceedings of the SAE International Off-Highway and Powerplant Conference*, Society of Automotive Engineers, Paper 901679.

Smith LH, Peeler RL, Bernd LH [1960]. Hydraulic fluid bulk modulus – Its effect on system performance and techniques for physical measurement. In *Proceedings of the 16th National Conference on Industrial Hydraulics*, Vol. 14, 179–196.

Sommer HT, Raze TL, Hart JM [1993]. The effects of optical material properties on particle counting results of light scattering and extinction sensors. In *Proceedings of the 10th
   International Conference on Fluid Power – The Future for Hydraulics*, MEP Publications,
   289–308.

Stecki JS, editor [1998]. *Total Contamination Control*, Fluid Power Net.

Stecki JS, editor [2002]. *Total Contamination Control*, Fluid Power Net.

Stewart HL [1979]. Fire-resistant hydraulic fluids. *Plant Eng.* 33, 157–160.

Suzuki K and Urata E [2002]. Cavitation erosion of materials for water hydraulics. In
   Power Transmission and Motion Control 2002, Professional Engineering Publications Ltd.,
   127–139.

Tikkanen S [2001]. Influence of line design on pump performance. In *Power Transmission
   and Motion Control 2001*, Professional Engineering Publications Ltd., 33–46.

Totten GE, Reichel J, Kling GH [1999]. Biodegradable fluids: A review. In *Proceedings of
   the 4th JHPS International Symposium on Fluid Power*, 285–290.

Tsai CP [1981]. Particle counting in water-based fluids using light-blockage-type automatic
   instruments. In *Proceedings of the 6th International Fluid Power Conference*, 87–94.

Urata E [1998]. Cavitation erosion in various fluids. In *Power Transmission and Motion Control 1998*, Professional Engineering Publications Ltd., 269–284.

Urata E [2002]. Evaluation of filtration performance of a filter element. In *Power Transmission and Motion Control 2002*, Professional Engineering Publications Ltd., 291–304.

Urata E [2005]. Notes on contamination control. In *Proceedings of the 6th JFPS International
   Symposium on Fluid Power*, 629–633.

Virvalo T, Makinen E, Vilenius M [1999]. On the damping of water hydraulic cylinder drives.
   In *Proceedings of the 4th JHPS International Symposium on Fluid Power*, 351–356.

Wang X, Han B, Xa H, Tang R [1999]. On the factors influencing the bubble content in
   air/hydro system. In *Proceedings of the 4th JHPS International Symposium on Fluid Power*,
   57–62.

Warring RH [1970]. *Fluids for Power Systems*, Trade and Technical Press.

Watton J [2007]. *Modelling, Monitoring and Diagnostic Techniques for Fluid Power Systems*,
   Springer-Verlag.

Watton J and Xue [1994]. A new direct-measurement method for determining fluid bulk
   modulus in oil hydraulic systems. In *Proceedings of FLUCOME '94*, pp. 543–548.

Wei K-X, Meng G, Zhu S-S [2004]. Fluid power control unit using electrorheological fluids.
   Int. J. Fluid Power 4(3), 49–54.

Yamaguchi A, Kazama T, Inoue K, Onoue J [2001]. Comparison of cavitation erosion test
   results between vibratory and cavitating jet methods. *Int. J. Fluid Power* 2(1), 25–30.

Yu J [1991]. Measurement of oil effective bulk modulus in hydraulic systems. *Chin. Fluid
   Power Eng.* No. 3, 46–48.

Zaun M [2006]. Design of cylinder drives based on electrorheological fluids. *Int. J. Fluid
   Power* 7(1), 7–14.

Zhu S, Wei K, Wang Q, Huang Y [2005]. The response performance of electrorheological
   fluids in a control flow field. *Int. J. Fluid Power* 6(3), 25–32.

3 Steady-State Characteristics of Circuit Components

3.1 Flow Through Pipes

3.1.1 The Energy Equation

Consider first a flow through a stream tube, as shown in Fig. 3.1.

Assuming ideal steady-state flow with no friction losses and constant density, it can then be shown that because the total energy must be constant, the energy equation simplifies to:

$$P_1 + \frac{1}{2}\rho U_1^2 + \rho g h_1 = \text{constant} = P_2 + \frac{1}{2}\rho U_2^2 + \rho g h_2, \qquad (3.1)$$

where pressures are absolute. Considering flow through a single pipe, it is unusual in fluid power to have a single pipe of significant length with a changing cross-sectional area, and so Eq. (3.1) becomes:

$$P_1 + \rho g h_1 = P_2 + \rho g h_2. \qquad (3.2)$$

Usually in high-pressure fluid power systems, the head effect is negligible, apart from pump inlet conditions, where the possibility of cavitation can exist. Assuming a steady-state flow rate Q then results in the simple flow-rate continuity equation:

$$Q = U_1 A_1 = U_2 A_2. \qquad (3.3)$$

For example, if a pump inlet is a distance H below the tank supply and the tank is at atmospheric pressure P_o, as shown in Fig. 3.2, then applying the energy equation gives the pump inlet pressure:

$$P_o + \rho g H = P_i. \qquad (3.4)$$

It is important that the pump inlet pressure not reach the fluid vapor pressure, which varies with temperature as discussed in Chapter 2, this being possible if the pump in Fig. 3.2 is drawing fluid from a tank positioned below its inlet. The inlet pressure would then be:

$$P_i = P_o - \rho g H. \qquad (3.5)$$

As will be shown later, this inlet pressure is depressed further in the presence of real pipe friction because of fluid viscosity effects, and the magnitude of the flow rate becomes important. This aspect is now considered.

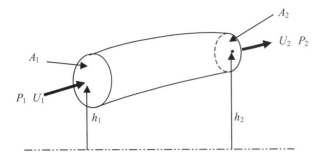

Figure 3.1. Flow through a stream tube.

3.1.2 Laminar and Turbulent Flow in Pipes; the Effect of Fluid Viscosity

If flow with a uniform velocity enters a pipe, then the effect of viscosity is to create a boundary layer that builds up from the wall, where the velocity must be zero, until fully developed flow is established, as shown in Fig. 3.3.

It is well established that as the flow rate is increased, then distinct flow regimes exist, as first established and published by Osborne Reynolds in 1883 (born in Ireland in 1842 and died in England in 1905). Initially, the fully developed flow velocity profile is parabolic, the regime known as laminar flow, and as the flow is increased, a transition stage occurs, following which the velocity profile adopts a flatter characteristic. The regime is known as turbulent flow, and the velocity profile has an associated, small, superimposed random turbulence component. This is shown schematically in Fig. 3.4 for profiles having the same mean velocity.

The mean velocity is closer to the maximum velocity for turbulent flow. Osborne Reynolds developed the concept of the Reynolds number Re to characterize these different flow regimes and embodied the fact that the pressure drop down a pipe, ΔP, varies typically with mean velocity, as shown in Fig. 3.5.

The transition between fully developed laminar flow and fully developed turbulent flow is "fuzzy" in the sense that there is no unique representation or equation to rigorously define it. However, it is clear that the relationship between pressure drop and mean velocity, or flow rate, varies from a linear form at low velocities to a nonlinear form at high velocities. This is more conveniently defined in terms of the Reynolds number Re.

3.1.3 The Navier–Stokes Equation

When analyzing laminar flow problems, the equations of motion in general x, y, z coordinates are represented by the general Navier–Stokes equation (after Lois M H

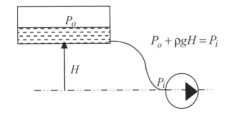

$$P_o + \rho g H = P_i$$

Figure 3.2. Pump inlet pressure.

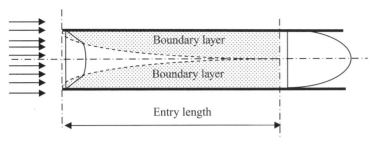

Figure 3.3. Establishment of a velocity profile in a pipe.

Navier, a French engineer, 1785–1836, and George G. Stokes, 1819–1903, a mathematician born in Ireland):

$$\rho \frac{DU}{Dt} = \rho F - \text{grad } p + \mu \nabla^2 U + \frac{\mu}{3} \text{grad}(\text{div } U),$$

$$\text{div } U = \nabla \cdot U = \frac{\partial u}{\partial x} + \frac{\partial v}{\partial y} + \frac{\partial w}{\partial z} \quad \text{grad } \phi = \nabla \phi = i \frac{\partial \phi}{\partial x} + j \frac{\partial \phi}{\partial y} + k \frac{\partial \phi}{\partial z},$$

(3.6)

where D/Dt represents the total differential and constant viscosity is assumed. For incompressible flow, the final term in (3.6) is zero and the three components are then given by Kay (1963):

$$\rho \left(\frac{\partial u}{\partial t} + u \frac{\partial u}{\partial x} + v \frac{\partial u}{\partial y} + w \frac{\partial u}{\partial z} \right) = \rho F_x - \frac{\partial p}{\partial x} + \mu \left(\frac{\partial^2 u}{\partial x^2} + \frac{\partial^2 u}{\partial y^2} + \frac{\partial^2 u}{\partial z^2} \right),$$

$$\rho \left(\frac{\partial v}{\partial t} + u \frac{\partial v}{\partial x} + v \frac{\partial v}{\partial y} + w \frac{\partial v}{\partial z} \right) = \rho F_y - \frac{\partial p}{\partial y} + \mu \left(\frac{\partial^2 v}{\partial x^2} + \frac{\partial^2 v}{\partial y^2} + \frac{\partial^2 v}{\partial z^2} \right), \quad (3.7)$$

$$\rho \left(\frac{\partial w}{\partial t} + u \frac{\partial w}{\partial x} + v \frac{\partial w}{\partial y} + w \frac{\partial w}{\partial z} \right) = \rho F_z - \frac{\partial p}{\partial z} + \mu \left(\frac{\partial^2 w}{\partial x^2} + \frac{\partial^2 w}{\partial y^2} + \frac{\partial^2 w}{\partial z^2} \right),$$

where u, v, w are the fluid velocity components in the x, y, z directions; F_x, F_y, F_z are the body forces in the x, y, z directions; and p is the pressure. In addition, flow continuity for incompressible flow gives:

$$\frac{\partial \rho}{\partial t} + \rho \text{ div } U = 0,$$

$$\frac{\partial u}{\partial x} + \frac{\partial v}{\partial y} + \frac{\partial w}{\partial z} = 0.$$

(3.8)

Considering the most common first two equations (3.7) and (3.8), the Navier–Stokes equations reduce to the two-dimensional (2D) form *for steady-state incompressible flow with no body forces*:

$$\rho \left(u \frac{\partial u}{\partial x} + v \frac{\partial u}{\partial y} \right) = -\frac{\partial p}{\partial x} + \mu \left(\frac{\partial^2 u}{\partial x^2} + \frac{\partial^2 u}{\partial y^2} \right),$$

$$\rho \left(u \frac{\partial v}{\partial x} + v \frac{\partial v}{\partial y} \right) = -\frac{\partial p}{\partial y} + \mu \left(\frac{\partial^2 v}{\partial x^2} + \frac{\partial^2 v}{\partial y^2} \right).$$

(3.9)

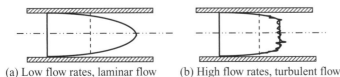

(a) Low flow rates, laminar flow   (b) High flow rates, turbulent flow

Figure 3.4. Laminar and turbulent velocity profiles for pipe flow.

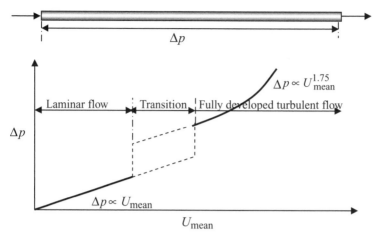

Figure 3.5. Pressure-drop variation with mean velocity for flow in a pipe.

For the case in which flow axial symmetry exists, then a cylindrical coordinate transformation gives:

$$\rho \left(u \frac{\partial u}{\partial x} + v \frac{\partial u}{\partial r} \right) = -\frac{\partial p}{\partial x} + \mu \left(\frac{\partial^2 u}{\partial x^2} + \frac{\partial^2 u}{\partial r^2} + \frac{1}{r} \frac{\partial u}{\partial r} \right),$$

$$\rho \left(u \frac{\partial v}{\partial x} + v \frac{\partial v}{\partial r} \right) = -\frac{\partial p}{\partial r} + \mu \left(\frac{\partial^2 v}{\partial x^2} + \frac{\partial^2 v}{\partial r^2} + \frac{1}{r} \frac{\partial v}{\partial r} - \frac{v}{r^2} \right). \tag{3.10}$$

3.1.4 Laminar Flow in a Circular Pipe

To apply the Navier–Stokes equation to incompressible steady-state flow through a smooth pipe, it is easier to use cylindrical coordinates (see Fig. 3.6). Only the first equation from (3.10) is required because this includes u velocity elements, whereas the co-equation includes v velocity elements, which are all zero:

$$\frac{dp}{dx} = \mu \left(\frac{d^2 u}{dr^2} + \frac{1}{r} \frac{du}{dr} \right) = \frac{\mu}{r} \frac{d}{dr} \left(r \frac{du}{dr} \right). \tag{3.11}$$

Integrating Eq. (3.11) twice with respect to r gives:

$$\frac{du}{dr} = \frac{r}{2\mu} \frac{dp}{dx} + \frac{C_1}{r} \rightarrow u = \frac{r^2}{4\mu} \frac{dp}{dx} + C_1 \ln(r) + C_2; \tag{3.12}$$

when $r = 0$, u must be finite so $C_1 = 0$;
when $r = r_0$, then $u = 0$, giving C_2;

$$u = -\frac{1}{4\mu} \frac{dp}{dx} \left(r_0^2 - r^2 \right). \tag{3.13}$$

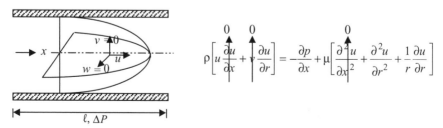

Figure 3.6. Laminar flow in a circular pipe.

Maximum velocity at the centerline:

$$u_{\max} = \frac{r_0^2}{4\mu} \frac{\Delta p}{\ell}. \tag{3.14}$$

Given that the pressure gradient:

$$\frac{dp}{dx} = -\frac{\Delta p}{\ell}, \tag{3.15}$$

$$u = \frac{1}{4\mu} \frac{\Delta p}{\ell} \left(r^2 - r_0^2\right). \tag{3.16}$$

The flow rate:

$$Q = \int_0^{r_0} u\, 2\pi r\, dr = \frac{\pi r_0^4}{8\mu} \frac{\Delta p}{\ell}, \tag{3.17}$$

and mean velocity:

$$u_{\text{mean}} = \frac{Q}{\pi r_0^2} = \frac{u_{\max}}{2}. \tag{3.18}$$

When considering laminar flow problems, it is common to specify the pressure–flow ratio as the resistance by analogy with electrical circuits in which pressure is analogous to voltage and flow rate is analogous to current. From Eq. (3.17),

$$\Delta p = RQ \rightarrow R = \frac{128\mu\ell}{\pi d^4}, \tag{3.19}$$

where d is the pipe diameter.

This is often referred to as the Hagen–Poiseuille equation (after GHL Hagen, a German engineer, 1797–1884, and JL Poiseuille, a French physicist, 1799–1869). The sensitivity of the pipe resistance to its diameter should be noted; hence, the need for a pipe to be correctly sized when flow rate and fluid viscosity are taken into account.

Worked Example 3.1

Considering flow through a pipe as previously analyzed, derive the basic differential equation for 2D flow.

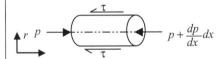

Worked Example 3.1

Consider symmetry about the centerline and equate forces:

$$\tau(2\pi r\, dx) + \pi r^2 \frac{dp}{dx} dx = 0 \rightarrow \tau = -\frac{r}{2} \frac{dp}{dx}.$$

Add Newton's law of viscosity and note that as r increases, the velocity decreases:

$$\tau = -\mu \frac{dp}{dx} \rightarrow \mu \frac{du}{dr} = \frac{r}{2} \frac{dp}{dx}.$$

3.1.5 The General Pressure-Drop Equation

The Reynolds number is defined as

$$R_e = \frac{U_{mean}d}{\nu}, \tag{3.20}$$

where d is the pipe diameter, U_{mean} is the mean velocity, and ν is the fluid kinematic viscosity. The transition region occurs for typically $2000 < Re < 4000$, the lower value often being used to define the upper limit for laminar flow. For laminar flow, as previously developed, the pressure-drop–flow-rate equation $\Delta P/Q$ is constant for a fixed pipe geometry and fluid viscosity as given by the Hagen–Poiseuille equation (3.19). However, when considering turbulent flow, it should be sensed from Fig. 3.5 that it is not possible to derive a pressure-drop–flow-rate equation in the way that has been done for fully developed laminar flow.

The main developments were made by means of postulated basic equations with constants derived from experimental data and evolved from the studies of Prandtl (Ludwig Prandtl, a German engineer, 1875–1953) and von Kármán (Theodore von Kármán, a Hungarian engineer, 1881–1963). Prandtl's mixing-length theory and the postulation of a 1/7 power law for the velocity distribution, combined with the Reynolds shear-stress postulation, lead to $U_{mean} = 0.82U_{max}$ but for a restricted range of Re. This is therefore of limited use in practice and to overcome this, a general equation is used to calculate pressure drop down a pipe by means of the use of the friction factor $4f$:

$$\Delta p = 4f \left(\frac{1}{2}\rho U_{mean}^2 \right) \left(\frac{\ell}{d} \right),$$

$$\text{laminar flow } 4f = \frac{64}{Re}, \quad Re < 2000, \tag{3.21}$$

$$\text{turbulent flow } 4f = \frac{0.316}{Re^{1/4}}, \quad Re > 3000.$$

Experimental work for flow in the transition region has led to useful equations such as the Colebrook–White formula, but it seems to have little use in fluid power. In addition, much work has been done on the effect of pipe roughness and the application of the Moody diagram to represent the full spectrum of Re and pipe condition or type. Again, pipe roughness tends to be a secondary issue when fluid power pressure-drop calculations are performed. The standard fluid mechanics textbooks can be used to further investigate the Moody diagram.

Worked Example 3.2

The inlet pipe to a pump is 20 mm in diameter and 2.5 m long. If the tank, at atmospheric pressure, is positioned 1 m above the pump inlet and the pump flow rate is 90 L/min when losses are taken into account, calculate the pump inlet pressure. Assume a high-water-content fluid having $\mu = 0.001$ N s/m², $\rho = 10^3$ kg/m³. At what temperature would the pump cavitate?

The solution to this problem is to consider both the energy equation and the pressure-drop equation. First, the Re number is calculated:

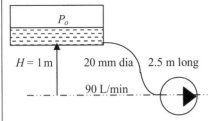

$H = 1\,m$ 20 mm dia   2.5 m long

P_o

90 L/min

Worked Example 3.2

$$\text{pump inlet conditions } U_{\text{mean}} = \frac{90 \times \dfrac{10^{-3}}{60}}{\dfrac{\pi \times 0.02^2}{4}} = 4.77 \text{ m/s},$$

$$\text{Re} = \frac{1000 \times 4.77 \times 0.02}{0.001} = 95,400.$$

Hence, the flow is turbulent. From earlier work, it is deduced that

$$\text{turbulent flow } 4f = \frac{0.316}{\text{Re}^{1/4}} = \frac{0.316}{95400^{1/4}} = 0.018,$$

$$\Delta p = 4f \left(\frac{1}{2} \rho U_{\text{mean}}^2 \right) \left(\frac{\ell}{d} \right) = 0.018 \left(\frac{1}{2} \times 1000 \times 4.77^2 \right) \left(\frac{2.5}{0.02} \right),$$

$$\Delta P = 0.26 \times 10^5 \text{ N/m}^2 (0.26 \text{ bar}).$$

The absolute inlet pressure at the pump is, therefore:

$$P_i = P_o + \rho g H - \Delta p = 10^5 + 0.1 \times 10^5 - 0.26 \times 10^5 \text{ N/m}^2,$$

$$P_i = 0.84 \times 10^5 \text{ N/m}^2 (0.94 \text{ bar}).$$

Considering the data on water vapor pressure given in Chapter 2, it can be seen that the temperature can increase to a value of around 90°C – that is, almost to the boiling point of water before cavitation occurs. The pressure drop that is due to pipe friction exceeds the pressure-head advantage of the raised tank.

Worked Example 3.3

A 400 MN press used for the forging of aerospace components by virtue of its size must have the axial piston pumps positioned some distance from the press and preferably below ground level. In such an application, the line diameter is 77 mm and the line length is 150 m. If the maximum pump flow rate is 4200 L/min per main cylinder and there are four cylinders, calculate the power dissipated during the pressing operation because of flow friction losses alone. Assume mineral oil having $\mu = 0.025$ Ns/m^2, $\rho = 860$ kg/m^3:

$$U_{\text{mean}} = \frac{4200 \times \dfrac{10^{-3}}{60}}{\dfrac{\pi \times 0.077^2}{4}} = 15 \text{ m/s}, \quad \text{Re} = \frac{860 \times 15 \times 0.077}{0.025} = 39,732.$$

The flow is turbulent, giving the friction factor as:

$$\text{turbulent flow } 4f = \frac{0.316}{\text{Re}^{1/4}} = \frac{0.316}{39732^{1/4}} = 0.0224,$$

$$\Delta P = 4f\left(\frac{1}{2}\rho U_{\text{mean}}^2\right)\left(\frac{\ell}{d}\right) = 0.0224\left(\frac{1}{2} \times 860 \times 15^2\right)\left(\frac{150}{0.077}\right),$$

$$\Delta P = 42.2 \times 10^5 \text{ N/m}^2 \text{ (42.2 bar)}.$$

The power dissipated down the four lines is $4 \times 0.07 \times (42.2 \times 10^5) = 11.82 \times 10^5$ W. The power dissipated is 1.18 MW.

This very large power loss that is due to pipe viscous friction losses alone indicates the power supply needed to operate a modern forging press of this size.

3.1.6 Temperature Rise in 3D Flow

For the numerical evaluation of internal flows – for example, flow between a pump barrel and its port plate – it is important to know the localized temperature because its value determines the local viscosity more significantly than other fluid properties. The energy equation is then required in conjunction with the Navier–Stokes equation. Considering the first law of thermodynamics for a fluid element gives (Kay, 1963):

$$\begin{aligned}&\text{rate of heat supply by condition } + \text{ rate of work by surface stresses}\\&= \text{rate of gain of internal energy.}\end{aligned} \tag{3.22}$$

Considering the energy–enthalpy equation and an ideal fluid then gives:

$$\rho J c_p \frac{DT}{Dt} = \frac{Dp}{Dt} + Jk\nabla^2 T + \Phi,$$

$$\nabla^2 T = \frac{\partial^2 T}{\partial x^2} + \frac{\partial^2 T}{\partial y^2} + \frac{\partial^2 T}{\partial z^2}, \tag{3.23}$$

where T is the temperature, k is the thermal conductivity, J is the mechanical equivalent of heat, and c_p is the fluid specific heat at constant pressure. Φ is the rate of work done by the stresses in distorting the fluid and is dissipated within the fluid as heat and is also referred to as the energy dissipation:

$$\Phi = \mu \begin{bmatrix} 2\left(\dfrac{\partial u}{\partial x}\right)^2 + 2\left(\dfrac{\partial v}{\partial y}\right)^2 + 2\left(\dfrac{\partial w}{\partial z}\right)^2 \\[2mm] + \left(\dfrac{\partial u}{\partial y} + \dfrac{\partial v}{\partial x}\right)^2 + \left(\dfrac{\partial v}{\partial z} + \dfrac{\partial w}{\partial y}\right)^2 + \left(\dfrac{\partial w}{\partial x} + \dfrac{\partial u}{\partial z}\right)^2 \\[2mm] - \dfrac{2}{3}(\text{div } u)^2. \end{bmatrix} \tag{3.24}$$

For steady-state conditions, then the temperature distribution within the fluid is given by Laplace's equations from (3.23) as:

$$\frac{\partial^2 T}{\partial x^2} + \frac{\partial^2 T}{\partial y^2} + \frac{\partial^2 T}{\partial z^2} = 0. \tag{3.25}$$

```
                                            ┌──────────────────────────┐
                                     ┌─────▶│     Update properties     │
                                     │      └──────────────────────────┘
                                     │                    │
                                     │                    ▼
                                     │      ┌──────────────────────────┐
                                     │      │ Solve the momentum equations │
                                     │      └──────────────────────────┘
                                     │                    │
                                     │                    ▼
                                     │   ┌────────────────────────────────────┐
                                     │   │ Solve the pressure-corrected (continuity) equation │
                                     │   │   Update pressure, face mass flow rate   │
                                     │   └────────────────────────────────────┘
                                     │                    │
                                     │                    ▼
                                     │   ┌────────────────────────────────────┐
                                     │   │ Solve energy, species, turbulence, and other │
                                     │   │            scalar equations          │
                                     │   └────────────────────────────────────┘
                                     │                    │
                                     │                    ▼
                                     └────◀ Convergence ? ──────▶ Stop
```

Figure 3.7. Flow diagram of the iterative process (www.ansys.com).

3.1.7 Computational Fluid Dynamics (CFD) Software Packages

CFD numerical analysis packages are now commonly used to model complex flow patterns within fluid power components and many other aspects of engineering, and now to a high degree of accuracy for both laminar and turbulent flow conditions. They are frequently used for both undergraduate projects work and postgraduate research work. The various notes and results in this book have been extracted from undergraduate and postgraduate projects undertaken by the author and J Thorpe, R Worthing, J Speedy, J Rhind-Tutt, and JM Haynes, using the ANSYS FLUENT software package. This applies governing integral equations for the conservation of mass, momentum, and other selected scalars such as turbulence to a reference domain. The domain is divided into discrete control volumes by a computational grid. The independent variables such as velocity, pressure, and conserved scalars are found by integration of the governing equations on the individual control volumes to construct algebraic equations for the dependent variables. Linearization of the discretized equations and solution of the resultant linear equation system will yield updated values of the dependent variables. A choice of numerical methods can be selected, the segregated solver or the coupled solver. Both solvers adopt a similar discretization process, and the iteration procedure is shown in Fig. 3.7.

GAMBIT was the former preprocessor for this software within which the geometries of the "reference domain" and "discrete control volumes" are generated, and it is possible to create 2D and 3D models containing structured and unstructured meshes. The discrete control volumes are created by splitting up the geometry using a mesh. Accuracy and stability of the model rely heavily on the quality of the mesh and there are tools available to measure parameters of mesh quality: the skewness and aspect ratio of each element can be calculated to evaluate the quality of the mesh. The former meshing technology within GAMBIT has now been integrated into ANSYS meshing.

The *mesh density* in important areas of the model should be increased because poor resolution in critical areas can dramatically affect the flow characteristics. Critical areas include shear areas, sudden expansion, and boundary layers. The FLUENT Users Guide states that no flow passage should be represented by fewer than five cells. The Users Guide recommends, when turbulent flow is modeled, that the regions where the mean flow changes dramatically should have a finer mesh because of the strong interaction between mean flow and turbulence. There are

two possible types of *skewness: size* and *angle*. Skewness calculates the difference between the size or angle of every cell compared with that of an equilateral cell of the same volume. The end result is in the form of a ratio between 0 and 1, where 0 is the optimum value for a quality mesh. Cells that have a skewness value nearing 1 can reduce the stability and accuracy of the mesh. Quadrilateral-type meshes will perform at a greater accuracy when cell vertex angles are close to $90°$ and $60°$ for triangular meshes. If very skewed elements do exist in the mesh, they can be tolerated if they are located in insignificant flow regions but can be damaging where dramatic flow changes occur. For 2D modeling, triangular and quadrilateral cell types are acceptable. For 3D modeling, tetrahedron, hexahedron, prism–wedge, and pyramid cells are used. Each of the different cell types can be arranged in a structured or unstructured format; unstructured meshes are used for more complex geometries because the benefits of using a complicated structured mesh would be lost. *When choosing a mesh type*, there are main issues to consider:

- *Setup time.* If models are of complex geometry, creating a structured mesh with quadrilateral and hexahedral cells can be very time-consuming and, in some cases, impossible; therefore, setup time for complex geometries is a major motivation for using unstructured grids with triangular and tetrahedral cells.
- *Computational expense.* To reduce run time, it is suggested that a triangular–tetrahedral mesh be used because it requires fewer cells but intelligently populated in important flow regions, compared with quadrilateral–hexahedral cells being forced into regions where it is not necessary.
- *Numerical diffusion* is a major source of error. Numerical diffusion is a problem in all practical numerical schemes because numerical diffusion arises from truncation errors that are a consequence of representing the fluid flow equations in discrete form. Techniques are available to reduce the amount of numerical diffusion. Aligning the mesh with flow can reduce numerical diffusion; however, the use of a triangular–tetrahedral-type mesh can never be aligned with the flow. With the use of a quadrilateral/hexahedral type mesh, it is possible to align flow and mesh but only for simple geometries such as flow in a long duct.

Turbulent flow is characterized by fluctuating velocities, and these fluctuations can be of small scale and high frequency. To model these small-scale fluctuations, an extremely dense mesh of impractical computational expense would be needed. Instead, the governing equations are time averaged to remove the small-scale changes in the velocity and momentum. This results in a set of equations that are less resource-consuming to solve. The new equations contain a set of unknown variables, and turbulence models are needed to determine these variables in terms of the known quantities. It is difficult to decide which turbulence model best suits a specific simulation, and it will depend on certain factors such as characteristics featured in flow, computational resources, and time available for simulation:

- *The k–ε turbulence model* is used frequently because of its reasonable accuracy, and the standard k–ε model is based on the transport equations for turbulent kinetic energy (k) and its dissipation rate (ε). The relative weakness in the standard k–ε model arises in flows where streamline curvature and swirl effects exist. This is related to its isotropic description of turbulence through the use of the turbulent viscosity μ_t.
- *The renormalization group (RNG) model* is part of two other variants of the standard k–ε model and has been used to derive equations from the

instantaneous Navier–Stokes equations; the resulting transport equations are similar to those of the standard k–ε model but with different constants and additional functions.

- *When the standard k–ε model is compared with the RNG k–ε model*, it is said that the RNG k–ε model would provide a more accurate solution because of the further refinement of the transport equations. In defense of the standard model, the RNG model is not as numerically stable as the standard k–ε model. The RNG model requires a lot more computational time and space.

- *The realizable k–ε model* is another variant of the standard k–ε model and derives from the combination of Boussinesq's hypothesis and the eddy-viscosity definition to give the transfer equations. The kinetic energy transfer equation remains the same; however, the dissipation model is refined and based on a dynamic equation of the mean-square vorticity fluctuation.

- *The Reynolds stress model (RSM)* discards the isotropic turbulent-viscosity hypothesis as used by the k–ε models. Instead, it uses a set of complex Reynolds stress transport equations derived from the Reynolds-averaged Navier–Stokes equations. There are now five additional transport equations for 2D applications and seven for 3D applications. This has a huge effect on computational time and space, increasing simulation time up to 60% per iteration and adding the need for further iterations. The RSM has the potential to outperform the previously discussed models because it accounts for the effects of streamline curvature, swirl, rotation, and rapid changes in strain energy in a more precise way. However, the modeling of the pressure-strain and dissipation-rate terms is particularly challenging.

Figure 3.8 shows a computer-aided design (CAD) drawing of an axial piston pump that has been modeled by use of CFD analysis and for time-varying operation, and as part of a research project supervised by the author. The geometry is complex,

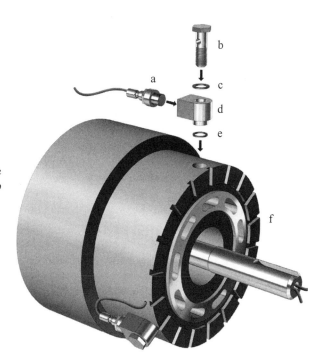

Figure 3.8. CAD drawing of part of the barrel part of an axial piston pump (Haynes, 2007).

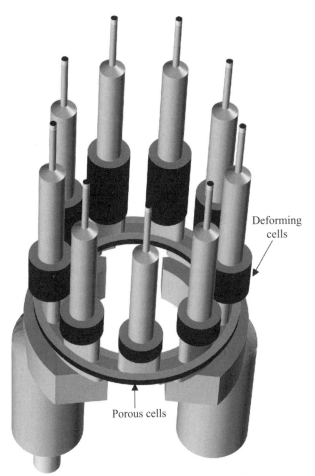

Deforming
cells   Figure 3.9. CFD pump meshed volumes
(Haynes, 2007).

Porous cells

involving time-varying piston chambers, flow through kidney slots, and flow across
the clearances at each slipper and between body and the rotating barrel.

Also shown is the way in which each kidney port internal pressure is experimentally measured: The pressure transducer signals are fed and retrieved by means of
a slip-ring data-acquisition system. The wiring aspect can be seen passing through
the center of the drive shaft, a connection then being made to the rotary slip-ring
system isolated from the drive shaft. It is difficult to visually present a 3D image
of the meshing because of the large number of cells used, but Fig. 3.9 shows some
information.

Figure 3.10 shows more details of the kidney slot interface mesh and the mesh
at the end of the piston.

The meshing required the use of deforming mesh and porous mesh techniques
to accommodate time-varying volumes and extremely fine leakage clearance volumes compared with those of normal-volume sizes. The geometry was produced
from separate volumes, sharing common faces. This enabled the mesh to move and
deform, replicating the changes in volumes that occur during pump operation. As
the pistons reciprocated, these volumes changed in volume accordingly. Hence, they
were specified as deforming zones. The circular face at each end of these volumes
was specified as a rigid body, but with a motion controlled by a user-defined function
(UDF). The velocity of each piston is dependent on its angular location; hence, each

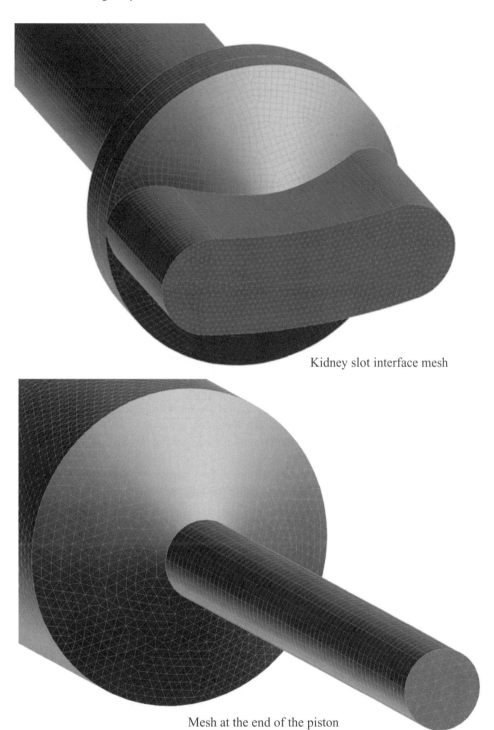

Kidney slot interface mesh

Mesh at the end of the piston

Figure 3.10. Part of the 3D meshing for an axial piston pump (Haynes, 2007).

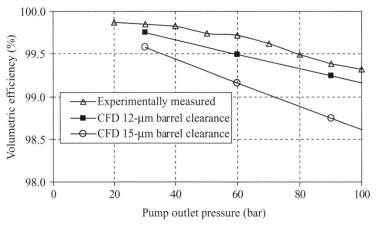

Figure 3.11. A comparison of measured and computed volumetric efficiency.

piston was assigned a UDF representative of this. Dynamic layering was used in this model to maintain mesh integrity during deformation. This meant that the volumes representing the pistons had to be meshed by a hexahedral meshing scheme. Volumes to the right of the moving face, depicting the volumes within the hollow pistons, were specified as rigid. These volumes were assigned a velocity equal to that of the moving face on the adjacent deforming zone. In addition to the deformation and velocity settings, all volumes representing the pistons were assigned a rotational boundary condition about the central pump axis. This, combined with the reciprocating motion of each piston, replicated the pumping dynamics of the pump. A good comparison between measured volumetric efficiency and CFD-computed volumetric efficiency was found when a port plate clearance of 12 μm was assumed, as shown in Fig. 3.11 (Haynes, 2007).

Further examples of CFD modeling are shown where appropriate in this book and when experimental validation is possible to some degree.

3.2 Restrictors, Control Gaps, and Leakage Gaps

3.2.1 Types

The very nature of fluid power component design means that there are very small clearances between moving parts, necessary to minimize leakages while maintaining adequate lubrication, and control areas necessary to either create pressure drops and/or direct a variable flow rate to different output ports. Hence, there is a need to determine pressure–flow-rate characteristics of such elements as accurately as possible, and here lies yet another fluid power issue: the diversity of control-area shapes and the variation of flow through them means that their pressure–flow-rate characteristics are not constant. Consider, for example, just a few restrictors shown in Fig. 3.12.

3.2.2 Orifice-Type Restrictors

The pressure-drop–flow-rate characteristic of such elements is usually derived from the ideal energy equation and then corrected for losses, usually by means of

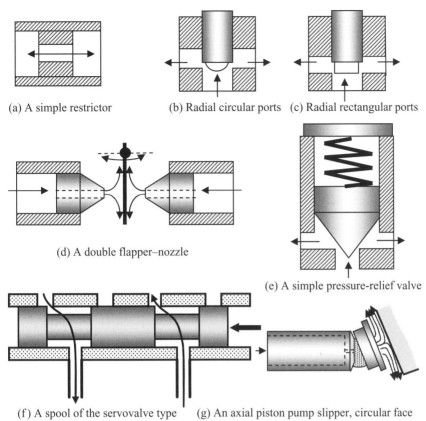

(a) A simple restrictor (b) Radial circular ports (c) Radial rectangular ports

(d) A double flapper–nozzle

(e) A simple pressure-relief valve

(f) A spool of the servovalve type (g) An axial piston pump slipper, circular face

Figure 3.12. Some flow restrictions met in fluid power components.

experimental testing. To illustrate the basic principle, consider a simple orifice in a pipe as shown in Fig. 3.13. It can be seen that the effect of fluid viscosity and the orifice geometry is to create both flow recirculation and a reduction in the effective cross-sectional area at the orifice exit from a_0 to a_c at the vena contracta. The pressure at the restriction, P_c, is virtually impossible to measure in real components; therefore, there is little to be gained by applying the energy equation between the upstream point and the orifice point, P_1 and P_c. However, in reality, the pressure beyond the orifice changes rapidly to the downstream pressure P_2, and this pressure may be used in the energy equation.

Figure 3.13. Flow through an orifice in a pipe.

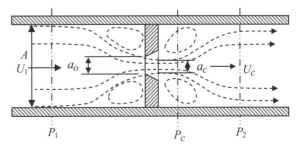

The ideal energy equation is first derived and then modified to take into account the vena contracta and recirculation losses:

$$P_1 + \frac{1}{2}\rho U_1^2 = P_2 + \frac{1}{2}\rho U_c^2, \tag{3.26}$$

$$Q = U_1 A = U_c a_c = U_c C_c a_o, \tag{3.27}$$

where C_c is the contraction coefficient relating the area at the vena contracta to the true orifice area. These two equations lead to the solution for the flow rate:

$$Q = C_d a_o \sqrt{\frac{2(P_1 - P_2)}{\rho}}, \tag{3.28}$$

$$\text{discharge coefficient } C_d = \frac{C_c}{\sqrt{1 - \left(\frac{C_c a_o}{A}\right)^2}}, \tag{3.29}$$

where both $C_d < 1.0$ and $C_c < 1.0$. Usually, the size of typical orifices and other restrictors used in fluid power components means that $a_o \ll A$ and, therefore, $C_d \approx C_c$. To take into account the recirculation losses, the discharge coefficient is modified to a flow coefficient to give the following equation:

$$Q = C_q a_o \sqrt{\frac{2(P_1 - P_2)}{\rho}}, \tag{3.30}$$

$$\text{flow coefficient } C_q = \frac{C_{rc} C_c}{\sqrt{1 - \left(\frac{C_c a_o}{A}\right)^2}}, \tag{3.31}$$

where C_{rc} is the recirculation loss coefficient. Determination of the coefficients C_{rc} and C_c can be done with a CFD package, although there are still some uncertain issues (e.g., the exact geometry mesh, real upstream and downstream conditions), and a great deal of effort is also placed on experimental determination. Application of Eq. (3.30) is therefore relatively easy in practice.

Worked Example 3.4

We may determine a measure of the significance of the ideal orifice equation by considering the flow through a cross-sectional area, 1 mm², with a pressure drop of 100 bar, using a typical mineral oil with $\rho = 860$ kg/m³:

$$\text{ideal flow rate } Q = a_o \sqrt{\frac{2(P_1 - P_2)}{\rho}} = 10^{-6} \sqrt{\frac{2 \times 10^7}{860}} = 0.152 \times 10^{-3} \text{ m}^3\text{/s},$$

$$Q = 9.12 \text{ L/min.}$$

It is now necessary to look at a few examples to illustrate typical orifice-type flow coefficients found in practice. In all cases, some way has to be found of plotting the appropriate flow coefficient as both geometry and flow conditions are changed, and it makes sense to choose the Reynolds number Re as the reference parameter.

To do this, it is necessary to define the characteristic dimension in Re because restrictors are not necessarily circular, as is obvious from the array of restrictors shown in Fig. 3.12. It is customary to define a hydraulic diameter given by:

$$\text{Re} = \frac{U_{\text{mean}} d_h}{\nu}, \quad \text{where } d_h = \frac{4 \times \text{flow area}}{\text{flow perimeter}}. \tag{3.32}$$

For a circular orifice this, of course, gives $d_h = d$, the orifice diameter. It should be noted that the concept of laminar and turbulent flow does not have the same Re implications as flow through a pipe. Some measured flow coefficients are shown in Fig. 3.14 for a circular orifice→short tube and for different flow rates and orifice sizes. These results are averaged from a large data set, experimentally determined, and the smallest orifice diameter is 2 mm. They show the complicated relationship between flow rate and geometry and how in practice these must be reasonably determined so that the correct flow coefficient is used.

At very low flow rates, the flow coefficient collapses whatever the orifice dimension ratio and a peak occurs as a true orifice geometry is used, typically $\ell/d < 0.5$.

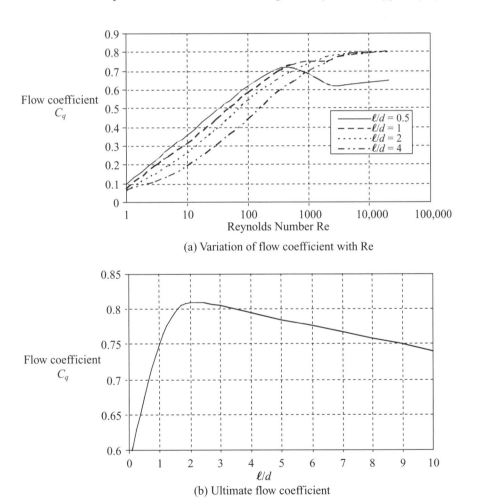

(a) Variation of flow coefficient with Re

(b) Ultimate flow coefficient

Figure 3.14. Flow coefficient for an orifice→short tube (adapted from Lichtarowicz et al., 1965).

For very large Re, the flow coefficient settles to its ultimate value, as shown in Fig. 3.14(b).

Worked Example 3.5

Determine the Re for the flow through an orifice of cross-sectional area 1 mm², with a pressure drop of 100 bar, using a typical mineral oil having $\rho = 860$ kg/m³ and $\mu = 0.025$ Ns/m², and an orifice length–diameter $= 0.5$.

Assume first ideal flow:

$$\text{ideal flow rate } Q = a_o \sqrt{\frac{2(P_1 - P_2)}{\rho}} = 0.152 \times 10^{-3} \text{m}^3/\text{s},$$

$$\text{mean flow rate } U_{\text{mean}} = \frac{0.152 \times 10^{-3}}{10^{-6}} = 152 \text{ m/s},$$

$$d_h = d \rightarrow \text{Re} = \frac{\rho U_{\text{mean}} d_h}{\mu} = \frac{(860) \times (152) \times (1.13 \times 10^{-3})}{0.025}$$

$$= 5909.$$

Now consider Fig. 3.14 and $\ell/d = 0.5$ to give the flow coefficient $C_q \approx 0.71$. Hence, recalculate to give:

$$Q = 0.71 \times (0.152 \times 10^{-3}) = 0.108 \times 10^{-3} \text{ m}^3/\text{s} \,(6.48 \text{ L/min}),$$
$$U_{\text{mean}} = 108 \text{ m/s},$$
$$\text{Re} = 4198 \rightarrow \text{ from Fig. 3.10, } C_q \approx 0.71, \text{ as previously assumed.}$$

So the actual orifice flow rate is actually 6.48 L/min compared with the ideal value of 9.12 L/min, a significant reduction when the appropriate flow coefficient is used.

Next, consider a PRV having eight circular radial ports that are simultaneously uncovered as the valve main spindle rises because of an increase in pressure below it and in excess of the initial pressure setting. This simplest of port geometries is shown in Fig. 3.15.

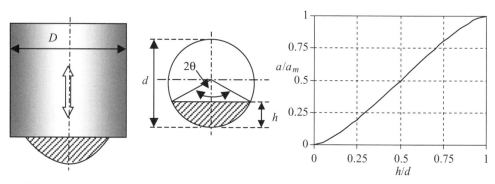

Figure 3.15. Circular radial port used in a PRV.

The projected cross-sectional area uncovered is given by:

$$\frac{a}{a_m} = \frac{\theta}{\pi} - \frac{\sin 2\theta}{2\pi}, \quad \cos \theta = 1 - 2\frac{h}{d}, \tag{3.33}$$

where a_m is the maximum circular cross-sectional area. Given the large flows generated by such small openings in fluid power components, as discussed earlier, some series approximations can be useful, and the most common ones are as follows:

$$\frac{a}{a_m} \approx 1.7 \left(\frac{h}{d}\right)^{1.5}, \quad 5.3\% \text{ error at } \frac{h}{d} = 0.15, \tag{3.34}$$

$$\frac{a}{a_m} \approx 1.7 \left(\frac{h}{d}\right)^{1.5} \left(1 - 0.3\frac{h}{d}\right), \quad 2.2\% \text{ error at } \frac{h}{d} = 0.15. \tag{3.35}$$

The details are as follows:

- port diameter $d = 4$ mm
- port length $\ell = 12.7$ mm, giving $\ell/d = 3.2$
- main spindle diameter $D = 12.7$ mm
- spindle lift before port is opened $x_0 = 5.5$ mm

When considering the calculation of Re, the hydraulic diameter is used as previously defined:

$$\text{Re} = \frac{U_{\text{mean}}d_h}{\nu}, \quad \text{where } d_h = \frac{4 \times \text{flow area}}{\text{flow perimeter}} = d\frac{(2\theta - \sin 2\theta)}{(2\theta + \sin 2\theta)}. \tag{3.36}$$

Measurements are shown in Fig. 3.16 and presented in two ways: one using Re as a reference, the other using flow rate as a reference. The results are presented for different pressure drops across the port, and it can be seen that a good approximation over the data range presented would be to assume $C_q \approx 0.55$, giving a possible error of $\pm 8\%$ over the operating range shown.

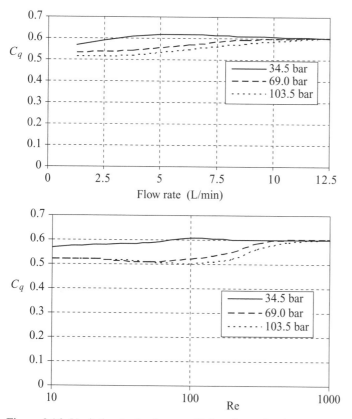

Figure 3.16. Variation in the flow coefficient for a circular port.

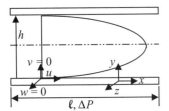

Figure 3.17. Flow between parallel plates.

3.2.3 Flow Between Parallel Plates

If flow between infinitely wide parallel plates is considered (see Fig. 3.17) and therefore has axial symmetry ($v = w = 0$), then consideration of (3.9) gives:

$$\rho\left[u\frac{\partial u}{\partial x}\overset{0}{} + v\frac{\partial u}{\partial y}\overset{0}{}\right] = -\frac{\partial p}{\partial x} + \mu\left[\frac{\partial^2 u}{\partial x^2}\overset{0}{} + \frac{\partial^2 u}{\partial y^2}\right],$$

$$\rho\left[u\frac{\partial v}{\partial x}\overset{0}{} + v\frac{\partial v}{\partial y}\overset{0}{}\right] = -\frac{\partial p}{\partial y} + \mu\left[\frac{\partial^2 v}{\partial x^2}\overset{0}{} + \frac{\partial^2 v}{\partial y^2}\overset{0}{}\right].$$

It then follows that:

$$\mu\frac{d^2 u}{dy^2} = \frac{dp}{dx}, \quad 0 = -\frac{dp}{dy}. \tag{3.37}$$

From the pressure gradient, the pressure does not vary with y and therefore can vary only with x down the gap. However, u is a function of y only and pressure can only decrease with x:

$$\frac{dp}{dx} = \text{constant, say} - \frac{\Delta P}{\ell}, \tag{3.38}$$

where ΔP is the total pressure drop across the gap of length ℓ. This gives:

$$\mu\frac{d^2 u}{dy^2} = -\frac{\Delta P}{\ell}. \tag{3.39}$$

Integrating twice and noting that at the centerline $du/dy = 0$ and at the plate position $y = 0, h$ and $u = 0$ gives:

$$u = \frac{\Delta P}{2\mu\ell}(h - y)y, \tag{3.40}$$

$$\text{the centerline velocity} \quad u_{\text{max}} = \frac{h^2 \Delta P}{8\mu\ell}. \tag{3.41}$$

If the plate width is w, then the total flow rate Q is:

$$Q = \int_0^h wu\,dy = \frac{wh^3 \Delta P}{12\mu\ell}, \tag{3.42}$$

$$u_{\text{mean}} = \frac{Q}{wh} = \frac{h^2 \Delta P}{12\mu\ell} = \frac{2}{3}u_{\text{max}}. \tag{3.43}$$

Hence, the velocity profile is parabolic, and note that the flow rate varies with h^3. It can be seen from Eq. (3.42) that the resistance is given by:

$$\Delta P = RQ \rightarrow \text{resistance } R = \frac{12\mu\ell}{wh^3}. \qquad (3.44)$$

Worked Example 3.6

Considering flow between a pair of parallel flat plates as previously analyzed, derive the basic differential equation for 2D flow.

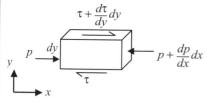

Worked Example 3.6

Consider a fluid volume and unit depth of plate and equate forces:

$$pdy + \left(\tau + \frac{d\tau}{dy}dy\right)dx = \left(p + \frac{dp}{dx}dx\right)dy + \tau dx,$$

$$\frac{d\tau}{dy} = \frac{dp}{dx}.$$

Add Newton's law of viscosity:

$$\tau = \mu\frac{du}{dy} \rightarrow \mu\frac{d^2u}{dy^2} = \frac{dp}{dx}.$$

This is as previously determined from the Navier–Stokes equation resulting in Eq. (3.37).

3.2.4 Flow Between Annular Gaps

Consider first concentric pipes, such as those shown in Fig. 3.18, where the gap width is h, the outer cylinder radius is r, and $h \ll r$. This type of clearance can exist around pistons or spools, although it is unlikely that the gap will be uniform in practice. It should be recalled from data in Chapter 2 and from typical flow calculations that clearances will be typically $<20\,\mu m$ and, hence, considerably smaller than typical component physical dimensions.

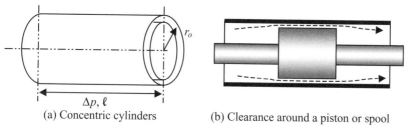

    (a) Concentric cylinders   (b) Clearance around a piston or spool

Figure 3.18. Flow between concentric cylinders.

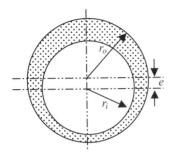

Figure 3.19. Flow between asymmetric cylinders.

Recall earlier work on the flow equation between a pair of flat parallel plates:

$$Q = \frac{wh^3 \Delta P}{12\mu\ell}, \quad \text{where the width } w \text{ is now } 2\pi r_o,$$

$$Q = \frac{\pi r_o h^3 \Delta P}{6\mu\ell}, \quad \text{resistance } R = \frac{6\mu\ell}{\pi r_o h^3}. \tag{3.45}$$

Again, the significance of clearance is noted; doubling the clearance increases the flow (leakage) by a factor of 8 for the same pressure drop. Next, consider *cylinder asymmetry* that will probably exist in practice as shown in Fig. 3.19, the clearances being exaggerated. The relative displacement is called the eccentricity e and, following some trigonometry and approximations that are due to the very small clearances being considered here, it can be shown that:

$$Q = \frac{\pi r_o (r_o - r_i)^3}{6\mu\ell} \left(1 + \frac{3}{2}\varepsilon^2\right) \Delta P, \tag{3.46}$$

$$\text{eccentricity ratio } \varepsilon = \frac{e}{(r_o - r_i)}, \tag{3.47}$$

where ℓ is the length of the assembly.

This is a highly significant finding because if the inner cylinder is just touching the outer cylinder, the flow rate is increased by a factor of 2.5 times the value with concentric cylinders, assuming the same pressure drop.

3.2.5 Flow Between an Axial Piston Pump Slipper and Its Swash Plate

Now consider an axial piston slipper connected to the end of a piston, as shown in Fig. 3.20. The motion of the piston around the swash plate can result in rotation of the piston and integral slipper. However, the experience of one manufacturer suggests that for a particular pump, the piston does not rotate as evident from wear at one point on the rear of the piston, even though there is motion between the piston and it barrel bore. In addition, assemblies in which the slipper is swaged around a spherical end, as shown in Fig. 3.20, can result in varying friction forces such that new slippers may also not rotate. The slipper predominantly acts as a hydrostatic

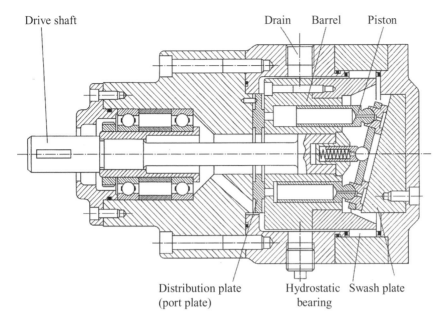

Drive shaft   Drain   Barrel   Piston

Distribution plate   Hydrostatic   Swash plate
(port plate)   bearing

(a) An in-line axial piston pump

Note: Supplied by N. Nervegna, Politecnico di Torino.

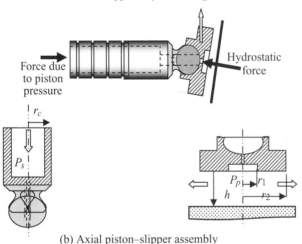

(b) Axial piston–slipper assembly

Figure 3.20. Consideration of the slipper in an axial piston pump.

bearing. In practice, the angle of the swash plate requires the slipper to also be at a similar angle that is influenced to a small extent by an additional small variation in force balance and associated dynamics. However, typically 97% of the force balance will be hydrostatic, the remainder being hydrodynamically associated with a very small additional tilt of the slipper. When the true hydrostatic force balance is computed, the slipper angle must be taken into account.

Consider, therefore, the configuration in which the slipper does not rotate and the gap is uniform. Also consider the central region of the slipper from where the pressure varies from a value slightly lower than supply pressure to tank pressure at the edge of the slipper. Adapting the equations for flow between parallel plates as given by Eqs. (3.37) and (3.40) gives:

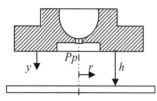

$$\mu \frac{d^2u}{dy^2} = \frac{dp}{dx} \rightarrow \mu \frac{d^2u}{dy^2} = \frac{dp}{dr}, \tag{3.48}$$

$$u = -\frac{1}{2\mu} \frac{dp}{dr} y(h - r), \tag{3.49}$$

$$Q = \int_0^h 2\pi r u\, dy = -\frac{\pi r h^3}{6\mu} \frac{dp}{dr}, \tag{3.50}$$

$$-\int_{r_1}^r \frac{1}{r} dr = \int_{P_p}^P \frac{\pi h^3 dp}{6\mu Q} \rightarrow \ln \frac{r_1}{r} = \frac{\pi h^3 (p - P_p)}{6\mu Q}. \tag{3.51}$$

Noting that at the edge of the slipper, $r = r_2$ and $p = 0$, then from Eq. (3.51) we have:

$$\ln \frac{r_2}{r_1} = \frac{\pi h^3 P_p}{6\mu Q} \rightarrow \text{resistance } R = \frac{P_p}{Q} = \frac{6\mu}{\pi h^3} \ln \left(\frac{r_2}{r_1} \right), \tag{3.52}$$

$$\frac{6\mu Q}{\pi h^3 P_p} = \frac{1}{\ln \left(\dfrac{r_2}{r_1} \right)}. \tag{3.53}$$

It then follows that the radial pressure distribution is given by:

$$\frac{p}{P_p} = 1 - \frac{\ln \left(\dfrac{r}{r_1} \right)}{\ln \left(\dfrac{r_2}{r_1} \right)}. \tag{3.54}$$

This radial pressure distribution is shown in Fig. 3.21.

    The total normal force acting on the slipper is then given by the addition of the contribution from the central pocket, plus the contribution from the logarithmic pressure decay from the pocket edge to the slipper edge. Neglecting the very small

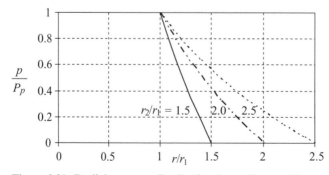

Figure 3.21. Radial pressure distribution for a slipper with a constant-gap width.

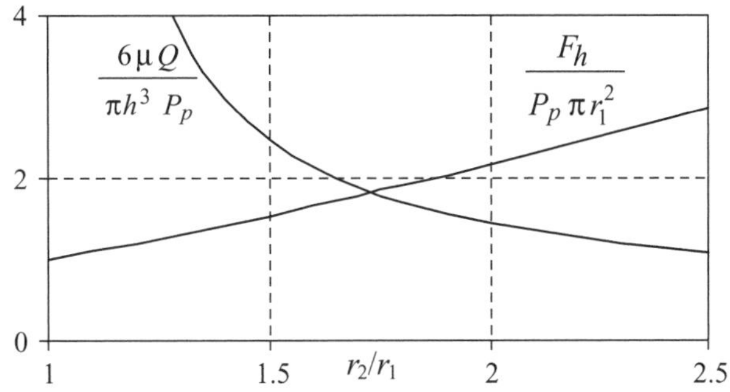

Figure 3.22. Flow-rate and total force functions for a slipper.

lubricating hole reduction effect, this becomes:

$$F_h = P_p \pi r_1^2 + \int_{r1}^{r2} 2\pi r p\, dr = \frac{P_p \pi r_1^2}{2 \ln\left(\frac{r_2}{r_1}\right)} \left[\left(\frac{r_2}{r_1}\right)^2 - 1 \right], \qquad (3.55)$$

$$\frac{F_h}{P_p \pi r_1^2} = \frac{\left[\left(\frac{r_2}{r_1}\right)^2 - 1 \right]}{2 \ln\left(\frac{r_2}{r_1}\right)}. \qquad (3.56)$$

Flow-rate equation (3.53) and total force equation (3.56) are shown in Fig. 3.22.

Clearly, increasing the slipper land ratio r_2/r_1 increases the hydrostatic force that can be achieved and also reduces the leakage flow rate that is highly sensitive to the clearance h adopted. It is now possible to consider the design approach to force balance achievement by equating the piston internal pressure force, $F_s = P_s \pi r_c^2$, to the slipper total force given by Eq. (3.56) when the resolved component of hydrostatic force along the piston axial axis is used. Also, it is necessary to consider the force balance ratio set by the designer and given by:

$$\text{force balance ratio } \alpha = \frac{F_h}{F_s} \quad \text{and usually set} > 1. \qquad (3.57)$$

There is also a negligible flow reaction force; therefore, neglecting the slipper orifice area gives:

$$\frac{r_2}{r_c} = \frac{r_2}{r_1} \sqrt{\frac{\alpha P_s}{P_p \cos \theta} \frac{2 \ln\left(\frac{r_2}{r_1}\right)}{\left[\left(\frac{r_2}{r_1}\right)^2 - 1 \right]}}. \qquad (3.58)$$

The swash-plate angle is θ. Assuming the case $\alpha P_s / P_p \cos \theta \approx 1$, not uncommon in practice, then Eq. (3.58) may be represented as shown in Fig. 3.23. This shows that if a particular piston diameter is selected from a pump displacement requirement, then increasing the slipper outer radius, effectively increasing r_2/r_c, must be

Table 3.1. *A comparison of three different piston designs,* $\alpha P_s / P_p \cos \theta = 1$

Piston in Fig. 3.24	r_2 (mm)	r_1 (mm)	r_c (mm)	r_2/r_1	r_2/r_c	r_2/r_c Theory	Error (%)
A	8.00	4.76	6.43	1.68	1.25	1.27	1.6
B	12.30	6.76	9.50	1.82	1.29	1.31	1.6
C	15.85	7.93	11.80	2.00	1.34	1.36	1.5

accompanied by a specific slipper radius ratio r_2/r_1. The solution shown in Fig. 3.23 will be slightly increased by typically less than 3% when the swash-plate angle effect is included.

Now consider some actual piston–slipper assemblies with a slipper design of the type being analyzed here and shown in Fig. 3.24. Data for the three pistons are shown in Table 3.1.

Interestingly, the orifice diameter for each slipper is remarkably similar and close to 1 mm, this also being true for much bigger slippers analyzed by the author. These comparisons with the approximation $\alpha P_s / P_p \cos \theta \approx 1$ show a very good correlation with errors less than 1.6%.

Table 3.1 shows that as the piston size increases, so does each ratio r_2/r_c and r_2/r_1. This effect is illustrated in Fig. 3.25.

It is now possible to outline the design approach to determine the operating characteristics for a hydrostatic slipper by considering the pressure-drop–flow characteristic for the slipper and the force balance characteristic. As also discussed earlier, the well-established orifice equation is used and the flow through the slipper orifice is equated to the flow across the slipper. Combining this with the force balance equation then gives:

$$C_q a_0 \sqrt{\frac{2(P_s - P_p)}{\rho}} = \frac{\pi h^3 P_p}{6\mu \ln\left(\dfrac{r_2}{r_1}\right)}, \tag{3.59}$$

$$\frac{\alpha P_s}{P_p \cos \theta} = \frac{\left(\dfrac{r_1}{r_c}\right)^2 \left[\left(\dfrac{r_2}{r_1}\right)^2 - 1\right]}{2 \ln\left(\dfrac{r_2}{r_1}\right)}. \tag{3.60}$$

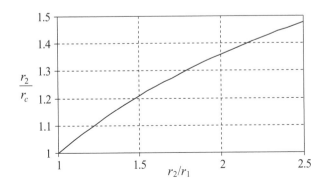

Figure 3.23. Solution for the slipper outer radius.

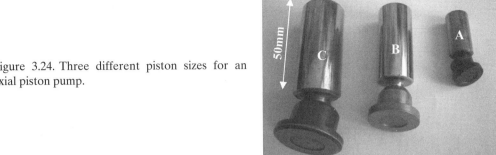

Figure 3.24. Three different piston sizes for an axial piston pump.

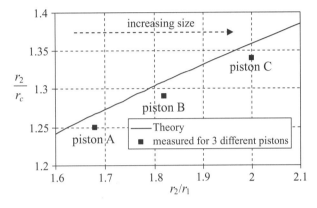

Figure 3.25. A comparison of three different piston designs.

Worked Example 3.7

An axial piston slipper is defined as follows:

Piston diameter $r_c = 11.4$ mm, swash-plate angle of $20°$.
Slipper radii $r_1 = 8$ mm, $r_2 = 16$ mm, slipper orifice diameter $r_o = 1$ mm.
Orifice coefficient of 0.61.
Mineral oil $\rho = 860$ kg/m^3, $\mu = 0.025$ N s/m^2.
$P_s = 210$ bar.
Calculate the slipper gap at the pressure side and the inlet side assuming an inlet pressure boosted to 5 bar and perfect force ratio balance.

(i) From Eq. (3.60) for perfect balance:

$$\frac{(1)(210)}{P_p(0.94)} = \frac{\left(\frac{8}{11.4}\right)^2 \left[\left(\frac{16}{8}\right)^2 - 1\right]}{2 \ln\left(\frac{16}{8}\right)} = 1.0657 \rightarrow P_p = 209.63 \text{ bar.}$$

The pressure drop across the orifice is 0.37 bar and the slipper pocket pressure is 209.63 bar. Evaluating flow equation (3.59) gives:

$$0.61(0.786 \times 10^{-6})\sqrt{\frac{2(210 - 209.63) \times 10^5}{860}} = \frac{\pi h^3 209.63 \times 10^5}{6(0.025) \ln\left(\frac{16}{8}\right)},$$

$$4.45 \times 10^{-6} = 6.34 \times 10^8 h^3 \rightarrow h = 19.1 \times 10^{-6} \text{ m} = 19.1 \text{ } \mu\text{m.}$$

The flow rate is $Q = 4.45 \times 10^{-6} \text{ m}^3/\text{s} \, (0.27 \text{ L/min})$.

Now check Re for the orifice to ensure that the correct flow coefficient has been used:

$$\text{orifice mean velocity } u = \frac{4.45 \times 10^{-6}}{0.786 \times 10^{-6}} = 5.66 \text{ m/s},$$

$$\text{Re} = \frac{\rho u d}{\mu} = \frac{860 \times 5.66 \times 0.001}{0.025} = 195.$$

This means that the selected flow coefficient is probably acceptable from Fig. 3.14.

(ii) Considering the pump inlet at a pressure of 5 bar now gives:

$$\frac{(1)(5)}{P_p(0.94)} = \frac{\left(\frac{8}{11.4}\right)^2 \left[\left(\frac{16}{8}\right)^2 - 1\right]}{2 \ln\left(\frac{16}{8}\right)} = 1.0657 \rightarrow P_p = 4.99 \text{ bar}.$$

The pressure drop across the orifice is now only 0.01 bar and the slipper pocket pressure is 4.99 bar. Evaluating flow equation (3.59) gives:

$$0.61(0.786 \times 10^{-6})\sqrt{\frac{2(0.01) \times 10^5}{860}} = \frac{\pi h^3 4.99 \times 10^5}{6(0.025) \ln\left(\frac{16}{8}\right)},$$

$$0.73 \times 10^{-6} = 0.15 \times 10^8 h^3 \rightarrow h = 36.5 \times 10^{-6} \text{ m} = 36.5 \, \mu\text{m}.$$

The flow rate is $Q = 0.73 \times 10^{-6} \text{ m}^3/\text{s} \, (0.044 \text{ Ls/min})$.

Now check Re for the orifice to ensure that the correct flow coefficient has been used:

$$\text{orifice mean velocity } u = \frac{0.73 \times 10^{-6}}{0.786 \times 10^{-6}} = 0.93 \text{ m/s},$$

$$\text{Re} = \frac{\rho u d}{\mu} = \frac{860 \times 0.93 \times 0.001}{0.025} = 32.$$

From Fig. 3.14, the flow coefficient should be lower and probably around 0.45. Recalculating then gives:

$$0.45(0.786 \times 10^{-6})\sqrt{\frac{2(0.01) \times 10^5}{860}} = \frac{\pi h^3 4.99 \times 10^5}{6(0.025) \ln\left(\frac{16}{8}\right)},$$

$$0.54 \times 10^{-6} = 0.15 \times 10^8 h^3 \rightarrow h = 33 \times 10^{-6} \text{ m} = 33 \, \mu\text{m}.$$

The flow rate is $Q = 0.54 \times 10^{-6} \text{ m}^3/\text{s} \, (0.032 \text{ Ls/min})$.

The recalculation results in just a small reduction in the estimated gap width with an associated reduction in flow rate. The new $\text{Re} = 24$, and this gives only a small reduction in the new estimated flow coefficient. What is clear is that as the piston moves from the high-pressure side to the low-pressure side, the gap width will significantly increase, although the flow loss will be significantly reduced.

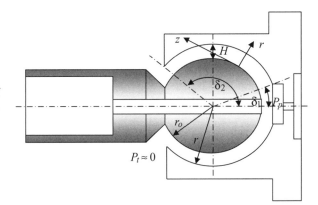

Figure 3.26. Flow between the gap of a ball and socket (gap highly exaggerated).

3.2.6 Flow Between a Ball and Socket

Now consider flow through the annular gap between a ball and a socket of the type used to connect an axial pump–motor slipper to its piston. The slipper is usually swaged onto the ball but, in some cases, for larger machines, this may not be the case and the slipper sits free and is held in position by reacting pressures and springs. The actual clearance may not be uniform around the ball in practice, but this analysis will assume a uniform clearance. Flow leakage through this type of gap is usually small compared with the slipper leakage; for this analysis, laminar flow will be assumed. Therefore, consider Fig. 3.26.

Following the analysis given in Bergada, Kumar, and Watton (2007), the gap area is given by:

$$ds = \int_0^H 2\pi(r_o + r)\sin\delta\, dr. \tag{3.61}$$

The flow rate is therefore given by:

$$dQ = \int_0^H u\, 2\pi(r_o + r)\sin\delta\, dr, \tag{3.62}$$

where u is the velocity at the elemental width dr across the gap.

This velocity profile has been defined and used in earlier examples for laminar flow and, for the geometry of Fig. 3.26, is given by:

$$u = -\frac{1}{\mu}\frac{dP}{dz}\frac{r}{2}(H - r). \tag{3.63}$$

The flow rate is then given by:

$$dQ = \int_0^H -\frac{1}{\mu}\frac{dP}{dz}\frac{r}{2}(H - r)2\pi(r_o + r)\sin\delta\, dr. \tag{3.64}$$

Also note that:

$$dz = \left(r_o + \frac{H}{2}\right)d\delta. \tag{3.65}$$

The flow rate is then evaluated using Eqs. (3.64) and (3.65) to give:

$$Q = \alpha P_p \frac{\pi H^3}{6\mu}, \quad \alpha = \frac{1}{\ln\left(\dfrac{\tan\delta_2/2}{\tan\delta_1/2}\right)}. \tag{3.66}$$

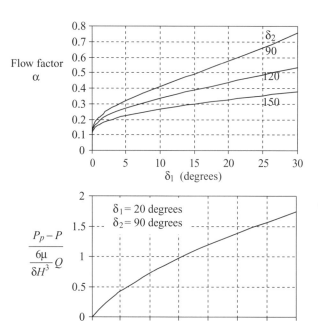

Figure 3.27. Leakage flow and pressure-drop characteristics for a ball and socket joint with a small uniform gap.

The pressure drop around the perimeter is given by:

$$\frac{P_p - P}{\frac{6\mu}{\pi H^3} Q} = \ln\left(\frac{\tan \delta/2}{\tan \delta_1/2}\right).$$

(3.67)

The leakage flow factor α and the nondimensional pressure drop are shown in Fig. 3.27. The maximum angle tends to be around $\delta_2 = 90°$. Increasing this angle reduces the leakage flow, at a fixed total pressure drop, because of the increased resistance of the longer flow path. Increasing the inlet angle δ_1 has a significant effect on leakage flow as the total angle δ_2 is reduced. Beyond an inlet angle of typically $\delta_1 = 5°$, the leakage flow increases almost linearly with this angle increase. The pressure-drop characteristic is not severely nonlinear; the one shown in Fig. 3.27 represents just one design condition, $\delta_1 = 20°$ and $\delta_2 = 90°$.

3.2.7 Flow Between Nonparallel Plates — Reynolds Equation

This problem is often referred to as a bearing hydrodynamic lubrication problem and may be studied by means of application of Reynolds equation. Consider a stationary thrust plate at a small angle relative to the reference boundary that has components of steady velocity U and v in the x and y directions, as shown in Fig. 3.28, in Cartesian coordinates.

Considering a fluid element within the wedge of general height h, then the difference between the inlet flow and the outlet flow is given by:

$$\delta Q = -\frac{\partial Q_x}{\partial x}\delta x - \frac{\partial Q_y}{\partial y}\delta y.$$

(3.68)

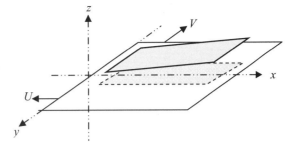

Figure 3.28. Reference axes for a fixed thrust plate and a moving boundary.

But, from previous work:

$$Q_x = -\frac{h^3}{12\mu}\frac{\partial p}{\partial x}, \quad Q_y = -\frac{h^3}{12\mu}\frac{\partial p}{\partial y}. \tag{3.69}$$

This leads to:

$$\delta Q = \left[\frac{\partial}{\partial x}\left(\frac{h^3}{12\mu}\frac{\partial p}{\partial x}\right) + \frac{\partial}{\partial y}\left(\frac{h^3}{12\mu}\frac{\partial p}{\partial y}\right)\right]\delta x\delta y. \tag{3.70}$$

The flow rate, because of the velocity gradient effect, is given by:

$$\delta Q = \frac{U}{2}\frac{\partial h}{\partial x}\delta x\delta y + \frac{V}{2}\frac{\partial h}{\partial y}\delta y\delta x = \left[\frac{U}{2}\frac{\partial h}{\partial x} + \frac{V}{2}\frac{\partial h}{\partial y}\right]\delta x\delta y. \tag{3.71}$$

For incompressible steady-state flow, the sum of Eqs. (3.70) and (3.71) must equal zero:

$$\frac{\partial}{\partial x}\left(\frac{h^3}{12\mu}\frac{\partial p}{\partial x}\right) + \frac{\partial}{\partial y}\left(\frac{h^3}{12\mu}\frac{\partial p}{\partial y}\right) + \frac{U}{2}\frac{\partial h}{\partial x} + \frac{V}{2}\frac{\partial h}{\partial y} = 0. \tag{3.72}$$

This is the Reynolds equation of lubrication. If it is assumed that the viscosity does not change significantly with position and, hence, with pressure, then Eq. (3.72) becomes:

$$\frac{\partial}{\partial x}\left(h^3\frac{\partial p}{\partial x}\right) + \frac{\partial}{\partial y}\left(h^3\frac{\partial p}{\partial y}\right) + 6\mu\left(U\frac{\partial h}{\partial x} + V\frac{\partial h}{\partial y}\right) = 0. \tag{3.73}$$

Now consider applying the Reynolds equation to a plane wedge bearing with uniform viscosity and a bearing height variation in x only; that is, no side leakage in the y direction, as illustrated in Fig. 3.29.

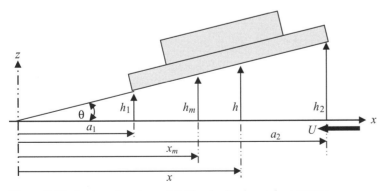

Figure 3.29. A plane bearing exhibiting hydrodynamic pressure.

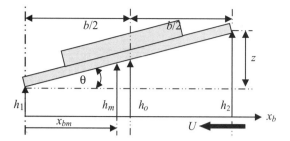

Figure 3.30. Coordinates with reference to the bearing.

Considering that the viscosity is assumed constant, there is no side leakage, the bearing angle is very small such that $\tan \theta \approx \theta$, and the fluid-film thickness varies only with x, then the Reynolds equation becomes:

$$\frac{\partial}{\partial x}\left(h^3 \frac{\partial p}{\partial x}\right) + 6\mu U \frac{\partial h}{\partial x} = 0. \tag{3.74}$$

Integrating with respect to x and defining $h = h_m$ when $(\partial p/\partial x) = 0$ and using the small-angle approximation $h \approx x\theta$ and $h_m \approx x_m\theta$ gives:

$$\frac{dp}{dx} = -\frac{6\mu U}{\theta^2}\left(\frac{1}{x^2} - \frac{x_m}{x^3}\right). \tag{3.75}$$

Integrating Eq. (3.75) to determine the pressure distribution and inserting the boundary conditions that $p = p_0$ at $x = a_1$ and also at $x = a_2$ then gives:

$$p = \frac{6\mu U}{\theta^2}\left(\frac{1}{x} - \frac{x_m}{2x^2} - C_1\right),$$
$$x_m = \frac{2a_1 a_2}{(a_1 + a_2)}. \tag{3.76}$$

Again, using the small-angle approximation $h_1 \approx a_1\theta$ and $h_2 \approx a_2\theta$, we may evaluate the constant C_1 in Eq. (3.76) to give:

$$p - p_0 = \frac{6\mu U}{(a_1 + a_2)\theta^2}\left[\frac{(a_1 + a_2)}{x} - \frac{a_1 a_2}{x^2} - 1\right]. \tag{3.77}$$

Using Eq. (3.76) then gives the maximum pressure p_m:

$$p_m - p_0 = \frac{3\mu U(a_1 - a_2)^2}{2(a_1 + a_2)a_1 a_2\theta^2}. \tag{3.78}$$

The load supported per unit width F is given by:

$$F = \int_{a_1}^{a_2}(p - p_0)dx = \frac{6\mu U}{\theta^2}\left[\ln\frac{a_2}{a_1} - \frac{2(a_2 - a_1)}{(a_1 + a_2)}\right]. \tag{3.79}$$

Now transfer the coordinate system to the bearing as shown in Fig. 3.30. The bearing is now located with respect to its trailing edge, the bearing width being b, and the relative height because of tilt is given by $z = (h_2 - h_1)$. The gap height at the bearing center is h_0.

The bearing pressure distribution then becomes:

$$\frac{p - p_0}{\left(\dfrac{6\mu U b}{h_0^2}\right)} = \frac{\overline{z}\,\overline{x}_b(1 - \overline{x}_b)}{2(1 + \overline{z}\,\overline{x}_b - 0.5\overline{z})^2} \quad \overline{x}_b = \frac{x_b}{b} \quad \overline{z} = \frac{z}{h_0}. \tag{3.80}$$

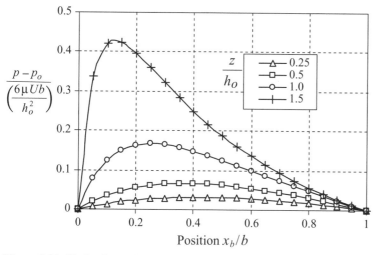

Figure 3.31. Hydrodynamic pressure distribution.

This pressure distribution is shown in Fig. 3.31.
The position for maximum pressure is then given by:

$$\frac{x_{bm}}{b} = \frac{1}{2}\left(1 - \frac{\bar{z}}{2}\right). \tag{3.81}$$

It can be seen that this position is always to the left of the bearing centerline and probably by a very small amount in fluid power applications with low values of $\bar{z}$; for example, a pump slipper. The theory also indicates that when the bearing trailing edge just touches the moving boundary, then the position for maximum pressure is at the trailing edge. Considering the mean force per unit length, F, given by Eq. (3.79), and using the coordinate transformation gives:

$$\frac{F}{\left(\frac{6\mu U b^2}{h_0^2}\right)} = \frac{1}{\bar{z}^2}\left[\ln\left(\frac{2 + \bar{z}}{2 - \bar{z}}\right) - \bar{z}\right]. \tag{3.82}$$

The force variation per unit length with geometry is shown in Fig. 3.32. It may be seen that the change in total force is almost proportional to the change in the geometry ratio z/h_0, particularly for small practical values of tilt; essentially, increasing the tilt gives a proportional increase in total force for a constant h_0.

Using a series expansion of the log term in Eq. (3.82) shows that for small values of z/h_0, then the slope of Fig. 3.32 is $1/12$. Therefore, the force per unit width of the bearing becomes:

$$F \approx \frac{\mu U b^2 z}{2h_o^3}, \quad z < 0.5h_o. \tag{3.83}$$

This shows that tilt, of course, must exist, but the force generated is more sensitive to changes in central clearance than tilt is. The flow rate per unit width of bearing is given by adding the component from the Reynolds equation to the entrained flow effect that is due to the velocity variation from zero at the bearing to U at the moving

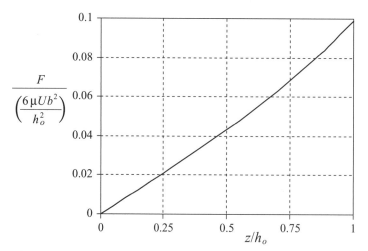

Figure 3.32. Variation in total force for a tilted plate and a sliding boundary.

boundary. This gives:

$$Q_x = \frac{Uh}{2} - \frac{h^3}{12\mu}\frac{dp}{dx}. \tag{3.84}$$

Considering the condition when $dp/dx = 0$ then gives the solution:

$$\frac{Q_x}{\left(\frac{Uh_o}{2}\right)} = 1 - \left(\frac{\bar{z}}{2}\right)^2. \tag{3.85}$$

For parallel faces, $\bar{z} = 0$, the flow rate is entirely entrained, as expected. The flow rate is reduced because of zero as the trailing edge just touches the moving boundary, $z = 2h_o$. The hydrodynamic effect on flow rate will probably be small for practical tilts in fluid power elements.

To account for side leakage in practice, the infinitely long bearing theory developed may be factored by a suitable constant. Early work showed that a bearing having an aspect ratio of 1 required the ideal load generated to be multiplied by a factor of 0.44 (Freeman, 1962; Kingsbury, 1931). The result of a recent finite-difference solution provided by a colleague, RWS Snidle, gave a factor varying between 0.42 for $\bar{z} = 0.1$ and 0.44 for $\bar{z} = 0.5$.

Worked Example 3.8

Consider Worked Example 3.7 concerning a pump slipper and use the following data:

> Piston diameter $r_c = 11.4$ mm.
> Slipper radii $r_1 = 8$ mm, $r_2 = 16$ mm.
> Slipper orifice diameter $r_o = 1$ mm.
> Orifice flow coefficient, 0.61.
> Viscosity $\mu = 0.025$ N s/m^2.
> Piston pressure $Ps = 210$ bar.
> Slipper peripheral velocity $U = 4.4$ m/s.
> Force provided at the piston end = 2.14 kN.

Now consider the case in which the slipper hydrostatic opposing force is 2.04 kN, giving a force balance of 95.3%. The remaining force of 100 N must be obtained by the slipper tilting and an approximate solution will be pursued.

Assume a slipper square-area approximation and corrected for side leakage. The square bearing ($b \times b$) has a dimension $b = \sqrt{\pi} r_2 = 28.36$ mm.

Also assume from the previous calculation that the clearance at the center of the slipper remains at $h_o = 19.1\ \mu$m. Then, applying a load factor of 0.43 to account for side leakage and also including the finite bearing length gives:

$$\frac{F}{\dfrac{(0.43)(b)6\mu U b^2}{h_o^2}} = \frac{1}{\bar{z}^2}\left[\ln\left(\frac{2+\bar{z}}{2-\bar{z}}\right) - \bar{z}\right]$$

$$= \frac{(100)(19.1 \times 10^{-6})^2}{(0.43)(28.36 \times 10^{-3})6(0.025)(4.4)(28.36 \times 10^{-3})^2} = 0.00556.$$

This gives a solution for $\bar{z} = (z/h_o) = 0.067 \rightarrow z = 1.28\ \mu$m.

This is a small displacement of $\pm 0.64\ \mu$m relative to the slipper central gap of 19.1 μm, giving a very small slipper angle when considering the actual slipper diameter of 32 mm (or 28.36 mm when considering the square-bearing approximation).

The direct addition of hydrostatic and hydrodynamic forces is not strictly correct because the actual flow characteristic has to be solved with both effects occurring. However, this example does serve to show that *extremely small slipper tilts will probably exist in practice.*

3.2.8 Flow Through Spool Valves of the Servovalve Type and the Use of a CFD Package for Analysis

Now consider flow through servovalve ports as illustrated schematically in Fig. 3.33. A servovalve is designed to be a proportional control valve; that is, the flow rate is proportional to the spool displacement at a fixed pressure differential. Hence, the control ports are rectangular in shape and machined onto the bush as indicated in Fig. 3.33(c). It can be readily appreciated that for both flow-entry and flow-exit ports, a high-velocity jet will be created because of the pressure differential across each port and the very small spool opening, typically $x < 0.5$ mm.

Further details on servovalve types and their performance are presented later in this chapter, but whatever the principle of operation, a pressure differential must be generated by the input control current. This pressure differential is used to move the spool to its required position; therefore, some form of feedback is required. In the design shown in Fig. 3.33, this feedback is achieved by the feedback wire that engages with a slot machined in the spool. The pressure differential is generated by the double flapper–nozzle amplifier stage by means of a pair of restrictors connected internally to the supply pressure port.

It is well-established practice that an orifice-type flow equation may be used to determine the flow characteristic of a servovalve port, provided an appropriate flow coefficient is selected. The spool diameter is 8 mm, and the distance between the active spool lands is 14.94 mm. Here, the latest development in CFD packages becomes invaluable for analyzing these complex flow paths, and a 3D study is

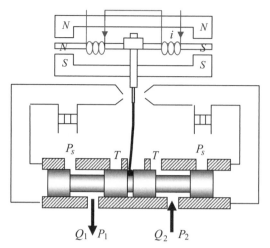

(a) Schematic of a force-feedback-type servovalve

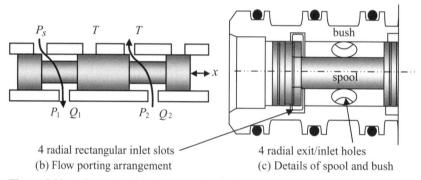

| 4 radial rectangular inlet slots | 4 radial exit/inlet holes |
| (b) Flow porting arrangement | (c) Details of spool and bush |

Figure 3.33. A force-feedback servovalve design.

highlighted. The manufacturer's 3D drawing must first be translated to a 3D mesh suitably refined around the port region of interest. In this study, the ANSYS FLUENT package was used.

For example, Fig. 3.34 shows "slices" taken across the port restriction and axially along the spool flow path indicating the mesh refinement needed for such problems. The 3D flow and pressure vectors are difficult to visualize in the absence of color; hence, only 2D slices give some meaningful information. Spool displacements from 0.1 mm to 0.5 mm were studied, resulting in 200,000–400,000 grid elements per port. Flow rate is computed together with mean pressures and shear forces acting on each spool face. Jet velocity "angles" may be derived in a number of ways, and flow coefficients are computed for different spool displacements and pressure differentials.

Figure 3.35 shows 2D slices at the inlet port and outlet port for a pressure drop of 100 bar; that is, a supply pressure of 200 bar for connected ports. The port openings are 0.5 mm, the maximum for the servovalve they represent. Figure 3.35 shows the different flow regimes that exist for the inlet and outlet ports. The inlet-port characteristic is dominated by a large recirculation zone, and the return-port characteristic is dominated by reattachment and an associated small recirculation zone.

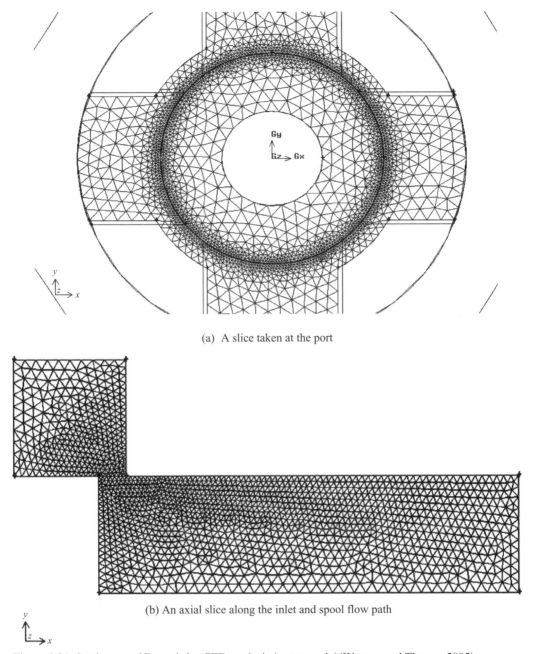

(a) A slice taken at the port

(b) An axial slice along the inlet and spool flow path

Figure 3.34. Setting up a 3D mesh for CFD analysis (not to scale)(Watton and Thorpe, 2005).

Very small openings have not been pursued here because of additional complexities of meshing and computation times, but it does seem that:

- the flow coefficients are similar at small openings and fall as the displacement approaches the closed position
- for increasing displacements, the supply-port flow coefficient is greater than the return-port flow coefficient

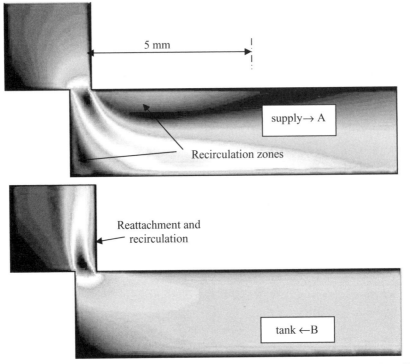

Figure 3.35. Velocity contour magnitudes, 2D slice, spool opening 0.5 mm $P_s = 200$ bar, ports connected, maximum velocity 155 m/s in both cases.

- the maximum variation in flow coefficient is only 0.05 over the range of conditions modeled
- the pressure drop has little effect on each flow coefficient for the supply pressures used

The mass flow rates are calculated from the converged results and the standard orifice-type flow-rate equation applied. This allows calculation of the flow coefficients for a range of openings and pressure drops. Some calculations are shown in Fig. 3.36 for spool openings between 0.1 mm and 0.5 mm.

It is possible to validate the CFD predictions by using the manufacturer's flow data. In this study, Star Hydraulics UK provided four new servovalves, and the average of the four rated characteristics provided by the manufacturer was used. The rated current produced a maximum spool displacement of 0.5 mm at the rated supply pressure of 70 bar. One of the flow characteristics is shown in Fig. 3.37.

With the ports connected and assuming equal port openings a_o:

$$Q = C_{q1}\, a_o \sqrt{\frac{2(P_s - P_1)}{\rho}} = C_{q2}\, a_o \sqrt{\frac{2P_2}{\rho}}, \tag{3.86}$$

and because $P_1 = P_2$, and defining:

$$Q = C_q\, a_o \sqrt{\frac{2P_s}{\rho}} \tag{3.87}$$

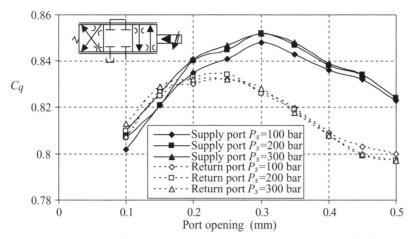

Figure 3.36. Flow coefficients for a servovalve computed with a 3D CFD package.

then the flow coefficient:

$$C_q = \frac{C_{q1}C_{q2}}{\sqrt{C_{q1}^2 + C_{q2}^2}}.$$ (3.88)

The average CFD data for each port are then used to calculate the "apparent" flow coefficient C_q with the ports connected, as given by Eq. (3.88). A comparison with the averaged measured values is shown in Fig. 3.38. The comparison suggests that the CFD predictions are good, bearing in mind the difficulty in deciding the correct experimental flow rate at small spool displacements. It can be seen from

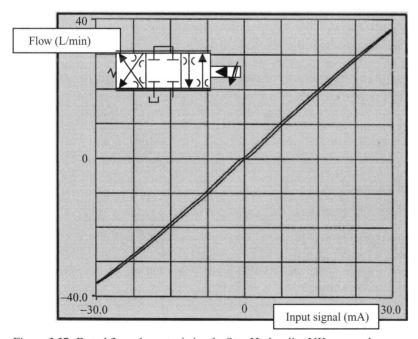

Figure 3.37. Rated flow characteristic of a Star Hydraulics UK servovalve.

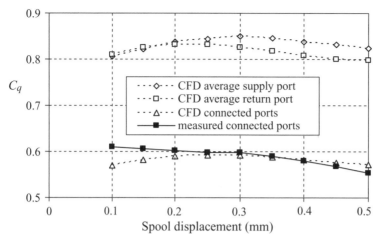

Figure 3.38. Servovalve flow coefficients, CFD and measured (Watton and Thorpe, 2005).

Fig. 3.38 that for small spool displacements, errors in graph interpretation become more significant as nonlinear effects become more dominant because of servovalve electrical hysteresis and the precise flow characteristic around the ostensibly closed condition.

Over the operating conditions studied, the supply-port average $C_{q1} = 0.83$ with a maximum error of $\pm 3\%$ and the return-port average $C_{q2} = 0.82$ with a maximum error of $\pm 2\%$. Therefore, it is reasonable to use a common, and constant, flow coefficient of $C_q = 0.83$ for design purposes.

3.2.9 Flow Characteristics of a Cone-Seated Poppet Valve

PRV poppets have traditionally had small cone seats because they are reliable from a leakage point of view and are dynamically more stable. To prevent leakage, the seats have a flat land, usually referred to as a chamfered land, and with lengths generally smaller than 2 mm. Such a configuration also helps to minimize permanent marks on the poppet. For the last 40 years or more, poppet valves have undergone a very small evolution, and the following study considers the way theory can be developed to obtain suitable design equations. As chamfered conical seats increase in length, and the clearance decreases, laminar flow occurs and some of the main characteristics of such valves then are as follows:

- improved stability
- sharp reduction of the vortex creation at the outlet and very small flow separation
- less prone to cavitation, resulting in improved reliability
- a linear pressure–flow characteristic, preferable from a control point of view
- the concept of discharge coefficient variation with flow rate does not exist because the flow is laminar
- the effect of temperature variation on the pressure–flow characteristic for laminar flow is well known and predictable
- flow-reaction-force effects could be negligible

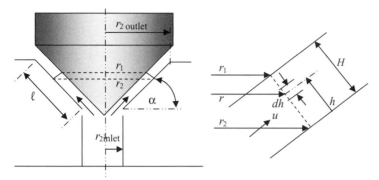

Figure 3.39. A cone-seated poppet valve.

Following the analysis by Bergada (Bergada and Watton, 2004), using Fig. 3.39 together with the earlier example of flow between parallel plates gives:

$$u = -\frac{1}{\mu}\frac{dp}{dx}\frac{h}{2}(H - h).$$ (3.89)

The flow rate through the restrictor gap will be:

$$Q = \int_o^H u\, 2\pi[r_2 - h\cos(90 - \alpha)]dh.$$ (3.90)

Integration of the flow-rate term then gives:

$$Q = -\frac{\pi}{\mu}\frac{dp}{dx}\left[r_2 H^3\frac{1}{6} - H^4\cos(90 - \alpha)\frac{1}{12}\right].$$ (3.91)

The pressure distribution along the seat is then given by:

$$\int_0^\ell \frac{dx}{r_2 H^3\frac{1}{6} - H^4\cos(90 - \alpha)\frac{1}{12}} = \int_{P_{\text{inlet}}}^{P_{\text{inlet}}} -\frac{\pi}{\mu}\frac{dp}{Q}.$$ (3.92)

The boundary conditions are as follows:

$$r_2 = x\cos\alpha + r_{2\text{inlet}},$$

$$x = 0, r = r_{2\text{inlet}}\ x = \ell, r = r_{2\text{outlet}},$$

$$\cos\alpha = \frac{r_{2\text{outlet}} - r_{2\text{inlet}}}{\ell} = \frac{r_2 - r_{2\text{inlet}}}{x}.$$ (3.93)

The pressure–flow characteristic may be obtained from (3.92):

$$Q = \frac{(P_{\text{inlet}} - P_{\text{outlet}})\dfrac{\pi H^3\cos\alpha}{6\mu}}{\ln\left[\dfrac{2\cos\alpha + \dfrac{(2r_{2\text{inlet}} - H\sin\alpha)}{\ell}}{\dfrac{(2r_{2\text{inlet}} - H\sin\alpha)}{\ell}}\right]}.$$ (3.94)

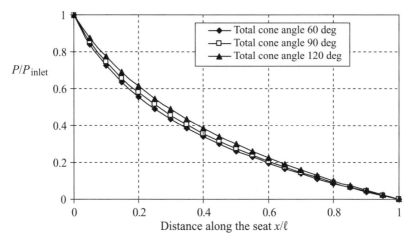

Figure 3.40. Theoretical pressure decay for a cone-seated poppet valve (Bergada and Watton, 2004).

If the pressure at a general distance x along the seat is given by P, then:

$$Q = \frac{(P_{inlet} - P)\dfrac{\pi H^3 \cos\alpha}{6\mu}}{\ln\left[\dfrac{\dfrac{2x\cos\alpha}{\ell} + \dfrac{(2r_{2inlet} - H\sin\alpha)}{\ell}}{\dfrac{(2r_{2inlet} - H\sin\alpha)}{\ell}}\right]}. \tag{3.95}$$

The pressure distribution may be completely defined along the seat as follows:

$$P = P_{inlet} - (P_{inlet} - P_{outlet})\frac{\ln\left[\dfrac{\dfrac{2x\cos\alpha}{\ell} + \dfrac{(2r_{2inlet} - H\sin\alpha)}{\ell}}{\dfrac{(2r_{2inlet} - H\sin\alpha)}{\ell}}\right]}{\ln\left[\dfrac{2\cos\alpha + \dfrac{(2r_{2inlet} - H\sin\alpha)}{\ell}}{\dfrac{(2r_{2inlet} - H\sin\alpha)}{\ell}}\right]}. \tag{3.96}$$

It will be seen that the pressure distribution does not vary linearly with position along the cone and the distribution also varies with the cone angle. For the type of seating being considered here, it is usual that $H\sin\alpha \ll 2r_{2inlet}$ and, therefore, the effect of the poppet clearance H on the pressure distribution is negligible.

Some results for pressure drop, in nondimensional form, are presented in Fig. 3.40 for a value $2r_{2inlet}/\ell = 0.2$ and for $H\sin\alpha \ll 2r_{2inlet}$. The outlet pressure is considered to be at atmospheric conditions. For this fixed clearance and seat length, increasing the cone angle increases the pressure-distribution magnitude, although the effect is very small.

To validate the theory, a test rig is considered with a seat length of $\ell = 30$ mm, an inlet diameter of 4 mm, and an angle $\alpha = 45°$, as shown in Fig. 3.41. Static pressure tappings, by means of brass inserts, were positioned in a spiral around the seated area to enable the pressure decay to be measured with sufficient spatial resolution.

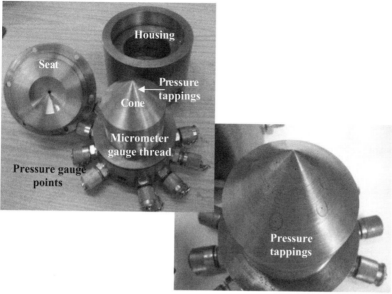

Figure 3.41. Experimental test rig for the cone-seated poppet valve.

Each tapping was connected to a pressure test gauge point at the top of the poppet by internal drilled connections.

The poppet position was adjusted by a micrometer gauge thread machined onto the assembly; such a fine thread actually distorts under pressure, creating an additional displacement. This is easily measured relative to the fixed base to create a calibration table, but it does make it difficult to obtain readings for a variable-pressure test; for example, to determine the pressure–flow-rate characteristic. For fixed-pressure tests, the clearance can easily be set to a predetermined value. The maximum thread movement measured was 15 μm for a pressure of 150 bar, although the distortion does not vary linearly with pressure.

A comparison between theory and measurement for the pressure distribution is shown in Fig. 3.42 for an angle $\alpha = 45°$. The pressure trend predicted is clearly demonstrated as is the experimental variation met in practice, particularly along the second half of the poppet face. This is primarily due to machining and mechanical assembly issues that make it almost impossible to obtain a constant, very small clearance along the seat and at different radial points around the poppet; concentricity of the poppet and seat cannot be assessed.

In keeping with previous work, the valve resistance R_c becomes, in nondimensional form:

$$\frac{\pi H^3 R_c}{6\mu} = \frac{\ln\left[\dfrac{2\cos\alpha + \dfrac{(2r_{2inlet} - H\sin\alpha)}{\ell}}{\dfrac{(2r_{2inlet} - H\sin\alpha)}{\ell}}\right]}{\cos\alpha}. \tag{3.97}$$

Because the term involving the clearance H is negligible in practice, it can be seen that the valve resistance term on the right-hand side of Eq. (3.97) is essentially

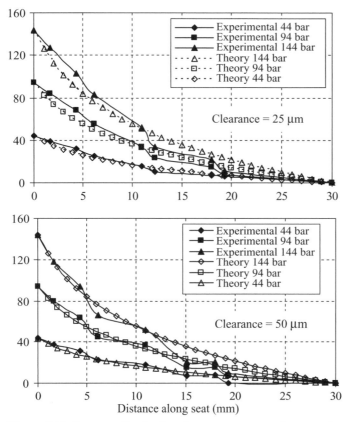

Figure 3.42. Pressure distribution down a cone-seated poppet face (Bergada and Watton, 2004).

constant for a particular geometry and realistic clearances. Hence, the valve resistance varies as expected in an inverse relationship with the characteristic dimension H^3.

3.2.10 A Double Flapper–Nozzle Device for Pressure-Differential Generation

Now consider a pair of flapper–nozzles in conjunction with a pair of orifices used to generate a pressure differential by small movements of the flapper positioned midway between the nozzles, as shown in Fig. 3.43.

The spool area is A and the spool velocity is U. The nozzle diameter is typically $d_n = 0.5$ mm, the flapper clearance in the midposition typically $x_{nm} = 0.03$ mm, and

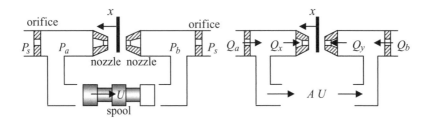

Figure 3.43. Schematic of a double flapper–nozzle amplifier used to move a spool (not to scale).

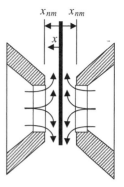

x_{nm} x_{nm}

x

Figure 3.44. Flapper–nozzle schematic.

the orifice diameter typically $d_o = 0.2$ mm. It is common for such a device to be used in servovalves that use force-feedback, the pressure differential generated being used to move the spool. It will be immediately clear from Fig. 3.43 that at the flapper midposition, often called the null position, the maximum leakage flow back to tank will exist, hence producing a small inherent power loss. As the flapper is moved to the left – in practice, by electromagnetic means – then pressure P_1 will increase and pressure P_2 will decrease, thus providing a pressure differential across the spool, which will then move unless restrained in some way. The flow loss and power loss will decrease as the flapper position is changed. To analyze the flapper–nozzle bridge, the conventional restrictor flow equations are appropriate and given by:

$$Q_a = Q_x + AU, \qquad Q_b = Q_y - AU; \tag{3.98}$$

$$Q_a = C_{qo}\, a_o \sqrt{\frac{2(P_s - P_a)}{\rho}}, \qquad Q_b = C_{qo}\, a_o \sqrt{\frac{2(P_s - P_b)}{\rho}}; \tag{3.99}$$

$$Q_x = C_{qn}\, a_{nx} \sqrt{\frac{2P_a}{\rho}}, \qquad Q_y = C_{qn}\, a_{ny} \sqrt{\frac{2P_b}{\rho}}. \tag{3.100}$$

To determine the nozzle effective flow area, it is assumed that the peripheral area is dominant because of the relative dimensions of the nozzle diameter–null clearance ratio of typically 15. Therefore, considering Fig. 3.44, it follows that:

$$a_o = \frac{\pi d_o^2}{4},$$
$$a_{nx} = \pi d_n(x_{nm} - x),$$
$$a_{ny} = \pi d_n(x_{nm} + x). \tag{3.101}$$

Considering the condition in which the spool motion is negligible, then the steady-state performance of the double flapper–nozzle amplifier may be derived from equating $Q_1 = Q_x$ and $Q_2 = Q_y$. This leads to:

$$\overline{P}_a = \frac{1}{1 + Z(1 - \overline{x})^2}, \quad \overline{P}_b = \frac{1}{1 + Z(1 + \overline{x})^2},$$

$$\overline{P}_a = \frac{P_a}{P_s}, \quad \overline{P}_b = \frac{P_b}{P_s} \quad \overline{x} = \frac{x}{x_{nm}},$$

$$Z = 16\left(\frac{C_{qn}}{C_{qo}}\right)^2 \left(\frac{d_n}{d_o}\right)^2 \left(\frac{x_{nm}}{d_o}\right)^2. \tag{3.102}$$

The pressure differential is then given by:

$$\overline{P}_a - \overline{P}_b = \frac{4Z\overline{x}}{[1 + Z(1 + \overline{x})^2][1 + Z(1 - \overline{x})^2]}. \tag{3.103}$$

At the null condition $\overline{x} = 0$ and $\overline{P}_a - \overline{P}_b = 0$; then:

$$\overline{P}_a = \overline{P}_b = \frac{1}{(1 + Z)},$$

$$\text{and the null gain } \frac{d(\overline{P}_a - \overline{P}_b)}{d\overline{x}} = \frac{4Z}{(1 + Z)^2}. \tag{3.104}$$

To obtain the maximum sensitivity at null, Eq. (3.104) leads to the maximum null gain when $Z = 1$, suggesting that the design should produce null pressures of $P_a = P_b = P_s/2$; that is, half supply pressure.

Considering the flow loss and the power loss, it follows that:

$$\overline{Q}_{\text{loss}} = \overline{W}_{\text{loss}} = \frac{(1 - \overline{x})}{\sqrt{1 + Z(1 - \overline{x})^2}} + \frac{(1 + \overline{x})}{\sqrt{1 + Z(1 + \overline{x})^2}}, \tag{3.105}$$

$$\overline{Q}_{loss} = \frac{Q_{\text{loss}}}{k_n} \quad \overline{W}_{\text{loss}} = \frac{W_{\text{loss}}}{P_s k_n} \quad k_n = C_{qn} \pi d_n x_{nm} \sqrt{\frac{2P_s}{\rho}}.$$

The pressure-differential–flapper-displacement characteristic and the power-loss and flow-loss characteristics are shown in Fig. 3.45 for a range of the design parameter Z.

If spool control is achieved with flapper displacements around the null condition, then a value of $0.5 < Z < 2$ is satisfactory, the suggested value of $Z = 1$ being ideal. Increasing Z does decrease the flow and power losses, which are both

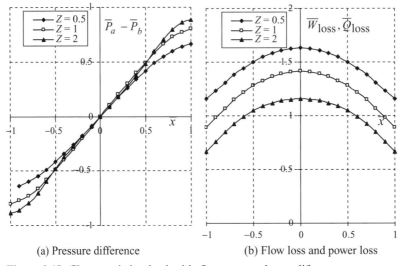

(a) Pressure difference   (b) Flow loss and power loss

Figure 3.45. Characteristic of a double flapper–nozzle amplifier.

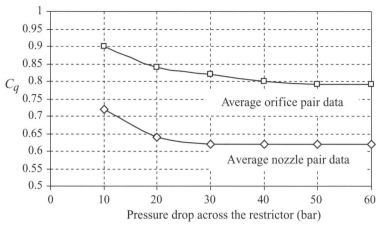

Figure 3.46. Measured flow coefficients for a double flapper–nozzle amplifier.

maximum at the null condition. The issue of design is therefore centered around the calculation of the design parameter Z, a value of $Z = 1$ being typical of that sought by servovalve manufacturers. From earlier work in this chapter, it is reasonable to select flow coefficients $C_{qn} \approx 0.6$ for the nozzle and $C_{qo} \approx 0.8$ for the orifice as a general design guide. For a more accurate analysis, it is necessary to have detailed flow characteristics for the nozzle pair and the orifice pair, which are matched by the manufacturer. The orifice diameter tends to vary between 0.15 and 0.4 mm, the nozzle diameter between 0.45 and 0.7 mm, the flapper clearance changing little around 0.03 mm. Some results measured by the author are shown in Fig. 3.46 using data typical of a large selection of servovalves of the force-feedback type.

It should be recalled that typical pressure drops will be around half supply pressure with a minimum therefore being in excess of 35 bar. From Fig. 3.46, it can be seen that the flow coefficients are reasonably constant at $C_{qn} \approx 0.62$ for the nozzle and $C_{qo} \approx 0.79$ for the orifice. The latest developments in sapphire-machining technology now ensure orifice dimensional consistency, drastic wear reduction, and batch quality such that diameter variation cannot be measured to any meaningful significance. A sapphire orifice is shown in Fig. 3.47.

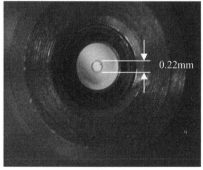
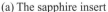

(a) The sapphire insert   (b) Orifice assembly

Figure 3.47. A sapphire orifice developed by Star Hydraulics Ltd. UK.

Worked Example 3.9

A flapper–nozzle amplifier is to be designed to have a maximum leakage flow loss of 0.5 L/min with a supply pressure of 210 bar. Assuming a fluid density of 860 kg/m^3 and given a flapper clearance of 0.03 mm, calculate:

 (i) the nozzle and orifice diameters, and
(ii) the power loss at null.

Solution for (i). Recall that:

$$Z = 16 \left(\frac{C_{qn}}{C_{qo}}\right)^2 \left(\frac{d_n}{d_o}\right)^2 \left(\frac{x_{nm}}{d_o}\right)^2 \to 1.$$

Select $C_{qn} \approx 0.6$ for each nozzle and $C_{qo} \approx 0.8$ for each orifice as a good starting point.

At null:

$$\overline{Q}_{loss} = \overline{W}_{loss} = \frac{2}{\sqrt{1+Z}} = 1.414, \quad \overline{Q}_{loss} = \frac{Q_{loss}}{k_n} \quad \overline{W}_{loss} = \frac{W_{loss}}{P_s k_n},$$

$$k_n = C_{qn}\pi d_n x_{nm}\sqrt{\frac{2P_s}{\rho}}.$$

Flow loss at null:

$$\frac{0.5 \times 10^{-3}}{60} = (1.414)(0.6)\pi(d_n \times 10^{-3})(x_{nm} \times 10^{-3})\sqrt{\frac{2(210 \times 10^5)}{860}}.$$

This gives $d_n x_{nm} = 0.014 \to d_n = 0.47$ mm:

$$16\left[\frac{0.6}{0.8}\right]^2 \frac{(d_n x_{nm})^2}{d_o^4} = 1 \quad 16\left[\frac{0.6}{0.8}\right]^2 \frac{(0.014)^2}{d_o^4} = 1 \, d_o = 0.2 \text{ mm}.$$

Solution for (ii). The null power loss:

$$W_e = 1.414 P_s k_n = (210 \times 10^5)(0.5 \times 10^{-3}/60),$$

$$W_e = 175 \text{ W}.$$

Sometimes servovalve manufacturers include a drain orifice at the flapper–nozzle region to provide a very small back pressure that tends to reduce flow instabilities and asymmetric flow effects. The addition of the drain orifice is shown schematically in Fig. 3.48, the drain flow returning to tank.

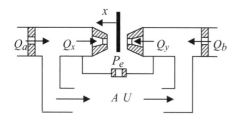

Figure 3.48. Flapper–nozzle amplifier with a drain orifice.

It is a reasonable assumption that the drain orifice equation has a similar flow characteristic as the other orifices and, for the spool at rest, the flow equation may be written as:

$$Q_e = Q_a + Q_b = Q_x + Q_y, \tag{3.106}$$

$$Q_e = \lambda C_{qo} d_o \sqrt{\frac{2 P_e}{\rho}}, \tag{3.107}$$

where Q_e is the drain flow and λ is the diameter scale factor compared with the existing orifice pair. The remaining flow equations are then modified to:

$$Q_a = C_{qo} \, a_o \sqrt{\frac{2(P_s - P_a)}{\rho}}, \qquad Q_b = C_{qo} \, a_o \sqrt{\frac{2(P_s - P_b)}{\rho}}, \tag{3.108}$$

$$Q_x = C_{qn} \, a_{nx} \sqrt{\frac{2(P_a - P_e)}{\rho}}, \qquad Q_y = C_{qn} \, a_{ny} \sqrt{\frac{2(P_b - P_e)}{\rho}}. \tag{3.109}$$

It is then a simple matter to show that the pressure differential is modified as follows:

$$\overline{P}_a - \overline{P}_b = \frac{4 Z \bar{x}(1 - \overline{P}_e)}{[1 + Z(1 + \bar{x}_n)^2][1 + Z(1 - \bar{x}_n)^2]}, \qquad \overline{P}_e = \frac{P_e}{P_s}. \tag{3.110}$$

Because the back pressure is of the order of $P_e \approx 1 - 2$ bar, then its effect on the pressure-differential characteristic is negligible.

3.2.11 The Jet Pipe and Deflector-Jet Fluidic Amplifier

Now consider alternative approaches to obtaining a pressure differential compared with the flapper–nozzle amplifier previously analyzed. The jet pipe amplifier and deflector-jet principles are shown in Fig. 3.49 and are allied to fluidic amplifier principles developed in the 1960s.

The supply pressure pipe pressure P_s produces a jet that divides equally between the two receivers, for both methods, when the unit is symmetrical. A small displacement of the pipe or deflector plate results in the flow rate increasing in one receiver, whereas the other receiver experiences a reduction in flow rate and by the same amount for an ideal device. Each flow rate is therefore converted to a static pressure; thus, a net pressure differential is experienced by the spool. To ensure that spool position is achieved, feedback is needed to control the jet source. This can be by means of spool-position electrical feedback for the jet pipe or by a feedback wire connected to the deflector plate. Fluidic devices are also shown in Fig. 3.49. The fluidic proportional amplifier, Fig. 3.49(c), uses lower-energy control jets to deflect the main jet, and the output pressure differential ΔP_o experiences a gain compared with the control pressure differential applied ΔP_c. The fluidic jet-deflection device, Fig. 3.49(d), creates a varying output pressure differential ΔP_o as the input jet is moved laterally. Whatever the approach selected, the pressure-differential–jet-deflection characteristic will be similar to a flapper–nozzle characteristic although possibly with an increased flow loss. One important advantage of these devices over flapper–nozzle devices is the tolerance to larger particles in the fluid; given normal filtration, they are less liable to blockage.

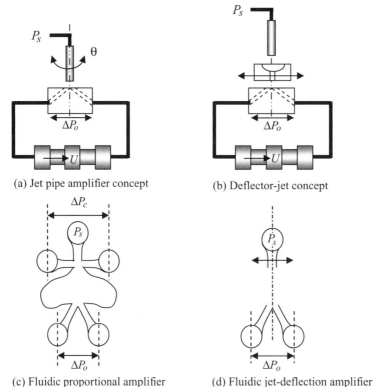

(a) Jet pipe amplifier concept   (b) Deflector-jet concept

(c) Fluidic proportional amplifier   (d) Fluidic jet-deflection amplifier

Figure 3.49. Schematics of the jet pipe and deflector-jet amplifiers.

To obtain a preliminary feel for the performance of a fluidic jet-deflector stage, consider the 2D free jet shown in Fig. 3.50.

The free jet issuing from the jet pipe is assumed here to initially have a constant velocity profile. It will experience an increasing cross-sectional area because of fluid boundary entrainment, and the pipe exit velocity will remain constant within the "potential core" of length x_0. Beyond this point, a normal distribution type of velocity profile will occur and the centerline velocity will then continually decrease with increasing distance. Clearly, the performance of these amplifiers depends on the distance between the pipe exit and the receivers, this distance significantly affecting pressure gain and maximum pressure–energy recovery. The velocity profile of a free jet has been studied in some detail (Abromovich, 1963; Kirshner and Katz, 1975),

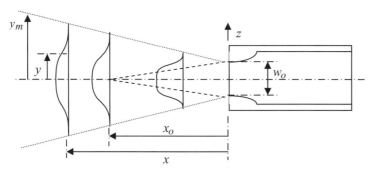

Figure 3.50. A submerged free jet.

Table 3.2. *Velocity profiles for fully established flow $x > x_0$*

Albertson et al. (1950)	$u = U_o \sqrt{\dfrac{x_o}{x}} \exp -\dfrac{\pi}{2} \left(\dfrac{x_o}{w_o}\right)^2 \left(\dfrac{z}{x}\right)^2$
Simson (1966)	$u = U_o \sqrt{\dfrac{x_o}{x}} \left[1 - \left(\dfrac{y}{y_m}\right)^{7/4}\right]^2$
Approximation to Simson	$u = U_o \sqrt{\dfrac{x_o}{x}} \left[1 - 3\left(\dfrac{y}{y_m}\right)^2 + 2\left(\dfrac{y}{y_m}\right)^3\right]$
U_o = centerline velocity	$y_m = 1.38 w_o \left(\dfrac{x}{x_o}\right), \quad x_o \approx 5.2 w_o$

and Table 3.2 shows some models that have been considered for the region beyond the potential core.

To analyze the jet-deflection characteristic, consider recovery of the jet energy in the two downstream receivers (diffusers) and lateral motion z of the supply jet. In reality, the jet-deflection angle may be neglected. Also assume that each receiver width is the same as the velocity profile half-width y_m. For fully developed jet flow, then the pressure differential, assuming a blocked load, is given by:

$$\frac{\Delta P_o}{P_s} = \int_0^{\bar{z}} \bar{u}^2 d\bar{y} + \int_0^{1-\bar{z}} \bar{u}^2 d\bar{y} - \int_{\bar{z}}^1 \bar{u}^2 d\bar{y}, \quad \bar{z} = \frac{z}{y_m}, \quad \bar{y} = \frac{y}{y_m}, \quad \bar{u} = \frac{u}{U_o},$$

$$\frac{\Delta P_o}{P_s} = 2 f(\bar{z}) + f(1 - \bar{z}) - f(1). \tag{3.111}$$

Using the approximation velocity profile from Table 3.2 then gives:

$$f(\bar{z}) = \frac{x_o}{x}(\bar{z} - 2\bar{z}^3 + \bar{z}^4 + 1.8\bar{z}^5 - 2\bar{z}^6 + 0.571\bar{z}^7). \tag{3.112}$$

Placing the receivers at the end of the potential core, $x = x_o$, gives the characteristic shown in Fig. 3.51. It will be seen that for this receiver placement at the end of the potential core, then the maximum pressure differential recovered is 2/3 of the

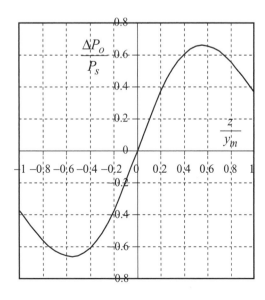

Figure 3.51. Pressure differential for a jet-deflection device, $x = x_o$.

supply pressure. Notice also that a maximum jet deflection of only half the receiver width may be utilized for a positive-gain characteristic.

For this analysis, each receiver width was set at half the total jet width:

$$\text{receiver width} = y_m = 1.38w_o \left(\frac{x}{x_o}\right) = 1.38w_o. \tag{3.113}$$

Therefore, each receiver width should be greater than the supply nozzle width. It should be noted that for a practical servovalve application, the supply nozzle may well be nearer to a short tube design, in which case the exit velocity profile will probably be parabolic rather than constant velocity. In this situation, the concept of a potential core may not be valid. However, the velocity profile distribution downstream will approach a shape similar to a normal distribution of the type used in this example. A preliminary study, beyond the scope of this section, has shown that a detailed 3D CFD model is essential to produce the pressure-differential characteristic of a fluidic amplifier whose total length may well be no greater than 10 mm. The maximum pressure recovery achieved will probably be low because of amplifier length constraints.

3.3 Steady-State Flow-Reaction Forces

3.3.1 Basic Concepts

Considering a flow direction change within a control volume, a change of momentum is experienced generally in the x and y directions; hence, a force is required to cause the momentum change. Hence, the rigid boundary experiences an equal and opposite force, termed the flow-reaction force. This is given by:

$$\text{fluid accelerating force} = \int_{A\text{outlet}} \rho V_j V dA - \int_{A\text{inlet}} \rho V_j V dA,$$

$$\text{Rigid-body flow-reaction force} = -\text{fluid accelerating force.} \tag{3.114}$$

In fluid power component problems, the dominant flow-reaction force is usually along the main flow direction along the component main axis.

3.3.2 Application to a Simple Poppet Valve

Consider Fig. 3.52, where the poppet movement creates an orifice area a and the inlet port has a cross-sectional-area A. The initial and relatively low mean velocity at the inlet has clearly increased in magnitude along the main axis as it leaves the exit port, which usually has a significantly smaller flow area.

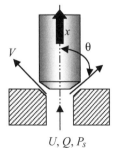

Figure 3.52. Flow through a simple poppet valve.

The force to cause the fluid acceleration is:

$$F_x = \rho Q V \cos\theta - \rho Q U$$
$$= \frac{\rho Q^2 \cos\theta}{a} - \frac{\rho Q^2}{A}$$
$$\approx \frac{\rho Q^2 \cos\theta}{a}. \tag{3.115}$$

From the orifice equation:

$$Q = C_q a \sqrt{\frac{2P_s}{\rho}}. \tag{3.116}$$

So, combining with Eq. (3.115) gives:

$$F_x \uparrow = \left(2C_q^2 \cos\theta\right) P_s a.$$

Therefore, the flow-reaction force is:

$$F_{rx} \downarrow = \left(2C_q^2 \cos\theta\right) P_s a. \tag{3.117}$$

This *acts to oppose the direction of the spool motion*. Because for this example the exit port area is proportional to the port opening – that is, poppet displacement – it can be deduced that this flow-reaction force has the characteristic of a resisting spring. Let d be the inlet port diameter; then:

$$a = \pi d x,$$
$$F_{rx} \downarrow = \left(2\pi C_q^2 \cos\theta\, P_s d\right) x = kx, \tag{3.118}$$
$$k = \text{equivalent spring stiffness}.$$

The ratio of flow-reaction force–inlet static force is given by:

$$\frac{F_{rx}}{P_s A} = \frac{2\pi C_q^2 \cos\theta\, P_s d x}{P_s \pi d^2/4} = 8C_q^2 \cos\theta \left(\frac{x}{d}\right). \tag{3.119}$$

So, considering reasonable values of $C_q = 0.7$ and $\theta = 60°$ gives:

$$\frac{F_{rx}}{P_s A} \approx 2\left(\frac{x}{d}\right). \tag{3.120}$$

For a poppet diameter of 8 mm, a lift of 0.25 mm and a supply pressure of 210 bar would give a flow rate of 58.4 L/min. The flow reaction resisting force would be 6.25% of the static force on the poppet.

3.3.3 Application to the Main Stage of a Two-Stage Pressure-Relief Valve

Consider the basic concept of a two-stage PRV of the type shown in Fig. 3.53.

The required pressure setting is obtained with the first-stage setting spring that has a stiffness much greater than the main-stage retaining spring. The first stage therefore requires a leakage path back to tank. The back pressure P_d is equal to supply pressure P_s, and the effect of the second-stage spring, and sometimes a small differential area, is to keep the main-stage poppet closed. When the system pressure force exceeds the first-stage setting, then the small poppet is displaced and a small flow is created through the first-stage variable orifice. Because of restrictors a and b, there is a rapid drop in pressure P_d, the main-stage poppet is displaced, and the

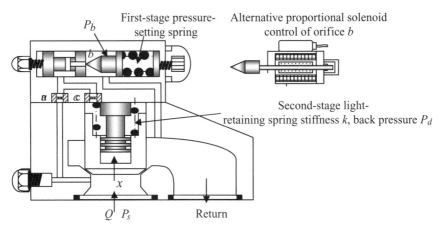

Figure 3.53. A two-stage PRV.

supply pressure is controlled close to its desired value. Note that the main poppet has a machined seating face inclined at an angle θ. The orifice c is called a damping orifice and helps to improve the dynamic motion of the main-stage poppet.

Considering the circuit results in the following steady-state equations for force balance across the main-stage poppet and flow through the main stage, where A is the main poppet cross-sectional area, the inlet diameter is d_i, and the poppet flow area is a_o:

$$(P_s - P_d)A = F_o + kx + \alpha 2 C_q^2 \cos\theta P_s a_o, \tag{3.121}$$

$$Q = C_q a_o \sqrt{\frac{2P_s}{\rho}}, \tag{3.122}$$

$$a_o = \pi d_i x \sin\theta. \tag{3.123}$$

The factor α is included to take into account the variation of flow-reaction force from the ideal theoretical value. These equations may be rearranged to give:

$$(P_s - P_d) = \frac{F_o}{A} + \frac{P_s}{A\sqrt{\dfrac{2P_s}{\rho}}}\left(\frac{k}{C_q P_s \pi d_i \sin\theta} + \alpha 2 C_q \cos\theta\right) Q. \tag{3.124}$$

Hence, plotting $(P_s - P_d)/Q$ then allows the unknown terms to be estimated for different supply pressures. The method requires the ability to measure the internal pressure P_d and also requires accurate transducers because the pressure difference is typically $(P_s - P_d) < 5\,\text{bar}$. For the valve tested, the following data apply:

spring stiffness $k = 1.76 \times 10^4$ N/m   fluid density $\rho = 850$ kg/m^3
main spool diameter $d_i = 22$ mm   main spool seat angle $\theta = 35°$

For this valve, the spring-resisting term is negligible compared with the flow-reaction force for realistic pressures, and this means that the method proposed allows evaluation only of the product αC_q. The results are shown in Fig. 3.54, the pressure-differential–flow-rate characteristic having only a restricted range of supply pressures possible because of the large flow rates created for a small value of $(P_s - P_d)$, the very advantage of this type of PRV.

The point-by-point calculation shows a small variation in αC_q as the flow rate increases at a fixed pressure. Using the average slopes gives a value of $\alpha C_q = 1.75$

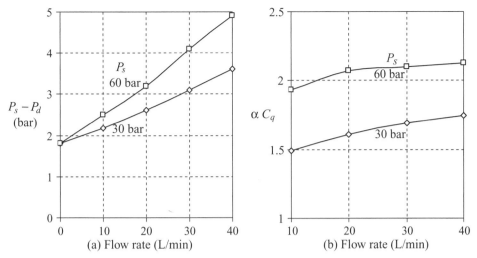

Figure 3.54. Measured characteristics of a two-stage PRV (Davies and Watton, 1993).

for a pressure of 30 bar and $\alpha C_q = 2.12$ for a pressure of 60 bar. What is clear is that whatever assumption is made for the flow coefficient C_q – for example, $C_q = 0.7$ – the value of α is probably 2.5–3. This means that the flow-reaction force is significantly higher than the value from simple momentum theory. Note also, the main spool position is controlled by the flow-reaction force, not by the spring force. Evaluating the spring precompression from Fig. 3.54(a) gives $F_o = 68.4$ N and, hence, a precompression of 3.9 mm.

3.3.4 Application to a Spool Valve

The problem with flow-reaction-force analysis is the correct choice of jet angle and jet velocity, and this was evident from the results in the previous example. In reality, the velocity profile across a flow-control port has a nonlinear distribution because it must be zero at each port boundary. Also, the flow path will be curved, often with reattachment to an adjacent boundary as illustrated in Fig. 3.35 for a spool-valve using CFD analysis. Consider, therefore, a spool-valve geometry typical for a servovalve, as shown in Fig. 3.55.

When the momentum-change theory is applied to evaluate the flow-reaction force, it is seen that the fluid passing through the inlet port has experienced an acceleration along the main spool axis. This also occurs at the return port. Because the annular velocity inside the servovalve, after the inlet port has been passed and before the return port has been entered, is very much smaller than each port jet velocity, then its momentum contribution may be neglected.

Figure 3.55. A servovalve spool.

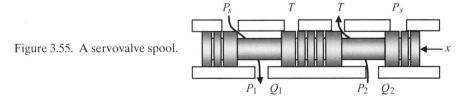

The net flow force that is due to fluid acceleration at both ports is given by:

$$F_x \approx (\rho\, Q V \cos\theta)_{\text{inlet}} - (\rho\, Q V \cos\theta)_{\text{return}}$$

$$\rightarrow$$

$$F_x = \left(\frac{\rho\, Q^2 \cos\theta}{a}\right)_{\text{inlet}} - \left(\frac{\rho\, Q^2 \cos\theta}{a}\right)_{\text{return}}. \tag{3.125}$$

Using the flow equation for each port then gives the equal and opposite flow-reaction force:

$$F_x = \left[2C_q^2 \cos\theta(P_s - P_1)a_o\right]_{\text{inlet}} - \left[2C_q^2 \cos\theta(P_2)a_o\right]_{\text{return}},$$

$$F_x = 2C_q^2 a_o[\cos\theta(P_s - P_1)_{\text{inlet}} - \cos\theta(P_2)_{\text{return}}], \tag{3.126}$$

where a_o is the orifice area and equal for the inlet port and return port. If the jet angles just happen to be equal, then:

$$F_x = 2C_q^2 a_o \cos\theta[P_s - P_{\text{load}}], \tag{3.127}$$

where the load pressure differential is defined as $P_{\text{load}} = P_1 - P_2$. The flow-reaction force is clearly at a maximum when the load pressure differential is zero.

Now consider the CFD analysis of flow through a servovalve spool, as outlined in Subsection 3.2.8, where flow coefficients were assessed. The analysis allows the flow-reaction force effect to be evaluated by use of the pressure data integrated around each spool annulus area. Then, by using the flow-reaction force equation and the previously calculated flow coefficients, we may determine a representative jet angle. The results, obtained with smoothed data, are shown in Fig. 3.56 for the inlet port and the return port.

The results show different trends associated with different flow regimes at the supply port and the return port, as previously discussed. It could be argued that the variation in jet angle is not significant in both cases, a midrange value being $\theta \approx 68°$ for the inlet port and $\theta \approx 67°$ for the return port. These values are close to the von Mises analytical solution limiting value of $\theta \approx 69°$.

Now consider using a visual approach in which the angle of the velocity vector having the greatest magnitude is selected to be representative of the dominant momentum contribution across the orifice. A vector evaluation software routine

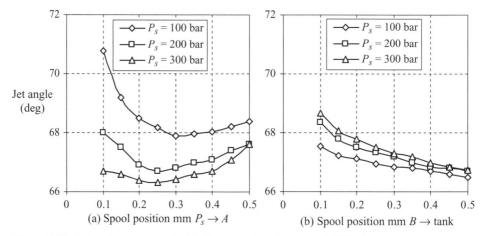

Figure 3.56. Jet angles computed with flow-reaction theory and CFD data.

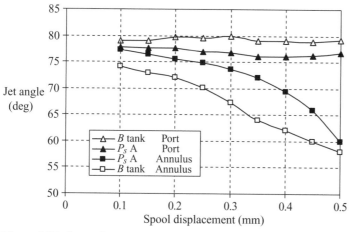

Figure 3.57. Jet angles computed with visual interpretation from a CFD analysis.

was used to do this, and this approximate method produced the results shown in Fig. 3.57 and shows a different picture compared with Fig. 3.56. The main flow path into and out of each orifice is set by the four peripheral slots termed "ports" and the remaining flow path area between the ports are termed "annulus." Visual interpretation of the jet angle shows the port values to be greater than the annulus values. However, the use of these angles for flow-reaction-force calculation using simple momentum-change theory would give misleading results.

This analysis of the two approaches suggests that a visual interpretation approach should be treated with caution and the maximum jet velocity angle may not be representative of the average momentum effect assumed in the simple theory.

3.3.5 Application to a Cone-Seated Poppet Valve

Consider the cone-seated poppet valve previously analyzed for its pressure and flow characteristic. The poppet is again shown in Fig. 3.58. This apparently simple device presents an analytical challenge because the velocity through the poppet clearance changes with distance. In addition, the velocity distribution at the inlet must be taken into account because it cannot be considered uniform, as developed by JM Bergada in Bergada and Watton (2004).

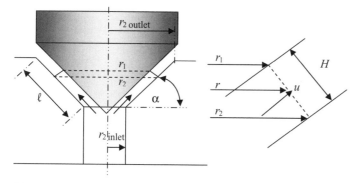

Figure 3.58. Schematic of the cone-seated poppet valve.

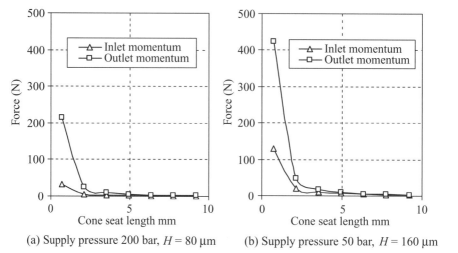

(a) Supply pressure 200 bar, $H = 80\,\mu m$ (b) Supply pressure 50 bar, $H = 160\,\mu m$

Figure 3.59. Steady-state flow reaction forces for a cone-seated poppet (Bergada and Watton, 2004).

A parabolic velocity distribution is assumed:

$$u_{\text{inlet pipe}} = u_{\max}\left[1 - \left(\frac{r}{r_{2\text{inlet}}}\right)^2\right]. \tag{3.128}$$

Hence, the inlet momentum term is:

$$\int_{A_{\text{inlet}}} \rho V_j V dA = \int_0^{r_{2\text{inlet}}} \rho u^2 2\pi r\,dr = \frac{4}{3}\rho\pi u_{\max}^2 r_{2\text{inlet}}^2. \tag{3.129}$$

It then follows that:

$$\int_{A_{\text{outlet}}} \rho V_j V dA = \int_0^H \rho u^2 \sin\alpha [r_{2\text{outlet}} - h\sin\alpha]2\pi\,dh$$

$$= \frac{\rho 2\pi\sin\alpha}{120\mu^2}\left[\frac{K_1^2(P_{\text{outlet}} - P_{\text{inlet}})^2}{K_3^2(\ell K_1 + K_2)^2}\right]\left[H^5 r_{2\text{outlet}} - \frac{H^6\sin\alpha}{2}\right], \tag{3.130}$$

$$K_1 = \frac{H^3\cos\alpha}{6},$$

$$K_2 = \frac{H^3 r_{2\text{inlet}}}{6} - \frac{H^4\sin\alpha}{12},$$

$$K_3 = \ln\left[\frac{2\ell\cos\alpha + 2r_{2\text{inlet}} - H\sin\alpha}{2r_{2\text{inlet}} - H\sin\alpha}\right]. \tag{3.131}$$

The net-flow reaction force is then given by subtracting Eq. (3.130) from Eq. (3.129), and each contribution is shown in Fig. 3.59 for the example poppet with $\alpha = 45°$ and an inlet radius of $r_{2\text{inlet}} = 2\,mm$. The flow-rate equation previously derived is first required to determine the appropriate inlet velocity profile maximum velocity $u_{\max}$. It can be seen from Fig. 3.59(a) that for a small clearance, the high resistance and, hence, reduced flow rate results in a relatively small net flow-reaction force, even with a large supply pressure. For a reduced supply pressure, shown in Fig. 3.59(b) but with an increased clearance, the flow rate is increased, which results in

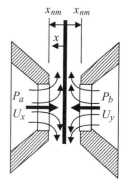

Figure 3.60. Flapper–nozzle schematic.

an increased net flow-reaction force. In both cases shown, each momentum contribution collapses rapidly as the seat length exceeds typically 3 mm. Hence, for seat lengths in excess of this value, the flow-reaction force may be neglected in comparison with the static inlet force for this poppet valve design. This leads to an important conclusion that flow-reaction force effects could be minimized if true cone seating is used for valve poppets.

3.3.6 Application to a Flapper–Nozzle Stage

Consider just the flapper–nozzle stage of the amplifier previously discussed in Section 3.2.10 and shown in Fig. 3.60. The forces acting on each side of the flapper may be written as:

$$F_x = \left(P_a + \frac{1}{2}\rho U_x^2 \right) a_n, \tag{3.132}$$

$$F_y = \left(P_b + \frac{1}{2}\rho U_y^2 \right) a_n, \tag{3.133}$$

$$\text{nozzle csa} = a_n = \frac{\pi d_n^2}{4},$$

$$Q_x = C_{qn} a_{nx} \sqrt{\frac{2P_a}{\rho}}, \quad a_{nx} = \pi d_n(x_{nm} - x),$$

$$Q_y = C_{qn} a_{ny} \sqrt{\frac{2P_b}{\rho}}, \quad a_{ny} = \pi d_n(x_{nm} + x),$$

$$U_x = \frac{Q_x}{a_n}, \quad U_y = \frac{Q_y}{a_n}. \tag{3.134}$$

Assuming equal flow coefficients for each nozzle, combining these equations gives the variation in net flapper force with flapper position as follows:

$$F_x - F_y = a_n \left\{ (P_a - P_b) + \frac{16 C_{qn}^2 x_{nm}^2}{d_n^2} \left[P_a \left(1 - \frac{x}{x_{nm}} \right)^2 - P_b \left(1 + \frac{x}{x_{nm}} \right)^2 \right] \right\}. \tag{3.135}$$

If the flapper–nozzle pair forms part of a pressure-differential-generating bridge incorporating a pair of orifices, as discussed in Section 3.2.10, then P_1 and P_2 may be expressed in terms of P_s, Z, and $\bar{x}$ to give:

$$\frac{(F_x - F_y)}{P_s a_n} = \frac{4\bar{x}(Z - \alpha)}{[1 + Z(1 + \bar{x}_n)^2][1 + Z(1 - \bar{x}_n)^2]},$$ (3.136)

$$Z = 16 \left(\frac{C_{qn}}{C_{qo}}\right)^2 \left(\frac{d_n}{d_o}\right)^2 \left(\frac{x_{nm}}{d_o}\right)^2, \quad \alpha = 16 C_{qn}^2 \left(\frac{x_{nm}}{d_n}\right)^2.$$

Considering the equation for pressure differential then gives:

$$\frac{(F_x - F_y)}{a_n} = \left(1 - \frac{\alpha}{Z}\right)(P_a - P_b),$$ (3.137)

$$(P_a - P_b) = \frac{4\bar{x}ZP_s}{[1 + Z(1 + \bar{x}_n)^2][1 + Z(1 - \bar{x}_n)^2]}.$$

Therefore, the flow-reaction force is proportional to the pressure-differential generated by the flapper–nozzle bridge. This means that for a modern servovalve design, the flow-reaction force may be considered approximately proportional to flapper displacement, particularly around the null condition. Because usually $\alpha \ll Z$, it can be deduced that Eq. (3.136) is identical to the pressure-differential equation. It also means that the maximum flow-reaction force occurs when the flapper is displaced to its maximum position, giving:

$$\frac{(F_x - F_y)_{max}}{P_s a_n} = \frac{4(Z - \alpha)}{(1 + 4Z)}.$$ (3.138)

This value is typically ≈ 0.8 for a servovalve. To ensure a reacting – hence, stabilizing – flow-reaction force, it is necessary that $Z > \alpha$ and gives:

$$16 \left(\frac{C_{qn}}{C_{qo}}\right)^2 \left(\frac{d_n}{d_o}\right)^2 \left(\frac{x_{nm}}{d_o}\right)^2 > 16 C_{qn}^2 \left(\frac{x_{nm}}{d_n}\right)^2,$$

$$\frac{d_n}{d_o} > \sqrt{C_{qo}} \approx 0.9.$$ (3.139)

This is absolutely guaranteed if the nozzle diameter is greater than the orifice diameter, which is the case in practice. For example, the servovalve previously discussed has values of $d_n = 0.47$ mm and $d_o = 0.22$ mm; hence, a ratio of ≈ 2.

3.4 Other Forces on Components

3.4.1 Static and Shear-Stress Components

The poppet valve previously analyzed again serves to illustrate the other forces that must be considered to complete the steady-state force balance. Normally, gravitational forces are neglected because component masses are usually small. The remaining forces are static, flow reaction, and shear forces, illustrated as shown in Fig. 3.61.

In practice, the shear-stress component will probably be small; for example, in the previous work discussed on a servovalve spool, the shear-stress component was found, through CFD analysis, to contribute $< 2\%$ of the total axial force.

Static force at "free" inlet and outlet:

$$\int_{A \text{ inlet}} P_i\, A_i,$$

$$\int_{A \text{ outlet}} P_o\, A_o.$$

Static along the cone and seat:

$$\int_{A \text{ cone}} p\, dA_j.$$

Shear stress along the cone surface:

$$\int_{A \text{ cone}} \tau\, dA_j.$$

Flow-reaction forces:

$$\int_{A \text{ inlet}} \rho V_j\, V\, dA_j,$$

$$\int_{A \text{ outlet}} \rho V_j\, V\, dA_j.$$

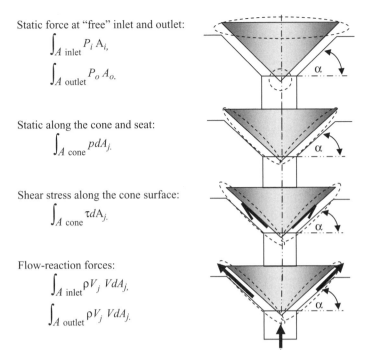

Figure 3.61. The various forces acting on a poppet.

3.4.2 Transient Flow-Reaction Forces

Although this chapter is concerned with steady-state characteristics, it is appropriate to include transient effects at this point because they are usually linked with flow-reaction forces. The transient force arises from the motion of a port opening–closing poppet or spool that creates a transient flow rate until the new steady-state conditions have been established. Consider Fig. 3.62.

Considering Newton's second law of motion and the spool annulus cross-sectional area a_s, an approximation to the net force to cause acceleration of a fluid slug of length ℓ through the valve is given by:

$$F_{\rightarrow} = \rho \ell a_s \frac{dV_1}{dt} - \rho \ell a_s \frac{dV_2}{dt}. \tag{3.140}$$

Hence, the transient flow-reaction force that resists motion is given by:

$$F_{\leftarrow} = \rho \ell \left(\frac{dQ_1}{dt} - \frac{dQ_2}{dt} \right). \tag{3.141}$$

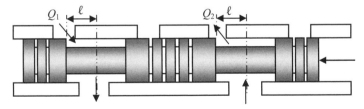

Figure 3.62. A spool-valve schematic.

Recalling the general flow equations discussed in Section 3.2.8, and assuming equal flow coefficients for an orifice area a_o that varies linearly with displacement x, we have:

$$Q_1 = C_q a_o \sqrt{\frac{2(P_s - P_1)}{\rho}}, \qquad Q_2 = C_q a_o \sqrt{\frac{2P_2}{\rho}}.$$

Letting $a_o = wx$, we have:

$$F = \rho \ell w C_q \frac{d}{dt}\left(x\sqrt{\frac{2(P_s - P_1)}{\rho}} - x\sqrt{\frac{2P_2}{\rho}} \right). \tag{3.142}$$

The transient net flow-reaction force then becomes a function of spool opening x and the rate of change of both P_1 and P_2 with time. This can be done only within a complete valve and circuit simulation, but often the transient flow-reaction term is neglected.

3.5 The Electrohydraulic Servovalve

3.5.1 Servovalve Types

Servovalves are the heart of modern fluid power control systems in the sense that they are the ideal interface between low-power electrical control signals and high-power pressure or flow output. This means that advanced electrical transducer technology may be combined with simple-to-advanced control theory, either by a dedicated digital controller or by a purpose-designed microcomputer. Figure 3.63 shows three examples from the many servovalves that are commercially available.

The term *electrohydraulic servovalve* has become synonymous with fast-acting high-precision proportional control valves, but they all have common control aspects, as follows:

- a power supply port, a tank return port, two load ports
- an electrical input
- an electrical-to-electromagnetic force-generating stage
- spool-valve pressure-differential actuation by means of the force-generating mechanism
- a mechanical or electrical mechanism for spool-valve positioning
- flow output from supply pressure to port A or port B and flow return from port B or port A back to tank

A variation on the servovalve theme is shown in Fig. 3.64 that represents a "proportional valve" in which the input electrical signal generates an electromagnetic force by means of a permanent magnet differential motor, which then directly moves the spool. Position control is achieved by an electrical position transducer whose signal is coupled to the integral electronic control circuit.

A particular advantage of this valve is the removal of the first hydraulic stage met in conventional servovalves and the possibility of a nozzle or orifice blockage that is due to oil contamination particles is eliminated. However, the response of the spool to a sudden requirement to move will be less than for a conventional servovalve, although improvements continue to be made. This may not be a serious issue in many applications.

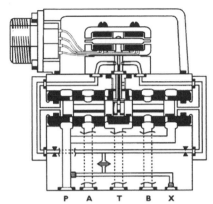

(a) Flapper–nozzle type using a force-feedback wire for spool-position control

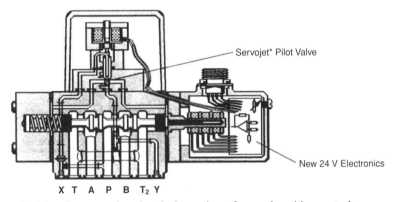

(b) A jet-pipe type using electrical transducer for spool-position control

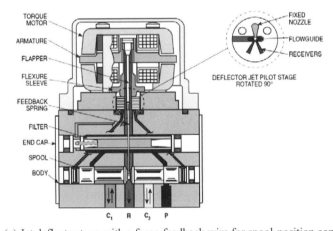

(c) Jet-deflector type with a force-feedback wire for spool-position control

Figure 3.63. Three types of servovalve manufactured by Moog Controls Ltd.

3.5.2 Servovalve Rating

When selecting a servovalve from a manufacturer's catalogue, the conventional approach is to select a suitable size by using the "rated" data provided. To understand this, it is necessary to return to the servovalve flow characteristics that are now expressed in terms of current input rather than spool displacement. Following

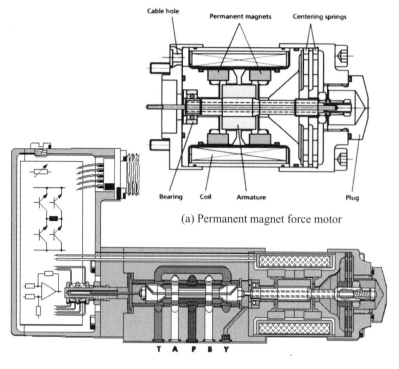

(a) Permanent magnet force motor

(b) Valve schematic

Figure 3.64. A direct-acting proportional control valve, Moog Controls Ltd.

Fig. 3.65, the two flow equations representing flow out of the servovalve and flow back through the servovalve are then given as follows, assuming critically lapped spool lands and a negligible return (tank) pressure:

$$Q_1 = k_f i \sqrt{P_s - P_1},$$
$$Q_2 = k_f i \sqrt{P_2}. \qquad (3.143)$$

A critically lapped spool means that each spool face, or land, just matches each port edge so that the smallest applied current will open the ports. *For the rated condition test, the two output ports are connected*, and therefore equating the two flow rates in Eq. (3.143) gives:

$$P_1 + P_2 = P_s. \qquad (3.144)$$

Hence, the sum of line pressures must equal the supply pressure, and this is a good experimental test to check that the port flow characteristics are matched. If the connected output ports have a restrictor in the line to create a load pressure differential P_{load}, then, by definition:

$$P_1 - P_2 = P_{\text{load}}. \qquad (3.145)$$

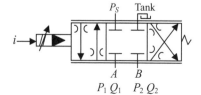

Figure 3.65. Servovalve symbol.

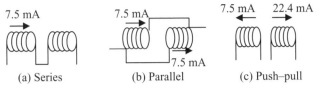

Figure 3.66. Coil connections to give equivalent effect at 15 mA.

Hence, from these two previous equations, it follows that:

$$Q_1 = Q_2 = Q = k_f i \sqrt{\frac{P_s - P_{\text{load}}}{2}}. \tag{3.146}$$

The rated condition test is then specified for a total load pressure drop of 70 bar at a particular current applied, and called the rated current, where:

$$\text{total valve pressure drop} = (P_s - P_1) + (P_2 - 0) \tag{3.147}$$
$$= P_s - P_{\text{load}} = 70 \text{bar}. \tag{3.148}$$

Therefore, from Eq. (3.139), the rated flow rate becomes:

$$Q_{\text{rated}} = k_f i_{\text{rated}} \sqrt{35 \, \text{bar}}. \tag{3.149}$$

Care must be taken in choosing the correct units when applying this equation. The two coils may be connected in one of three ways – series, parallel, and push–pull – as shown in Fig. 3.66. The disadvantage of the series connection is that operation ceases if the coil connection is broken.

Worked Example 3.10

A servovalve is rated at 38 L/min with a rated current of 15 mA applied.
  The valve flow constant k_f is given by:

$$Q = k_f i \sqrt{35},$$
$$38 = k_f 15 \sqrt{35}, \qquad k_f = 0.428.$$

So, for example, at a supply pressure of 210 bar with a current of 10 mA and no load pressure differential, the flow rate will be:

$$Q = k_f i \sqrt{\frac{P_s - P_{\text{load}}}{2}} = 0.428 \times 10 \sqrt{\frac{210 - 0}{2}} = 43.86 \, \text{L/min}.$$

3.5.3 Flow Characteristics, Critically Lapped Spool

The characteristic for a critically lapped spool has already been discussed in the previous subsection, and this characteristic is pursued further here. The flow equation for a servovalve with a critically lapped spool, and the output ports connected, has been previously derived:

$$Q_1 = Q_2 = Q = k_f i \sqrt{\frac{P_s - P_{\text{load}}}{2}}. \tag{3.150}$$

The flow characteristic for a positive current is shown in the following example, noting that for negative currents, the characteristic is reflected about both axes.

Worked Example 3.11

The flow–pressure characteristic of a critically lapped servovalve is shown for a positive current and with the ports connected by a restrictor valve. The supply pressure $P_s = 100$ bar. Determine the flow constant k_f.

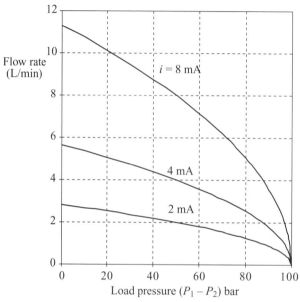

Worked Example 3.11

The flow-rate equation is:

$$Q = k_f i \sqrt{\frac{P_s - P_{\text{load}}}{2}},$$

so select the part of the characteristic where the load pressure is zero to give $Q = k_f i \sqrt{P_s/2}$:

$$\frac{\partial Q}{\partial i} = k_f \sqrt{\frac{P_s}{2}} \approx \frac{11.2}{8} = 1.4, \quad k_f = \frac{1.4}{\sqrt{P_s/2}} = \frac{1.4}{\sqrt{50}} = 0.2.$$

For both performance and control studies, it is common to define the flow gain and the pressure gain as follows:

$$\text{flow gain,} \quad K_{qi} = \frac{\partial Q}{\partial i} = k_f \sqrt{\frac{P_s - P_{\text{load}}}{2}}, \tag{3.151}$$

$$\text{pressure gain,} \quad K_{qp} = \frac{\partial Q}{\partial P_{\text{load}}} = -\frac{k_f i}{2\sqrt{2(P_s - P_{\text{load}})}}. \tag{3.152}$$

Using the original flow equation gives:

$$\text{flow gain,} \quad K_{qi} = \frac{\partial Q}{\partial i} = \frac{Q}{i}, \tag{3.153}$$

$$\text{pressure gain,} \quad K_{qp} = \frac{\partial Q}{\partial P_{\text{load}}} = -\frac{Q}{2(P_s - P_{\text{load}})}. \tag{3.154}$$

The pressure sensitivity K_{pi} is defined as the relationship between pressure-differential and applied current. Recall that:

$$\left(\frac{\partial Q}{\partial i}\right)\left(\frac{\partial i}{\partial P_{\text{load}}}\right)\left(\frac{\partial P_{\text{load}}}{\partial Q}\right) = -1,$$

$$K_{qi}\frac{1}{K_{pi}}\frac{1}{K_{qp}} = -1,$$

$$\text{pressure sensitivity,} \; K_{pi} = \frac{\partial P_{\text{load}}}{\partial i} = -\frac{K_{qi}}{K_{qp}} = \frac{2(P_s - P_{\text{load}})}{i}. \tag{3.155}$$

This suggests that the pressure sensitivity $K_{pi} \to \infty$ as $i \to 0$. This is not the case in practice because each ostensibly critically lapped land will have some extremely small inherent leakage path associated with it. This could be across the perimeter of a land because of clearance between the spool outer diameter and the spool bush, or sometimes deliberately introduced by machining a small underlap that is sometimes specified by the user. Whatever the origin, spool underlap creates damping around the null position, as will be shown later, and has a stabilizing influence. Therefore, the pressure-sensitivity measured characteristic is a good test of how accurately the critical lands and the spool–bushing have been manufactured.

3.5.4 Servovalve with Force Feedback

This type of servovalve is shown in Fig. 3.63(a), and there are several variations on this design that still retain a feedback wire. Consider the electromagnetic stage, often called the torque motor, coupled to the flapper and torque wire, as shown in more detail in Fig. 3.67.

The operation of the torque motor and flapper stage is then as follows:

- The flapper armature has two coils positioned at each end and may be connected in series, parallel, or push–pull. The interaction between the magnetic field generated along the axis of the coils and the poles of the two permanent magnets creates a torque and, hence, rotation of the flapper.
- The flapper is secured to the servovalve body by a flexible support called the flexure tube, and its rotational stiffness is important in the control of the flapper displacement at the pair of nozzles. The center of rotation of the flapper is not necessarily at the base of the flexure tube.
- The effect of displacing the flapper to the left is to create a pressure differential that moves the spool to the right, causing a rotation of the torque wire coupled to the spool. This creates a feedback torque reacting against the electromagnetic torque generated, and the flapper rapidly moves to its central position. This is very close to the flapper rest condition, as a result of the design, with a resulting displacement of the spool.
- The end of the torque wire has an accurately machined ball attached that is matched to the slot in the spool.

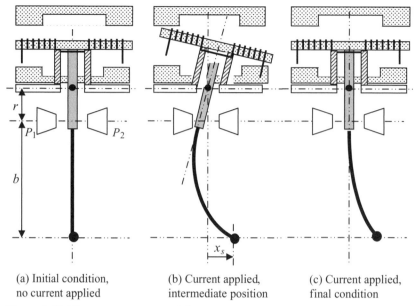

(a) Initial condition,   (b) Current applied,   (c) Current applied,
no current applied   intermediate position   final condition

Figure 3.67. The torque motor and flapper–nozzle stage of a force-feedback servovalve (not to scale and deflections highly exaggerated).

- The maximum spool displacement is typically 0.5 mm at the rated condition. If the input current is suddenly applied, then the total time for the spool to move to its required position will be typically 10–50 ms.

The torque generated by the torque motor is proportional to the current applied but also has to overcome a torque generated by virtue of the fact that rotation of the armature occurs within a magnetic field. Hence,

$$\text{generated torque,} \quad T = k_t i + k_m \theta, \tag{3.156}$$

where θ is the rotation of the armature and flapper. For a more detailed study of this characteristic, see Urata (2004). The resisting torque consists of the flexure tube torque, the flow-reaction-force effect at the flapper, and the wire torque effect from its location in the spool slot that has moved a distance x_s. This gives:

$$\text{resisting torque,} \quad T = k_a \theta + (P_a - P_b) a_n r + k y (r + b), \tag{3.157}$$

where k_a is the flexure tube rotational stiffness, a_n is the nozzle cross-sectional area, $(P_a - P_b)$ is the pressure differential generated by the flapper–nozzle stage, and k is the lateral stiffness of the wire evaluated at the spool slot position. The total deflection of the wire, y, is given by:

$$y = x_s + (r + b)\theta. \tag{3.158}$$

The displacement of the flapper at the nozzles is given by:

$$x = r\theta. \tag{3.159}$$

Because flapper operation is designed to be around the central, null, position from earlier work, the pressure differential generated may be simply written as:

$$(P_a - P_b) = \left(\frac{x}{x_{nm}}\right) P_s. \tag{3.160}$$

Spool displacement is then determined from the force balance across the spool, which is dominated by the feedback wire force and the spool flow-reaction force:

$$(P_a - P_b)a_s = ky + 2C_q^2 a_o \cos\theta\,[P_s - P_{\text{load}}], \qquad (3.161)$$

where a_s is the spool end cross-sectional area, the spool orifice area $a_0 = wx_s$ for rectangular ports having an area gradient w, $P_{\text{load}} = P_1 - P_2$. Combining these equations then gives the relationship between spool displacement and input current as follows:

$$x_s = \frac{(1-\alpha)k_t i}{\beta\dfrac{(k+k_{fr})x_{nm}}{r\,P_s a_s} + k(r+b)(1-\alpha)}, \qquad (3.162)$$

$$\alpha = \frac{k(r+b)x_{nm}}{r\,P_s a_s}, \qquad (3.163)$$

$$\beta = k_a - k_m + k(r+b)^2 + \frac{a_n r^2 P_s}{x_{nm}}, \qquad (3.164)$$

$$k_{fr} = 2C_q^2 w \cos\theta(P_s - P_{\text{load}}). \qquad (3.165)$$

The spool displacement will be proportional to input current provided that the denominator of Eq. (3.162) is positive. The flow-reaction equivalent stiffness k_{fr} will probably be much smaller than the wire stiffness k, so that the effect of load pressure differential P_{load} may not present a problem. In practice, $\alpha \ll 1$ and can be neglected. Notice also the destabilizing magnetic constant $-k_m$, the magnitude of which can be varied during manufacture, the process known as detuning. In particular, β can be detuned to a very small value by magnetically increasing k_m, and Eq. (3.162) then becomes:

$$x_s \approx \frac{k_t i}{k(r+b)}. \qquad (3.166)$$

The input electrical torque is balanced by the wire feedback torque because of spool position, and the flapper will return to its central position between the nozzles.

3.5.5 Servovalve with Spool-Position Electrical Feedback

This type of servovalve can operate without wire force feedback as previously analyzed, but it is usual to leave this feedback mechanism in place to produce spool centering if the electrical-position-feedback transducer fails. Also, the use of electrical feedback means that the torque motor stage can be detuned more than usual and the feedback wire stiffness also reduced. Bearing in mind the previous analysis for the wire feedback-force case, approximation Eq. (3.166), a block diagram for position feedback can be constructed as shown in Fig. 3.68.

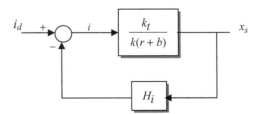

Figure 3.68. Feedback control of a servovalve spool with force and position feedback.

The current signal from the position transducer, having a gain of H_i mA/mm, is subtracted from demand current i_d and the error current is then used to drive the torque motor. The relationship between spool displacement and demand current may then be written as:

$$\frac{x_s}{i_d} = \frac{1}{\dfrac{k(r+b)}{k_t} + H_i}. \tag{3.167}$$

Therefore, if:

$$\frac{k(r+b)}{k_t} \ll H_i,$$

then:

$$\frac{x_s}{i_d} \approx \frac{1}{H_i}. \tag{3.168}$$

Spool position now depends on only the position transducer gain and, therefore, this is a desirable design. From Eq. (3.167), this is helped by reducing the wire stiffness k in addition to applying the electromagnetic detuning process inherent in this design. Note that this analysis has neglected dynamic behavior, and care must be taken that the design approach does not destabilize the valve.

3.5.6 Flow Characteristics, Underlapped Spool

Figure 3.69 shows a spool that is symmetrically underlapped; that is, all lands are machined back by the same very small amount.

The underlap is indicated by u and for a servovalve, it is equivalent to a current i_u, the current needed to just close off the underlap as the spool moves. Considering the flow equations then gives:

$$Q_1 = k_f(i_u + i)\sqrt{P_s - P_1} - k_f(i_u - i)\sqrt{P_1}, \tag{3.169}$$

$$Q_2 = k_f(i_u + i)\sqrt{P_2} - k_f(i_u - i)\sqrt{P_s - P_2}. \tag{3.170}$$

These equations are valid only within the underlap region $-i_u < i < i_u$.

If the ports are connected such that $Q_1 = Q_2 = Q$, then the same condition for pressures occurs as for the critically lapped spool:

$$P_1 + P_2 = P_s. \tag{3.171}$$

So, again defining:

$$P_{\text{load}} = P_1 - P_2, \tag{3.172}$$

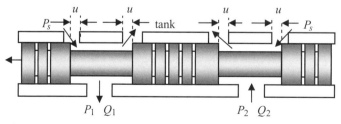

Figure 3.69. A servovalve symmetrically underlapped spool (underlap exaggerated).

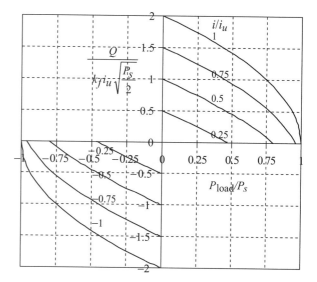

Figure 3.70. Flow characteristic for a servovalve underlapped spool.

we may write the general flow equation as:

$$Q = k_f(i_u + i)\sqrt{\frac{P_s - P_{\text{load}}}{2}} - k_f(i_u - i)\sqrt{\frac{P_s + P_{\text{load}}}{2}}. \tag{3.173}$$

The nondimensional form of this equation is particularly useful and may be written as:

$$\overline{Q} = \frac{Q}{k_f i_u \sqrt{\dfrac{P_s}{2}}} = (1 + \overline{i})\sqrt{1 - \overline{P}_{\text{load}}} - (1 - \overline{i})\sqrt{1 + \overline{P}_{\text{load}}},$$

$$\overline{i} = \frac{i}{i_u} \qquad \overline{P}_{\text{load}} = \frac{P_{\text{load}}}{P_s}. \tag{3.174}$$

This characteristic is shown in Fig. 3.70.

It is now useful to consider the *blocked-load* test in which the output ports are blocked, the current is varied within the underlap region, and the pressures P_1 and P_2 are measured. From Eqs. (3.169), (3.170), and (3.171), setting each flow to zero gives:

$$\overline{P}_1 = \frac{(1 + \overline{i})^2}{2(1 + \overline{i}^2)}, \quad \overline{P}_2 = \frac{(1 - \overline{i})^2}{2(1 + \overline{i}^2)}, \quad \overline{P}_{\text{load}} = \frac{2\overline{i}}{(1 + \overline{i}^2)}, \tag{3.175}$$

$$\overline{P}_1 = \frac{P_1}{P_s}, \quad \overline{P}_2 = \frac{P_2}{P_s}, \quad \overline{P}_{\text{load}} = \frac{P_{\text{load}}}{P_s}.$$

This blocked-load characteristic is shown in Fig. 3.71.

It follows that at the null condition:

$$\frac{d\overline{P}_{\text{load}}}{d\overline{i}}\bigg|_{\overline{i}=0,\overline{P}_{\text{load}}=0} = 2 \rightarrow \frac{dP_{\text{load}}}{di} = \frac{2P_s}{i_u}. \tag{3.176}$$

This is the pressure sensitivity K_{pi}, and its determination from the blocked-load test and use of Eq. (3.176) allow the underlap to be experimentally determined. The pressure sensitivity $\neq \infty$ as is the case for a critically lapped spool, and it can be predetermined to improve system damping, when connected to a real load actuator. However, it does produce an additional flow loss – hence, a power loss.

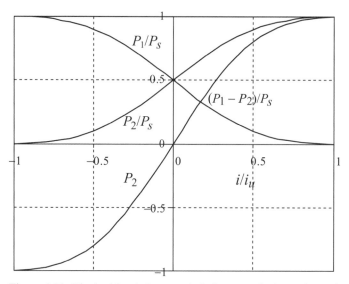

Figure 3.71. Blocked-load characteristic for an underlapped spool.

The leakage flow back to tank can also be measured, albeit probably a very small value, under blocked-load conditions. This is given by:

$$Q_{\text{leak}} = k_f(i_u - i)\sqrt{P_1} + k_f(i_u + i)\sqrt{P_2}. \tag{3.177}$$

Inserting values for pressure then gives:

$$\overline{Q}_{\text{leak}} = \frac{Q_{\text{leak}}}{k_f i_u \sqrt{2P_s}} = \frac{(1 - \bar{i}^2)}{\sqrt{1 + \bar{i}^2}}. \tag{3.178}$$

This flow leakage loss characteristic is shown in Fig. 3.72.

The maximum flow loss occurs at the spool midposition, and the region of underlap can be validated experimentally by determining the current at which the leakage flow approaches zero. For this test, the servovalve flapper–nozzle-stage

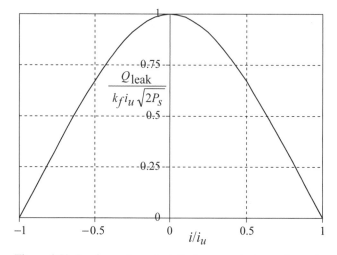

Figure 3.72. Leakage characteristic for an underlapped spool.

leakage flow or jet-pipe-stage leakage flow will be present in the measurement and will also vary slightly with input current within the underlap region.

Now considering the servovalve flow and pressure gains with the *ports connected*, it follows from Eq. (3.173) that:

$$\text{flow gain, } K_{qi} = \frac{\partial Q}{\partial i} = k_f \sqrt{\frac{P_s - P_{\text{load}}}{2}} + k_f \sqrt{\frac{P_s + P_{\text{load}}}{2}}, \tag{3.179}$$

$$\text{pressure gain, } K_{qp} = \frac{\partial Q}{\partial P_{\text{load}}} = -\frac{k_f(i_u + i)}{2\sqrt{2(P_s - P_{\text{load}})}} - \frac{k_f(i_u - i)}{2\sqrt{2(P_s - P_{\text{load}})}}. \tag{3.180}$$

Considering then the null condition gives:

$$\text{flow gain, } K_{qi} = \frac{\partial Q}{\partial i} = 2k_f \sqrt{\frac{P_s}{2}}, \tag{3.181}$$

$$\text{pressure gain, } K_{qp} = \frac{\partial Q}{\partial P_{\text{load}}} = -\frac{k_f i_u}{\sqrt{2P_s}}. \tag{3.182}$$

Notice that both the flow gain and pressure gain at the null condition are twice the values for a critically lapped spool as given by Eqs. (3.151) and (3.152). The flow gain doubling at null in particular is useful because the conventional rating test will show this if the spool is underlapped. An actual servovalve measurement is shown in Fig. 3.73.

The supply pressure chosen was to achieve a reasonable flow-measuring accuracy by use of a gear-type flow meter with a range of ± 16 L/min. Although the accuracy of measurement is at its limit when the servovalve current approaches the underlap region, it will be deduced that this servovalve has an underlap equivalent to $i_u \approx \pm 0.33$ mA.

3.6 Positive-Displacement Pumps and Motors

3.6.1 Flow and Torque Characteristics of Positive-Displacement Machines

A positive-displacement pump or motor has a number of moving elements, such as the gear, vane, and axial piston types like those shown in Fig. 3.74, that transfer the fluid from one port to another.

There are other types, such as screw, internal gear, ring gear, and radial piston, but in all cases they are defined by an important parameter known as the *machine displacement D*, which has the units of cubic meters per radian, quite simply the volume of fluid required per radian of revolution. Therefore, if the machine speed is ω rad/s, then the flow rate is given by:

$$Q = D\omega \quad \text{m}^3/\text{s}. \tag{3.183}$$

For the ideal machine, the input power must equal the output power. Therefore, if the flow rate is Q m^3/s at a pressure of P N/m^2 and the torque is T N m at a speed of ω rad/s, then:

$$\text{fluid power } (PQ) = \text{mechanical power } (T\omega),$$

$$T = \frac{QP}{\omega} = DP \quad \text{N m}. \tag{3.184}$$

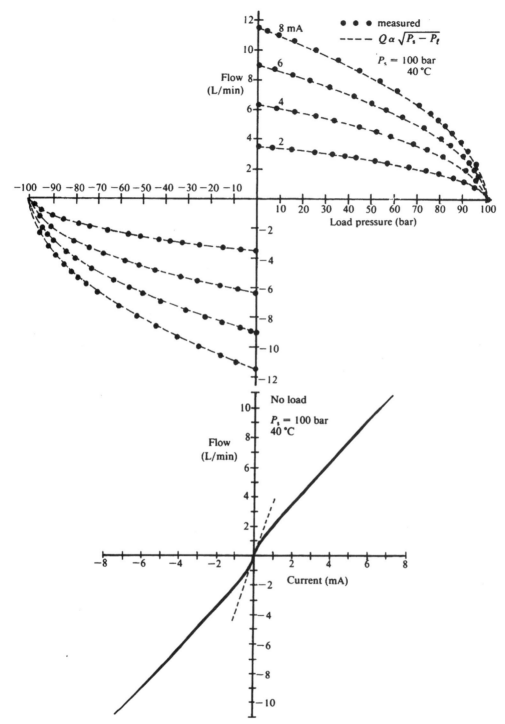

Figure 3.73. Flow characteristic for a servovalve with ports connected, $P_s = 100$ bar, ISO 32 mineral oil at 40°C.

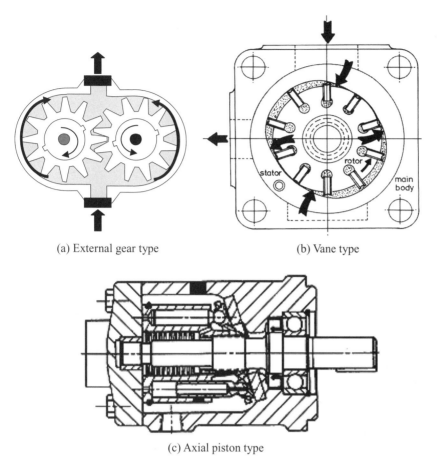

(a) External gear type   (b) Vane type

(c) Axial piston type

Figure 3.74. Some common positive-displacement machines.

In other words, the flow and torque are uniquely defined by the displacement D once the machine speed and working pressure is known. A pump is usually unidirectional and draws the fluid from a tank or by means of a boost pump and, in both cases, at a relatively low pressure compared with that at the delivery port. A motor can be bidirectional – for example, when driving a mobile machine wheel – and both the inlet and outlet ports may be pressurized. It then follows from Fig. 3.75 that:

Figure 3.75. Pump and motor positive-displacement machines.

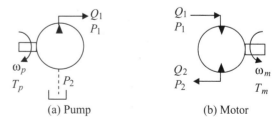

(a) Pump   (b) Motor

pump   motor

$$Q_1 = D_p \omega_p, \qquad Q_2 = Q_1 = D_m \omega_m, \qquad (3.185)$$

$$T_p = D_p(P_1 - P_2), \qquad T_m = D_m(P_1 - P_2). \qquad (3.186)$$

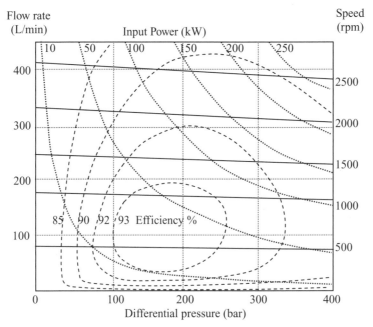

Figure 3.76. A typical manufacturer's performance plot for a pump.

In practice, there are fluid and mechanical losses so the ideal equations must be modified. A pump output is load pressure, and a motor output is speed and load torque. Therefore, the ideal flow and torque equations are modified as follows:

$$\text{pump} \qquad\qquad\qquad \text{motor}$$

$$Q_1 = D_p\omega_p - Q_{1\text{loss}}, \qquad Q_1 = D_m\omega_m + Q_{1\text{loss}}, \qquad (3.187)$$

$$Q_2 = D_p\omega_p + Q_{2\text{loss}}, \qquad Q_2 = D_m\omega_m - Q_{2\text{loss}},$$

$$T_p = D_p(P_1 - P_2) + T_{\text{loss}}, \qquad T_m = D_m(P_1 - P_2) - T_{\text{loss}}. \qquad (3.188)$$

Flow losses associated with the inlet pressure are different from those associated with the outlet pressure for reasons of both flow continuity and the different values of each pressure. For a pump, only the output flow-rate variation under load is usually required. Each loss term cannot be uniquely defined because of the variety of different designs and types of machines commercially available. They depend on speed, clearances, flow-path special designs, seal types, materials, the fluid used, pressure, and so on, but in practice often show dominating linear relationships apart from very low-speed cases for motors. This often means that some simple tests can be performed, given appropriate instrumentation, to determine the dominant loss terms if data are not supplied by the manufacturer. However, the understanding of the effect of detailed design changes on losses is quite complex, as indicated from examples considered in this book. A typical manufacturer's performance plot for a pump is shown in Fig. 3.76.

These performance characteristics are particularly useful for determining the best operating condition for an application. For example, if the drive speed is 1500 rpm and close to a common electric motor drive speed of 1440 rpm, then the best efficiency is gained by running the pump at a pressure of around 200 bar. However, significant variations in pressure of ±100 bar around the optimum will reduce the efficiency only by 1%–2%. Note also that as the drive speed is reduced, the flow-rate leakage loss is reduced at the same working pressure.

3.6.2 Geometrical Displacement of a Positive-Displacement Machine

Theoretically, it is possible to determine the displacement of a positive-displacement machine providing the variation of displacement with position of each displacing element is known. Consider pumping action, in which volumes decrease with time; then, the theoretical flow rate Q_t is given by:

$$Q_t = -\sum_{i=1}^{n} \frac{dV_i}{dt} = -\omega \sum_{i=1}^{n} \frac{dV_i}{d\theta},$$

$$Q_t = D_g\omega, \quad D_g = -\sum_{i=1}^{n} \frac{dV_i}{d\theta}, \tag{3.189}$$

where D_g is the theoretical geometrical displacement, n is the number of pumping elements appropriate, and V_i is the corresponding volume that varies with angular position.

For a *gear pump*, Fig. 3.74(a), the displaced volume variation from two counter-rotating gears may be determined from the varying gear geometry. Note that fluid is trapped by a gear tooth at the inlet and carried around with it, depositing it to the outlet. It is not drawn through the central meshing part of the gear combination. Gear pumps are the simplest, offer the lowest cost of positive-displacement pump designs, are often manufactured from aluminum, and are ideal as boost pumps that usually do not require large pressures and/or flow rates. They usually have an even number of gear teeth, and the mean flow rate will have a superimposed flow ripple that is due to the repetitive pumping contribution of each tooth. Gear-tooth design is actually a complex process, with design standards varying from country to country. To get a feel for some of the basic issues, consider a pair of gears as shown in Fig. 3.77. The base circle may be considered as the radius whereby the tooth involute form is achieved by unwinding a taught string wrapped around the base circle (Mabie and Reinholz, 1987). For a gear pump, each gear is identical and the meshing point is where the two involute forms just touch, given by point P in Fig. 3.77.

The line through point P and tangent to the two base circles defines the pressure angle ψ. For further details, see Ivantysyn and Ivantysynova (2002), where it is shown that by considering the tooth contact displacement interval:

$$\text{displacement,} \quad D = 2b\pi \left(\frac{2r_p}{n}\right)^2 \omega \left[n + 1 - \frac{\pi^2 \cos^2 \psi}{12}\right], \tag{3.190}$$

where b is the gear width and n is the even number of teeth.

The flow ripple ratio is given by:

$$\frac{\delta Q}{Q} = \frac{\pi^2 \cos^2 \psi}{4\left(n + 1 - \dfrac{\pi^2 \cos^2 \psi}{12}\right)}. \tag{3.191}$$

Hence, for a typical pressure angle of $\psi = 20°$, the flow ripple becomes:

$$\frac{\delta Q}{Q} = \frac{2.18}{[n + 0.27]}. \tag{3.192}$$

This illustrates that a large flow ripple can be generated by a gear pump, compared with an axial piston pump, unless a large number of teeth are used. However, the

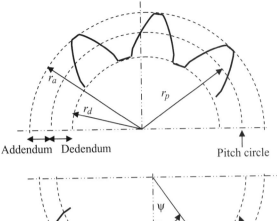

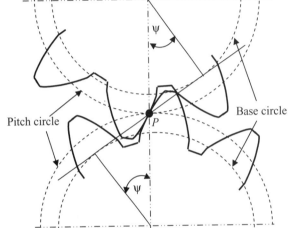

Figure 3.77. Meshing involute gears applicable to a gear pump.

displacement is then reduced so either the speed must be increased or the width of the gears or pitch radius increased.

A *vane pump*, Fig. 3.74(b), operation relies on a precise stator geometry design, and it is this stator, or cam ring, contour variation with angular position that creates the volume variation with angular position. Figure 3.78 shows a vane pump geometry measured by the author, who obtained it by rotating an accurate electronic-position transducer within and around the stator using a low-speed electric motor drive assembly. The motor has a gear box with two outputs, one driving a pivoted stylus with a ball end touching the cam ring and the other end touching the position transducer. The other gear output drives a precision rotary potentiometer to measure the angle turned during operation.

Consider the pump operation with reference to Figs. 3.74 and 3.78. Fluid is collected at the inlet port by the vane and then carried to the precompression zone. At this point, the package of fluid is trapped between the current vane and the vane ahead for a short part of the precompression zone. Because the stator radius is decreasing with increasing angular rotation, the trapped package of fluid is compressed, ideally, to match the load pressure at the outlet port. This can be achieved at only one pressure setting, for a fixed pump speed, which is that designed for the pump. Important aspects of its operation are as follows:

- It will be noted that the outlet-port radius change is significantly higher than within the precompression zone to give the pump its displacement value and thus its flow-delivery characteristic. A constant rate of radius change throughout

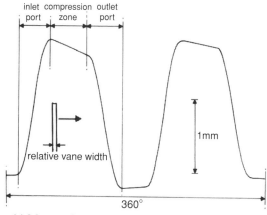

(a) Measured geometry variation with angular position

Figure 3.78. Measured stator geometry of a fixed displacement vane pump.

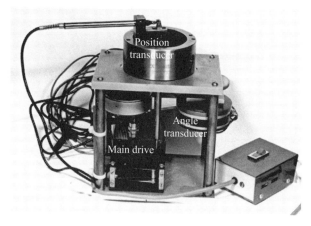

(b) Motor drive and measuring transducer

the pumping cycle would give such a high precompression that the pump could suffer from intolerably–high pressure fluctuations.

- The symmetrical shape of the stator profile ensures that a similar pumping action is also taking place at the other outlet port.
- A further design feature of this and many other positive-displacement pumps, particularly axial piston types, is the use of timing grooves at the entrance to and the exit from the precompression zone. These timing grooves – in this case, v-slots – are used to aid the transition of pressure between each region. In particular, it is necessary to ensure that the pressure achieved at the end of precompression is matched to the load pressure as smoothly as possible. The timing grooves help in this respect but only at the design pressure and speed. At other operating conditions, any resulting pressure differential across the timing grooves will cause a flow of fluid in the appropriate direction either into or out of the precompression zone.
- Each vane undertakes radial motion during its rotation, and this motion within the slots is a further crucial design issue in terms of both friction and leakage minimization.

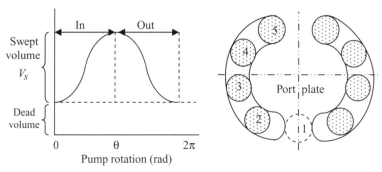

Figure 3.79. Motion of a piston within an axial piston machine.

- Because the flow-delivery process is one in which vanes repetitively enter and leave the delivery port, the flow rate must contain a ripple characteristic superimposed on the mean flow.

It will be clear from considerations for the gear pump that it is a complex matter, if not impossible, to develop a generic equation for displacement and ripple for a balanced vane pump because of the complex design of the stator geometry. However, experience shows that the vane-pump ripple is lower than the gear-pump ripple.

Consider now an *axial piston pump* as shown in Fig. 3.74(c). Each piston has a slipper assembly that moves tangentially around the swash plate because of the rotation of the pump barrel. Consider then the position of a piston as it makes one revolution (Fig. 3.79). The swash plate is at an angle to create a flow rate and, therefore, the displaced volume of one piston varies because of its axial motion within the barrel during one revolution. This piston instantaneous volume V is given by:

$$V = V_s \frac{(1 - \cos\theta)}{2}. \qquad (3.193)$$

For n pistons, the total swept volume V_t is given by:

$$\frac{V_t}{V_s} = \sum_{i=1}^{n} \frac{1 - \cos\left[\theta + (i-1)\frac{2\pi}{n}\right]}{2}. \qquad (3.194)$$

Considering Fig. 3.79 for a nine-piston pump, only pistons 1–5 are considered, with piston 1 just entering the port-plate kidney slot. The total volume variation with position, given by Eq. (3.194), then depends on the angle at which piston 5 leaves the kidney slot. In this example, it is assumed that piston 5 rotates 20° before it is fully closed off from the kidney port. The summation continues until 40° of rotation has been reached, at which point the volume variation repeats itself. Now consider the flow rate generated by the velocity effect from each piston. From Eqs. (3.189) and (3.194), the total flow rate is given by

$$Q = \frac{V_s \omega}{2} = \sum_{i=1}^{n} \sin\left[\theta + (i-1)\frac{2\pi}{n}\right]. \qquad (3.195)$$

Now define a geometrical displacement D_g defined as:

$$D_g = \frac{nV_s}{2\pi} = \frac{nd^2\ell}{8} \text{ m}^3/\text{rad}, \qquad (3.196)$$

where d is the piston diameter and ℓ is its stroke.

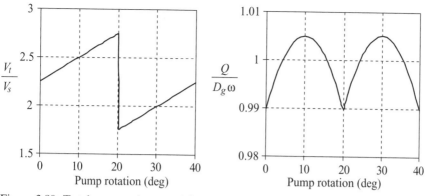

Figure 3.80. Total swept volume and flow rate for a 9-piston pump.

Equation (3.195) then becomes:

$$\frac{Q}{D_g\omega} = \frac{\pi}{n}\sum_{i=1}^{n}\sin\left[\theta + (i-1)\frac{2\pi}{n}\right].$$ (3.197)

The volume change and the flow-rate change with angular position are shown in Fig. 3.80 for a nine-piston pump.

The total swept-volume variation with angular position is remarkably linear between peaks and results in:

$$\text{maximum} = \frac{(n+2)}{4}, \quad \text{minimum} = \frac{(n-2)}{4}, \quad \text{mean} = \frac{n}{4}.$$ (3.198)

The total swept volume changes by ±22% about the mean for a nine-piston machine. Notice also that the flow rate has an inherent ripple that is remarkably close to a sine wave at a frequency of (n × machine speed) but rectified, even for the three-piston case. Frequency-spectrum analyses of pump ripple therefore often show a frequency component of $(2n \times$ machine speed) that is due to interpretation of this rectified sine wave. The peak-to-peak value of the ripple is given by the following exact solution:

$$\frac{\delta Q}{D_g\omega} = \frac{\pi}{n}\tan\frac{\pi}{2n}, \quad n \text{ even,}$$

$$\frac{\delta Q}{D_g\omega} = \frac{\pi}{2n}\tan\frac{\pi}{4n}, \quad n \text{ odd.}$$ (3.199)

Table 3.3 shows the ripple variation for different numbers of pistons.

Notice that the pump ripple for an odd number of pistons is the same as if the odd number were doubled to an even number, and it is for this reason that an axial piston pump usually has an odd number of pistons. For a nine-piston pump, the geometric ripple is 1.53% of the mean flow rate.

Table 3.3 *Geometric ripple for an axial piston pump*

n	3	4	5	6	7	8	9	10
$\frac{\delta Q}{D_g\omega}$	0.14	0.325	0.05	0.14	0.0253	0.078	0.0153	0.05

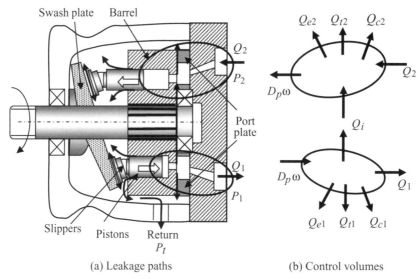

(a) Leakage paths   (b) Control volumes

Figure 3.81. Schematic of an axial piston pump.

3.6.3 Flow Losses for an Axial Piston Machine

An *axial piston pump* is considered here because it has a variety of flow losses that are similar in origin to other positive-displacement machines. Figure 3.81 shows a schematic of an axial piston pump indicating various flow leakage paths that contribute to the total flow loss.

To determine the flow into the pump and out of the pump, it is necessary to consider suitable flow-control volumes within which the flow-loss terms interact, as shown in Fig. 3.81(b). Flow continuity may then be applied to each control volume to determine the output flow-rate and the input flow-rate balance.

The difference between the ideal displaced flow rate and the actual flow rate is due to the following main issues:

- cross-port leakage Q_i
- external leakages across components and to the pump casing Q_e
- losses that are due to fluid compression Q_c
- timing-groove losses Q_t

The pump may have a pressure-boosted inlet to minimize cavitation for large flow rates, so the inlet line pressure P_2 may not be at tank return pressure P_t. Some pumps have a "floating" port plate to provide the flow path between the external pipe connection point and the pistons in the rotating barrel. The face of the body seen by the pistons is kidney-shaped, as indicated in Fig. 3.79, and these inlet and outlet kidney slots may be cast directly into the pump body or on the floating port plate. The very nature of an axial piston pump design results in leakages across faces, lands, and orifices, in addition to other fluid flow effects that are due to compressibility and timing grooves. Therefore, considering Fig. 3.81 gives the following flow-continuity equations:

$$\text{outlet,} \quad Q_1 = D_p\omega - Q_i - Q_{e1} - Q_{c1} - Q_{t1}, \tag{3.200}$$

$$\text{inlet,} \quad Q_2 = D_p\omega - Q_i + Q_{e2} + Q_{c2} + Q_{t2}. \tag{3.201}$$

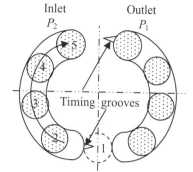

Figure 3.82. The use of timing grooves.

Considering practical output flow-rate measurements, the overall leakage flow characteristic of an axial piston pump varies in a sufficiently linear way with increasing pressure, and this suggests that individual leakage terms are dominated by a linear pressure-drop–flow characteristic. This is assumed here.

Cross-port leakage occurs through the clearance between the rotating barrel inner face and the opposing face and is given by:

$$Q_i = \frac{P_1 - P_2}{R_i}. \tag{3.202}$$

External losses occur through the clearance between the rotating barrel outer face and the opposing face. Each piston will also have a very small flow loss across its perimeter and from the piston chamber to case, and a much more significant flow loss through the slipper lubricating hole and across the slipper face. These three elements of flow loss are lumped together as a single term:

$$Q_{e1} = \frac{P_1}{R_{e1}}, \quad Q_{e2} = \frac{P_2}{R_{e2}}. \tag{3.203}$$

Fluid compressibility losses are due to the fact that each line, for the general case, is pressurized and results in an effective compressibility flow rate. Considering the fluid compressibility equation and the ideal displaced flow rate, it is assumed that this effect may be characterized as:

$$Q_{c1} = D_p\omega\left(\frac{P_1}{\beta}\right), \quad Q_{c2} = D_p\omega\left(\frac{P_2}{\beta}\right), \tag{3.204}$$

where β is the fluid effective bulk modulus.

Timing-groove flow losses is the final term being considered here and are connected with the timing grooves that usually exist at the ends of each kidney port, as shown in Fig. 3.82. Note that a pump may have only one timing groove.

Each small timing groove has a cross-sectional area that varies from zero at its apex to a maximum at the entry point of the kidney slot. The piston that is moving toward its bottom dead center (piston 5 in Fig. 3.82) will then begin to move in the opposite direction toward its top dead center as its role changes from drawing fluid in to pushing fluid out. To improve the pressure change from a low inlet value P_2 to full pump pressure P_1, the groove is designed to smooth this transition. The performance of such a timing groove is optimum around a particular design speed and pressure but rarely achieves its desired effect perfectly; consequently, it creates a small backflow of fluid. Hence, the pump output flow rate experiences a small decrease when averaged over one half-cycle. The second timing groove operates on a similar principle, allowing a smooth transition from full pump pressure P_1 to low inlet pressure P_2; for example, as shown for piston 1 in Fig. 3.82. For the purpose of this analysis, these flow rates will be treated as losses and assumed to be proportional to the appropriate pressure:

$$Q_{t1} = \frac{P_1}{R_{t1}}, \quad Q_{t2} = \frac{P_2}{R_{t2}}. \tag{3.205}$$

Considering flow-continuity equations (3.200) and (3.201) then gives:

$$\text{inlet,} \quad Q_2 = D_p\omega - \frac{(P_1 - P_2)}{R_i} + \frac{P_2}{R_{e2}} + D_p\omega\left(\frac{P_2}{\beta}\right) + \frac{P_2}{R_{t2}}, \tag{3.206}$$

$$\text{outlet,} \quad Q_1 = D_p\omega - \frac{(P_1 - P_2)}{R_i} - \frac{P_1}{R_{e1}} - D_p\omega\left(\frac{P_1}{\beta}\right) - \frac{P_1}{R_{t1}}. \tag{3.207}$$

Collecting common terms then gives:

$$\text{inlet,} \quad Q_2 = D_p\omega - \frac{P_1}{R_i} + P_2\left(\frac{D_p\omega}{\beta} + \frac{1}{R_i} + \frac{1}{R_{e2}} + \frac{1}{R_{t2}}\right), \tag{3.208}$$

$$\text{outlet,} \quad Q_1 = D_p\omega - P_1\left(\frac{D_p\omega}{\beta} + \frac{1}{R_i} + \frac{1}{R_{e1}} + \frac{1}{R_{t1}}\right) + \frac{P_2}{R_i}. \tag{3.209}$$

If the "true" flow losses are lumped together, and it is reasonable to assume that $R_{e1} = R_{e2} = R_{ext}$ and $R_{t1} = R_{t2} = R_{tim}$, then the flow-continuity equations become:

$$\text{inlet,} \quad Q_2 = D_p\omega - \frac{P_1}{R_i} + P_2\left(\frac{1}{R_e} + \frac{1}{R_i}\right), \tag{3.210}$$

$$\text{outlet,} \quad Q_1 = D_p\omega - P_1\left(\frac{1}{R_e} + \frac{1}{R_i}\right) + \frac{P_2}{R_i}, \tag{3.211}$$

$$\frac{1}{R_e} = \frac{D_p\omega}{\beta} + \frac{1}{R_{ext}} + \frac{1}{R_{tim}}. \tag{3.212}$$

The evaluation of the various loss parameters depends on the measurements possible in practice. It is important to note that the difference between the input and the output flow rate is not the same as the case drain leakage:

$$Q_2 - Q_1 = \frac{P_1 + P_2}{R_e}, \tag{3.213}$$

$$Q_{drain} = \frac{P_1 + P_2}{R_{ext}}. \tag{3.214}$$

Not all the components of pump external resistance can be determined from the measurement of case drain leakage and output flow rate. It is important to be aware of the changes to the four components of pump resistance as the displacement, speed, or both is changed. During experimental testing, it is also important to measure the speed of the pump drive because a small change under load could lead to incorrect deductions regarding flow losses. A measured set of output flow-rate characteristics from a pump set in the author's laboratory is shown in Fig. 3.83 for the main in-line axial piston pump and a vane pump used for lower-pressure circuits. Both pumps are connected to the same electric motor drive running at 1440 rev/min. The inlet line from the overhead supply tank is connected to both pumps by a large-diameter inlet pipe, and the inlet pressure is negligible.

Figure 3.83. Flow characteristics of an axial piston pump and a vane pump.

Considering that the output flow only is usually measured:

$$\text{outlet,} \quad Q_1 = D_p\omega - P_1\left(\frac{1}{R_e} + \frac{1}{R_i}\right) + \frac{P_2}{R_i},$$

$$Q_1 = D_p\omega - \frac{P_1}{R_p}, \quad \frac{1}{R_p} = \left(\frac{1}{R_e} + \frac{1}{R_i}\right). \tag{3.215}$$

These data suggest pump output resistances of $R_p \approx 6 \times 10^{10}\,\mathrm{N\,m^{-2}/m^3\,s^{-1}}$ for the axial piston pump and $R_p \approx 10 \times 10^{10}\,\mathrm{N\,m^{-2}/m^3\,s^{-1}}$ for the smaller vane pump.

Worked Example 3.12

Data for an axial piston pump tested by the author in his laboratory for use with a 95/5 oil-in-water emulsion for a steel mill continuous-caster unit application are shown. The inlet is boosted by a centrifugal pump to a pressure of typically 3.5 bar. The output and case drain flow rates are measured.

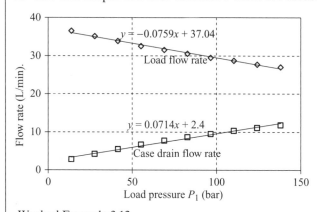

Worked Example 3.12

Also shown are the least-squares fits to the data. These indicate a flow rate of 37.04 L/min when $P_1 \to 0$:

$$\text{outlet,} \quad Q_1 = D_p\omega - P_1\left(\frac{1}{R_e} + \frac{1}{R_i}\right) + \frac{P_2}{R_i},$$

$$\text{case drain,} \quad Q_{\mathrm{drain}} = \frac{P_1 + P_2}{R_{\mathrm{ext}}},$$

$$\frac{1}{R_e} = \frac{D_p\omega}{\beta} + \frac{1}{R_{\mathrm{ext}}} + \frac{1}{R_{\mathrm{tim}}}.$$

The measurement accuracy at very low pressures becomes an issue, and it seems only sensible to use the slope of the case drain characteristic rather than the intercept predicted as $P_1 \to 0$. In addition, the drain flow rate shows a weak second-order effect at low pressures.

Considering the linear part gives:

$$\frac{1}{R_{\text{ext}}} \approx \frac{0.0714 \times 10^{-3}}{60 \times 10^5} = 0.119 \times 10^{-10} \ \text{N}\,\text{m}^{-2}/\text{m}^3\,\text{s}^{-1}.$$

Given that the cross-port leakage resistance R_i is usually much greater than the external resistance R_e, then from the load flow characteristic at $P_1 \to 0$,

$$Q_1 \approx D_p\omega \to D_p = \frac{37.04 \times 10^{-3}}{60 \times 150.8} = 4.09 \times 10^{-6} \text{m}^3/\text{rad}.$$

This compares with the computed geometrical displacement given by:

$$D_g = \frac{nd^2\ell}{8} = \frac{9 \times (0.0127)^2(0.0223)}{8} = 4.05 \times 10^{-6}\text{m}^3/\text{rad}.$$

The slope of the load flow-rate characteristic is given by:

$$\frac{1}{R_e} + \frac{1}{R_i} = \frac{0.0759 \times 10^{-3}}{60 \times 10^5} = 0.127 \times 10^{-10}.$$

Recall that:

$$\frac{1}{R_e} = \frac{D_p\omega}{\beta} + \frac{1}{R_{\text{ext}}} + \frac{1}{R_{\text{tim}}},$$

$$\frac{1}{R_e} + \frac{1}{R_i} = \frac{D_p\omega}{\beta} + \frac{1}{R_{\text{ext}}} + \frac{1}{R_{\text{tim}}} + \frac{1}{R_i}.$$

Assuming a fluid effective bulk modulus $\beta = 1.4 \times 10^9 \text{N/m}^2$ and inserting data for compressibility resistance and R_{ext} then gives:

$$\frac{D_p\omega}{\beta} + \frac{1}{R_{\text{ext}}} + \frac{1}{R_{\text{tim}}} + \frac{1}{R_i} = 0.127 \times 10^{-10},$$

$$0.004 \times 10^{-10} + 0.119 \times 10^{-10} + \frac{1}{R_{\text{tim}}} + \frac{1}{R_i} = 0.127 \times 10^{-10},$$

$$\frac{1}{R_{\text{tim}}} + \frac{1}{R_i} = 0.004 \times 10^{-10}.$$

Considering next an *axial piston motor*, we may directly apply results obtained for a pump, recalling that in many applications both lines will be pressurized. From Eqs. (3.210) and (3.211), it follows that:

$$Q_1 = D_m\omega + P_1\left(\frac{1}{R_e} + \frac{1}{R_i}\right) - \frac{P_2}{R_i}, \tag{3.216}$$

$$Q_2 = D_m\omega + \frac{P_1}{R_i} - P_2\left(\frac{1}{R_e} + \frac{1}{R_i}\right). \tag{3.217}$$

Hence, if a constant-speed test can be arranged and the motor driven – for example, by coupling the motor to a servovalve – then the resistance terms R_e and R_i may be determined with the following derivations from Eqs. (3.216) and (3.217):

$$Q_1 - Q_2 = \frac{(P_1 + P_2)}{R_e}, \tag{3.218}$$

$$\frac{Q_1 + Q_2}{2} = D_m\omega + (P_1 - P_2)\left(\frac{1}{R_i} + \frac{1}{2R_e}\right). \tag{3.219}$$

If the flow-rate difference is constant, then the external resistance R_e can be derived from Eq. (3.218) if the sum of line pressures is constant. With a servovalve controlling the motor, this will not be the case because of motor torque losses. However, tests show that the sum of line pressures is above 90% of the maximum possible, particularly at increasing pressure differentials and speeds. A set of flow-rate measurements taken on a motor coupled to a servovalve, and undertaken by the author, is shown in Fig. 3.84.

It is clear from the measured characteristics that the flow difference is sufficiently constant, indicating that it is probably insensitive to line pressures, given the large changes that occur over the test conditions. The average flow difference for the complete set of data is 0.47 L/min. Note the limited lower-pressure differential range that is due to motor torque losses, to be discussed later, but dominated by stiction–coulomb friction. However, for realistic loads, the sum of line pressures approaches the maximum possible of 100 bar, the supply pressure to the servovalve. A typical sum of pressure for calculations can be taken to be 95 bar.

Worked Example 3.13

Considering Fig. 3.84, determine the motor's resistance values.

(i) To determine R_e, it can be seen that the difference between the input flow rate and the output flow rate is sufficiently constant for a range of pressure differentials and speeds, allowing Eq. (3.218) to be used:

$$Q_1 - Q_2 = \frac{(P_1 + P_2)}{R_e}.$$

The flow-rate difference is remarkably constant, ≈ 0.47 L/min for a range of pressure differentials and speeds. The sum of line pressures is ≈ 95 bar with sufficient accuracy, and therefore:

$$R_e = \frac{95 \times 10^5}{\dfrac{0.47 \times 10^{-3}}{60}} = 1.21 \times 10^{12}\,\mathrm{N\,m^{-2}/m^3\,s^{-1}}.$$

(ii) To determine R_i, from Eq. (3.219):

$$\frac{Q_1 + Q_2}{2} = D_m\omega + (P_1 - P_2)\left(\frac{1}{R_e} + \frac{1}{2R_i}\right).$$

Hence, the inverse of the slope of the (mean flow rate)/(pressure differential) graph gives:

$$\left(\frac{1}{R_i} + \frac{1}{2R_e}\right) \approx 0.692 \times 10^{-12},$$

$$\left(\frac{1}{R_i} + 0.413 \times 10^{-12}\right) \approx 0.692 \times 10^{-12} \rightarrow R_i = 3.58 \times 10^{12}\,\text{N m}^{-2}/\text{m}^3\,\text{s}^{-1}.$$

For this motor, the external leakage resistance is smaller, although not negligible, than the cross-port leakage resistance, as might be expected primarily because of slipper leakage.

The motor displacement can be determined from the mean flow equation. By considering the extrapolated data to the zero-pressure-differential condition, the mean flow intercepts can be determined and plotted against speed. The results here give a remarkably linear characteristic. At a speed of 710 rpm, the best straight-line fit gives a mean flow of 7.5 L/min:

$$D_m = \frac{7.5 \times 10^{-3}}{60 \left(\dfrac{710 \times 2\pi}{60}\right)} = 1.68 \times 10^{-6}\,\text{m}^3/\text{rad}.$$

3.6.4 Torque Losses for an Axial Piston Machine

For a *pump* running at ostensibly a constant speed, the input torque from the drive must overcome torque losses to provide the output hydraulic torque. From earlier work in this chapter, it will be recalled that:

$$T_p = D_p(P_1 - P_2) + T_{\text{loss}}. \tag{3.220}$$

The torque loss characteristic depends on the particular type of pump being considered, although there are some common terms that are due to viscous friction and stiction–coulomb friction. A torque loss that is due to fluid viscosity arises from the fact that there are rotating components surrounded by the working fluid, and this torque loss is assumed to be proportional to pump speed. Stiction friction arises from the small torque necessary to cause the pump shaft to just turn. This then falls to the coulomb friction value as the shaft is rotating. Stiction friction obviously becomes more important for motors around zero speed and will be discussed later. For a pump at a particular speed and pressure, it is sufficient for this analysis to use the following general equation:

$$T_p = D_p(P_1 - P_2) + B_v\omega_p + T_{sc}, \tag{3.221}$$

where B_v is the viscous friction coefficient and T_{sc} is the stiction–coulomb friction torque loss function.

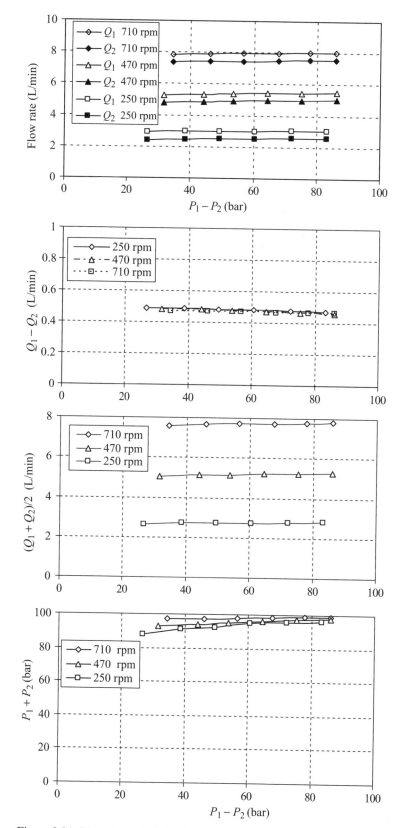

Figure 3.84. Flow characteristic of an in-line axial piston motor when coupled to a servovalve. ISO 32 mineral oil at 40°C (Watton, 2006).

For axial piston machines, the running value of friction is affected by the load pressure because of the orientation of the pistons and the swash plate, and its mathematical form is not precise. Previous work – for example, Hibi and Ichikawa (1975) – and other studies suggest that this friction effect can be considered proportional to the displacement and the pressure differential $P_1 - P_2$. Therefore, Eq. (3.221) for an axial piston pump becomes:

$$T_p = D_p(P_1 - P_2) + B_v\omega_p + \alpha D_p(P_1 - P_2) + T_c,$$
$$T_p = (1 + \alpha)D_p(P_1 - P_2) + B_v\omega_p + T_c. \tag{3.222}$$

Here, T_c is the coulomb friction running torque. Therefore, it might be expected that a practical measurement of shaft torque against pressure differential will overestimate the pump displacement, as evident from Eq. (3.222).

Worked Example 3.14

Worked Example 3.12 considered an axial piston pump test using a 95/5 oil-in-water emulsion. The inlet is boosted by a centrifugal pump to a pressure of typically 3.5 bar. Now consider the measured torque characteristic.

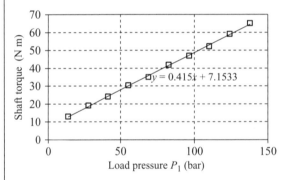

Worked Example 3.14

Also shown is the best straight-line fit to the data. Recall the torque equation (3.222):

$$T_p = (1 + \alpha)D_p (P_1 - P_2) + B_v\omega_p + T_c,$$
$$T_p = (1 + \alpha)D_p P_1 + \underbrace{[B_v\omega_p + T_c - (1 + \alpha)D_p P_2]}_{\text{cannot be separated from this test}}.$$

From the graph, $[B_v\omega_p + T_c - (1 + \alpha)D_p P_2] \approx 7.15\,\text{N m}$. From the slope of $Tp/(1 + \alpha)D_p P_1$,

$$(1 + \alpha)D_p = \frac{0.415}{10^5} = 4.15 \times 10^{-6}\ \text{m}^3/\text{rad}.$$

From torque:

$$(1 + \alpha)D_p = 4.15 \times 10^{-6}\ \text{m}^3/\text{rad}.$$

From flow rate:

$$D_p = 4.09 \times 10^{-6}\ \text{m}^3/\text{rad}.$$

From piston geometry:

$$D_g = 4.05 \times 10^{-6} \text{ m}^3/\text{rad}.$$

This suggests that $\alpha \approx 0.02$.

Now consider the *torque losses for a motor* and particularly the most common axial piston type. The analysis for a pump has to be modified because a motor will operate at all speeds, including very low speeds around zero. Equation (3.222) now becomes:

$$T_m = (1 - \alpha)D_m(P_1 - P_2) - B_v\omega_m - T_{sc}, \tag{3.223}$$

where T_{sc} is now a nonlinear term embracing stiction and coulomb friction. A measured torque loss $[\alpha D_m(P_1 - P_2) + B_v\omega_m + T_{sc}]$ characteristic is shown in Fig. 3.85 for a range of speeds and pressure differentials.

A similar characteristic holds for speeds in the reverse direction. This is the same motor whose flow characteristics were shown in Fig. 3.84 and analyzed in Worked Example 3.13. The coulomb friction value is $T_c \approx 1.7$ N m and equivalent to a pressure differential of 10 bar because the motor displacement is $D_m = 1.68 \times 10^{-6}$ m^3/rad. The stiction value is typically doubled and has an equivalent pressure differential of $T_s \approx 20$ bar. This is often termed the *break-out pressure*. The effect of pressure differential is evident, as discussed earlier, together with the viscous friction effect. For this motor, $\alpha \approx 0.05$ using the midspeed range and pressure differentials between 50 and 100 bar. The viscous friction coefficient is given by a typical value of $B_v \approx 0.02$ N m/rad s^{-1}.

For computer-modeling purposes, the stiction–friction characteristic can be approximated by:

$$T_{sc} = T_c + (T_s - T_c)e^{-N/N_{ref}}, \tag{3.224}$$

where N is the motor speed and N_{ref} is selected to give a sufficiently acceptable representation of the measured characteristic; for example, 20 rpm in Fig. 3.85. To complete the calculations, the torque–pressure-differential characteristic is needed at different motor speeds, Fig. 3.86 being representative of the present motor.

It can be seen that the slope of each of the measured torque characteristics is less than the ideal value as anticipated, the average least-squares estimate giving a slope of $(1 - \alpha)D_m = 1.6 \times 10^{-6}$ m^3/rad . Assuming the expected no-loss value to be

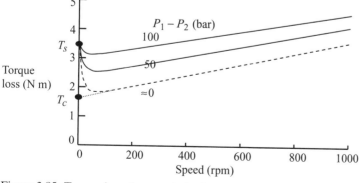

Figure 3.85. Torque loss characteristics for an axial piston motor.

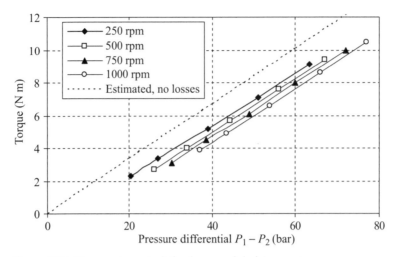

Figure 3.86. Torque characteristics for an axial piston motor.

1.68×10^{-6} m^3/rad, then $\alpha \approx 0.05$, similar to the value deduced from the flow loss characteristic in Worked Example 3.13.

3.6.5 Machine Efficiency — Axial Piston Pump

The overall efficiency is the product of volumetric efficiency and mechanical efficiency, and defined as follows:

(i) *Volumetric efficiency*

$$\eta_v = \frac{\text{output hydraulic flow rate}}{\text{input mechanically} - \text{generated flow rate}},$$

$$= \frac{D_p\omega_p - \dfrac{P_1}{R_p} + \dfrac{P_2}{R_i}}{D_p\omega_p},$$

$$= 1 - \frac{P_1}{D_p\omega_p R_p} + \frac{P_2}{D_p\omega_p R_i}. \tag{3.225}$$

Assume that the inlet pressure effect can be neglected and define a leakage pressure-loss term as follows:

$$\text{leakage pressure loss,} \quad P_q = D_p\omega_p R_p, \tag{3.226}$$

$$\eta_v = 1 - \frac{P_1}{P_q}. \tag{3.227}$$

(ii) *Mechanical efficiency*

$$\eta_m = \frac{\text{output hydraulic torque}}{\text{input mechanical torque}},$$

$$= \frac{D_p(P_1 - P_2)}{(1+\alpha)D_p(P_1 - P_2) + B_v\omega_p + T_c}, \tag{3.228}$$

$$= \frac{1}{(1+\alpha) + \dfrac{(B_v\omega_p + T_c)}{D_p(P_1 - P_2)}}.$$

Assume that the inlet pressure effect can be neglected and define a friction pressure-loss term as follows:

$$\text{Friction pressure loss,} \quad P_f = \frac{(B_v \omega_p + T_c)}{D_p}, \tag{3.229}$$

$$\eta_m = \frac{1}{(1+\alpha) + \dfrac{P_f}{P_1}}. \tag{3.230}$$

(iii) *The overall pump efficiency* is then given by:

$$\eta_{\text{pump}} = \eta_v \eta_m = \frac{1 - \dfrac{P_1}{P_q}}{(1+\alpha) + \dfrac{P_f}{P_1}},$$

$$(1+\alpha)\eta_{\text{pump}} = \frac{\overline{P}_1 - \overline{P}_1^2}{\overline{P}_1 + \dfrac{1}{k_p}}, \tag{3.231}$$

$$\overline{P}_1 = \frac{P_1}{P_q}, \qquad k_p = \frac{(1+\alpha)P_q}{P_f} = \frac{(1+\alpha)D_p^2 \omega_p R_p}{(B_v \omega_p + T_c)}. \tag{3.232}$$

For large and desirable values of k_p, the efficiency curve becomes asymptotic to the straight line defined by $(1+\alpha)\eta_{\text{pump}} = 1 - \overline{P}_1$. Increasing k_p is better achieved by increasing the displacement that is due to the squared effect indicated in Eq. (3.232).

A plot of pump efficiency variation with load pressure is shown in Fig. 3.87 for different values of k_p and illustrates that maximum efficiency occurs along the line $(1+\alpha)\eta_{\text{pump}} = 1 - \overline{P}_1$.

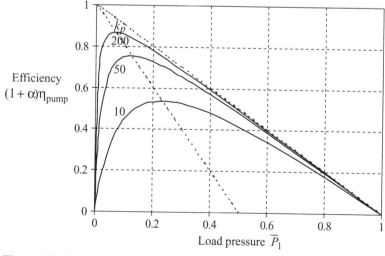

Figure 3.87. Total efficiency of an axial piston pump.

The condition for maximum efficiency is obtained by differentiating Eq. (3.231) with respect to load pressure, and this gives the solution:

$$\overline{P}_1 = \frac{\sqrt{1+k_p}-1}{k_p},$$

$$\text{maximum } (1+\alpha)\eta_{\text{pump}} = 1 - 2\overline{P}_1 \qquad (3.233)$$

$$= \frac{(\sqrt{1+k_p}-1)^2}{k_p}.$$

A feel for the effect of k_p on the pressure for maximum efficiency can be obtained by noting that if k_p is large, then a good approximation becomes:

$$\overline{P}_1 = \frac{\sqrt{1+k_p}-1}{k_p} \approx \frac{1}{\sqrt{k_p}},$$

$$P_1 = \sqrt{\frac{P_q P_f}{(1+\alpha)}} = \sqrt{\frac{\omega_p R_p (B_v \omega_p + T_c)}{(1+\alpha)}}. \qquad (3.234)$$

The pressure for maximum efficiency, therefore, will probably not be affected significantly by increasing the pump displacement, assuming the other loss terms remain the same. In fact, increasing the displacement also increases k_p, making it more likely that Eq. (3.234) applies, as mentioned earlier.

Worked Example 3.15

Considering Examples 3.12 and 3.14, loss coefficients were established for an axial piston pump operating with a 95/5 oil-in-water emulsion. These were found to be as follows:

output resistance, $\quad R_p = 7.9 \times 10^{10} \, \text{N m}^{-2}/\text{m}^3\text{s}^{-1}$

displacement, $\quad D_p = 4.07 \times 10^{-6} \, \text{m}^3/\text{rad}, \quad$ pump speed $\omega_p = 150.8 \, \text{rad/s}$,

friction loss $(B_v \omega_p + T_c) = 7.15 \, \text{N m}, \quad$ friction loss coefficient $\alpha = 0.02$,

$$P_q = D_p \omega_p R_p, \quad P_f = \frac{(B_v \omega_p + T_c)}{D_p},$$

$$k_p = \frac{(1+\alpha)P_q}{P_f}, \quad \eta_{\text{pump}} = \frac{1}{(1+\alpha)} \frac{(\overline{P}_1 - \overline{P}_1^2)}{(\overline{P}_1 + \frac{1}{k_p})}.$$

This results in:

$$P_q = (4.07 \times 10^{-6})(150.8)(7.9 \times 10^{10}) = 485 \, \text{bar},$$

$$P_f = \frac{7.15}{4.07 \times 10^{-6}} = 15.2 \, \text{bar},$$

$$k_p = \frac{1.02 \times 485}{15.2} = 32.55,$$

$$\eta_{\text{pump}} = \frac{0.98 \left(\overline{P}_1 - \overline{P}_1^2\right)}{(\overline{P}_1 + 0.031)}, \quad \overline{P}_1 = \frac{P_1 \, (\text{bar})}{485}.$$

A comparison between the measure efficiency and the theoretical efficiency is shown.

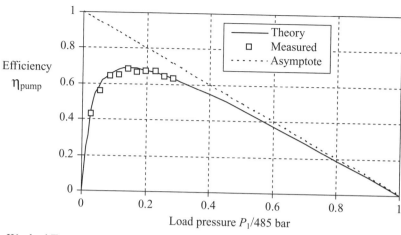

Worked Example 3.15

The maximum efficiency occurs at a pressure given by:

$$\overline{P}_1 = \frac{\sqrt{1 + k_p} - 1}{k_p} = \frac{\sqrt{1 + 32.55} - 1}{32.55} = 0.147,$$

$$P_1 = 0.147 \times 485 = 71.4\,\text{bar}.$$

The data scatter suggests that the condition for maximum efficiency probably matches that predicted with the pump performance model.

3.6.6 Machine Efficiency — Axial Piston Motor

Considering the previous work on an axial piston pump and the parameter definitions, then the equations for the same machine now acting as a motor become as follows:

(i) *Volumetric efficiency*

For a motor, there becomes an issue of defining the flow rate because both lines are usually pressurized. The mean flow rate defined in Eq. (3.219) therefore will be used for this analysis:

$$\frac{Q_1 + Q_2}{2} = D_m \omega_m + \frac{(P_1 - P_2)}{R_m},$$

$$\frac{1}{R_m} = \frac{1}{R_i} + \frac{1}{2R_e}, \tag{3.235}$$

$$\eta_v = \frac{\text{output mechanically} - \text{generated flow rate}}{\text{input hydraulic flow rate}}$$

$$= \frac{D_m \omega_m}{D_m \omega_m + \dfrac{(P_1 - P_2)}{R_m}}$$

$$= \frac{1}{1 + \dfrac{(P_1 - P_2)}{D_m \omega_m R_m}}. \tag{3.236}$$

Now define the leakage pressure loss:

$$P_q = D_m \omega_m R_m \tag{3.237}$$

$$\eta_v = \frac{1}{1 + \dfrac{(P_1 - P_2)}{P_q}}. \tag{3.238}$$

(ii) *Mechanical efficiency*

$$\eta_m = \frac{\text{output mechanical torque}}{\text{input hydraulic torque}}$$

$$= \frac{(1 - \alpha)D_m(P_1 - P_2) - (B_v \omega_m + T_c)}{D_m(P_1 - P_2)}$$

$$= (1 - \alpha) - \frac{(B_v \omega_m + T_c)}{D_m(P_1 - P_2)}. \tag{3.239}$$

Now, again, define a friction pressure loss term:

$$P_f = \frac{(B_v \omega_m + T_c)}{D_m}. \tag{3.240}$$

The mechanical efficiency then becomes:

$$\eta_m = (1 - \alpha) - \frac{P_f}{(P_1 - P_2)}. \tag{3.241}$$

(iii) *The overall motor efficiency* is then given by:

$$\eta_{\text{motor}} = \eta_v \eta_m = \frac{(1 - \alpha) - \dfrac{P_f}{(P_1 - P_2)}}{1 + \dfrac{(P_1 - P_2)}{P_q}},$$

$$\frac{\eta_{\text{motor}}}{(1 - \alpha)} = \frac{\overline{P}_\ell - \dfrac{1}{k_m}}{\overline{P}_\ell + \overline{P}_\ell^2},$$

$$\overline{P}_\ell = \frac{(P_1 - P_2)}{P_q} \quad k_m = \frac{(1 - \alpha)P_q}{P_f} = \frac{(1 - \alpha)D_m^2 \omega_m R_m}{(B_v \omega_m + T_c)}. \tag{3.242}$$

For large and desirable values of k_m, the efficiency curve becomes asymptotic to the straight line defined by:

$$\frac{\eta_{\text{motor}}}{(1 - \alpha)} = \frac{1}{(1 + \overline{P}_\ell)},$$

and maximum efficiency occurs along the line:

$$\frac{\eta_{\text{motor}}}{(1 - \alpha)} = \frac{1}{(1 + 2\overline{P}_\ell)}. \tag{3.243}$$

A plot of motor efficiency variation with pressure differential is shown in Fig. 3.88.

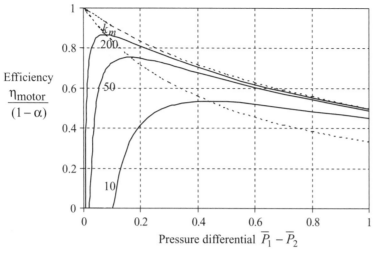

Figure 3.88. Total efficiency of an axial piston motor.

The condition for maximum efficiency is obtained by differentiating (3.242) with respect to pressure differential, and this gives the solution:

$$\overline{P}_\ell = \frac{\sqrt{1 + k_m} + 1}{k_m}$$

$$\text{maximum,} \quad \frac{\eta_{motor}}{(1 - \alpha)} = \frac{1}{(1 + 2\overline{P}_\ell)}$$

$$= \frac{(\sqrt{1 + k_m} - 1)^2}{k_m}. \tag{3.244}$$

These equations are similar to those for a pump, and it follows that for real machine data, the maximum efficiency condition can be approximated by:

$$\overline{P}_\ell = \frac{\sqrt{1 + k_m} + 1}{k_m} \approx \frac{1}{\sqrt{k_m}},$$

$$P_\ell = \sqrt{\frac{P_q P_f}{(1 - \alpha)}} = \sqrt{\frac{\omega_m R_m (B_v \omega_m + T_c)}{(1 - \alpha)}}. \tag{3.245}$$

An interesting conclusion may be drawn from this analysis for a pump and a motor if it is assumed that the displacements are equal, friction is dominated by coulomb friction, and α is negligible. The ratio of pressures at maximum efficiency is then given by:

$$\frac{P_{1pump}}{(P_1 - P_2)_{motor}} \approx \sqrt{\frac{k_m}{k_p}} \approx \sqrt{\frac{\omega_m R_m}{\omega_p R_p}} = \sqrt{\frac{\omega_m}{\omega_p} \frac{\left(1 + \dfrac{R_i}{R_e}\right)}{\left(1 + \dfrac{R_i}{2R_e}\right)}}. \tag{3.246}$$

Therefore, if the speeds are equal, the load pressure for maximum efficiency of a pump will be higher than the differential pressure for a motor. In practice, the absolute maximum efficiency for a motor can also occur at a lower speed than for the same machine acting as a pump.

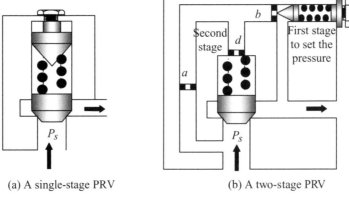

(a) A single-stage PRV   (b) A two-stage PRV

Figure 3.89. Schematics of a PRV.

3.7 Pressure-Relief Valve Pressure–Flow Concepts

The basic principle of a PRV operation is that the system pressure force acts against a variable resisting force, usually a spring mechanism, until a set force, equivalent to a set pressure, has been reached. Any pressure increase above this "cracking pressure" P_c will cause the main poppet to open, allowing flow relief to regulate the pressure. It is usually assumed that flow across the main poppet follows Bernoulli's equation. Figure 3.89 shows the principle of operation of a single-stage PRV and a two-stage PRV.

   The two-stage PRV is discussed in Section 3.3.3, the cracking pressure being set by the first-stage spring, which is much stiffer than the second-stage spring behind the poppet. The smallest movement of the first-stage poppet creates a pressure behind the main poppet at a value between tank pressure and system pressure. This produces a smaller increase from the cracking pressure during operation than with the simple single-stage PRV because the system pressure is opposed by the weak second-stage spring plus the small flow-reaction force. Actually, the flow-reaction force can be important for this two-stage PRV, as discussed in Section 3.3.3. For the single-stage PRV, the pressure increase during operation is directly affected by the stiffness of the single main stiff spring plus the small flow-reaction force. The pressure–flow characteristic of PRVs can vary in terms of the pressure-drop–flow-rate characteristic when manufacturers' data are considered. A good design has a small pressure increase as flow is passed through it. Because the poppet opening is usually controlled by a main-stage spring, to a greater or lesser extent, then the characteristic can be generically defined. Once the pressure has exceeded the cracking pressure P_c, then the net pressure difference creates the poppet opening x. Flow is assumed to obey Bernoulli's equation, and the flow area is assumed to be proportional to poppet opening x. The flow rate through the PRV is then given by:

$$Q = K(P - P_c)\sqrt{P} \quad P > P_c. \tag{3.247}$$

So, increasing the flow area, by increasing K, reduces the valve resistance at the opening point and at all other operating points. It will be obvious from Eq. (3.247) that because the flow rate rapidly increases when the pressure is typically no greater than 20 bar above the cracking pressure, then the square-root effect is only small. Figure 3.90 shows the characteristic of two different two-stage PRVs.

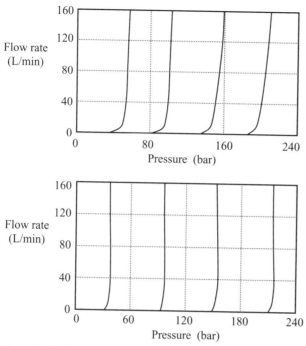

Figure 3.90. Flow/pressure characteristics of two different two-stage PRVs.

It can be seen that each characteristic shows a nonlinear effect around the crack-ing point, but the characteristic becomes remarkably linear over the higher-pressure range, which is still only a small value above the cracking pressure. Manufacturers' data for some single-stage PRVs show a near-linear characteristic over the entire pressure range achievable above the cracking pressure.

The valve resistance just as it opens; that is, when $P = P_c$, is given by:

$$R_{rv} = \left.\frac{dP}{dQ}\right|_{P=P_c} = \frac{2}{K\sqrt{P_c}}. \tag{3.248}$$

A good feature is that even if the flow characteristic obeys the Bernoulli form given by Eq. (3.247), then the valve resistance decreases by only 13% as the pressure increases 10% beyond the cracking pressure. This has some importance when the dynamics of a circuit containing a PRV is considered.

3.8 Sizing an Accumulator

When considering the calculation of changing gas volumes within an accumulator, it is usually assumed that slowly changing processes are considered as isothermal, whereas rapidly changing processes are considered as adiabatic; that is:

$$PV^n = \text{constant},$$

$$n = 1 \text{ isothermal}, \quad n = 1.4 \text{ adiabatic}. \tag{3.249}$$

Actually, n changes with the rate of charging, as shown in Watton and Xue (1995), and the exponent n can actually exceed 1.4, the tests described sometimes giving $n = 1.6$ for the largest volume of 4 L used from the 1 L, 2 L, 4 L set. In addition, the

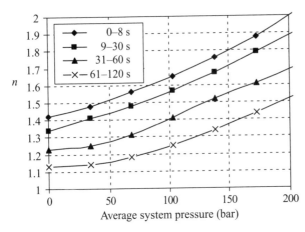

n

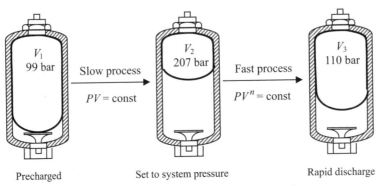

Figure 3.91. The variation in gas exponent n (rearranged from *Parker Engineers Design Book*, Vol. 1, 2001).

exponent n also varies with both the charging time and the pressure level, as shown in Fig. 3.91.

Consider the problem of deciding an accumulator size if 4 L of oil is to be delivered when a circuit, operating at 207 bar, discharges to a pressure of 110 bar in 0.8 s. It is assumed here that the operating pressure is achieved by a slow charging process such that the ideal isothermal process $n = 1$ may be assumed. From Fig. 3.91, and assuming that the discharging process has a mean pressure of 158.5 bar, take $n = 1.82$. The accumulator bladder volume states are shown in Fig. 3.92.

Figure 3.92. The system process of charging and discharging.

(i) Precharge the accumulator to 90% of the lowest pressure used:

$$P_1 = 0.9 \times 110 = 99 \text{ bar}.$$

(ii) Slow isothermal charging to working pressure:

$$P_1 V_1 = P_2 V_2,$$
$$99 V_1 = 207 V_2 \rightarrow V_1 = 2.09 V_2.$$

(iii) Fast discharging:

$$P_2 V_2^{1.82} = P_3 V_3^{1.82},$$
$$207 V_2^{1.82} = 110 V_3^{1.82} \rightarrow V_3 = 1.41 V_2.$$

(iv) But the required volume to be discharged is:

$$V_3 - V_2 = 4\,L,$$
$$1.41V_2 - V_2 = 4 \rightarrow V_2 = 9.76\,L,$$
$$V_1 = 20.5\,L.$$

Select a 21-L accumulator.

3.9 Design of Experiments

Whether or not a computer simulation or experimental testing is undertaken, it is often useful to determine trends in performance around a particular operating condition. If a simple linear model, for example, can be determined with sufficient accuracy for changes about the chosen operating condition as a number of important parameters are varied, then this can lead to a rapid assessment of potential design improvements. The design of experiments, in the present context, is an ordered approach to parameter changes such that minimum computation is necessary to determine the functional relationship. It is based on the matrix method of least squares, which actually can represent either linear or nonlinear functional relationships.

For example, with n design variables $x_1, x_2, \ldots, x_n$ and the design output y, then a linear combination would be as follows:

$$y = b_0 + b_1 x_1 + b_2 x_2 + \cdots + b_n x_n. \tag{3.250}$$

Because each combination of design variables leads to one experiment, then there will be N experiments resulting in the following matrix form:

$$\begin{bmatrix} y^1 \\ y^2 \\ y^3 \\ \vdots \\ y^n \end{bmatrix} = \begin{bmatrix} 1 & x_1^1 & x_2^1 \ldots x_n^1 \\ 1 & x_1^2 & x_2^2 \ldots x_n^2 \\ 1 & x_1^3 & x_2^3 \ldots x_n^3 \\ \vdots & \vdots & \vdots \\ 1 & x_1^N & x_2^N \ldots x_n^N \end{bmatrix} \begin{bmatrix} b_0 \\ b_1 \\ b_2 \\ \vdots \\ b_n \end{bmatrix} \tag{3.251}$$

In general terms, a times-series is represented in matrix notation as follows:

$$\mathbf{Y} = \mathbf{A}\boldsymbol{\beta}. \tag{3.252}$$

The least-squares solution for the unknown coefficients $\boldsymbol{\beta}$ is:

$$\boldsymbol{\beta} = (\mathbf{A}^T\mathbf{A})^{-1}\mathbf{A}^T\mathbf{Y}. \tag{3.253}$$

The error vector $\mathbf{e}$ is defined as:

$$\mathbf{e} = \mathbf{Y} - \mathbf{A}\boldsymbol{\beta}. \tag{3.254}$$

The variance σ^2 is given by:

$$\sigma^2 = \mathbf{e}^T\mathbf{e}/(N - n). \tag{3.255}$$

$(N - n)$ is referred to as the number of degrees of freedom. The variance of estimates is given by:

$$V(\boldsymbol{\beta}) = \sigma^2(\mathbf{A}^T\mathbf{A})^{-1}. \tag{3.256}$$

Considerable computational effort can be gained if the design variables are expressed in nondimensional values and set to specific values of $-1, 0, +1$ because the matrix $(\mathbf{A}^T\mathbf{A})^{-1}$ can be transformed into diagonal form. However, the number of experiments is then fixed.

Example

Average noise measurements are made on a pump in which both the pressure and speed may be varied. Determine a best-estimate linear relationship given the following experimental conditions:

Speed	x_1	1500 rpm	1000 rpm	500 rpm
Pressure	x_2	250 bar	200 bar	150 bar

Nondimensionalize with respect to a reference speed of 1000 rpm and a reference pressure of 200 bar as follows:

$$x \to \frac{(x - x_{\text{ref}})}{(x_{\text{max}} - x_{\text{min}})/2}. \tag{3.257}$$

This gives:

		-1	0	$+1$
Speed	x_1	1500 rpm	1000 rpm	500 rpm
Pressure	x_2	250 bar	200 bar	150 bar

Then, design the experiments by using all combinations as follows:

$$\mathbf{A} = \begin{bmatrix} & x_1 & x_2 \\ 1 & 1 & 1 \\ 1 & 1 & 0 \\ 1 & 1 & -1 \\ 1 & 0 & 1 \\ 1 & 0 & 0 \\ 1 & 0 & -1 \\ 1 & -1 & 1 \\ 1 & -1 & 0 \\ 1 & -1 & -1 \end{bmatrix}. \tag{3.258}$$

It is then an easy matter to show that:

$$\mathbf{A}^T\mathbf{A} = \begin{bmatrix} 9 & 0 & 0 \\ 0 & 6 & 0 \\ 0 & 0 & 6 \end{bmatrix}; \tag{3.259}$$

$$\beta_0 = \frac{1}{9}[y_1 + y_2 + y_3 + y_4 + y_5 + y_6 + y_7 + y_8 + y_9] \pm \sigma/\sqrt{9},$$

$$\beta_1 = \frac{1}{6}[(y_1 + y_2 + y_3) - (y_7 + y_8 + y_9)] \pm \sigma/\sqrt{6}, \tag{3.260}$$

$$\beta_2 = \frac{1}{6}[(y_1 + y_4 + y_7) - (y_3 + y_6 + y_9] \pm \sigma/\sqrt{6}.$$

The nine measured noise data for the nine different parameter combinations given by Eq. (3.258) are as follows:

$$\text{Noise level dB(A) } \mathbf{Y} = \begin{bmatrix} 88.6 \\ 85.8 \\ 84.1 \\ 87.2 \\ 84.9 \\ 82.8 \\ 86.2 \\ 83.8 \\ 81.5 \end{bmatrix}.$$

Application of Eq. (3.260) then gives:

$$\beta_0 = 85 \pm 0.066,$$
$$\beta_1 = 1.17 \pm 0.081,$$
$$\beta_2 = 2.17 \pm 0.081.$$

The pump noise level is then given by the following linear approximation:

$$\text{Noise level} = 85 + 1.17 \frac{(N - 1000)}{500} + 2.17 \frac{(P - 200)}{50} dB(\mathbf{A}).$$

Clearly, increasing the pressure has a more significant effect on noise level than increasing speed for this particular machine.

3.10 References and Further Reading

Abromovich GN [1963]. *The Theory of Turbulent Jets*, MIT, 1963.

Albertson ML et al. [1950]. Diffusion of submerged jets. *Trans. Am. Soc. Civil Eng.* 115.

Baylet V, O'Doherty T, Watton J [1993]. Reversed flow characteristic of a cone-seated poppet valve. In *Proceedings of the 5th International Symposium on Refined Flow Modelling and Turbulence Measurements*, Presses Ponts et Chaussees, 867–874.

Bergada JM, Kumar S, Watton J [2007]. Towards an analytical solution for axial piston pump leakage and output flow ripple. Presented at FLUCOME '07, Tallahassee, FL.

Bergada JM and Watton J [2002]. A direct leakage flow rate calculation method for axial pump grooved pistons and slippers, and its evaluation for a 5/95 fluid application. In *Proceedings of the 5th International Japan Fluid Power Society Symposium on Fluid Power*, 259–264.

Bergada JM and Watton J [2002]. Axial piston pump slipper balance with multiple lands. In *Proceedings of the ASME International Mechanical Engineering Congress and Exposition, IMECE 2002*. American Society of Mechanical Engineers, Vol. 2, Paper 39338.

Bergada JM and Watton J [2004]. A direct solution for flow rate and force along a cone-seated poppet valve for laminar flow conditions. *Proc. Inst. Mech. Eng.* 218, 197–210.

Bergada JM and Watton J [2005]. Force and flow through hydrostatic slippers with grooves. In *Proceedings of the 8th International Symposium on Fluid Control, Measurement and Visualization, FLUCOME 2005*, China Aerodynamics Research Society, Paper 240.

Bergemann M [1990]. Noise problems of hydraulic piston pumps with odd and even numbers of cylinders. In *Proceedings of the BHRA 9th International Symposium on Fluid Power*, Scientific and Technical Information, 235–248.

Borghi M, Milani M, Paoluzzi R [2000]. Stationary axial flow forces analysis on compensated spool valves. *Int. J. Fluid Power* 1(1), 17–28.

Borghi M, Milani M, Paoluzzi R [2005]. Influence of notch shape and number on the metering characteristics of hydraulic spool valves. *Int. J. Fluid Power* 6(2), 5–18.

Brault F [1986]. The use of hydraulic accumulators in energy storage. In *Proceedings of the 7th International Fluid Power Symposium*, British Hydrodynamics Research Association 48/1–10.

Brookes CA, Fagan MJ, James RD, McConnachie J [1996]. The selection and performance of ceramic components in a sea water pump. In *Proceedings of the 3rd JHPS International Symposium on Fluid Power*, Japan Fluid Power Society, 3–12.

Burton R, Ruan J, Ukrainetz P [2003]. Hydraulic bridge for pressure control in a *p-q* multiple line segment control valve. *Int. J. Fluid Power* 4(3), 5–16.

Chapple PJ and Dorey RE [1986]. The performance comparison of hydrostatic piston motors: factors affecting their application and use. In *Proceedings of the 7th BHRA Fluid Power Symposium*, British Hydrodynamics Research Association, 1–8.

Chenvisuwat T, Park S, Kitagawa A [2002]. Development of a poppet-type brake pressure control valve for a friction brake of rolling stock. In *Proceedings of the 5th JFPS International Symposium on Fluid Power*, Japan Fluid Power Society, Vol. 3, 733–738.

Crabtree AB, Manring ND, Johnson RE [2005]. Pressure measurements for translating hydrostatic thrust bearings. *Int. J. Fluid Power* 6(3), 19–24.

Davies RM and Watton J [1993]. Some Practical Considerations regarding the Dynamic Performance of Proportional Relief Valves. 10th International Fluid Power Conference, Brugge, Mechanical Engineering Publications Ltd, 199–218.

de Pennington A, 't Mannetji JJ, Bell R [1974]. The modelling of electrohydraulic control valves and its influence on the design of electrohydraulic drives. *Inst. Mech. Eng. J. Mech. Eng. Sci.* 16, 196–204.

Del Vescovo G and Lippolis A [2003]. Three-dimensional analysis of flow forces on directional control valves. *Int. J. Fluid Power* 4(2), 15–24.

Di Rito G [2007]. Experiments and CFD simulations for the characterisation of the orifice flow in a four-way servovalve. *Int. J. Fluid Power* 8(2), 37–46.

Dobchuk J, Burton R, Nikiforuk P [2007]. A modified turbulent orifice equation approach for modelling valves of unknown configuration. *Int. J. Fluid Power* 8(3), 25–30.

Dong X and Ueno H [1999]. Flows and flow characteristics of spool valve. In *Proceedings of the 4th JHPS International Symposium on Fluid Power*, Japan Fluid Power Society, 51–56.

Douglas JF, Gasiorek JM, Swaffield JA, Jack LB [2005]. *Fluid Mechanics*. Pearson, Prentice-Hall.

Edge KA [1999]. Designing quieter hydraulic systems – Some recent developments and contributions. In *Proceedings of the 4th JHPS International Symposium on Fluid Power*, Japan Fluid Power Society, 3–27.

Edge K, Burrows C, Lecky-Thomson N [1996]. Modelling of cavitation in a reciprocating plunger pump. In *Proceedings of the 3rd JHPS International Symposium on Fluid Power*, Japan Fluid Power Society, 473–478.

Fairhurst M and Watton J [2001]. CFD analysis of a pump pressure compensator operating with a water-based fluid. In *Power Transmission and Motion Control*, Professional Engineering Publications Ltd., 21–32.

Fisher MJ. [1962]. A theoretical determination of some characteristics of a tilted hydrostatic slipper bearing. British Hydromechanics Research Association, Rep. RR 728.

Fluent Inc. [2006]. *FLUENT Users Guide*.

Freeman P [1962]. *Lubrication and Friction*, Pitman.

Fu X, Du X, Zoul J, Ji H, Ryu S, Ochiai M [2007]. Characteristics of flow through throttling valve undergoing a steep pressure gradient. *Int. J. Fluid Power* 8(1), 29–38.

Gao D [2004]. Investigation of flow structure inside spool valve with FEM and PIV methods. *Int. J. Fluid Power* 5(1), 51–66.

Gao H, Fu X, Yang H [2002]. Numerical and experimental investigation of cavitating flow in oil hydraulic ball valve. In *Proceedings of the 5th JHPS International Symposium on Fluid Power*, 923–928.

Gilardino L, Mancò S, Nervegra N, Viotto F [1999]. An experience in simulation: The case of a variable displacement axial piston pump. In *Proceedings of the 4th JHPS International Symposium on Fluid Power*, Japan Fluid Power Society, 85–91.

Gordic D, Babic M, Jovicic N [2004]. Modelling of spool position feedback servovalves. *Int. J. Fluid Power* 5(1), 37–50.

Green WL [1970]. The poppet valve-flow force compensation. In *Proceedings of the International Fluid Power Conference*, British Hydrodynamics Research Association, Paper 2, pp. S1–S6.

Green WL [1973]. The effects of discharge times on the selection of gas charged accumulators. In *Proceedings of the 3rd International Fluid Power Symposium*, British Hydrodynamics Research Association, pp. D1/1–15.

Hagiwara T [1962]. Studies on the characteristics of radial flow nozzles (1st report). *Bull. JSME* 5, 656–663.

Handroos H and Halme J [1996]. Semi-empirical model of a counterbalance valve. In *Proceedings of the 3rd JHPS International Symposium on Fluid Power*, Japan Fluid Power Society, 525–530.

Harris RM, Edge KA, Tilley DG [1993]. Predicting the behaviour of slipper pads in swashplate-type axial piston pumps. In *Proceedings of the ASME Winter Annual Meeting*, American Society of Mechanical Engineers, 1–9.

Harris RM, Edge KA, Tilley DG [1996]. Predicting the behaviour of slipper pads in swashplate-type axial piston pumps. *ASME J. Dyn. Syst. Meas. Control* 118(1), 41–47.

Hayase T, Shirai A, Xia Y, Hayashi S [2002]. Fundamental consideration on numerical analysis of unsteady flow through spool valves. In *Proceedings of the 5th JHPS International Symposium on Fluid Power*, Japan Fluid Power Society, 929–934.

Haynes JM [2007]. Axial piston pump leakage modelling and measurement. Ph.D. thesis, Cardiff University, School of Engineering, UK.

Henri DH, Hollerbach JM, Nahvi A [1998]. An analytical and experimental investigation of a jet pipe controlled electropneumatic actuator. *IEEE Trans. Robot. Automation* 14, 601–611.

Hibi A and Ichikawa T [1975]. Torque performance of hydraulic motor in whole operating condition from start to maximum speed and its mathematical model. In *Proceedings of the BHRA 4th International Fluid Power Symposium*, British Hydromechanics Research Association B3, 29–38.

Hooke CJ and Kakoullis YP [1978]. The lubrication of slippers on axial piston pumps. In *Proceedings of the 5th International Fluid Power Symposium*, British Hydromechanics Research Association, B2, 13–26.

Hooke CJ and Kakoullis YP [1981]. The effects of centrifugal load and ball friction on the lubrication of slippers in axial piston pumps. In *Proceedings of the 6th International Fluid Power Symposium*, British Hydromechanics Research Association, 179–191.

Hooke CJ and Kakoullis YP [1983]. The effects of nonflatness on the performance of slippers in axial piston pumps. *Proc. Inst. Mech. Eng.* 197 C, 239–247.

Hooke CJ and Li KY [1988]. The lubrication of overclamped slippers in axial piston pumps centrally loaded behaviour. *Proc. Inst. Mech. Eng.* 202(CH), 287–293.

Hooke CJ and Li KY [1989]. The lubrication of slippers in axial piston pumps and motors. The effect of tilting couples. *Proc. Inst. Mech. Eng.* 203 C, 343–350.

Hu Y and Liu D. [2000]. Static characteristics of relief valve with pilot G-π bridge hydraulic resistances network. In *Proceedings of the 5th JFPS International Symposium on Fluid Power*, 739–744.

Iboshi N and Yamaguchi A [1982]. Characteristics of a slipper bearing for swash plate type axial piston pumps and motors, theoretical analysis. *Bull. JSME* 25, 1921–1930.

Iboshi N and Yamaguchi A [1983]. Characteristics of a slipper bearing for swash plate type axial piston pumps and motors, experimental. *Bull. JSME* 26, 1583–1589.

Ichiyanagi T, Kojima E, Edge KA [1997]. The fluid borne noise characteristics of a bent axis motor established using the "Secondary Source" method. In *Proceedings of the 5th Scandinavian Conference on Fluid Power*, Linkoping University, 123–138.

Ijas M and Virvalo T [2002]. Problems in using an accumulator as a pressure damper. In *Power Transmission and Motion Control*, Professional Engineering Publications Ltd., 277–289.

Ito K, Kiyoshi I, Keiji S [1996]. Visualisation and detection of cavitation in V-shaped groove type valve plate of an axial piston pump. In *Proceedings of the 3rd JHPS International Symposium on Fluid Power*, Japan Fluid Power Society, 67–72.

Iudicello F and Mitchell D [2002]. CFD modelling of the flow in a gerotor pump. In *Power Transmission and Motion Control*, Professional Engineering Publications Ltd., 53–64.

Ivantysyn J and Ivantysynova M [2002]. *Hydrostatic Pumps and Motors*, Akademia Books International.

Ivantysynova M [1999]. A new approach to the design of sealing and bearing gaps of displacement machines. In *Proceedings of the 4th JHPS International Symposium on Fluid Power*, 45–50.

Ivantysynova M and Huang C [2002]. Investigation of the gap flow in displacement machines considering elastohydrodynamic effects. In *Proceedings of the 5th JFPS International Symposium on Fluid Power*, 219–229.

Ivantysynova M and Lasaar R [2004]. An investigation into micro and macrogeometric design of piston/cylinder assembly of swash plate machines. *Int. J. Fluid Power* 5(1), 23–36.

Jen YM and Lee CB [1993]. Influence of an accumulator on the performance of a hydrostatic drive with control of the secondary unit. *Proc. Inst. Mech. Eng.* 207, 173–184.

Johnston DN, Edge KA, Vaughan ND [1991]. Experimental investigation of flow and force characteristics of hydraulic poppet and disk valves. *Proc. Inst. Mech. Eng.* 205, 161–171.

Kakoulis YP [1977]. Slipper lubrication in axial piston pumps. M.Sc. thesis, University of Birmingham, UK.

Karmel AM [1986]. A study of the internal forces in a variable-displacement vane pump – Part 1: A theoretical analysis. *Trans. ASME. J. Fluids Eng.* 108, 227–232.

Karmel AM [1986]. A study of the internal forces in a variable-displacement vane pump – Part 2: A parametric study. *Trans. ASME J. Fluids Eng.* 108, 233–237.

Kay JM [1963]. *An Introduction to Fluid Mechanics and Heat Transfer*, Cambridge University Press.

Kazama T [2005]. Numerical simulation of a slipper model for water hydraulic pumps/motors in mixed lubrication. In *Proceedings of the 6th JFPS International Symposium on Fluid Power*, Japan Fluid Power Society, 7–10.

Kingsbury A [1931]. On problems in the theory of film lubrication with an experimental method of solution. *Trans. Am. Soc. Mech. Eng.* 53, 59.

Kirshner JM and Katz S [1975]. *Design Theory of Fluidic Components*, Academic.

Kobayashi, S, Hirose, M, Hatsue, J, Ikeya M [1988]. Friction characteristics of a ball joint in the swash-plate type axial piston motor. In *Proceedings of the 8th International Symposium on Fluid Power*, British Hydromechanics Research Association J2–565–592.

Kobayashi S and Ikeya M [1990]. The structural analysis of piston balls and hydrostatic slipper bearings in swash-plate type axial piston motors. In *Proceedings of the BHRA 9th International Symposium on Fluid Power*, British Hydromechanics Research Association, 19–32.

Koc E and Hooke CJ [1996]. Investigation into the effects of orifice size, offset and overclamp ratio on the lubrication of slipper bearings. *Tribol. Int.* 29, 299–305.

Koc E and Hooke CJ [1997]. Considerations in the design of partially hydrostatic slipper bearings. *Tribol. Int.* 30, 815–823.

Koc E, Hooke CJ, Li KY [1992]. Slipper balance in axial piston pumps and motors. *Trans. ASME J. Tribol.* 114, 766–772.

Koivula T, Ellman A, Vilenius M [1999]. The effect of oil type on flow and cavitation properties in orifices and annular clearances. In *Power Transmission and Motion Control*, Professional Engineering Publications Ltd., 151–165.

Konami S and Nishiumi T [1999]. *Hydraulic Control Systems* [in Japanese], TDU.

Kosodo H, Nara M, Kakehida S, Imanari Y [1996]. Experimental research about pressure-flow characteristics of V-notch. In *Proceedings of the 3rd JHPS International Symposium on Fluid Power*, Japan Fluid Power Society, 73–78.

Kozuma F, Arita T, Tsuda H [2005]. Development of energy-saving power steering. In *Proceedings of the 6th JHPS International Symposium on Fluid Power*, Japan Fluid Power Society, 297–300.

Kreith F and Eisenstadt R [1957]. Pressure drop and flow characteristics of short capillary tubes at low Reynolds numbers. *Trans. Am. Soc. Mech. Eng.* July, 1070–1078.

Lee SY and Blackburn JF [1952]. Contributions to hydraulic control-steady state axial forces on control valve pistons. *Trans. Am. Soc. Mech. Eng.* August, 1005–1011.

Lichtarowicz A, Duggins RK, Markland E [1965]. Discharge coefficients for incompressible non-cavitating flow through long orifices. *J. Mech. Eng. Sci.* 7(2), 210–219.

Li KY and Hooke CJ [1991]. A note on the lubrication of composite slippers in water-based axial piston pumps and motors. *Wear* 147, 431–437.

Mabie HM and Reinholz CF [1987]. Mechanisms and Dynamics of Machinery. John Wiley & Sons.

Maiti R, Saha R, Watton J [2002]. The static and dynamic characteristics of a pressure relief valve with a proportional controlled pilot stage. *Proc. Inst. Mech. Eng.* 216, 143–156.

Maiti R, Surawattanawan P, Watton J [1999]. Performance prediction of a proportional solenoid control pressure relief valve. In *Proceedings of the 4th JHPS International Symposium on Fluid Power*, Japan Fluid Power Society, 321–326.

Manco S, Nervegna N, Gilardino L [2002]. Advances in the simulation of axial piston pumps. In *Proceedings of the 5th JFPS International Symposium on Fluid Power*, 251–258.

Manring ND and Dong Z [2004]. The impact of using a secondary swash plate angle within an axial piston pump. *Trans. ASME J. Dyn. Syst. Meas. Control* 126, 65–74.

Manring ND and Kasaragadda SB [2003]. The theoretical flow ripple of an external gear pump. *Trans. ASME J. Dyn. Syst. Meas. Control* 205, 396–404.

Manring ND and Zhang Y [2001]. The improved volumetric efficiency of an axial piston pump utilizing a trapped volume design. *Trans. ASME J. Dyn. Syst. Meas. Control* 123, 479–487.

Martin MJ and Taylor R [1978]. Optimised port plate timing for an axial piston pump. In *Proceedings of the 5th BHRA Fluid Power Symposium*, British Hydromechanics Research Association, B5–51–66.

Massey B [1989]. *Mechanics of Fluids*, Taylor & Francis.

Masuda K and Ohuchi H [1996]. Noise reduction of a variable piston pump with even number of cylinders. In *Proceedings of the 3rd JHPS International Symposium on Fluid Power*, Japan Fluid Power Society, 91–96.

McCandlish D and Dorey R [1981]. Steady-state losses in hydrostatic pumps and motors. In *Proceedings of the 6th International Fluid Power Symposium*, British Hydromechanics Research Association, C3/133–144.

Mills RD [1969]. Numerical solutions of viscous flow through a pipe orifice at low Reynolds numbers. *Proc. Inst. Mech. Eng. J. Mech. Eng. Sci.* 11, 168–174.

Mokhtarzadeh-Dehghan MR, Ladommatos N, Brennan TJ [1997]. Finite element analysis of flow in a hydraulic pressure valve. *App. Math. Model.* 21, 437–445.

Nakada T and Ikebe Y [1981]. Measurement of the axial flow force on a spool valve. *J. Fluid Control* 13(3), 29–40.

Nakanishi T, Hayashi S, Hayase T, Shirai A, Jotatsu M, Kawamoto H [1999]. Numerical simulation of water hydraulic relief valve. In *Proceedings of the 4th JHPS International Symposium on Fluid Power*, Japan Fluid Power Society, 555–560.

Nakano K [1992]. Experimental study for the compensation of axial flow force in a spool valve. *J. Fluid Control* 21, 7–26.

Nakayama Y and Boucher RF [2002]. *Introduction to Fluid Mechanics*, Butterworth Heinemann, reprinted with revisions.

Nigro FEB, Strong AB, Alpay SA [1978]. A numerical study of the laminar viscous incompressible flow through a pipe orifice. *Trans. ASME. J. Fluids Eng.* 100, 467–472.

Nikiforuk PN, Ukrainetz PR, Tsai SC [1968]. Detailed analysis of a two-stage four-way electrohydraulic flow control valve. *Proc. Inst. Mech. Eng. Sci.* 10, 133–140.

Olems L [2000]. Investigations of the temperature behaviour of the piston cylinder assembly in axial piston pumps. *Int. J. Fluid Power* 1(1), 27–38.

Palumbo A, Paoluzzi R, Borghi M, Milani M [1996]. Forces on a hydraulic spool valve. In *Proceedings of the 3rd JHPS International Symposium on Fluid Power*, Japan Fluid Power Society, 543–548.

Park S-H, Kitigawa A, Kawashima M [2004]. Water hydraulic high-speed solenoid valve. Part 1: Development and static behaviour. *Proc. Inst. Mech. Eng.* 218, 399–409.

Petherick PM, Birk AM [1991]. State of the art review of pressure relief valve design, testing and modelling. *Pressure Vessel Technol.* 113, 46–54.

Rampen WHS and Salter SH [1990]. The digital displacement pump. In *Proceedings of the BHRA 9th International Symposium on Fluid Power,* British Hydromechanics Association, 33–45.

Scharf S and Murrenhoff H [2000]. Measurement of friction forces between piston and bushing of an axial piston displacement unit. *Int. J. Fluid Power* 6(1), 7–17.

Schlosser WMJ [1961]. Mathematical model for displacement pumps and motors. *Hydraulic Power Transmission,* April, 252–257 and 269; May, 324–328.

Seet GL, Penny JET, Foster K [1985]. Applications of a computer model in the design and development of a quiet vane pump. *Proc. Inst. Mech. Eng.* 199, B4, 247–253.

Shearer JL [1960]. Resistance characteristics of control valve orifices. In *Proceedings of the IMechE Symposium on Recent Mechanical Engineering Developments in Automatic Control,* Institute of Mechanical Engineering, 35–41.

Simson AK [1966]. Gain characteristics of subsonic pressure-controlled proportional fluid jet amplifiers. *ASME. J. Basic Eng.* June.

Somashekhar SH, Singaperumal M, Kumar RK [2006]. Modelling the steady-state analysis of a jet pipe electrohydraulic servo valve. *Proc. Inst. Mech. Eng.* 220, 109–130.

Somashekhar SH, Singaperumal M, Kumar RK [2007]. Mathematical modelling and simulation of a jet pipe electrohydraulic flow control valve. *Proc. Inst. Mech. Eng.* 221, 365–382.

Stone JA [1960]. Discharge coefficients and steady-state flow forces for hydraulic poppet valves. *Trans. ASME,* 144–154.

Stecki J and Matheson P [2005]. Adavances in automotive hydraulic drives. In *Proceedings of the 6th JHPS International Symposium on Fluid Power,* Japan Fluid Power Society, 664–669.

Suzuki K and Urata E [2002]. Experimental study on hydrostatic supports of water hydraulic valves. In *Proceedings of the 5th JHPS International Symposium on Fluid Power,* Japan Fluid Power Society, 177–180.

Takahashi K and Ishizawa S [1989]. Viscous flow between parallel disks with time varying gap width and central fluid source. In *Proceedings of the JHPS International Symposium on Fluid Power,* Japan Fluid Power Society, 407–414.

Takahashi K and Tsukiji T [1982]. The unsteady jet from the metering orifice of a spool valve. *Bull. JSME* 25, 576–582.

Takahashi T, Yamashina C, Miyakawa S [1999]. Development of water hydraulic proportional control valve. In *Proceedings of the 4th JHPS International Symposium on Fluid Power,* Japan Fluid Power Society, 549–554.

Takahashi K and Yu K [2002]. Analysis of some characteristics of hydraulic control valves using a streamline coordinate method. In *Proceedings of the 5th JHPS International Symposium on Fluid Power,* Japan Fluid Power Society, 917–922.

Takashima M et al. [1996]. Development of high-performance components for pollution-free water hydraulic system. In *Proceedings of the 3rd JHPS International Symposium on Fluid Power,* Japan Fluid Power Society, 49–54.

Takenaka T, Yamane R, Iwamizu T [1964]. Thrust of the disk valves. *Bull. Japan Society of Mechanical Engineering* 7, 558–566.

Tanaka K, Nakahara T, Kyogoku K [1993]. Piston rotation and frictional forces between piston and cylinder of piston and motor. In *Proceedings of the 2nd JHPS International Symposium on Fluid Power,* Japan Fluid Power Society, 235–240.

Tanaka K, Nakahara T, Kyogoku K [2003]. Half-frequency whirl of pistons in axial piston pumps under mixed lubrication. *Proc. Inst. Mech. Eng. J. Tribol.* 217, 93–102.

Tesar V and Watton J [1988]. Hydrodynamics of an idealised poppet valve. In *Proceedings of FLUCOME '88 Conference,* H S Stephens & Associates, Bedford, England, 36–39.

Thoma JU [1969]. Mathematical models and effective performance of hydrostatic machines and transmissions. *Hydraulic Pneumatic Power* 15, 179.

Totten GE, Kling GH, Reichel J [1999]. Development of hydraulic pump performance standards: an overview of current activities. In *Proceedings of the 4th JHPS International Symposium on Fluid Power,* Japan Fluid Power Society, 63–67.

Tsukiji T, Soshino M, Yonezawa Y [1993]. Numerical and experimental flow visualisation of unsteady flow in a spool valve. In *Proceedings of the 2nd JHPS International Symposium on Fluid Power,* Japan Fluid Power Society, 379–384.

Tsuta T, Iwamoto T, Umeda T [1999]. Combined dynamic response analysis of a piston-slipper system and lubricants in hydraulic piston pump. In *Emerging Technologies in Fluids, Structures and Fluid/Structure Interactions.* American Society of Mechanical Engineering, Vol. 396, 187–194.

Ueno H, Okajima A, Muromiya Y [1993]. Visualisation of cavitating flow and numerical simulation of flow in a poppet valve. In *Proceedings of the 2nd JHPS International Symposium on Fluid Power,* Japan Fluid Power Society, 385–390.

Urata E [1969]. Thrust of poppet valve. *Bull. JSME* 12, 1099–1109.

Urata E [2004]. One-degree-of-freedom model for torque motor dynamics. *Int. Journal of Fluid Power* 5(2), 35–42.

Vacca A and Cerutti M [2007]. Analysis and optimization of a two-way valve using response surface methodology. *Int. J. Fluid Power* 8(3), 43–58.

Vaughan ND, Johnston DN, Edge KA [1992]. Numerical simulation of fluid flow in poppet valves. *Proc. Inst. Mech. Eng.* 206, 119–127.

Washio S, Nakamura Y, Yu Y [1991]. Static characteristics of a piston-type relief valve. *Proc. Inst. Mech. Eng. J. Mech. Eng. Sci.,* 213, 231–239.

Watton J [1983]. Steady-state and dynamic flow characteristics of positive displacement pumps using laser Doppler anemometry. In *Proceedings of the International Conference on Optical Techniques in Process Control,* British Hydromechanics Research Association, 165–178.

Watton J [1987]. The effect of drain orifice damping on the performance characteristics of a servovalve flapper/nozzle stage. *ASME J. Dyn. Syst. Meas. Control* 109, 19–23.

Watton J [1988]. The design of a single-stage relief valve with directional damping. *J. Fluid Control* 18(2), 22–35.

Watton J [1988]. The effect of servovalve underlap on the accuracy and dynamic response of single-rod actuator position control systems. *J. Fluid Control* 18(3), 7–24.

Watton J [2006]. An explicit design approach to determine the optimum steady state performance of axial piston motor speed drives. IMechE Journal of Systems and Control Engineering, 12, 131–144.

Watton J [2007]. *Modelling Monitoring and Diagnostic Techniques for Fluid Power Systems,* Springer.

Watton J and Thorp J [2005]. Flow characteristics of a servovalve using a 3D CFD analysis. FLUCOME'05, 8th International Symposium on Fluid Control, Measurement and Visualisation. China Aerodynamics Research Society, CD-Rom Paper 13–14.

Watton J and Bergada JM [1994]. Progress towards an understanding of the pressure/flow characteristics of a servovalve two flapper/double nozzle flow divider using CFD modelling. In *Proceedings of FLUCOME '94,* ENSAE Toulouse, France, 47–51.

Watton J and Nelson RJ [1993]. Evaluation of an electrohydraulic forge valve behaviour using a CAD package. *Appl. Math. Model.* 17, 355–368.

Watton J and Watkins-Franklin KL [1990]. The transient pressure characteristic of a positive displacement vane pump. *Proc. Inst. Mech. Eng. J. Power and Energy* 204, 269–275.

Watton J and Xue Y [1995]. Identification of fluid power component behaviour using dynamic flow rate measurement. Proc Institution of Mechanical Engineers, Part C, *Journal of Mechanical Engineering Science,* Vol. 209, 179–191.

Weixiang S, Songnian L, Sihua G [1990]. A new technique for steady-state flow force compensation in spool valves. *Proc. Inst. Mech. Eng. J. Process Mech. Eng.* 204, 7–14.

Wieczoreck U and Ivantysynova M [2000]. CASPAR-A computer-aided design tool for axial piston machines. In *Proceedings of the Power Transmission Motion and Control International Workshop, PTMC 2000*, Professional Engineering Publishing, UK, 113–126.

Wieczoreck U and Ivantysynova M [2002]. Computer-aided optimization of bearing and sealing gaps in hydrostatic machines: the simulation tool CASPAR. *Int. J. Fluid Power* 3(1), 7–20.

Winkler B, Mikota G, Scheid R, Marhartsgruber B [2003]. Modelling and simulation of the elastohydrodynamic behaviour of sealing gaps. In *Proceedings of the 1st International Conference in Fluid Power Technology, Methods for Solving Practical Problems in Design and Control*, Fluid Power Net Publications, 155–164.

Wu D, Burton R, Schoenau G [2002]. An empirical discharge coefficient model for orifice flow. *Int. J. Fluid Power* 3(3), 13–18.

Wu D, Burton R, Schoenau G, Bitner D [2003]. Modelling of orifice flow rate at very small openings. *Int. J. Fluid Power* 4(1), 31–40.

Xin F, Hong J, Ryu S [2005]. Investigation into pressure distribution of spool valves with notches. In *Proceedings of the 6th JFPS International Symposium on Fluid Power*, Japan Fluid Power Society, 635–639.

Zarotti GL and Nervegna N [1981]. Pump efficiencies approximation and modelling. In *Proceeding of the 6th International Fluid Power Symposium*, British Hydromechanics Research Association, C4/145–164.

Zarotti LG and Paoluzzi R [1993]. Triple controls of variable displacement pumps. In *Proceedings of the 2nd JHPS International Symposium on Fluid Power*, Japan Fluid Power Society, 215–220.

Zhang R, Alleyne AG, Presetiawan EA [2004]. Performance limitations of a class of two-stage electrohydraulic flow valves. *Int. J. Fluid Power* 5(2), 47–53.

4 Steady-State Performance of Systems

4.1 Determining the Power Supply Pressure Variation during Operation for a Pump–PRV–Servovalve Combination: A Graphical Approach

First, consider a pump supply circuit to illustrate some calculation issues that arise for even this simple circuit, which is shown in Fig. 4.1. The PRV cracking pressure is set at 150 bar.

The flow–pressure characteristics of each component is shown, where the units are flow in liters per minute, pressure in bar, and current in milliamperes. The steady-state flows and the flow-continuity equations are:

$$Q_p = 40 - 0.03P,$$
$$Q_{rv} = 0.25(P - 150)\sqrt{P}, \quad P > 150\,\text{bar},$$
$$Q_{sv} = 0.18i\sqrt{P - P_{\text{load}}},$$
$$Q_p = Q_{rv} + Q_{sv}. \tag{4.1}$$

The pressure differential across the servovalve output ports is P_{load}, as required by the load actuator.

Nonlinear equation (4.1) in pressure P is not difficult to solve numerically, but an insight into the effect of the circuit interconnection can be gained by a graphical plot of Eq. (4.1). This is done in Fig. 4.2 for a servovalve load pressure differential of $P_{\text{load}} = 0$ and for servovalve currents of 0, 10, and 20 mA.

As the current is increased, the load flow is increased and the supply pressure P would be expected to fall. Figure 4.2 is drawn by plotting Q_p and the sum $(Q_{rv} + Q_{sv})$. The intersection of the two curves gives the operating pressure P.

For zero servovalve current, the pump flow rate matches the PRV flow rate to give a pressure of 161 bar.

As the current is increased to 10 mA and then to 20 mA, the pressure falls to 155 bar and then to 105 bar. Note that for a servovalve current of 20 mA, the PRV is not in operation because its cracking pressure is set at 150 bar. Such a graphical approach is quick to implement and gives a better visual impact than a purely mathematical numerical solution.

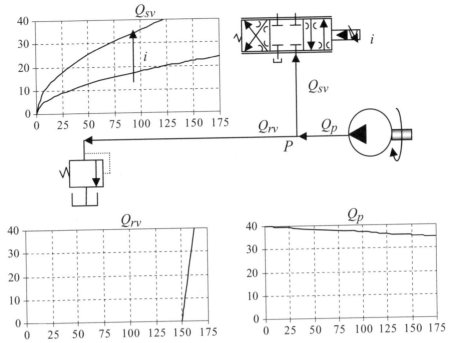

Figure 4.1. A power supply subcircuit linking a pump–PRV–servovalve.

4.2 Meter-Out Flow Control of a Cylinder

Figure 4.3 shows the system with the flow-control valve and check valve positioned such that load runaway is avoided. The lowering speed is set to 0.1 m/s.

The following data apply:

Pump flow rate $= 45$ L/min, load mass $M = 500$ kg
Cylinder bore diameter $= 76.2$ mm $\rightarrow A_1 = 4.56 \times 10^{-3}$ m^2
Cylinder rod diameter $= 44.3$ mm $\rightarrow A_2 = 3.20 \times 10^{-3}$ m^2, $\gamma = A_1/A_2 = 1.425$
Pressure drop across the directional valve at full pump flow $= 8$ bar

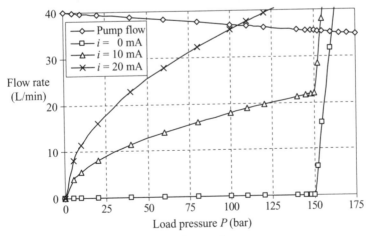

Figure 4.2. Graphical solution for the supply pressure.

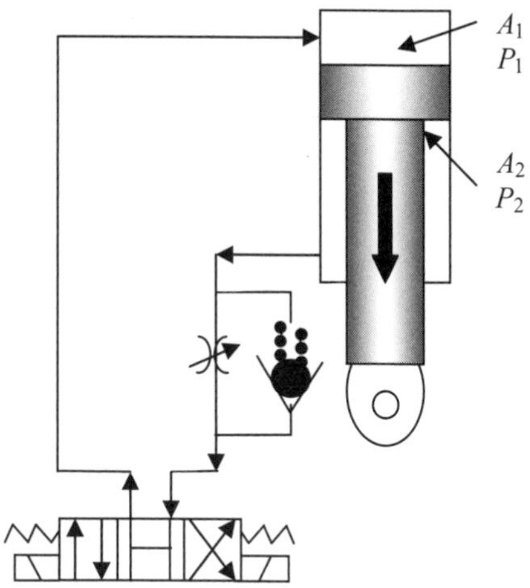

Figure 4.3. Meter-out flow control of an actuator lowering.

Flow-control valve setting requires a 20-bar pressure drop
Pressure drops across the check valve and down the lines may be neglected
Required retracting velocity $= 0.1$ m/s

(i) Lowering
The flow-controlled flow rates are calculated as:

$$Q_{in} = A_1\, U = (4.56 \times 10^{-3})(0.1) = 4.56 \times 10^{-4}\ \text{m}^3/\text{s}\ (27.36\ \text{L/min}),$$
$$Q_{out} = A_2\, U = (3.20 \times 10^{-3})(0.1) = 3.20 \times 10^{-4}\ \text{m}^3/\text{s}\ (19.20\ \text{L/min}).$$

So, *the flow-control valve is set at 19.20 L/min.*

It is assumed that the directional-valve pressure drop across any port is proportional to $(\text{flow})^2$. Starting at the tank end of the system, the pressure at the directional-valve return port is $(8)(19.2/45)^2 = 1.46$ bar.

Because the flow-control valve requires a minimum pressure drop of 20 bar, the annulus pressure $= (1.46 + 20) = P_2 = 21.46$ *bar.*

The steady-state force equation is:

$$P_2 A_2 = P_1 A_1 + Mg,$$
$$P_1 = P_2 \frac{A_2}{A_1} - \frac{Mg}{A_1} = \frac{21.46}{1.425} - \frac{500 \times 9.81}{(4.56 \times 10^{-3})(10^5)} \qquad (4.2)$$
$$= 4.3\ \text{bar}.$$

Adding the pressure drop across the directional-valve supply port gives a supply pressure $= (8)(27.36/45)^2 + 4.3 = P_s > 7.26$ *bar.* This is a very low pressure that is due to the load mass effect on the force equation.

(ii) Lifting
The check valve now allows flow-control bypass, as shown in Fig. 4.4.

To determine the system pressures, assume that the PRV is set to a high value and that all the pump flow is delivered to the actuator. Hence, $Q_{in} = 45$ L/min and $Q_{out} = (45)(1.425) = 64.13$ L/min.

The pressure drop across the directional valve is 8 bar at the supply ports and $(8)(64.13/45)^2 = 16.25$ bar at the return ports. So, $P_1 = 16.25$ *bar.*

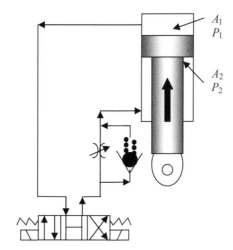

Figure 4.4. Actuator lifting with flow control bypassed.

The steady-state force equation is:

$$P_2 A_2 = P_1 A_1 + Mg,$$

$$P_2 = P_1 \frac{A_1}{A_2} + \frac{Mg}{A_2} = (16.25)(1.425) + \frac{500 \times 9.81}{(3.20 \times 10^{-3})(10^5)}$$

$$= 38.49 \, bar. \tag{4.3}$$

The supply pressure is now determined by adding the pressure drop of 8 bar across the directional-valve supply port to give $P_s > 46.49 \, bar$. This is higher than the requirement for lowering, so it is deduced that the supply pressure should be set to a value greater than 46.48 bar.

The lifting speed is $(45 \times 10^{-3}/60)/(3.20 \times 10^{-3} \, m^2) = 0.23$ m/s, a factor of 2.3 times the lowering speed. If the lifting speed is unacceptable, then it will be necessary to also control the flow in the lifting direction – for example, with a flow-control valve placed in the other line.

4.3 A Comparison of Counterbalance-Valve and an Overcenter-Valve Performances to Avoid Load Runaway

Figure 4.5 shows the circuit with the appropriate valve place in the appropriate line to prevent load runaway. The overcenter valve is piloted by the pressure from the other line when lowering is attempted.

The system is to perform a pressing operation; data are as follows:

Cylinder bore diameter = 80 mm → $A_1 = 0.005$ m^2
Cylinder rod diameter = 60 mm → $A_2 = 0.0028$ m^2 $\gamma = A_1/A_2 = 1.79$
Load mass equivalent to 5 kN
Press force required = 100 kN
Losses may be neglected for a first-order estimate of pressures

(i) Counterbalance valve

Static pressure when the load is stationary:
$P_2 = (5000)/(0.0028) = 17.8$ bar.
Set counterbalance pressure to $(1.3)(17.8) = P_{cb} = 23$ bar.

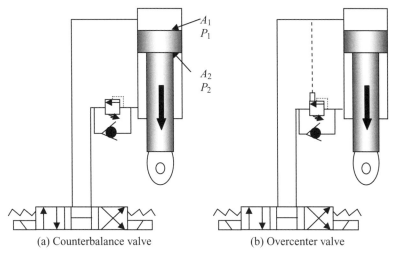

(a) Counterbalance valve   (b) Overcenter valve

Figure 4.5. Actuator runaway control.

Pressure on full bore to overcome counterbalance $= 23/1.79 \approx 13$ bar.
To press, consider the force equation:

$$P_1 A_1 + Mg = P_{cb} A_2 + F,$$
$$P_1 = \frac{(F - Mg)}{A_1} + P_{cb}\frac{A_2}{A_1} = \frac{(100{,}000 - 5000)}{(0.005)(10^5)} + 13 = 203 \text{ bar} \quad (4.4)$$

(ii) Overcenter valve

Pilot pressure to open the valve $=$ counterbalance setting$/2 = 23/2 = 11.5$ bar.
    It is assumed that when the overcenter valve is piloted, the annulus pressure is now zero; during pressing, it is deduced from the preceding that:

$$P_1 = \frac{(F - Mg)}{A_1} + 0 = \frac{(100{,}000 - 5000)}{(0.005)(10^5)} = 190 \text{ bar}. \quad (4.5)$$

It can be seen that the pressure needed is less than when a counterbalance valve is used. Recall also that in each case, the supply pressure is more than adequate to operate each valve.

Worked Example 4.1

Consider a lifting cylinder with flow control for both lifting and lowering. This is achieved with a flow-control valve within a four-check-valve bridge, as shown in the diagram. Therefore, the lifting and lowering speeds are the same. However, the pump has to supply a flow rate of 60 L/min when lowering and flow regeneration can be used to reduce this flow-rate requirement by recirculating return flow.

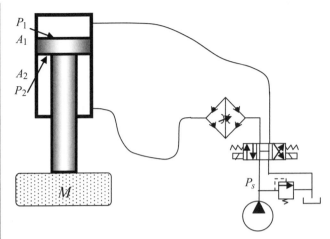

Worked Example 4.1

The lifting section is remote from the power supply unit; therefore, the system has long flexible hoses between the pump unit and the actuator. This example therefore considers pressure losses down each pipe together with directional-valve and check-valve pressure losses. Data are as follows:

Check-valve flow setting = 30 L/min
$A_1 = 0.1$ m^2, $A_2 = 0.05$ m^2, load mass $M = 20{,}000$ kg
Lines 20 m long, 13 mm diameter
Fluid viscosity $\mu = 0.025$ N s/m^2
Directional-valve port pressure drop   2 bar at 30 L/min
   8 bar at 60 L/min
Check-valve pressure drop   3 bar at 30 L/min
The load velocity in both directions is:

$$U = \frac{30 \times 10^{-3}/60}{0.05} = 0.01 \text{ m/s}.$$

Lowering

For the inlet line at 60 L/min, the pressure drop is:

$$\Delta p_{\text{inletline}} = \frac{128\mu\ell}{\pi d^4}Q = \frac{128(0.025)(20)}{\pi(0.013)^4(10^5)}10^{-3} \text{ bar} = 7.13 \text{ bar}.$$

The pressure drop across the directional-valve supply ports is 8 bar:

$$P_1 = P_s - (8 + 7.13) = P_s - 15.13.$$

For the return line at 30 L/min, the pressure drop is $7.13/2 = 3.57$ bar.
The pressure drop across two check valves is 6 bar.
Allow at least 20-bar pressure drop across the flow-control valve.
The pressure drop across the directional-valve return ports is 2 bar:

$$P_2 = 3.57 + 6 + 20 + 2 = 31.57 \text{ bar}.$$

Load force balance

$$P_1 A_1 + Mg = P_2 A_2 \rightarrow P_1 = P_2 \frac{A_2}{A_1} - \frac{Mg}{A_1},$$

$$P_s - 15.13 = \frac{31.57}{2} - \frac{20000(9.81)}{0.1(10^5)} = 15.79 - 19.62 = -3.83 \text{ bar},$$

$$P_s = 11.3 \, bar.$$

So, any sensible supply pressure will allow the load to lower at the set speed.

Lifting

The individual pressure drops are exactly the same as for lowering, the absolute values now being:

$$P_1 = (7.13 + 8) = 15.13, \text{ bar}$$
$$P_2 = P_s - (2 + 6 + 20 + 3.57) = (P_s - 31.57) \text{ bar}.$$

So, the load force equation now becomes:

$$P_1 = P_2 \frac{A_2}{A_1} - \frac{Mg}{A_1} \rightarrow 15.13 = \frac{(P_s - 31.57)}{2} - 19.62,$$

$$P_s = 101.07 \, bar.$$

A supply pressure above this value will create a larger pressure drop across the flow-control valve and is acceptable.

4.4 Drive Concepts

There are many ways of controlling motors and linear actuators, and Fig. 4.6 illustrates conceptually the most common drives using pumps and motors.

If the fluid is passed back to the tank from the motor, rather than being returned back to the pump, then it is called an open circuit, as shown in Figs. 4.6(a) and 4.6(c). If fluid is returned back to the pump, then it is known as a closed circuit, shown in Figs. 4.6(b) and 4.6(d), and can utilize a smaller tank. If machines are used with case drains, a boost pump is also necessary to make up the flow leakage back to tank; this will be discussed later. Figure 4.6(e) illustrates a power transfer unit (PTU) used to transfer fluid from the healthy power supply side to the other failing power supply side. It requires that two machines be coupled together, and both must able to operate as either a pump or a motor. The displacements must also be different and, therefore, the displacement of one machine must be changed to a value depending on whether it is acting in the pump mode or in the motor mode. When the pressure differential across the PTU is within the friction pressure range, then it will not rotate and power transfer ceases. Figure 4.7 shows a variety of drives using a servovalve to control the fluid direction of flow and may be used with either motors or cylinders.

Accurate speed and position control can be obtained using a servovalve with feedback, although motor speed feedback control may also be achieved by

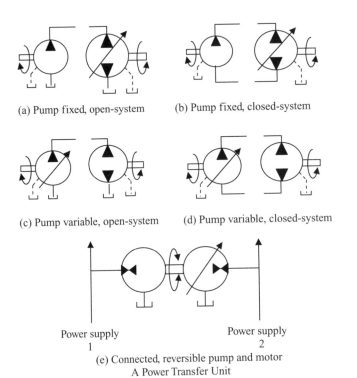

(a) Pump fixed, open-system

(b) Pump fixed, closed-system

(c) Pump variable, open-system

(d) Pump variable, closed-system

Power supply
1

Power supply
2

(e) Connected, reversible pump and motor
A Power Transfer Unit

Figure 4.6. Some different ways of combining pumps and motors.

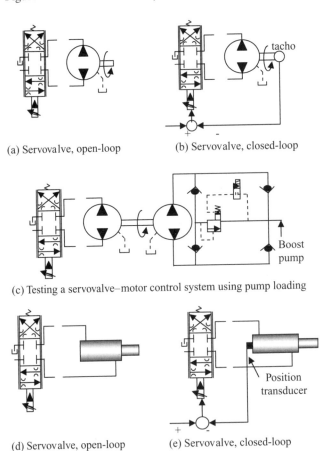

(a) Servovalve, open-loop

(b) Servovalve, closed-loop

tacho

(c) Testing a servovalve–motor control system using pump loading

Boost
pump

(d) Servovalve, open-loop

(e) Servovalve, closed-loop

Position
transducer

Figure 4.7. Motor and cylinder drives using a servovalve.

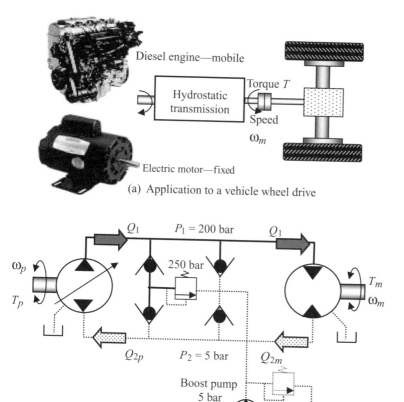

Diesel engine—mobile

Torque T

Hydrostatic transmission

Speed

ω_m

Electric motor—fixed

(a) Application to a vehicle wheel drive

Q_1 $P_1 = 200$ bar Q_1

ω_p 250 bar

T_p

T_m

ω_m

Q_{2p} $P_2 = 5$ bar Q_{2m}

Boost pump
5 bar

(b) Hydraulic circuit

Figure 4.8. A hydrostatic drive using a variable-displacement pump.

controlling a pump swash plate with an electrically operated actuator. Of course, in practice, other components are needed to ensure safe operation and system viability, as will be seen later.

4.5 Pump and Motor Hydraulically Connected: A Hydrostatic Drive

Hydrostatic drives, or hydrostatic transmissions (Fig. 4.8), are used in mobile applications, thus giving effectively an infinite gear ratio in both directions of motor–wheel rotation. The wheel motors will be axial piston or radial piston, with some variations, and it is easy to have separate wheel control if required.

The drive line is at the high pressure of 200 bar, and the return line is boosted to a low pressure of 5 bar. The protection PRV is set higher than the expected maximum working pressure; in this case, 250 bar. Normal operation therefore allows fluid to circulate from the pump to the motor and then back to the pump inlet, with boost pump flow added to make up system fluid losses.

The check-valve bridge ensures that this normal flow circulation occurs until the high pressure exceeds the main PRV setting of 250 bar; for example, because of stalling of the wheel motor during operation. The high-pressure line is then connected to the low-pressure line by the PRV that dissipates some energy while allowing recirculation of flow back to the inlet of the pump. The whole process is reversible because of the symmetry of the check-valve bridge.

The pump and motor flow rates are as previously developed in Chapter 3:

Pump   Motor

$$\text{outlet } Q_1 = D_p\omega_p - \frac{P_1}{R_p} + \frac{P_2}{R_i} \quad \rightarrow \quad \text{inlet } Q_1 = D_m\omega_m + \frac{P_1}{R_m} - \frac{P_2}{R_i}$$

$$\text{inlet } Q_2 = D_p\omega_p - \frac{P_1}{R_i} + \frac{P_2}{R_p} \quad \leftarrow \quad \text{outlet } Q_2 = D_m\omega_m + \frac{P_1}{R_i} - \frac{P_2}{R_m} \quad (4.6)$$

$$\frac{1}{R_p} = \left(\frac{1}{R_e} + \frac{1}{R_i}\right)_p, \qquad\qquad \frac{1}{R_m} = \left(\frac{1}{R_e} + \frac{1}{R_i}\right)_m,$$

$$\frac{1}{R_{ep}} = \left(\frac{D_p\omega_p}{\beta} + \frac{1}{R_{ext}} + \frac{1}{R_{tim}}\right)_p \qquad \frac{1}{R_{em}} = \left(\frac{D_m\omega_m}{\beta} + \frac{1}{R_{ext}} + \frac{1}{R_{tim}}\right)_m.$$

Equating the pump output flow rate and the motor inlet flow rate then gives:

Pump   Motor

$$D_p\omega_p - \frac{P_1}{R_p} + \frac{P_2}{R_i} = D_m\omega_m + \frac{P_1}{R_m} - \frac{P_2}{R_i}, \qquad (4.7)$$

$$D_m\omega_m = D_p\omega_p - P_1\left(\frac{1}{R_p} + \frac{1}{R_m}\right) + \frac{2P_2}{R_i}.$$

Considering the torque equations for the pump and the motor gives

$$T_p = (1 + \alpha)\,D_p(P_1 - P_2) + B_v\omega_p + T_{scp}, \qquad (4.8)$$

$$T_m = (1 - \alpha)\,D_m(P_1 - P_2) - B_v\omega_m - T_{scm}. \qquad (4.9)$$

Rearrangement of the flow rate and torque equations leads to a complicated set of functions, but the transmission performance may be evaluated with the following steps:

Step 1

$$\frac{1}{R_1} = \left(\frac{1}{R_p} + \frac{1}{R_m} - \frac{2}{R_i}\right), \qquad \frac{1}{R_2} = \left(\frac{1}{R_p} + \frac{1}{R_m}\right),$$

$$\frac{\omega_m}{\omega_p} = \frac{\dfrac{D_p}{D_m} - \dfrac{T_m}{\omega_p R_2(1-\alpha)D_m^2} - \left[\dfrac{P_2}{\omega_p R_1 D_m} + \dfrac{T_{scm}}{\omega_p R_2(1-\alpha)D_m^2}\right]}{\left[1 + \dfrac{B_v}{(1-\alpha)D_m^2 R_2}\right]}. \qquad (4.10)$$

Step 2

$$P_1 = P_2 + \frac{T_m}{(1-\alpha)D_m} + \frac{B_v\omega_p}{(1-\alpha)D_m}\left(\frac{\omega_m}{\omega_p}\right) + \frac{T_{scm}}{(1-\alpha)D_m}. \qquad (4.11)$$

Step 3

$$\frac{T_m}{T_p} = \frac{(1-\alpha)(P_1 - P_2) - \frac{B_v\omega_p}{D_m}\left(\frac{\omega_m}{\omega_p}\right) - \frac{T_{scm}}{D_m}}{(1+\alpha)(P_1 - P_2)\left(\frac{D_p}{D_m}\right) + \frac{B_v\omega_p}{D_m} + \frac{T_{scp}}{D_m}}. \tag{4.12}$$

Step 4

The transmission efficiency is given by:

$$\eta_{\text{tran}} = \frac{T_m\omega_m}{T_p\omega_p}. \tag{4.13}$$

The calculation will proceed by first establishing the displacement ratio D_p/D_m and then varying the load torque T_m. Typical calculations are covered in the following examples.

Worked Example 4.2: A Hydrostatic Transmission Performance

Assume identical machine designs such that $R_p = R_m$:

Maximum pump displacement $D_p = 4 \times 10^{-6}$ m^3/rad
Pump speed $\omega_p = 1440$ rpm
Motor displacement $D_m = 4 \times 10^{-6}$ m^3/rad
$R_e = 0.5 \times 10^{12}$ N m^{-2}/m^3 s^{-1} and dominated by external leakage R_{ext}
$R_i = 2 \times 10^{12}$ N m^{-2}/m^3 s^{-1}
$B_v = 0.02$ N m/rad s^{-1}, $\alpha = 0.05$, $T_{scp} = T_{scm} = 6$ N m
$P_2 = 5$ bar (boost)

It follows that:

$$R_p = R_m = 0.4 \times 10^{12}\,\text{N m}^{-2}/\text{m}^3\,\text{s}^{-1},$$

$$R_1 = 0.25 \times 10^{12}\,\text{N m}^{-2}/\text{m}^3\,\text{s}^{-1}, \quad R_2 = 0.2 \times 10^{12}\,\text{N m}^2/\text{m}^3\,\text{s}^{-1},$$

$$\frac{\omega_m}{\omega_p} = \frac{\frac{D_p}{D_m} - \frac{T_m}{458} - 0.0164}{1.0066},$$

$$P_1 = 5 + 2.63\,T_m + 7.94\left(\frac{\omega_m}{\omega_p}\right) + 15.79\,(\text{bar}),$$

$$\frac{T_m}{T_p} = \frac{0.95(P_1 - 5) - 7.54\left(\frac{\omega_m}{\omega_p}\right) - 15}{1.05(P_1 - 5)\left(\frac{D_p}{D_m}\right) + 7.54 + 15}.$$

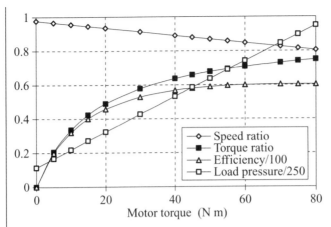

Worked Example 4.2(a)

To determine the flow make-up needed from the boost pump, consider a load pressure $P_1 = 200$ bar. This occurs at a motor load torque of 65.6 N m, gives a motor–pump speed of 0.835, and a transmission efficiency of 60.2%.

Ideal pump flow rate $D_p\omega_p = (4 \times 10^{-6})(150.8) = 36.19$ L/min
Motor ideal flow rate $0.835\, D_p\omega_p = 30.22$ L/min

The various flows are shown on the circuit diagram.

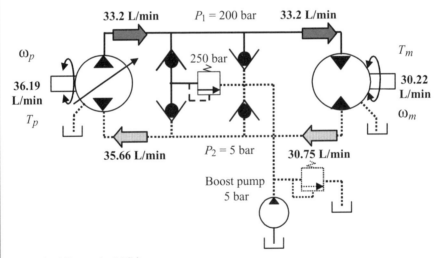

Worked Example 4.2(b)

Therefore, the make-up flow necessary from the boost pump is:

$$Q_{\text{make-up}} = 35.66 - 30.75 = 4.91 \text{ L/min}.$$

To create flow through the boost PRV and thus create the boost pressure, then set the boost pump flow rate greater than this value.

Worked Example 4.3: A Hydrostatic Drive

Load torque = 200 N m
No-load pump flow = 60 L/min
Pump and motor leakage resistance = 2.5×10^{11} N m^{-2}/m^3 s^{-1}
Motor torque loss = $9 + 0.04\omega_m$ N m
Negligible return pressure

(i) Calculate the motor displacement to give a maximum speed of 200 rpm.
(ii) Determine the line pressure.

200 rpm = 20.95 rad/s
Motor torque equation:

$$D_m P = T + T_{losses}$$
$$D_m P = 200 + 9 + 0.04\omega_m$$
$$D_m P = 210 \ N_m{}^*$$

Flow continuity between pump and motor: $Q_{po} - P/R = D_m \ \omega_m + P/R$,
$$10^{-3} = 20.95 D_m + 0.8 \times 10^{-11} P *$$

Hence, solve* for displacement and pressure:

$20.95 D_m^2 - 10^{-3} D_m + 1.68 \times 10^{-9} = 0$
let $y = 10^6 D_m$, $0.021 y^2 - y + 1.68 = 0$,
$y = 45.9$ or 1.74.
$y = 45.9$ gives $D_m = 45.9 \times 10^{-6}$ m^3/rad and $P = 45.8$ bar.
$y = 1.74$ gives $D_m = 1.74 \times 10^{-6}$ m^3/rad and $P = 1207$ bar.
So, choose the first solution:

$$D_m = 45.9 \times 10^{-6} \ m^3/\bar{r}ad, \quad P = 45.8 \ bar.$$

4.6 Pump and Motor Shaft Connected: A Power Transfer Unit (PTU)

This approach is used to ensure that pressure is maintained in a circuit in the event of an unacceptable or unexplained drop in pressure and seems to be particularly sought in aerospace applications. Another existing healthy circuit is used to supply flow to the faulty circuit in a manner that attempts to restore the faulty circuit pressure to the best possible. This, of course, requires that sufficient flow be available from the appropriate healthy circuit and also a power transfer mechanism. The power transfer approach is shown in Fig. 4.9.

For the purpose of example, in the figure, healthy circuit a provides make-up flow rate to faulty circuit b via the PTU, the left-hand side of the PTU acting as a motor and the right-hand side of the PTU acting as a pump. PTU operation can be either unidirectional or bidirectional. The PTU rotates because the net pressure differential across the motor unit is greater than the net pressure differential across the pump unit. This creates a torque unbalance beyond the friction value when one circuit pressure changes due to a fault condition. In practice, a variable-displacement axial piston machine is used together with a bent-axis fixed-displacement piston machine. The machine acting as a motor must have a displacement greater than

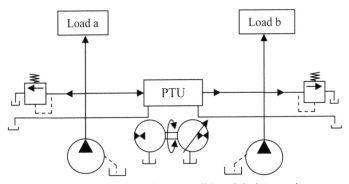

Figure 4.9. Power transfer using reversible axial piston units.

the machine acting as a pump. Consequently, as the direction of power transfer is changed, the swash-plate stroke of the variable displacement machine must also be changed. Consider Figure 4.10.

Power transfer is from left to right because of a pressure drop in supply line 2 relative to supply line 1. Therefore, the machine at the left-hand side is acting as a motor and the machine at the right-hand side is acting as a pump. Applying the previously established flow rate and torque equations for a motor and a pump, and assuming similar loss characteristics for each machine, gives:

$$\text{PTU flow rates,} \quad \text{motor,} \quad Q_1 = D_m\omega + \frac{P_1}{R_{mp}}, \tag{4.14}$$

$$\text{pump,} \quad Q_2 = D_p\omega - \frac{P_2}{R_{mp}}, \tag{4.15}$$

$$\text{PTU torque,} \quad D_m P_1 - D_p P_2 = B_v\omega + T_{sc}. \tag{4.16}$$

Here, the viscous coefficient B_v and the friction torque T_{sc} is the sum for both machines. Considering each identical supply pump,

$$Q_{p1} = Q_{po} - \frac{P_1}{R_{ps}}, \tag{4.17}$$

$$Q_{p2} = Q_{po} - \frac{P_2}{R_{ps}}. \tag{4.18}$$

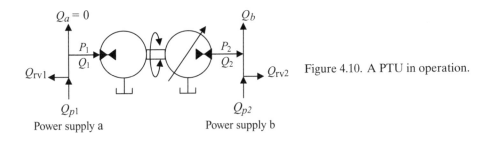

Figure 4.10. A PTU in operation.

Assuming that both identical PRVs are in operation on both power supply sides, then a linear flow-rate–pressure-drop characteristic may be used to give:

$$Q_{rv1} = \frac{P_1 - P_{rv}}{R_v} \quad P_1 > P_{rv}, \tag{4.19}$$

$$Q_{rv2} = \frac{P_2 - P_{rv}}{R_v} \quad P_2 > P_{rv}, \tag{4.20}$$

where R_v is the resistance of the PRV. For the purpose of identifying unique features of a PTU performance, *a suddenly demanded flow rate Q_b at side 2 is used* and the load flow rate at side 1 is zero. Therefore, flow continuity on each side from Fig. 4.10 gives:

$$Q_{p1} = Q_1 + Q_{rv1}, \tag{4.21}$$

$$Q_{p2} = Q_b - Q_2 + Q_{rv2}. \tag{4.22}$$

Inserting the flow equations defined in the previous equations then gives:

$$P_1 = R_t \left(Q_{po} - D_m\omega + \frac{P_{rv}}{R_v} \right), \tag{4.23}$$

$$P_2 = R_t \left(Q_{po} + D_p\omega + \frac{P_{rv}}{R_v} - Q_b \right), \tag{4.24}$$

$$\frac{1}{R_t} = \frac{1}{R_{ps}} + \frac{1}{R_{mp}} + \frac{1}{R_v}. \tag{4.25}$$

Also, because it has already been stated that a PTU performs better if the pump displacement is less than the motor displacement, then let:

$$D_p = \varepsilon D_m, \quad \varepsilon < 1. \tag{4.26}$$

Combining the torque and flow equations and the displacement ratio equation then gives:

$$\overline{\omega} = \frac{(\overline{P}_{rv} + 1)(1 - \varepsilon) + \varepsilon\overline{Q}_b - \overline{T}_{sc}}{(1 + \varepsilon^2 + \overline{B}_v)}, \tag{4.27}$$

$$\overline{P}_1 = \frac{P_1}{P_{rv}} = \frac{R_t}{R_v} \left[\frac{(1 - \overline{\omega})}{\overline{P}_{rv}} + 1 \right], \tag{4.28}$$

$$\overline{P}_2 = \frac{P_2}{P_{rv}} = \frac{R_t}{R_v} \left[\frac{(1 + \varepsilon\overline{\omega} - \overline{Q}_b)}{\overline{P}_{rv}} + 1 \right], \tag{4.29}$$

$$\frac{R_v}{R_t} = \frac{R_v}{R_{ps}} + \frac{R_v}{R_{mp}} + 1, \quad \overline{\omega} = \frac{D_m\omega}{Q_{po}} \quad \overline{T}_{sc} = \frac{T_{sc}}{D_m R_t Q_{po}},$$

$$\overline{P}_{rv} = \frac{P_{rv}}{R_v Q_{po}}, \quad \overline{Q}_b = \frac{Q_b}{Q_{po}}, \quad \overline{B}_v = \frac{B_v}{R_t D_m^2}. \tag{4.30}$$

The individual pressures are defined with respect to the PRV cracking pressure P_{rv} so that a check can be made on whether a particular design causes any pressure to fall below the PRV setting.

PTU operation ceases when the driving differential pressure is within the friction dead-band, which is defined as:

$$\overline{P}_1 - \varepsilon\overline{P}_2 = \pm\frac{\overline{T}_{sc}}{\overline{P}_{rv}}. \tag{4.31}$$

Worked Example 4.4: Characteristics Determination

Consider that both machines have the following data:

PTU machines, $R_{mp} = 10^{12}$ N m^{-2}/m^3 s^{-1}
Supply pumps, $R_{ps} = 10^{12}$ N m^{-2}/m^3 s^{-1}
PRVs, $R_v = 0.25 \times 10^{10}$ N m^{-2}/m^3 s^{-1}
PRV cracking pressure, $P_{rv} = 210$ bar
Supply pumps no-load flow rate, $Q_{po} = 24$ L/min
Fixed-displacement machine displacement, $D_m = 4 \times 10^{-6}$ m^3/rad
PTU total friction torque, $T_{sc} = 12$ N m
PTU total viscous friction coefficient, $B_v = 0.04$ N m/rad s^{-1}

$$\frac{1}{R_t} = \frac{1}{R_{ps}} + \frac{1}{R_{mp}} + \frac{1}{R_v} = 2 \times 10^{-12} + 400 \times 10^{-12} = 402 \times 10^{-12},$$

$$R_t \cong R_v = 0.25 \times 10^{10}\,\text{N m}^{-2}/\text{m}^3\,\text{s}^{-1}.$$

The PRV resistance is clearly dominant. For no-load flows from each power supply, the pressure increase across each PRV is given by:

$$R_v Q_{po} = (0.25 \times 10^{10})(0.4 \times 10^{-3}) = 10\,\text{bar}.$$

So, if the PRV cracking pressure is 210 bar, the operating pressure for no-load is not 220 bar but ≈ 219 bar when the PTU losses are included:

$$\frac{T_{sc}}{D_m} = \frac{12}{4 \times 10^{-6}} = 30\,\text{bar}, \quad \overline{T}_{sc} = \frac{T_{sc}}{D_m R_t Q_{po}} = \frac{30}{10} = 3,$$

$$\overline{P}_{rv} = \frac{P_{rv}}{R_v Q_{po}} = \frac{210}{10} = 21, \quad \overline{B}_v = \frac{B_v}{R_t D_m^2} = \frac{0.04}{(0.25 \times 10^{10})(4 \times 10^{-6})^2} = 1.$$

To fully understand how a PTU operates, it is necessary to consider particular conditions of speed and pressure and their probable ranges.

The Condition for Zero Speed

From Eq. (4.27), this occurs when the following condition is satisfied:

$$\varepsilon = \frac{\overline{P}_{rv} + 1 - \overline{T}_{sc}}{\overline{P}_{rv} + 1 - \overline{Q}_b}. \tag{4.32}$$

The Condition for Each Pressure to Fall to Its PRV Setting

To understand the pressure behavior, it is noted that because it is probable that $R_t \approx R_v$, then the pressures are given by:

$$\frac{P_1}{P_{rv}} \approx \frac{(1 - \overline{\omega})}{\overline{P}_{rv}} + 1, \quad \frac{P_2}{P_{rv}} \approx \frac{(1 + \varepsilon\overline{\omega} - \overline{Q}_b)}{\overline{P}_{rv}} + 1. \tag{4.33}$$

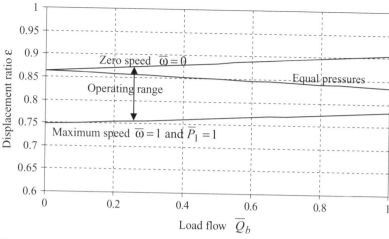

Figure 4.11. Determination of PTU displacement ratio range.

Therefore, the condition for each pressure to fall to the PRV setting is given by:

$$\frac{P_1}{P_{rv}} = 1 \quad \text{when} \quad \overline{\omega} = 1.$$

This gives $\varepsilon^2 + \varepsilon(\overline{P}_{rv} + 1 - \overline{Q}_b) - (\overline{P}_{rv} - \overline{B}_v - \overline{T}_{sc}) = 0,$　　(4.34)

$$\frac{P_2}{P_{rv}} = 1 \text{ when } 1 + \varepsilon\overline{\omega} = \overline{Q}_b \text{ and not possible.}$$　　(4.35)

The Condition for Equal Pressures

It seems good design sense to aim for equal pressures in each circuit because by definition, they must then be above the PRV setting. This condition is achieved when:

$$\overline{P}_1 = \overline{P}_2 \quad \text{when} \quad \overline{\omega} = \frac{\overline{Q}_b}{(1 + \varepsilon)},$$

$$\varepsilon^2(\overline{P}_{rv} + 1) + \varepsilon(\overline{T}_{sc} - \overline{Q}_b) + \overline{T}_{sc} + \overline{Q}_b(1 + \overline{B}_v) - \overline{P}_{rv} - 1 = 0.$$　　(4.36)

Considering Worked Example 4.4 data, it was deduced that $\overline{T}_{sc} = 3$, $\overline{P}_{rv} = 21$, $\overline{B}_v = 1$. The operating conditions are shown in Fig. 4.11 for various load flows.

From Fig. 4.11, the following points may be observed:

- The displacement ratio range is restricted overall to typically $0.75 < \varepsilon < 0.91$ but depends on the load flow rate to be supplied.
- For equal-pressure operation, the displacement ratio variation is not highly significant as the load flow changes, suggesting in this example that a value of $\varepsilon \approx 0.85$ would be acceptable. Considering the zero-load flow point, it follows from Eq. (4.32), and recalling that probably $\overline{P}_{rv} \gg 1$, that a good starting design is met by selecting a displacement difference given by:

$$D_m - D_p \approx \frac{T_{sc}}{P_{rv}}.$$　　(4.37)

This is an interesting result in that it shows that the friction level directly affects the displacement ratio required, but a change in PRV setting will also require a change in displacement ratio.

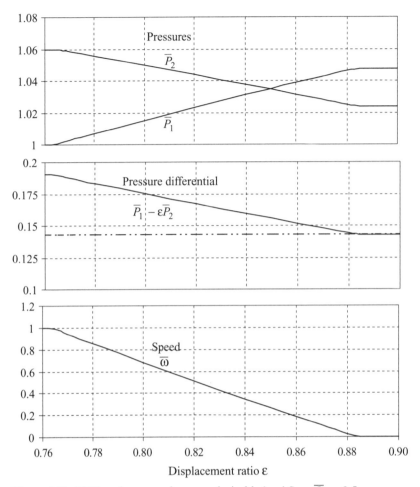

Figure 4.12. PTU performance for an undesirable load flow $\overline{Q}_b = 0.5$.

Figure 4.12 shows the effect of changing the displacement ratio for a particular undesirable load flow of $\overline{Q}_b = 0.5$. The following points may be observed:

- A displacement ratio beyond 0.88 will not allow the PTU to rotate and power transfer will not exist.
- As the displacement ratio decreases below 0.88, the pressure differential increases and the PTU speed increases. The supply pressure P_1 continually decreases and the pressure to be compensated P_2 increases as required.
- As the displacement ratio decreases further, the speed rises to its maximum when the supply pressure P_1 reaches the PRV setting and the analysis is then invalid.
- A lower displacement ratio is therefore better to provide the driving pressure differential beyond the friction dead-band.
- The PTU speed decreases linearly with increasing displacement ratio.

Consider data from aircraft applications and, in particular, the aerospace-recommended practice document (SAE ARP1280, Society of Automotive Engineers International, Warrendale, PA, 2007). Table 4.1 shows some of the data presented.

Table 4.1. *Data taken on PTUs used in different aircraft (SAE, 2007)*

Aircraft	D_m (10^{-6} m³/s)	D_p (10^{-6} m³/s)	Displacement ratio ε	Motor pressure (bar)	Pump pressure (bar)
		Unidirectional, fixed displacement			
DC-10/ND-11	1.23	1.13	0.92	200	179
757	4.00	3.65	0.91	172	150
Gulfstream 11	1.73	1.57	0.91	207	200
A-300	4.00	3.46	0.87	207	207
767	0.25	0.21	0.84	112	86
727, 747	0.25	0.21	0.84	207	207
737	0.81	0.63	0.78	169	166
		Bidirectional, variable displacement			
DC-10/MD-11	5.00	4.46–5.50	0.89–1.10	207	193
C-17A	3.15	2.62–3.66	0.83–1.16	275	255
A-320	2.10	1.57–2.62	0.75–1.25	207	200

The displacement ratio can vary from $0.78 \leq \varepsilon \leq 0.92$ for unidirectional operation with fixed displacements and from $0.75 \leq \varepsilon \leq 0.89$ for bidirectional operation with variable displacements. The theory presented gives results typical of this range with $0.75 \leq \varepsilon \leq 0.88$. For bidirectional operation, a swash-plate adjusting control system is required for changing the displacement ratio of the in-line axial piston unit, depending on the direction of power transfer (Fig. 4.13).

The displacement ratio would preferably be placed around the neutral position, or within the friction dead-band, when no PTU action is needed. Hydromechanical control is usually adopted and integral with the variable displacement unit to change the swash-plate position in the correct direction, as determined by the appropriate load condition.

4.7 Servovalve–Motor Open-Loop and Closed-Loop Speed Drives

4.7.1 Open-Loop Control

A servovalve coupled to a motor is a common method of motor speed control in practice, in which both load pressure and speed change during operation. It is also

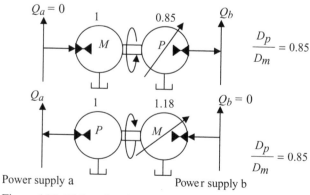

Figure 4.13. Bidirectional operation of a PTU with displacement control.

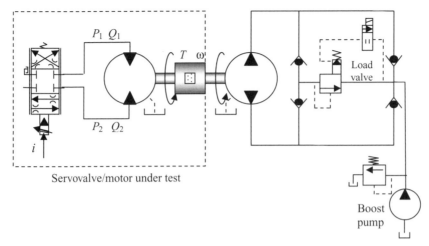

Figure 4.14. A servovalve–motor open-loop drive under test.

a particularly useful approach to determine the motor flow and leakage character-istics by varying the load and, hence, the motor pressure differential, and main-taining the motor speed constant. Initial work on pump–motor loss characteristics has developed essentially by means of extensive experimental work combined with a pragmatic approach to the forms and mathematical representation of each flow and torque loss term. Figure 4.14 shows the open-loop circuit and the pump–relief valve/check-valve bridge–flow make-up load circuit that is able to load the motor in either direction of rotation.

Both pressure gauges and pressure transducers were used to measure the load pressures P_1 and P_2, and gear-type flow meters were used to measure the load flow rates Q_1 and Q_2. A torque–speed transducer unit with electronic display was con-nected between the motor and load pump using a flexible coupling, and load was applied by switching in the load valve and manually adjusting the variable PRV set-ting. Thus, loading the pump loads the motor, and the complete performance can be measured. The advantage of having a load valve with integral solenoid off-loading is that the dynamic performance can be assessed by rapidly switching the load on and off at any load pressure setting.

Chapter 3 considered a critically lapped servovalve and the motor system being evaluated here and analyzed the steady-state characteristics in detail. The equations for the servovalve and motor are now brought together:

$$Q_1 = k_f i \sqrt{P_s - P_1} = D_m \omega + P_1 \left(\frac{1}{R_e} + \frac{1}{R_i} \right) - \frac{P_2}{R_i}, \qquad (4.38)$$

$$Q_2 = k_f i \sqrt{P_2} = D_m \omega + \frac{P_1}{R_i} - P_2 \left(\frac{1}{R_e} + \frac{1}{R_i} \right), \qquad (4.39)$$

$$D_m(P_1 - P_2) = T_m + T_{\text{losses}}. \qquad (4.40)$$

The losses combine all the pressure, speed, and friction terms previously discussed for a motor:

$$T_{\text{losses}} = \alpha D_m(P_1 - P_2) + B_v + T_{sc}. \qquad (4.41)$$

The equations as they stand may only be solved numerically. As discussed in Chapter 3, the sum of line pressures is close to supply pressure, particularly as the motor load is increased. Therefore, the use of a mean flow-rate equation based on the assumption that $P_1 + P_2 \approx P_s$ is sufficiently accurate (Watton, 2006):

$$\frac{(Q_1 + Q_2)}{2} \approx k_f i \sqrt{\frac{P_s}{2}} \sqrt{1 - \frac{(P_1 - P_2)}{P_s}}. \tag{4.42}$$

From Chapter 3, or Eqs. (4.38) and (4.39), the mean flow rate evaluated for the motor becomes:

$$\frac{(Q_1 + Q_2)}{2} = D_m \omega + \frac{(P_1 - P_2)}{R_m}, \tag{4.43}$$

$$\frac{1}{R_m} = \frac{1}{R_i} + \frac{1}{2R_e}.$$

Equating Eqs. (4.42) and (4.43) then gives:

$$k_f i \sqrt{\frac{P_s}{2}} \sqrt{1 - \frac{(P_1 - P_2)}{P_s}} = D_m \omega + \frac{(P_1 - P_2)}{R_m}. \tag{4.44}$$

Defining the no-load speed $\omega(0)$ when $P_{\text{load}} = P_1 - P_2 = 0$ and using nondimensional notation then gives the expression for motor speed as:

$$\overline{\omega} = \sqrt{1 - \overline{P}_{\text{load}}} - \alpha \overline{P}_{\text{load}},$$

$$\overline{\omega} = \frac{\omega}{\omega(0)}, \quad \omega(0) = \frac{k_f i \sqrt{\frac{P_s}{2}}}{D_m}, \quad \overline{P}_{\text{load}} = \frac{(P_1 - P_2)}{P_s}, \quad \alpha = \frac{P_s}{D_m R_m \omega(0)}. \tag{4.45}$$

This then allows the derivation of other important system properties:

$$\text{efficiency, } \eta = (\overline{P}_{\text{load}} - \overline{T}_{\text{losses}}) \left(1 - \frac{\alpha \overline{P}_{\text{load}}}{\sqrt{1 - \overline{P}_{\text{load}}}} \right), \tag{4.46}$$

$$\text{power transfer to motor } \overline{W}_m = \overline{P}_{\text{load}} \sqrt{1 - \overline{P}_{\text{load}}} + \varepsilon, \tag{4.47}$$

$$\overline{T}_{\text{losses}} = \frac{T_{\text{losses}}}{P_s D_m}, \quad \overline{W}_m = \frac{W_m}{P_s D_m \omega(0)}, \quad \varepsilon = \frac{P_s}{2 D_m \omega(0) R_e}. \tag{4.48}$$

Consider the following data:

- Shell Tellus ISO 32 mineral oil at 40°C
- $D_m = 1.68 \times 10^{-6}$ m^3/rad, $P_s = 100$ bar
- $R_e = 1.28 \times 10^{12}$ N m^{-2}/m^3 s^{-1}, $R_i = 4.2 \times 10^{12}$ N m^{-2}/m^3 s^{-1}
- $R_m = 1.59 \times 10^{12}$ N m^{-2}/m^3 s^{-1}
- Motor torque losses $\overline{T}_{\text{losses}} = 0.12$ at 234 rpm and 0.18 at 903 rpm

These values barely change with load pressure at the two test conditions. It is deduced that $\alpha = 35.7/N(0)$, $\varepsilon = 22.2/N(0)$, where $N(0)$ is the no-load speed (in revolutions per minute).

A comparison between the approximate theory and measurements is shown in Fig. 4.15, from which the usefulness of the approximate theory may be deduced. The speed behavior and the power-transferred predictions are particularly good, with the efficiency comparisons improving with increased load pressure differential. Over the working range of the drive, the theory is sufficient for design purposes.

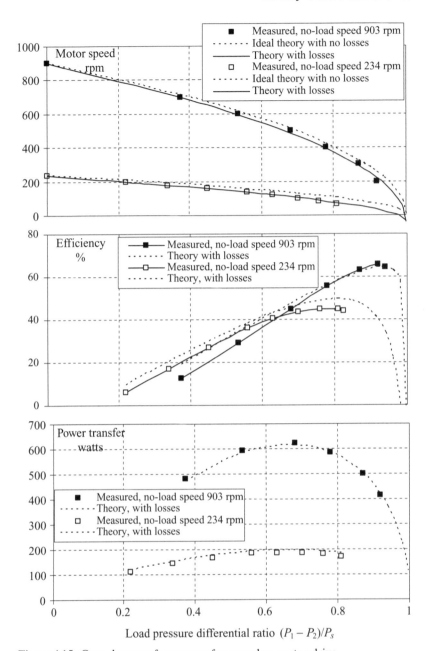

Figure 4.15. Open-loop performance of servovalve–motor drive.

4.7.2 Closed-Loop Control

Now consider the closed-loop drive with speed feedback as shown in Fig. 4.16.
    The two defining equations now become:

$$k_f i \sqrt{\frac{P_s - P_{\text{load}}}{2}} = D_m \omega + \frac{P_{\text{load}}}{R_m}, \tag{4.49}$$

$$i = G_a(V_d - H_t \omega), \qquad P_{\text{load}} = P_1 - P_2. \tag{4.50}$$

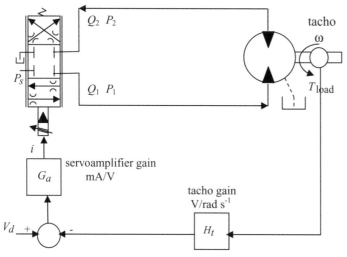

Figure 4.16. A servovalve–motor closed-loop drive.

Recalling that a reference speed $\omega(o)$ is defined as the speed when the pressure difference $P_{\text{load}} = 0$, then the closed-loop speed may be written as:

$$\frac{\omega}{\omega(0)} = \overline{\omega} = \frac{(1 + K)\sqrt{1 - \overline{P}_{\text{load}}} - \alpha \overline{P}_{\text{load}}}{1 + K\sqrt{1 - \overline{P}_{\text{load}}}},$$

$$K = \frac{k_f G_a H_t}{D_m}\sqrt{\frac{P_s}{2}}, \qquad \overline{P}_{\text{load}} = \frac{P_1 - P_2}{P_s}, \qquad \alpha = \frac{P_s}{D_m R_m \omega(0)}. \qquad (4.51)$$

Figure 4.17 shows the theory and experimental comparisons for the system with speed feedback.

Because of the minimum pressure differential available in practice, the no-load speed is difficult to assess for each test condition. A reference condition at a common speed of 500 rpm and a common load pressure of 30 bar has been used in Fig. 4.17, resulting in slightly different no-load speeds and, hence, slightly different values of α. The improved closed-loop characteristic can be clearly seen over

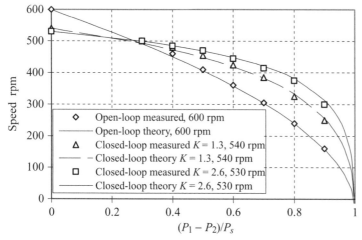

Figure 4.17. Closed-loop performance of a servovalve–motor drive and a comparison with its open-loop performance.

the lower load pressure range; however, at the higher gain of $K = 2.6$, noticeable yet small oscillations in pressures occurred, indicating that the condition for closed-loop instability was being approached. This is a dangerous condition, resulting in piston slipper bounce on the swash plate and clearly audible.

Efficiency and power transfer to motor may then be derived as:

$$\text{efficiency, } \eta = \frac{(\overline{P}_{\text{load}} - \overline{T}_{\text{losses}})\left(1 - \dfrac{\alpha\overline{P}_{\text{load}}}{(1+K)\sqrt{1 - \overline{P}_{\text{load}}}}\right)}{\left[1 + \dfrac{\alpha K\overline{P}_{\text{load}}}{(K+1)}\right]}, \tag{4.52}$$

$$\text{power transfer to motor, } \overline{W}_m = \overline{P}_{\text{load}}\left(\frac{(1+K)\sqrt{1 - \overline{P}_{\text{load}}} - \alpha\overline{P}_{\text{load}}}{1 + K\sqrt{1 - \overline{P}_{\text{load}}}}\right)$$
$$+ \alpha\overline{P}_{\text{load}}^2 + \varepsilon. \tag{4.53}$$

The condition for maximum efficiency with respect to the load pressure differential is given implicitly by:

$$\alpha^2\overline{P}_{\text{load}}^2\left[\sqrt{1 - \overline{P}_{\text{load}}} - \frac{K(\overline{P}_{\text{load}} - \overline{T}_{\text{losses}})}{2}\right]$$
$$- \alpha(1+K)\left\{2\overline{P}_{\text{load}}(1 - \overline{P}_{\text{load}}) + (\overline{P}_{\text{load}} - \overline{T}_{\text{losses}})\left[K(1 - \overline{P}_{\text{load}})^{3/2} + 1 - \frac{\overline{P}_{\text{load}}}{2}\right]\right\}$$
$$+ (1+K)^2(1 - \overline{P}_{\text{load}})^{3/2} = 0. \tag{4.54}$$

This cannot be represented in a general graphical way because of the various parameters K, $\overline{T}_{\text{losses}}$, α required.

It is sometimes useful to determine the condition to transfer maximum power to the load with respect to the load pressure differential. This is given, also implicitly, by:

$$\alpha = \left[\frac{1+K}{K}\right]\left[\frac{2K(1 - \overline{P}_{\text{load}})^{3/2} - (3\overline{P}_{\text{load}} - 2)}{\overline{P}_{\text{load}}(5 - 4\overline{P}_{\text{load}}) - 4K\overline{P}_{\text{load}}(1 - \overline{P}_{\text{load}})^{3/2}}\right]. \tag{4.55}$$

For the lossless case, then $\alpha = 0$ and (4.55) becomes:

$$\overline{P}_{\text{load}} = \frac{2}{3} + \frac{2}{3}K(1 - \overline{P}_{\text{load}})^{3/2}. \tag{4.56}$$

For the open-loop case, $K = 0$, the solution reduces to one that is well known whereby maximum power is transferred to the motor when the load pressure differential is two-thirds the supply pressure. Validation of closed-loop efficiency and power transfer for specific loss characteristics is shown in Fig. 4.18.

In general, the maximum efficiency and power transfer to the motor both occur at a higher pressure differential than for the open-loop system. However, the restriction of the theoretical assumptions at a very high pressure differential should be recalled.

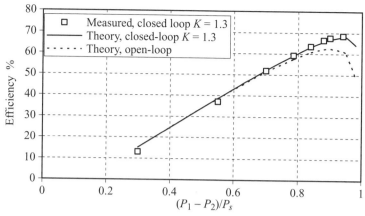

(a) Efficiency $\overline{T}_{\text{losses}} = 0.14$, $K = 1.3$, no-load speed 548 rpm

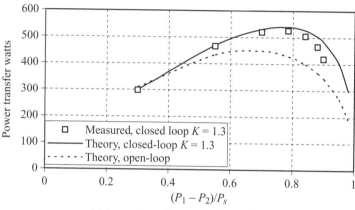

(b) Power transfer, no-load speed 548 rpm

Figure 4.18. Closed-loop efficiency and power transfer for a servovalve–motor drive.

4.8 Servovalve–Linear Actuator

Consider the basic open-loop system for an actuator with a single rod. The established servovalve flow equations, assuming a critically lapped spool, may be used to generate two flow-continuity equations for the system and for two cases of extending and retracting.

4.8.1 Extending

See Fig. 4.19 for an illustration of the discussion in this subsection.
Equating flow rates gives:

$$Q_1 = k_f i \sqrt{P_s - P_1} = A_1 U. \tag{4.57}$$

$$Q_2 = k_f i \sqrt{P_2} = A_2 U. \tag{4.58}$$

The force equation is:

$$P_1 A_1 - P_2 A_2 = F. \tag{4.59}$$

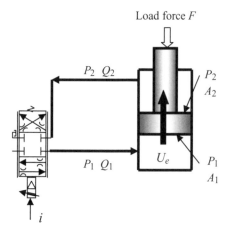

Figure 4.19. A servovalve-controlled actuator open-loop system, extending.

The pressures may be then determined by eliminating the velocity from Eqs. (4.57) and (4.58) and combining with Eq. (4.59) to give:

$$P_1 = \frac{P_s + \gamma^2 P_{\text{load}}}{(1 + \gamma^3)}, \qquad P_2 = \frac{\gamma P_s - P_{\text{load}}}{(1 + \gamma^3)}, \qquad (4.60)$$

$$U_e = \frac{k_f i}{A_2} \sqrt{\frac{\gamma P_s - P_{\text{load}}}{(1 + \gamma^3)}}, \qquad P_{\text{load}} = \frac{F}{A_2}, \qquad \gamma = \frac{A_1}{A_2}. \qquad (4.61)$$

It will be deduced that motion stops when $F = P_s A_1$ with $P_1 = P_s$ and $P_2 = 0$.

4.8.2 Retracting

See Fig. 4.20 for an illustration of the discussion in this subsection.
Equating flow rates gives:

$$Q_1 = k_f i \sqrt{P_1} = A_1 U. \qquad (4.62)$$

$$Q_2 = k_f i \sqrt{P_s - P_2} = A_2 U. \qquad (4.63)$$

The force equation is:

$$P_1 A_1 - P_2 A_2 = F. \qquad (4.64)$$

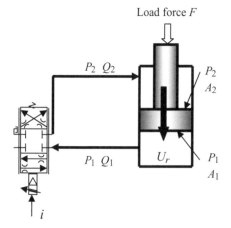

Figure 4.20. A servovalve-controlled actuator open-loop system, retracting.

The pressures may be then determined by eliminating the velocity from Eqs. (4.62) and (4.63) and combining with Eq. (4.64) to give:

$$P_1 = \frac{\gamma^2 P_s + \gamma^2 P_{\text{load}}}{(1+\gamma^3)}, \qquad P_2 = \frac{\gamma^3 P_s - P_{\text{load}}}{(1+\gamma^3)}, \tag{4.65}$$

$$U_r = \frac{k_f i}{A_2} \sqrt{\frac{P_s + P_{\text{load}}}{(1+\gamma^3)}}, \qquad P_{\text{load}} = \frac{F}{A_2}, \qquad \gamma = \frac{A_1}{A_2}. \tag{4.66}$$

It will be deduced that motion stops when $F = -P_s A_2$, with $P_2 = P_s$ and $P_1 = 0$. This condition, therefore, occurs for the load, acting to cause runaway.

Worked Example 4.5

A servovalve is rated at 38L/min at a rated current of 15 mA and a total valve pressure drop of 70 bar. The servovalve is coupled to a cylinder that is used to raise a platform as shown, and the platform mass may be assumed to act at the platform center. Data are as follows:

platform mass = 2000 kg
actuator bore = 50-mm diameter
actuator rod diameter = 35 mm
supply pressure = 140 bar

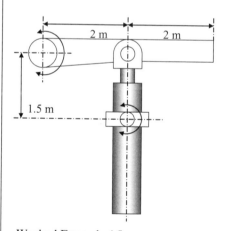

Worked Example 4.5

Determine the actuator speed when lifting if a servovalve current of 5 mA is applied when:

(i) the platform is horizontal,
(ii) the platform is just vertical, and also
(iii) estimate the actuation time.

First determine the servovalve flow constant k_f and other data:

$$Q_{rated} = k_f i_{rated} \sqrt{\frac{70\,bar}{2}}, \qquad 38 = k_f 15\sqrt{35}, \qquad k_f = 0.428,$$

$$A_1 = \frac{\pi \times 50^2 \times 10^{-6}}{4} = 1.96 \times 10^{-3}\,m^2,$$

$$A_2 = \frac{\pi(50^2 - 35^2) \times 10^{-6}}{4} = 1.00 \times 10^{-3}\,m^2, \quad \gamma = \frac{A_1}{A_2} = 1.96.$$

(i) *Platform horizontal*: $F = 2000\,g$ N, $P_{load} = F/A_2 = 196.2$ bar;

$$P_1 = \frac{P_s + \gamma^2 P_{load}}{(1 + \gamma^3)} = \frac{140 + 3.84 \times 196.2}{8.53} = 104.7\,bar,$$

$$P_2 = \frac{\gamma P_s - P_{load}}{(1 + \gamma^3)} = \frac{1.96 \times 140 - 196.2}{8.53} = 9.2\,bar,$$

$$Q_2 = k_f i \sqrt{P_2} = 0.428 \times 5\sqrt{9.2} = 6.49\,L/min,$$

$$U_e = \frac{Q_2}{A_2} = \frac{6.49 \times 10^{-3}}{60 \times 10^{-3}} = 0.108\,m/s.$$

(ii) *Platform vertical*: $F = 0$, $P_{load} = 0$;

$$P_1 = \frac{P_s}{(1 + \gamma^3)} = \frac{140}{8.53} = 16.4\,bar,$$

$$P_2 = \frac{\gamma P_s}{(1 + \gamma^3)} = \frac{1.96 \times 140}{8.53} = 32.2\,bar,$$

$$Q_2 = k_f i \sqrt{P_2} = 0.428 \times 5\sqrt{32.2} = 12.14\,L/min,$$

$$U_e = \frac{Q_2}{A_2} = \frac{12.14 \times 10^{-3}}{60 \times 10^{-3}} = 0.202\,m/s.$$

The actuator speed has almost doubled as the platform rotates to its vertical position.

(iii) *Actuation time*: The total travel of the actuator is 2.53 m; assume a first approximation to estimate the time for motion by considering a mean speed:

$$U_{mean} \rightarrow \frac{(0.108 + 0.202)}{2} = 0.155\,m/s,$$

$$t \rightarrow \frac{2.53}{0.155} = 16.32\,s.$$

However, the time will be between limits of 12.52 s and 23.43 s when either the maximum or minimum speeds are considered.

4.8.3 A Comparison of Extending and Retracting Operations

Selecting an arbitrary area ratio $\gamma = 2$ allows Fig. 4.21 to be constructed, illustrating how the pressures vary with load force for both extending and retracting operations. Both pressures are referenced to supply pressure.

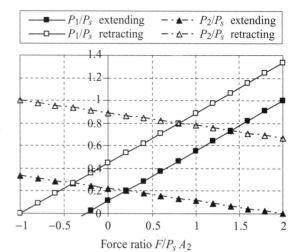

Figure 4.21. Variation in actuator pressure, area ratio $\gamma = A_1/A_2 = 2$.

It can be seen that both actuator pressures are greater for the retracting case than for the extending case. The ratio extending velocity divided by retracting velocity U_e/U_r is shown in Fig. 4.22, from which where it is deduced that the extending velocity is always greater than the retracting velocity.

An interesting solution occurs for the case when the extending velocity is equal to the retracting velocity for the same servovalve current. From Eqs. (4.61) and (4.66), this occurs when:

$$\gamma P_s - P_{\text{load}} = P_s + P_{\text{load}},$$
$$F = \frac{P_s A_{\text{rod}}}{2}, \tag{4.67}$$

where $A_{\text{rod}} = A_1 - A_2$ is the cross-sectional area of the actuator rod.

Therefore, the flow gain of the open-loop system can be made effectively the same for both directions of motion, thus aiding closed-loop control of a single-rod actuator by ensuring that Eq. (4.67) is satisfied. Also, the power transfer to the load is the same in both directions if Eq. (4.67) is satisfied. In addition, the power transfer increases as the area ratio $\gamma > 1$ increases for the same servovalve current and supply pressure. Of course, the power transfer is zero for a double-rod actuator because equal velocities in both directions at the same servovalve current can be achieved only for zero-load $F = 0$.

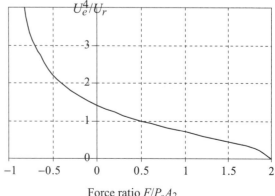

Figure 4.22. Variation in the velocity ratio U_e/U_r, area ratio $\gamma = A_1/A_2 = 2$.

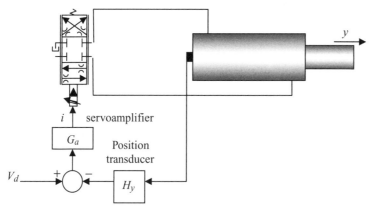

Figure 4.23. Closed-loop electrohydraulic position control.

Notice also from Fig. 4.21 that the pressure differential across the actuator can change sign as the load is increased. Any leakage present across the actuator piston will therefore also change the direction of action. The condition for $P_1 > P_2$ occurs when:

$$P_1 > P_2,$$

$$\text{extending,} \quad \frac{F}{P_s A_2} > \frac{(\gamma - 1)}{(\gamma^2 + 1)},$$

$$\text{retracting} \quad \frac{F}{P_s A_2} > \frac{\gamma^2(\gamma - 1)}{(\gamma^2 + 1)}. \tag{4.68}$$

Clearly, the load force is highest for the retracting condition.

4.9 Closed-Loop Position Control of an Actuator by a Servovalve with a Symmetrically Underlapped Spool

Consider a closed-loop position control system for the general case of a single-rod actuator, as shown in Fig. 4.23.

Recall the flow equations for a symmetrically underlapped spool, as discussed in Chapter 3 and the load force equation:

$$Q_1 = k_f(i_u + i)\sqrt{P_s - P_1} - k_f(i_u - i)\sqrt{P_1}, \tag{4.69}$$

$$Q_2 = k_f(i_u + i)\sqrt{P_2} - k_f(i_u - i)\sqrt{P_s - P_2}, \tag{4.70}$$

$$P_1 A_1 - P_2 A_2 = F. \tag{4.71}$$

Under closed-loop position control, $Q_1 = Q_2 = 0$; from Eqs. (4.69) and (4.70), this leads to the conclusion that the sum of pressures must be equal to supply pressure:

$$\overline{P}_1 + \overline{P}_2 = 1, \tag{4.72}$$

$$\overline{P}_1 = \frac{(1 + \overline{F})}{(1 + \gamma)}, \quad \overline{P}_2 = \frac{(\gamma - \overline{F})}{(1 + \gamma)}, \tag{4.73}$$

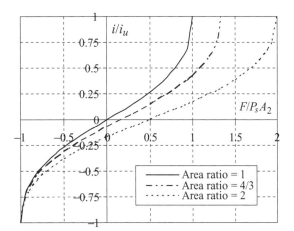

Figure 4.24. Servovalve current under position control with spool symmetrical underlap.

$$\bar{i} = \frac{\sqrt{1 + \overline{F}} - \sqrt{\gamma - \overline{F}}}{\sqrt{1 + \overline{F}} + \sqrt{\gamma - \overline{F}}}, \tag{4.74}$$

$$\overline{P}_1 = \frac{P_1}{P_s}, \quad \overline{P}_2 = \frac{P_2}{P_s}, \quad \overline{F} = \frac{F}{P_s A_2}, \quad \bar{i} = \frac{i}{i_u} = \frac{G_a H_y y_{error}}{i_u}. \tag{4.75}$$

It then follows that the position error can be driven to zero from Eq. (4.74), providing the load force satisfies the condition:

$$\overline{F} = \frac{(\gamma - 1)}{2} \rightarrow F = \frac{P_s A_{rod}}{2}, \tag{4.76}$$

where $A_{rod} = A_1 - A_2$ is the actuator rod cross-sectional area. For this optimum load condition, it then follows that both pressures are equal:

$$P_1 = P_2 \rightarrow \frac{P_s}{2}. \tag{4.77}$$

Figure 4.24 shows the current error for different values of area ratio γ.
    Some conclusions therefore may be drawn as follows:

- The supply pressure may be matched to the load to produce a zero position error, providing the load is constant and the supply pressure is within an acceptable range for both servovalve operation and system response.
- Higher loads can be positioned with zero error and will be aided by an increased area ratio. The sensitivity of steady-state error to variations in load is improved for increased area ratios.
- It was shown in Chapter 3 that spool underlap provides system damping that is due to the leakage resistance at the centralized position. The dynamic performance therefore may be better designed to some extent.
- At optimum conditions, the pressure differential across the actuator piston will be close to zero, thus minimizing leakage across the piston.

Now consider a practical example of this position control system with a variable load force and with supply pressure adaption to maintain zero position error as the

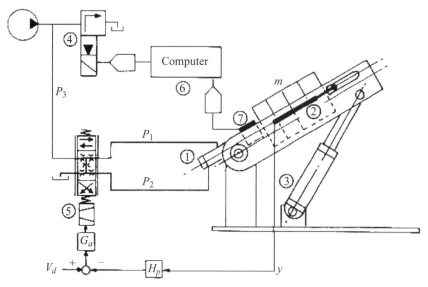

Figure 4.25. Position control with a variable load force and supply pressure adaption.

load force is changing. The supply pressure adjustment is achieved with a proportional PRV that is controlled electronically. The position control system is assembled within a frame that may be rotated in its vertical plane by a second linear actuator, as shown in Fig. 4.25.

The load variation on position control actuator (1) is generated by second actuator (3), the force is detected by load cell (7), and the voltage signal is fed through a digital-to-analog converter (DAC) within a microcomputer (6). A simple algorithm is then written to send a voltage via a DAC to set supply pressure (4) to match the load. Position transducer (2) provides the feedback signal for servovalve (5). System data for the position control system are as follows:

> tilt variable over a 58° angle, load mass = 72 kg, load variable up to $F = 599$ N
> actuator bore 25.4 mm diameter, rod 15.88 mm diameter, area ratio $\gamma = 1.64$
> servoamplifer gain $G_a = 58$ mA/V
> position transducer gain $H_p = 0.96$ mV/mm, load cell gain 1.9 mV/N

The proportional PRV supply pressure–voltage characteristic and the servovalve blocked flow pressure–current characteristic around the closed position are shown as Fig. 4.26. The computer control signal used to control the PRV was approximated by:

$$P_s = 21 - 11.4V + 4.4V^2. \tag{4.78}$$

The servovalve spool was symmetrically underlapped by the servovalve manufacturer to a value of 0.3 mm, equivalent to $i_u = 6$ mA. The lifting cylinder was actuated slowly over a period of 2 s such that system dynamics were negligible and the position control system position was recorded with and without supply pressure adaption. Measurements are shown in Fig. 4.27 as the load force is changed from 599 N to 242 N.

The effect of supply pressure on steady-state error is clear from Fig. 4.27, as expected. For this system, a servovalve error current of 1 mA is equivalent to a position error of 18 mm. This can be reduced, of course, by increasing the servoamplifier

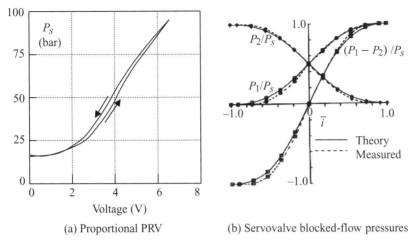

(a) Proportional PRV   (b) Servovalve blocked-flow pressures

Figure 4.26. Characteristics of the PRV and the servovalve.

gain G_a, but then the potential for closed-loop stability must be considered under normal operating conditions for closed-position control.

4.10 Linearization of a Valve-Controlled Motor Open-Loop Drive: Toward Intelligent Control

Now consider a drive such as a servovalve or proportional-valve motor drive. This was extensively covered earlier, from which it is clear that the valve's inherent square-root flow characteristic causes the speed to reduce with increasing load torque applied to the motor. However, the use of common programmable electronic controllers means that this flow characteristic can be easily and significantly linearized to improved open-loop behavior. By way of a real example, a proportional control valve is considered with an axial piston motor and for unidirectional speed control only. The system and control concept are shown in Fig. 4.28.

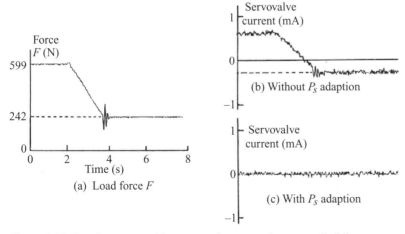

(a) Load force F

(b) Without P_s adaption

(c) With P_s adaption

Figure 4.27. Steady-state position error of a servovalve-controlled linear actuator with servovalve spool underlap (Watton and Al-Baldawi, 1991).

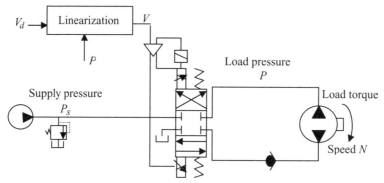

Figure 4.28. An open-loop motor drive with proportional-valve linearization.

The proportional-valve spool design is unique in that the return line pressure is held at a very low and constant pressure. The flow-rate characteristic is therefore defined for this example in terms of the load pressure P.

A Moog programmable servocontrol card was used to modify the applied voltage V_d such that the actual voltage V applied to the proportional valve improves the flow characteristic. A pressure transducer is required for monitoring the load pressure for valve linearization. The flow characteristic of the proportional flow-control valve was obtained experimentally and found to be of the following form:

$$Q = (aV - b)\sqrt{P_s - P}, \tag{4.79}$$

where $a \approx 2.48$ and $b \approx 2.25$ to give Q L/min with a pressure differential $(P_s - P)$ bar.

It is desirable to linearize this characteristic such that it becomes:

$$Q = aV_d\sqrt{P_{\text{ref}}}, \tag{4.80}$$

where P_{ref} is a chosen reference pressure, not necessarily supply pressure P_s. Therefore, the flow rate is now just proportional to applied voltage V_d.

Rearranging these two equations then gives:

$$V = V_d\sqrt{\frac{P_{\text{ref}}}{P_s - P}} + \frac{b}{a}. \tag{4.81}$$

If the reference pressure is chosen to be the supply pressure, then:

$$V = V_d\sqrt{\frac{1}{1 - P/P_s}} + \frac{b}{a}. \tag{4.82}$$

The linearization process requires a multiplication factor on V_d that becomes increasingly large as the load pressure P increases toward supply pressure. Therefore, in practice, the output voltage will have a saturation limit and linearization cannot be achieved over the full pressure range. This is not a serious restriction in practice because the load pressure P should not be designed to be close to supply pressure. Some measured results are shown in Fig. 4.29 for the minimum load speed of 515 rpm and a supply pressure of $P_s = 100$ bar.

The conventional square-root variation in speed is evident for the normal system, linearization resulting in a dramatic improvement with increasing load pressure. At the same time, the overall energy efficiency is inherently improved, the

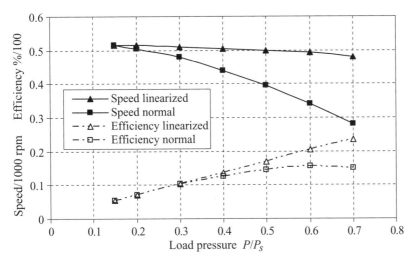

Figure 4.29. Improving the speed characteristic of an open-loop motor drive by flow-control valve linearization.

actual values being low because of the low motor speed in this example. In fact, the system efficiency now continually increases with increasing load pressure.

Further improvements in performance can be made by noting that for this system, the maximum efficiency is obtained approximately over a wide operating range by driving the pressure differential across the proportional valve to typically 20 bar. As the load pressure changes, this requires the supply pressure to be adjusted to maintain the constant pressure drop. This can be achieved in practice by using a proportional PRV. The supply pressure is set by the control voltage V_s and is relatively insensitive to the flow rate through the PRV. A measured characteristic for the valve used in this example is shown in Fig. 4.30.

It can be seen that, in practice, there is no need to determine the flow rate through the valve because a single defining characteristic may be used for supply

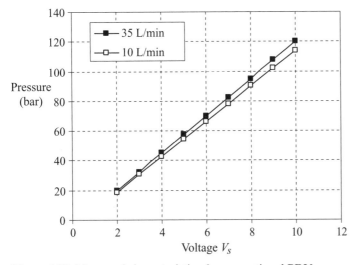

Figure 4.30. Measured characteristic of a proportional PRV.

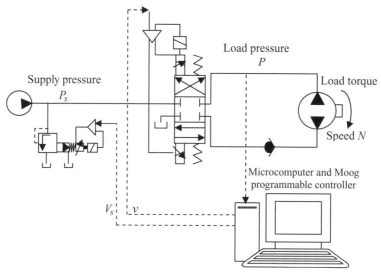

Figure 4.31. Intelligent control of an open-loop motor drive with flow-control valve linearization and supply pressure adaption.

tracking control purposes. The equations used for the pressure differential across the flow-control valve and the supply pressure are therefore given by:

$$\begin{aligned}
\text{flow-control valve fixed-pressure drop,} \quad & P_s - P = 20 \, \text{bar}, \\
\text{PRV characteristic,} \quad & P_s = -P_{so} + k_s V_s.
\end{aligned} \tag{4.83}$$

This gives the PRV control signal as follows:

$$V_s = \frac{P + 20 + P_{so}}{k_s}, \tag{4.84}$$

where $P_{so} \approx -5$ bar and $k_s \approx 12.2$ bar/V. As a consequence of this supply pressure adaption, the system efficiency is further improved, particularly at low load conditions. The PRV control voltage is easily implemented on the Moog programmable control card used in this example. The modified open-loop control concept for steady-state speed control is shown in Fig. 4.31.

Care must be taken to ensure a minimum operating pressure. This example also has considered only steady-state performance and, in practice, system dynamics must be considered. Dynamic stabilization is considerably improved with signal filtering together with dynamic pressure feedback compensation. However, this does reduce the system frequency bandwidth. Note, however, that the system is still open-loop, and true feedback control by use of measured speed is not used. For further details, particularly on the additional dynamic aspects, see Davies and Watton (1995).

A complete set of measured steady-state characteristics is shown in Fig. 4.32 with both flow-control valve linearization and supply pressure adaption.

The improvement in steady-state performance is clear and obtained by use of standard industrial components in common use. The improvement in efficiency with increasing load is highly significant when combined with a speed characteristic with negligible droop.

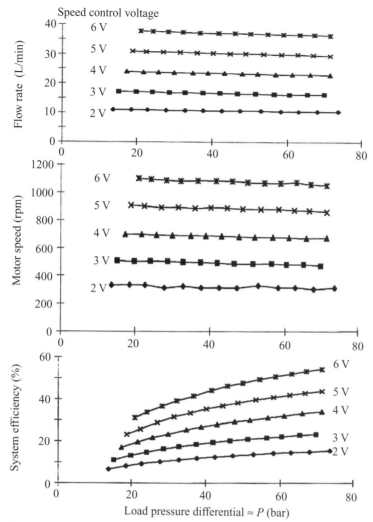

Figure 4.32. Intelligent steady-state control of an open-loop motor drive with flow-control valve linearization and supply pressure adaption.

4.11 References and Further Reading

Backe W [1993]. Recent research projects in hydraulics. In *Proceedings of the 2nd JHPS International Symposium on Fluid Power*, Japan Fluid Power Society, 1–27.

Bick DE, Dowty Rotol Ltd. [1979]. Power transfer unit, U.S. Patent 4,168,652.

Boehringer WE, Little J, Rothi RD, Westland CJ, McDonnell Douglas Corporation [1975]. Reciprocating transfer pump, U.S. Patent 3,890,064.

Boehringer WE, McDonnell Douglas Corporation [1981]. Hydraulic power transfer unit, U.S. Patent 4,286,927.

Chapple PJ and Dorey RE [1968]. The performance comparison of hydrostatic piston motors: factors affecting their application and use. In *Proceedings of the 7th BHRA Fluid Power Symposium*, British Hydromechanics Research Association, 1–8.

Davies RM and Watton J [1995]. Intelligent control of an electrohydraulic motor drive system. *J. Mechatronics* 5, 527–540.

Ebert H [1962]. Hydrostatic axial piston fluid transmission, U.S. Patent 3,052,098.

McGowan PT, Allied Signal Inc. [1987]. Power transfer apparatus, European Patent 0,280,532,B1.

McGowan PT, The Garrett Corporation [1979]. Fluid motors and pumps, European Patent 0,015,127.

SAE [2007]. Aerospace-recommended practice SAE ARP1280, reaffirmed 2002–07, Society of Automotive Engineers International, Warrendale, PA.

Watton J [1989]. Closed-loop design of an electrohydraulic motor drive using open-loop steady-state characteristics. *J. Fluid Control* 20 (1), 7–30.

Watton J [2006]. An explicit design approach to determine the optimum steady-state performance of axial piston motor speed drives. *Proc. Inst. Mech. Eng. J. Syst. Control Eng.* 220, 131–143.

Watton J and Al-Baldawi RA [1991]. Performance optimisation of an electrohydraulic control system with load-dependent supply pressure. *Proc. Inst. Mech. Eng.* 205, 175–189.

Zarotti GL and Nervegna N [1981]. Pump efficiencies: approximation and modelling. In *Proceedings of the 6th BHRA Fluid Power Symposium*, British Hydromechanics Research Association, 145–164.

5 System Dynamics

5.1 Introduction

The preceding chapters considered the steady-state behavior of common fluid power elements and systems. In reality, fluid power systems handle significant moving masses, and the combination of this with fluid compressibility results in system dynamics that usually cannot be neglected. In addition, individual components such as PRVs require a finite time to accommodate flow-rate changes. This also applies, for example, to a servovalve that again requires a finite time to change its spool position in response to a change in applied current. The combination of these issues means that the design of both open-loop and closed-loop control systems must take into account these dynamic issues. In particular, a closed-loop control system will almost certainly become unstable as system gains are increased because of such dynamic effects. Instability can lead to disastrous consequences if severe pressure oscillations occur. Instability in axial piston motor speed control systems, for example, can result in severe repetitive lifting and impact of the pistons on the swash plate.

Consider the design of a servoactuator that forms one of four to be used to provide the "road" input to the wheels of a vehicle sitting on a rig commonly called a "four-poster." Figure 5.1 shows one of the servoactuators and a block diagram of the position control system.

Determining the dynamic performance of the position control system only is relatively straightforward once the important dynamic features have been identified. The servovalve is manufactured to be critically lapped (four were donated by Star Hydraulics UK for the vehicle four-poster test rig at Cardiff University). The servoactuator has an integral position transducer and a PC is used to control each actuator by means of National Instruments DAQ technology. Data are as follows:

Load mass, 80 kg
Actuator stroke = 100 mm, position transducer gain $H_p = 0.1$ V/mm
Double-rod actuator bore diameter = 25.4 mm, rod diameter = 19.05 mm

The dynamic performance may be assessed by demanding a sudden change in position from center position to 5 mm and then back to the center position, as shown in Fig. 5.2.

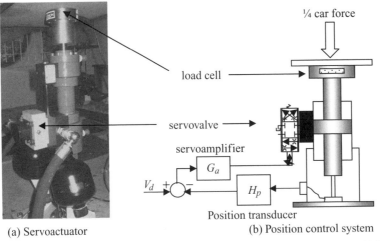

(a) Servoactuator   (b) Position control system

Figure 5.1. A servoactuator, one of four forming part of a four-poster vehicle test rig (Cardiff University School of Engineering).

Some important points emerge from this preliminary analysis:

- The general response is damped for the servoamplifier gain chosen.
- There is a small oscillation around the central position, $y = 0$, as evident from the velocity graph.
- The oscillation is not apparently significant from the position graph.
- Maximum transient flow rates may be estimated, thus helping the selection of the power supply pump.
- The existence of a small servovalve spool underlap and actuator friction will assist in dampening the small oscillations.
- The oscillations are not attributable to the servovalve but rather to the way it is used in a feedback-control-system sense. Making the position response faster will eventually lead to closed-loop instability.

Clearly, an accurate analysis of the system dynamic behavior prior to purchasing components can be a valuable asset, the two important issues that arise being:

- the validity of the model used
- the technique used for the system analysis

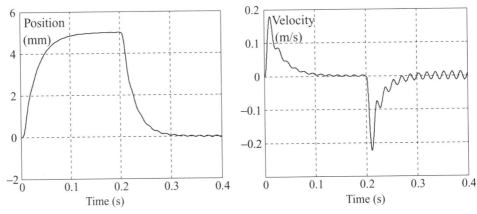

Figure 5.2. The dynamic response of the position control system.

Figure 5.3. Mass flow into and out of a control volume.

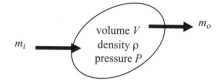

What is of crucial importance to fluid power systems dynamic modeling is the effect of fluid compressibility and moving masses, and these aspects are considered next.

5.2 Mass Flow-Rate Continuity

Consider an arbitrary control volume as shown in Fig. 5.3.
   Mass flow continuity then gives:

$$m_i - m_o = \frac{d}{dt}(\rho V). \tag{5.1}$$

Then, redefining mass flow rate by considering the appropriate volumetric flow rates gives:

$$\rho Q_i - \rho Q_o = \rho \frac{dV}{dt} + V \frac{d\rho}{dt}. \tag{5.2}$$

Using the definition of fluid bulk modulus β, together with the equation of state, the following equations are added:

$$\text{bulk modulus, } \beta = -\frac{dP}{dV/V}, \tag{5.3}$$

$$\text{equation of state, } \rho V = \text{constant} \rightarrow \frac{d\rho}{\rho} = -\frac{dV}{V}. \tag{5.4}$$

Equation (5.2) then becomes:

$$Q_i - Q_o = \frac{dV}{dt} + \frac{V}{\beta}\frac{dP}{dt}. \tag{5.5}$$

Equation (5.5) is used to analyze fluid power circuits for nonsteady conditions. Thus, the difference between the input flow rate Q_i and the output flow rate Q_o is equal to the rate of change of the boundary volume, plus an additional flow rate that is due to fluid compressibility and the existence of a pressure that is varying with time. For steady-state conditions, the compressibility term is zero, but there could still be a moving boundary such as a piston moving with a constant velocity.

5.3 Force and Torque Equations for Actuators

Considering a linear actuator and a motor (Fig. 5.4), the applied hydraulic force or moment has to overcome losses and the load requirement:

$$\text{linear actuator, } P_1 A_1 - P_2 A_2 = F_{\text{losses}} + F_{\text{load}} + M\frac{dU}{dt}, \tag{5.6}$$

$$\text{motor, } D_m(P_1 - P_2) = T_{\text{losses}} + T_{\text{load}} + J\frac{d\omega}{dt}, \tag{5.7}$$

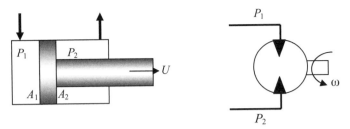

Figure 5.4. Forces applied to hydraulic actuators.

where M is the load mass, J is the rotary inertia, F_{load} and T_{load} are the load constant force and torque, and D_m is the motor displacement. The force and torque losses have a similar characteristic combining stiction–friction and viscous damping, as shown in Fig. 5.5, sometimes referred to as the *Stribeck curve*.

The stiction component of friction is often called the "breakaway" component, which then falls to a lower Coulomb friction level when motion occurs. For a motor, this characteristic does depend to some extent on the motor pressure differential, although the general shape is the same. The viscous friction component is usually assumed to be proportional to speed. The force equations then become:

$$\text{linear actuator,} \quad P_1 A_1 - P_2 A_2 = F_{cf} + B_v U + F_{load} + M\frac{dU}{dt}, \tag{5.8}$$

$$\text{motor,} \quad D_m(P_1 - P_2) = T_{cf} + B_v \omega + T_{load} + J\frac{d\omega}{dt}, \tag{5.9}$$

where F_{cf} and T_{cf} are the stiction–Coulomb friction losses and B_v is the viscous friction loss coefficient.

Worked Example 5.1

A motor is required to accelerate from zero to 500 rpm in 2 s with the following machine parameters:

Load torque $T = 80$ N m, Coulomb friction $T_{cf} = 10$ N m
Displacement $D_m = 6 \times 10^{-6}$ m³/rad, rotary inertia $J = 0.5$ kg m²
Viscous coefficient $B_v = 0.04$ N m/rads^{-1}
Flow losses may be neglected

(i) Calculate the pressure differential required.
(ii) If the motor is driven at a constant speed of 500 rpm, calculate the power and flow rate needed.

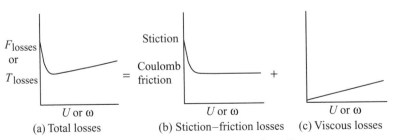

Figure 5.5. Components of force and torque losses for a linear actuator or motor.

Solution to (i):

$$D_m(P_1 - P_2) = T + (T_f + B_v\omega) + J\frac{d\omega}{dt},$$

$$6 \times 10^{-6}(P_1 - P_2) = 80 + (10 + 0.04 \times 52.37) + 0.5\frac{52.37}{2},$$

$$6 \times 10^{-6}(P_1 - P_2) = 80 + 10 + 2.09 + 13.09 = 105.18 \text{ N m},$$

$$(P_1 - P_2) = 175 \text{ bar}.$$

Solution to (ii):

$$\text{the maximum torque} = 105.18 \text{ N m}$$

$$\text{power} = T\omega = (105.18)(52.37) = 5.51 \text{ kW},$$

$$Q = D_m\omega = (6 \times 10^{-6})(52.37)$$

$$= 0.314 \times 10^{-3} \text{ m}^3/\text{s},$$

$$Q = 18.8 \text{ L/min}.$$

5.4 Solving the System Equations, Computer Simulation

A fluid power circuit mathematical model is a collection of nonlinear flow equations associated with components together with flow continuity and force equations applicable to components and actuators. The equations must be brought together in a form suitable for computer simulation by either a dedicated numerical solution code or a commercial simulation software package. The MATLAB Simulink package is commonly used for the simulation of any dynamical system and contains a comprehensive library of mathematical functions. The approach requires re-arrangement of the differential equations such that the highest differential is placed at the left-hand side of the equation by itself. The process of integration then allows the block diagram to be completed once all the equations have been included. For example, consider the simple system shown in Fig. 5.6 to illustrate some basic principles:

$$\text{flow continuity,} \quad Q_i - Q_o = \frac{dV}{dt} + \frac{V}{\beta}\frac{dP}{dt}, \tag{5.10}$$

$$Q = AU + \frac{V}{\beta}\frac{dP}{dt}, \tag{5.11}$$

$$\text{force,} \quad PA = Mg + B_v U + M\frac{dU}{dt}, \tag{5.12}$$

$$\text{volume,} \quad V = V(0) + Ay = A[y(0) + y]. \tag{5.13}$$

Rearranging Eqs. (5.11), (5.12), and (5.13) then gives:

$$\frac{dP}{dt} = \frac{\beta}{V}[Q - AU], \tag{5.14}$$

$$P = \int \frac{dP}{dt}dt = \int \frac{\beta}{V}[Q - AU]\,dt, \tag{5.15}$$

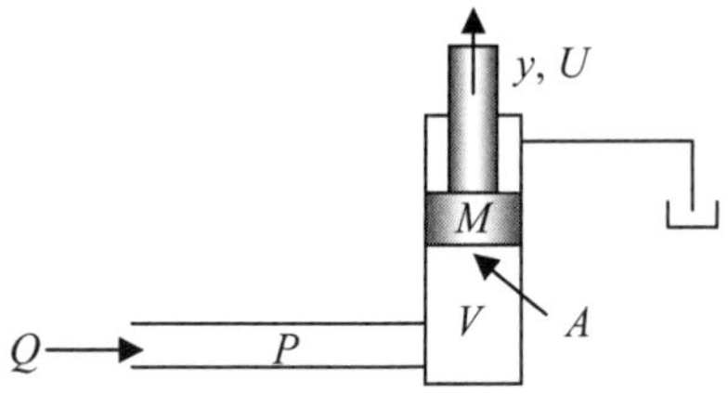

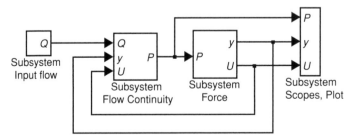

Figure 5.6. A simple circuit for lifting a load mass M.

$$\frac{\beta}{V} = \frac{\beta}{V(0)} \frac{1}{\left[1 + \dfrac{y}{y(0)}\right]}, \tag{5.16}$$

$$\frac{dU}{dt} = \frac{1}{M}[PA - Mg - B_v U], \tag{5.17}$$

$$U = \int \frac{dU}{dt}\,dt = \int \frac{1}{M}[PA - Mg - B_v U]\,dt,$$

$$y = \int U\,dt. \tag{5.18}$$

Data are as follows:

   Moving mass $M = 2000\,\text{kg}$, piston cross-sectional area $A = 0.004\,\text{m}^2$
   Input volume equivalent displacement $y(0) = 0.2\,\text{m}$, actuator stroke $= 0.8\,\text{m}$
   Velocity damping coefficient $B_v = 8 \times 10^4\,\text{N/ms}^{-1}$
   Input flow rate $Q = 24\,\text{L/min}$, fluid bulk modulus $\beta = 1.4 \times 10^9\,\text{N/m}^2$

The steady-state velocity will be:

$$U_{ss} = \frac{24 \times 10^{-3}}{60 \times 0.004} = 0.1\,\text{m/s}.$$

The pressure to overcome steady-state viscous friction will be:

$$\frac{(8 \times 10^4)(0.1)}{0.004} = 20 \times 10^5\,\text{N/m}^2 = 20\,\text{bar}.$$

The pressure to hold the load mass will be:

$$\frac{(2 \times 10^3 g)}{0.004} = 49.05 \times 10^5\,\text{N/m}^2 = 49.05\,\text{bar}.$$

Therefore, the total steady-state pressure will be 69.05 bar. The block diagram is then constructed as shown in Fig. 5.7. The left-hand side of the block diagram starts with the input Q, and the diagram is completed by the process of integration from left to right by means of the flow-continuity equation and the force equation.

Figure 5.7. MATLAB Simulink block diagram using subsystems.

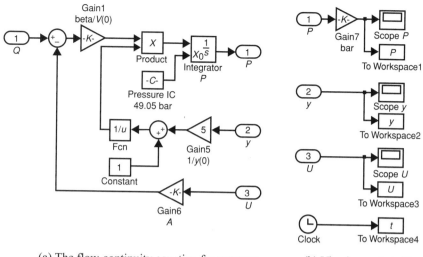

(a) The flow-continuity equation for pressure

(b) Viewing and plotting

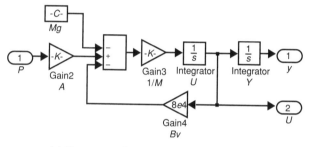

(c) Force equation for velocity and position

Figure 5.8. Subsystems for the lifting circuit simulation.

Each subsystem contains the appropriate mathematical models, inputs, and plotting facilities, as shown in Fig. 5.8.

It will be deduced from the data using steady-state theory that it will take 8 s for the load to be raised 0.8 m assuming a constant speed of 0.1 m/s. Even though dynamics exists at the beginning of the motion, because of the sudden application of the flow rate, these effects are very fast and have little effect on this total motion time based on steady-state velocity $U_{ss} = Q/A$. This is illustrated in Fig. 5.9.

The effect of fluid compressibility is to filter any oscillatory characteristic when position is observed or measured, and this is typical for linear actuator systems. This is particularly evident when the oscillatory component period is much lower than the total motion time. However, when the start-up or shutoff dynamic characteristic is examined in more detail, other interesting features appear, as shown in Fig. 5.10.

It can be seen that the short initial period of about 0.2 s is highly oscillatory with the pressure increasing to more than 110 bar compared with the steady-state value of 69 bar. The speed eventually settles to the steady-state value of 0.1 m/s, as anticipated. It is this importance of dynamic effects that demands analysis to determine whether or not that short-term unsatisfactory behavior will occur and can preferably be determined before the system is built and operated. A further advantage of

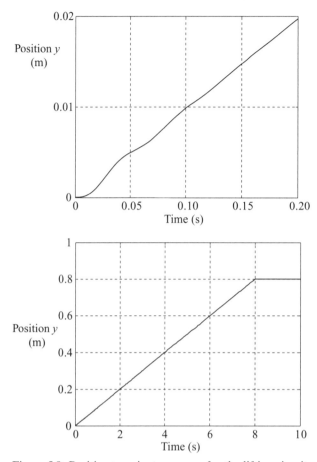

Figure 5.9. Position transient response for the lifting circuit.

computer simulation is that damping techniques can be introduced in an attempt to minimize oscillations.

5.5 Differential Equations, Laplace Transforms, and Transfer Functions

5.5.1 Linear Differential Equations

The system considered in the previous section has one linear differential equation for force and a nonlinear differential equation for flow-rate continuity, the nonlinear component being due to the actuator volume change with time. However, the actual volume change is negligible over the transient part of the operation, typically a 10% increase. Hence, if this variation in volume is neglected, then the system consists of two linear first-order differential equations:

$$\text{flow continuity, } Q = AU + \frac{V}{\beta}\frac{dP}{dt}, \tag{5.19}$$

$$\text{force, } PA = Mg + B_v U + M\frac{dU}{dt}, \tag{5.20}$$

$$\text{volume, } V = V(0) + Ay = \cong V(0). \tag{5.21}$$

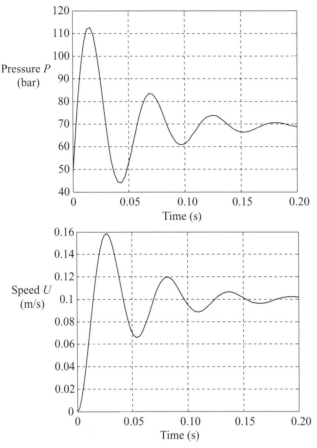

Figure 5.10. Pressure and speed transient response of the lifting circuit at the beginning of the 8-s motion time.

Equations (5.19) and (5.20) can now be combined to give:

$$U + \frac{V(0)B_v}{\beta A^2}\frac{dU}{dt} + \frac{V(0)M}{\beta A^2}\frac{d^2U}{dt^2} = \frac{Q}{A},$$

$$U + \frac{2\zeta}{\omega_n}\frac{dU}{dt} + \frac{1}{\omega_n^2}\frac{d^2U}{dt^2} = U_{ss}.$$

(5.22)

This is a second-order linear differential equation and written in standard second-order notation, where U_{ss} is the steady-state value of U, ζ is the damping ratio, and ω_n is the undamped natural frequency that would exist if no damping were present. The complementary component of the solution is given by:

$$U + \frac{2\zeta}{\omega_n}\frac{dU}{dt} + \frac{1}{\omega_n^2}\frac{d^2U}{dt^2} = 0 \text{ and try a solution } U = Ae^{\lambda t}:$$

$$1 + \frac{2\zeta}{\omega_n}\lambda + \frac{1}{\omega_n^2}\lambda^2 = 0 \rightarrow \lambda^2 + 2\zeta\omega_n\lambda + \omega_n^2 = 0,$$

(5.23)

$$\lambda = -\zeta\omega_n \pm \omega_n\sqrt{\zeta^2 - 1},$$

$$U = A_1 e^{-\zeta\omega_n t + \omega_n\sqrt{\zeta^2 - 1}t} + A_2 e^{-\zeta\omega_n t - \omega_n\sqrt{\zeta^2 - 1}t}.$$

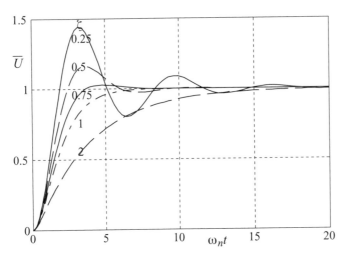

Figure 5.11. Step response of a second-order system.

The response therefore contains a decaying exponential term together with a term that can also decay with time or be oscillatory, depending on the value of ζ. Equation (5.23) is called the *characteristic equation* in control terminology, and *it is assumed that all the coefficients of the polynomial in λ are positive*. It should be obvious that for the system response to be stable, in the sense that the output must settle down to the state demanded, then *the roots of the characteristic equation should have negative real parts*. A linear second-order system, therefore, has the following properties:

$$
\begin{aligned}
&\text{oscillatory, underdamped for } \zeta < 1,\\
&\text{critically damped for } \zeta = 1,\\
&\text{over damped for } \zeta > 1.
\end{aligned}
\tag{5.24}
$$

From, Eq. (5.23), the actual damped frequency of oscillation, ω_d, is given by:

$$
\omega_d = \omega_n \sqrt{1 - \zeta^2}.
\tag{5.25}
$$

Consider again the second-order linear differential equation:

$$
\begin{aligned}
&U + \frac{2\zeta}{\omega_n}\frac{dU}{dt} + \frac{1}{\omega_n^2}\frac{d^2U}{dt^2} = U_{ss},\\
&\overline{U} + 2\zeta\frac{d\overline{U}}{d(\omega_n t)} + \frac{d^2\overline{U}}{d(\omega_n t)^2} = 1, \text{ where } \overline{U} = \frac{U}{U_{ss}}.
\end{aligned}
\tag{5.26}
$$

Some transient step responses are shown in Fig. 5.11 for different values of damping ratio ζ.

It is good sense in practice to aim for a damping ratio of $0.7 < \zeta < 1$. For the lifting system in the previous section, it follows that:

$$
\begin{aligned}
&\omega_n^2 = \frac{\beta A^2}{V(0)M}, \quad \frac{2\zeta}{\omega_n} = \frac{V(0)B_v}{\beta A^2},\\
&\omega_n = \sqrt{\frac{\beta A}{My(0)}}, \quad \zeta = \frac{B_v}{2}\sqrt{\frac{y(0)}{\beta M A}}.
\end{aligned}
\tag{5.27}
$$

For the example, the initial condition $y(0) = 0.2\,\text{m}$ and at the end condition $y(0) = 1.0\,\text{m}$. Recalling for this example that $M = 2000\,\text{kg}$, $B_v = 8 \times 10^4\,\text{N/m s}^{-1}$, $A = 0.004\,\text{m}^2$, $\beta = 1.4 \times 10^9\,\text{N/m}^2$, then:

when motion suddenly starts	when motion suddenly stops
$\omega_n = 118.32\,\text{rad/s}\,(18.83\,\text{Hz})$, $\zeta = 0.17$	$\omega_n = 52.92\,\text{rad/s}\,(8.42\,\text{Hz})$, $\zeta = 0.38$
$\omega_d = 18.56\,\text{Hz} \rightarrow \text{period} = 0.054\,\text{s}$	$\omega_d = 7.79\,\text{Hz} \rightarrow \text{period} = 0.13\,\text{s}$

The period of $0.054\,\text{s}$ agrees with the results shown in Fig. 5.10 at the start of motion, which also illustrates highly oscillatory motion and is explained by the low damping ratio $\zeta = 0.17$. Damping can be increased by reducing the actuator cross-sectional area A, but the undamped natural frequency will be reduced. If the differential equation is first-order – for example, if moving mass is considered negligible – then:

$$U + \frac{V(0)B_v}{\beta A^2}\frac{dU}{dt} = \frac{Q}{A}, \tag{5.28}$$

$$\overline{U} + \tau\frac{d\overline{U}}{dt} = 1, \tag{5.29}$$

where τ is called the time constant. The solution to Eq. (5.29) for a suddenly applied input (step input) is exponential in form and obtained by direct integration of Eq. (5.29). Assuming zero initial conditions then gives:

$$\overline{U} = 1 - e^{-t/\tau}. \tag{5.30}$$

The time constant τ may be experimentally determined by measuring the time for the response to achieve 63.2% of its final value; that is, when $\overline{U} = 0.632$ – which, of course, occurs when $t = \tau$.

5.5.2 Nonlinear Differential Equations, the Technique of Linearization for Small-Signal Analysis

Nonlinear differential equations are common, for example, when a servovalve is included or any other restrictor-type component is used that has a nonlinear area variation with displacement. If the ability of linear differential equations to give an insight into system dynamic behavior is to be utilized, then nonlinear terms must be changed to linear terms given that this will introduce some limitations. The process is called *linearization* or *small-signal analysis*. The linearized coefficients for a servovalve were discussed in Chapter 3, and this will be used here again, but in more detail, as a starting point. Consider the previous simple example but now including a servovalve as the input control element, as shown in Fig. 5.12.

The flow-continuity equation now becomes:

$$Q = k_f i\sqrt{P_s - P} = AU + \frac{V}{\beta}\frac{dP}{dt}. \tag{5.31}$$

Now *consider linearizing this equation* to deduce the dynamic performance close to a particular operating condition. The issue is the servovalve equation, the flow being a function of two variables: current i and pressure P. If small variations about the

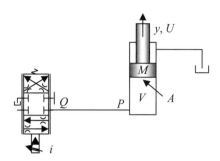

Figure 5.12. A simple circuit for lifting a load mass M.

steady-state operating point $Q(0)$, $U(0)$, $i(0)$, $P(0)$ are considered, then:

$$Q(0) + \delta Q = A[U(0) + \delta U] + \frac{V}{\beta}\frac{d[P(0) + \delta P]}{dt}. \tag{5.32}$$

But, the following is true:

$$Q(0) = AU(0), \quad \frac{dP(0)}{dt} = 0, \tag{5.33}$$

and, therefore:

$$\delta Q = A\delta U + \frac{V}{\beta}\frac{d\delta P}{dt}. \tag{5.34}$$

The small variation in servovalve flow rate about the operating point is obtained from the Taylor expansion:

$$\delta Q = \left.\frac{\partial Q}{\partial i}\right|_{i(0),P(0)}\delta i + \left.\frac{\partial Q}{\partial P}\right|_{i(0),P(0)}\delta P, \tag{5.35}$$

$$\delta Q = k_i\delta i - k_p\delta P. \tag{5.36}$$

It will be seen that the two partial differential components in Eqs. (5.35) and (5.36) are the linearized coefficients of the servovalve at the operating point discussed in Chapter 3 and for this example are given by:

$$\text{flow gain,} \quad k_i = \frac{\partial Q}{\partial i} = k_f\sqrt{P_s - P(0)} = \frac{Q(0)}{i(0)}, \tag{5.37}$$

$$\text{pressure gain,} \quad k_p = -\frac{\partial Q}{\partial P} = \frac{k_f i(0)}{2\sqrt{P_s - P(0)}} = \frac{Q(0)}{2[P_s - P(0)]}. \tag{5.38}$$

Completing the system model by adding the linearized force equation gives:

$$\text{force,} \quad PA = Mg + B_v U + M\frac{dU}{dt}, \tag{5.39}$$

$$\delta PA = B_v\delta U + M\frac{d\delta U}{dt},$$

$$\text{volume,} \quad V = V(0) + Ay \cong V(0). \tag{5.40}$$

Bringing system equations (5.34), (5.36), and (5.39) together gives:

$$k_i\delta i - k_p\delta P = A\delta U + \frac{V(0)}{\beta}\frac{d\delta P}{dt}, \tag{5.41}$$

$$\delta PA = B_v\delta U + M\frac{d\delta U}{dt}. \tag{5.42}$$

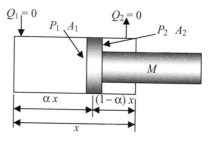

Figure 5.13. Sudden closure of a linear actuator with negligible damping.

The system linearized differential equation then becomes:

$$\delta U \left(1 + \frac{k_p B_v}{A^2}\right) + \left[\frac{k_p M}{A^2} + \frac{V(0)B_v}{\beta A^2}\right] \frac{d\delta U}{dt} + \frac{V(0)M}{\beta A^2}\frac{d^2\delta U}{dt^2} = \frac{k_i}{A}\delta i. \tag{5.43}$$

This is again a second-order, linearized differential equation now relating actuator speed to servovalve applied current and is *strictly applicable only to small variations about the particular operating point*. Rearranging this equation into standard second-order form then gives:

$$\delta U + \frac{\left[\dfrac{k_p M}{A^2} + \dfrac{V(0)B_v}{\beta A^2}\right]}{\left(1 + \dfrac{k_p B_v}{A^2}\right)}\frac{d\delta U}{dt} + \frac{\dfrac{V(0)M}{\beta A^2}}{\left(1 + \dfrac{k_p B_v}{A^2}\right)}\frac{d^2\delta U}{dt^2} = \frac{\dfrac{k_i}{A}\delta i}{\left(1 + \dfrac{k_p B_v}{A^2}\right)}, \tag{5.44}$$

$$\delta U + \frac{2\zeta}{\omega_n}\frac{d\delta U}{dt} + \frac{1}{\omega_n^2}\frac{d^2\delta U}{dt^2} = \delta U_{ss}.$$

The damping ratio ζ, the undamped natural frequency ω_n, and the damped frequency ω_d may then be determined by equating terms. It is useful to note that in practice,

- often the denominator term $\dfrac{k_p B_v}{A^2} \ll 1$ and can be neglected,
- for an operating condition $i(0) = 0$, then the pressure gain $k_p = 0$ and the flow gain k_i is finite for a critically lapped spool. The left-hand side of differential equation (5.44) is then identical to that determined for an idealized flow input, Eq. (5.22); that is, the critically lapped servovalve provides no damping around the closed condition.

5.5.3 Undamped Natural Frequency of a Linear Actuator

Consider a single-rod actuator, with no damping, and with the flow suddenly cut off, as shown in Fig. 5.13.

Because the undamped natural frequency is being considered, it is necessary to consider only load mass (load inertia) in the force equation:

$$P_1 A_1 - P_2 A_2 = M\frac{dU}{dt}. \tag{5.45}$$

Flow continuity on both sides gives:

$$\text{inlet side,} \quad 0 = A_1 U + \frac{V_1}{\beta}\frac{dP_1}{dt}, \tag{5.46}$$

$$\text{outlet side,} \quad 0 = A_2 U - \frac{V_2}{\beta}\frac{dP_2}{dt}. \tag{5.47}$$

Table 5.1. *Condition for minimum undamped natural frequency*

Area ratio γ	1.00	1.25	1.50	1.75	2.00	2.25
Length ratio α	0.50	0.53	0.55	0.57	0.59	0.60
	Double-rod	Single-rod				

Differentiating Eq. (5.45) and combining with Eqs. (5.46) and (5.47) gives:

$$U + \frac{M}{\left(\dfrac{A_1^2 \beta}{V_1} + \dfrac{A_2^2 \beta}{V_2} \right)} \frac{d^2 U}{dt^2} = 0. \tag{5.48}$$

It will be deduced that the volumes on either side are effectively two springs in series having a combined *hydraulic stiffness k*, given by:

$$k = \frac{A_1^2 \beta}{V_1} + \frac{A_2^2 \beta}{V_2}. \tag{5.49}$$

The undamped natural frequency is therefore given by:

$$\omega_n = \sqrt{\frac{\dfrac{A_1^2 \beta}{V_1} + \dfrac{A_2^2 \beta}{V_2}}{M}} = \sqrt{\frac{k}{m}}. \tag{5.50}$$

Given that $V_1 = A_1 \alpha x$ and $V_2 = A_2(1 - \alpha)x$ and also $\gamma = A_1/A_2$, then:

$$\omega_n = \sqrt{\frac{\beta A_2}{Mx} \left[\frac{\gamma}{\alpha} + \frac{1}{(1 - \alpha)} \right]}. \tag{5.51}$$

This clearly has a minimum because it varies from ∞ when $\alpha = 0$ to ∞ when $\alpha = 1$.

Differentiating Eq. (5.51) then gives the condition for minimum natural frequency as follows:

$$\omega_n \text{ minimum occurs when } \alpha = \frac{\sqrt{\gamma}}{1 + \sqrt{\gamma}}. \tag{5.52}$$

Table 5.1 shows this variation with area ratio.

It can be seen that, as expected, a double-rod actuator has its minimum natural frequency with equal volumes on either side. For a single-rod actuator, it can also be seen that when the area ratio is increased from 1 to 2, then the position for minimum natural frequency changes from 0.5 to 0.59, an increase of only 18%.

Worked Example 5.2

A double-rod cylinder has a design stroke of 0.5 m, and its internal diameter has to be determined. One design requirement is that the undamped natural frequency should be in excess of 100 Hz when the load mass $M = 200\,\text{kg}$, the effective fluid bulk modulus is $\beta = 0.9\,\text{GN/m}^2$, and the rod diameter is half the bore diameter, $d = D/2$.

Determine the cylinder bore diameter.
Choosing the midposition for the lowest natural frequency gives:

$$\omega_n^2 = \frac{2\beta}{V}\frac{A^2}{M} \rightarrow \sqrt{\frac{2\beta}{V}\frac{A^2}{M}} > 2\pi \times 100 \text{ rad/s},$$

$$\frac{2\beta}{(A \times \text{half stroke})}\frac{A^2}{M} > (200\pi)^2 \rightarrow A > \frac{(200\pi)^2 M \times \text{half-stroke}}{2\beta},$$

$$A > \frac{(200\pi)^2 200(0.25)}{2(0.9 \times 10^9)} = 0.011,$$

$$\frac{\pi}{4}\left[D^2 - \left(\frac{D}{2}\right)^2\right] > 0.011 \rightarrow \frac{3\pi D^2}{16} > 0.011,$$

$$D > 137 \text{ mm}.$$

5.5.4 Laplace Transforms and Transfer Functions

The use of differential equations and linearized differential equations now opens up a powerful analysis avenue to aid the understanding of the dynamic behavior of fluid power circuits. This is considerably further aided by the use of Laplace transforms that change linear differential equations from the time domain to the s domain, where s is the Laplace operator. The Laplace transform of a time-varying function $f(t)$ is defined as:

$$\mathcal{L}f(t) = F(s) = \int_0^\infty f(t)e^{-st}dt. \tag{5.53}$$

It is now simply a matter of applying this transform to each element of the differential equation, and the use of Table 5.2 greatly aids this process.

As an example, consider the previously considered lifting system consisting of a flow input Q, a fixed volume, and a linear actuator of mass M with the return connected to the tank, Fig. 5.14. The two equations were developed as follows:

$$Q = AU + \frac{V(0)}{\beta}\frac{dP}{dt}, \tag{5.54}$$

$$PA = Mg + B_v U + M\frac{dU}{dt}. \tag{5.55}$$

Now consider taking the Laplace transforms of these two linear differential equations by using Table 5.2. Also assume that the volume $V(0)$ that does not change

Figure 5.14. A simple circuit for lifting a load mass M.

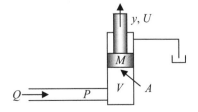

Table 5.2. *Some common Laplace transforms*

	$f(t)$	$F(s)$		$f(t)$	$F(s)$
1	Unit impulse	1	11	$\cos \omega t$	$\dfrac{s}{(s^2 + \omega^2)}$
2	Step h	$\dfrac{h}{s}$	12	$1 - \cos \omega t$	$\dfrac{\omega^2}{s(s^2 + \omega^2)}$
3	t^n n $= 1,2,3,-$	$\dfrac{n!}{s^{n+1}}$	13	$e^{-at} \sin \omega t$	$\dfrac{\omega}{(s + a)^2 + \omega^2}$
4	e^{-at}	$\dfrac{1}{(s + a)}$	14	$e^{-at} \cos \omega t$	$\dfrac{(s + a)}{(s + a)^2 + \omega^2}$
5	$t e^{-at}$	$\dfrac{1}{(s + a)^2}$	15	$\dfrac{d}{dt} f(t)$	$s F(s) - f(0)$
6	$\dfrac{1}{(n - 1)!} t^{n-1} e^{-at}$	$\dfrac{1}{(s + a)^n}$	16	$\dfrac{d^2}{dt^2} f(t)$	$s^2 F(s) - s f(0) - \dfrac{df(0)}{dt}$
7	$\dfrac{1}{a}(1 - e^{-at})$	$\dfrac{1}{s(s + a)}$	17	Delay $f(t - T)$	$e^{-sT} f(t)$
8	$\dfrac{1}{a^2}[at - (1 - e^{-at})]$	$\dfrac{1}{s^2(s + a)}$	18	Pulse function, magnitude h, duration T	$\dfrac{h}{s}(1 - e^{-sT})$
9	$\dfrac{1}{(b - a)}[e^{-at} - e^{-bt}]$, $b \neq a$	$\dfrac{1}{(s + a)(s + b)}$	19	Final-value theorem	$\underset{t \to \infty}{\text{Lim}} \; f(t) = s F(s)]_{s=0}$
10	$\sin \omega t$	$\dfrac{\omega}{(s^2 + \omega^2)}$	20	Initial-value theorem	$\underset{t \to 0}{\text{Lim}} \; f(t) = s F(s)]_{s=\infty}$

significantly over the transient response:

$$Q(s) = AU(s) + \frac{V(0)}{\beta}[s P(s) - P(0)], \tag{5.56}$$

$$P(s)A = Mg + B_v U(s) + M[s U(s) - U(0)]. \tag{5.57}$$

It can be seen that the process of taking Laplace transforms, with the inherent integration limits, has changed the differential equations to algebraic equations. This now allows the two first-order equations to be combined in a purely algebraic manner, allowing the establishment of transfer functions, a powerful tool for dynamic analysis and control studies. Combining algebraic equations (5.56) and (5.57) gives the following system transfer function about a steady-state operating point:

$$\begin{matrix} \text{Input} & \text{External} & \text{Initial} & \text{Conditions} \\ \downarrow & \downarrow & \downarrow & \downarrow \end{matrix} \tag{5.58}$$

$$U(s) = \frac{\dfrac{Q(s)}{A} - s\dfrac{V(0)Mg}{\beta A^2} + s\dfrac{V(0)U(0)}{\beta A^2} + \dfrac{V(0)P(0)}{A\beta}}{1 + \dfrac{V(0)B_v}{\beta A^2}s + \dfrac{V(0)M}{\beta A^2}s^2}.$$

For determining the complete solution, initial conditions $U(0)$ and $P(0)$ are required. However, if the response to input flow rate Q only is required, then the transfer function becomes:

$$U(s) = \frac{\dfrac{Q(s)}{A}}{1 + \dfrac{V(0)B_v}{\beta A^2}s + \dfrac{V(0)M}{\beta A^2}s^2} \tag{5.59}$$

$$= \frac{\dfrac{Q(s)}{A}}{1 + \dfrac{2\zeta}{\omega_n}s + \dfrac{s^2}{\omega_n^2}}.$$

Comparing this transfer function with the original differential equation, it can be seen that a mechanistic approach is to simply replace the differential operator $(d/dt) \to s$ if the relationship between the output and input only is required. Considering the earlier brief introduction to the solution of linear differential equations, (5.23), it can be seen that the *denominator of the transfer function is the characteristic equation when equated to zero*; in this case,

$$1 + \frac{2\zeta}{\omega_n}s + \frac{s^2}{\omega_n^2} = 0. \tag{5.60}$$

It can be seen that this is equivalent to replacing λ from the original differential equation complementary solution with the Laplace operator s.

To determine the load position, velocity is integrated. Recalling the Laplace transform symbol for integration, $1/s$, gives:

$$y(s) = \frac{U(s)}{s} = \frac{\dfrac{Q(s)}{A}}{s\left(1 + \dfrac{2\zeta}{\omega_n}s + \dfrac{s^2}{\omega_n^2}\right)}. \tag{5.61}$$

The relationship between a system output and a system input gives rise to the common notation for a transfer function as follows:

$$\frac{\text{output}}{\text{input}} = \frac{y(s)}{Q(s)} = G(s) = \frac{K}{s\left(1 + \dfrac{2\zeta}{\omega_n}s + \dfrac{s^2}{\omega_n^2}\right)}. \tag{5.62}$$

Transfer functions such as $G(s)$ can be directly inserted into MATLAB Simulink by standard blocks. This can speed up the simulation assembly process, particularly for components such as a servovalve whose dynamic performance is often expressed in transfer function form by the manufacturer; for example, $\zeta = 1$ and $\omega_n = 110\,\text{Hz}$. Note, however, that such parameters can change as the magnitude of the input current to the servovalve changes. This is particularly important if frequency-response analysis is to be used.

5.6 The Electrical Analogy

Previous work has considered the case in which a line and actuator volume can be "lumped" together, and the approach is sometimes referred to as a *lumped-*

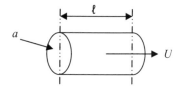

Figure 5.15. A fluid element subject to motion.

parameter analogy. However, in many applications, the lines connecting components could be extremely long and, in reality, are more accurately defined by the wave equations, often referred to as a distributed-parameter analysis. An approximation to the more accurate model of a line can be made with lumped approximations, and the electrical analogy is a useful mechanism for understanding this approach. Consider linear characteristics and a slug of fluid, cross-sectional area a and length ℓ, as shown in Fig. 5.15.
Assume the analogy:

$$\text{pressure } P \rightarrow \text{voltage } V,$$

$$\text{flow rate } Q \rightarrow \text{current } I.$$

(i) *Fluid resistance.* The pressure drop Δp, along the fluid element for *laminar flow*, is given by:

$$\Delta p = \frac{128\,\mu\ell}{\pi d^4} Q \rightarrow \Delta V = RI, \tag{5.63}$$

hydraulic resistance $\rightarrow$ electrical resistance.

(ii) *Fluid compressibility.*

$$\Delta Q = \frac{V}{\beta}\frac{dP}{dt} \rightarrow \Delta I = C\frac{dV}{dt}, \tag{5.64}$$

fluid compressibility $\rightarrow$ electrical capacitance.

(iii) *Fluid inertia.* The pressure drop that is due to fluid acceleration is given by:

$$\Delta pa = \rho\ell a\frac{dU}{dt},$$

$$\Delta p = \frac{\rho\ell}{a}\frac{dQ}{dt} \rightarrow \Delta V = L\frac{dI}{dt}, \tag{5.65}$$

fluid–mechanical mass $\rightarrow$ electrical inductance.

It is therefore concluded that for a fluid element, the analogies are given by:

$$\text{resistance} \quad R = \frac{128\,\mu\ell}{\pi d^4},$$

$$\text{inductance} \quad L = \frac{\rho\ell}{a}, \tag{5.66}$$

$$\text{capacitance} \quad C = \frac{V}{\beta}.$$

Applying this to an actuator, when considering mass and viscous friction effects, then gives:

$$\text{linear actuator,} \quad P_1 A_1 - P_2 A_2 = B_v U + M\frac{dU}{dt},$$

$$P_1\gamma - P_2 = \frac{B_v}{A_1 A_2}Q + \frac{M}{A_1 A_2}\frac{dQ}{dt}, \tag{5.67}$$

$$Q = U A_1 \quad \gamma = \frac{A_1}{A_2}.$$

$$\text{motor,} \quad D_m(P_1 - P_2) = B_v\omega + J\frac{d\omega}{dt},$$

$$P_1 - P_2 = \frac{B_v}{D_m^2}Q + \frac{J}{D_m^2}\frac{dQ}{dt}, \tag{5.68}$$

$$Q = D_m\omega.$$

It then follows that the electrical analogies for actuators are:

$$\text{linear actuator: viscous resistance,} \quad R_v = \frac{B_v}{A_1 A_2}, \text{inductance,} \quad L = \frac{M}{A_1 A_2}, \tag{5.69}$$

$$\text{motor: viscous resistance,} \quad R_v = \frac{B_v}{D_m^2}, \text{inductance} \quad L = \frac{J}{D_m^2}. \tag{5.70}$$

When considering whether or not to include *line dynamics* in a system model, it is noted that a blocked line with no friction losses will have a traveling-wave frequency of oscillation given by:

$$f = \frac{C_o}{2\ell}, \quad \text{velocity of sound in the fluid,} \quad C_o = \sqrt{\frac{\beta}{\rho}}. \tag{5.71}$$

If a system is expected to have a frequency component that is comparable with this frequency, line dynamics must be modeled with some accuracy. One approximation to the solution to the wave equation, to be discussed later, is to consider a lumped approximation to the true distributed-parameter solution. The next issue is how to distribute R, L, C in the line and how many "lumps" should be used. For example, Fig. 5.16 shows a two-lump approximation using a pair of π networks.

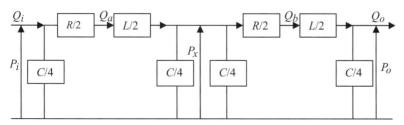

Figure 5.16. A line dynamics approximation using lumped π elements for laminar mean flow.

The set of equations using this approximation, and working from left to right, may then be written as follows:

$$Q_i - Q_a = \frac{C}{4}\frac{dP_i}{dt},$$

$$P_i - P_x = \frac{R}{2}Q_a + \frac{L}{2}\frac{dQ_a}{dt},$$

$$Q_a - Q_b = \frac{C}{2}\frac{dP_x}{dt}, \qquad (5.72)$$

$$P_x - P_o = \frac{R}{2}Q_b + \frac{L}{2}\frac{dQ_b}{dt},$$

$$Q_b - Q_o = \frac{C}{4}\frac{dP_o}{dt}.$$

These equations can be resolved only when the input and the output pressure–flow relationships have been included to close the solution. The solution approach, however, is common to other systems, flow differences are integrated to evaluate pressure, and pressure differences are integrated to evaluate flow rates. Experience has shown that it is better to rearrange the pressure-drop equations into individual terms and then integrate to determine the pressure.

If the mean flow rate through a pipe is turbulent, then the lumped analogy may be used with the "resistance" replaced with the nonlinear pressure-drop equation:

$$\Delta p = k_t Q^{1.75}, \qquad k_t = \frac{0.24\rho \ell v^{0.25}}{d^{4.75}}. \qquad (5.73)$$

Considering the line as a two-lump model then allows the equations previously described to now be written:

$$Q_i - Q_a = \frac{C}{4}\frac{dP_i}{dt},$$

$$P_i - P_x = \frac{k_t}{2}Q_a^{1.75} + \frac{L}{2}\frac{dQ_a}{dt},$$

$$Q_a - Q_b = \frac{C}{2}, \qquad (5.74)$$

$$P_x - P_o = \frac{k_t}{2}Q_b^{1.75} + \frac{L}{2}\frac{dQ_b}{dt},$$

$$Q_b - Q_o = \frac{C}{4}\frac{dP_o}{dt}.$$

The nonlinear pressure-drop terms will considerably increase damping compared with the incorrect use of a laminar-flow-type linear characteristic.

Worked Example 5.3

For a line with a load restrictor having a linear pressure–flow characteristic, determine the transfer function relating the output flow rate to the input flow rate using a single π lumped approximation for line dynamics. Determine the undamped natural frequency if the load resistance is negligible.

(i) Construct the system using a termination load resistance R_L.

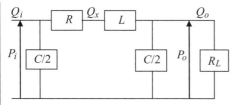

Worked Example 5.3

The equations, when Laplace transformed with zero initial conditions, then become:

time domain s domain

$$Q_i - Q_x = \frac{C}{2}\frac{dP_i}{dt} \rightarrow Q_i(s) - Q_x(s) = \frac{sC}{2}P_i(s)$$

$$P_i - P_o = RQ_x + L\frac{dQ_x}{dt} \rightarrow P_i(s) - P_o(s) = (R + sL)Q_x(s)$$

$$Q_x - Q_o = \frac{C}{2}\frac{dP_o}{dt} \rightarrow Q_x(s) - Q_o(s) = \frac{sC}{2}P_o(s)$$

$$P_o = R_L Q_o \rightarrow P_o(s) = R_L Q_o(s)$$

Rearranging then gives:

$$\frac{Q_o(s)}{Q_i(s)} = \frac{1}{1 + s\left(CR_L + \dfrac{CR}{2}\right) + s^2\left(\dfrac{LC}{2} + \dfrac{C^2 RR_L}{4}\right) + s^3 \dfrac{LC^2 R_L}{4}}.$$

(ii) If the load resistance is negligible, $R_L = 0$, then:

$$\frac{Q_o(s)}{Q_i(s)} = \frac{1}{1 + s\dfrac{CR}{2} + s^2 \dfrac{LC}{2}}, \qquad \omega_n = \sqrt{\frac{2}{LC}} = \frac{\sqrt{2}C_o}{\ell}$$

where C_o is the velocity of sound in the pipe and ℓ is the pipe length.

5.7 Frequency Response

Now consider the behavior of a system if a sinusoidal signal is applied to its input; for example, if a sinusoidal current is applied to a servovalve–motor drive as shown in Fig. 5.17.

Assuming a linear system, initially the output will fluctuate and then also settle down to a steady-state sine wave, but with a phase difference and amplitude change compared with the input. The phase difference and amplitude of the output will change as the frequency changes. It can be shown by Laplace transform theory that for steady-state conditions, the amplitude ratio relating the output sinusoid to the input sinusoid and the phase angle relating the output sinusoid to the input sinusoid are given by:

$$\text{amplitude ratio, } \frac{N}{I} = |G(s)|_{s=j\omega}, \tag{5.75}$$

$$\text{phase angle, } \phi = \angle G(s)|_{s=j\omega}. \tag{5.76}$$

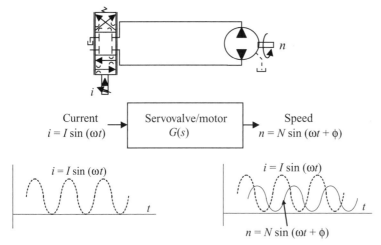

Figure 5.17. Frequency response of a servovalve–motor drive.

That is, the transfer function is evaluated as a complex vector with real and imaginary parts:

$$G(j\omega) = A + jB, \tag{5.77}$$

$$\text{amplitude ratio, } \frac{N}{I} = \sqrt{A^2 + B^2}, \tag{5.78}$$

$$\text{phase angle, } \phi = \tan^{-1} B/A. \tag{5.79}$$

The frequency response can therefore be represented as a single-graph polar plot, similar to an Argand diagram, or as a logarithmic plot, using two graphs to display magnitude and phase angle. Polar plots tend not to be used in practice because the amplitude ratio and phase of a transfer function can convey more information if plotted separately and in logarithmic coordinates. Such a logarithmic representation is known as a *Bode diagram*. The Bode diagram amplitude ratio is converted to decibels (dB) as follows:

$$\text{amplitude ratio (dB)} = 20 \log \frac{N}{I} = 20 \log |G(s)|_{s=j\omega}. \tag{5.80}$$

The amplitude ratio diagram and the phase angle diagram are both plotted against frequency expressed on a logarithmic basis. When the magnitude is considered, it will be evident that common transfer functions will have asymptotic approximations over some parts of the frequency range, and this can help in deciding the probable form of a transfer function when experimentally measured. Note also that a product of transfer functions becomes additive when a Bode diagram amplitude ratio is considered on a logarithmic basis.

For example, consider a transfer function representing the relationship between the sinusoidal variation in motor speed and the applied servovalve current, and given by

$$\frac{n(s)}{i(s)} = G(s) = \frac{10}{(1 + 0.05s)(1 + 0.01s)}. \tag{5.81}$$

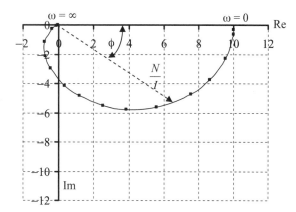

Figure 5.18. Polar plot of a gain and two first-order transfer functions.

For frequency response, change to the frequency domain by means of the transformation $s = j\omega$:

$$\frac{n(j\omega)}{i(j\omega)} = G(j\omega) = \frac{10}{(1 + 0.05\,j\omega)(1 + 0.01\,j\omega)}. \tag{5.82}$$

The amplitude ratio and phase angle are given by:

$$\frac{N}{I} = \frac{10}{\sqrt{[1 + (0.05\omega)^2][1 + (0.01\omega)^2]}}, \tag{5.83}$$

$$\phi = -\tan^{-1} 0.05\omega - \tan^{-1} 0.01\omega. \tag{5.84}$$

Therefore, as the frequency is increased from 0 to ∞, the amplitude ratio varies from 10 to 0 and the phase angle varies from 0 to $-180°$. This frequency response is shown as a polar (vector) plot in Fig. 5.18 as the frequency is increased from zero toward a high value.

Now consider the frequency response plotted as a Bode diagram. It may be readily deduced from Eq. (5.81) that the overall transfer function of the servovalve–motor drive has three individual components – one gain and two individual first-order transfer functions, as follows:

$$10, \quad \frac{1}{(1 + 0.05s)}, \quad \frac{1}{(1 + 0.01s)}. \tag{5.85}$$

Because the magnitude is converted to decibels, each component is added on a log scale. The gain becomes 20 log 10 = 20 dB. Each *first-order lag* function may be considered in general terms, with a time constant τ, as follows:

$$20 \log\frac{1}{(1 + s\tau)} \rightarrow 20 \log\frac{1}{(1 + j\omega\tau)}, \text{ the magnitude being}$$

$$\rightarrow 0\,\text{dB at low frequency}, \tag{5.86}$$

$$\rightarrow 20 \log\frac{1}{\omega\tau} \text{ at high frequency.}$$

Therefore, at high frequency, the slope is -20 dB/decade change in frequency. At a frequency $\omega\tau = 1 \rightarrow \omega = \frac{1}{\tau}$, known as the *break frequency*, the magnitude is 0 dB. Therefore, the magnitude of a first-order lag transfer function has a zero gain (in decibels) from low frequency up to the break frequency and will be asymptotic to a slope of -20 dB/decade as the frequency is increased. This slope is constructed from the break frequency. In the demonstration example, there are two first-order transfer function components; therefore, there are two break frequencies of 20 rad/s and 100 rad/s. The magnitude (amplitude ratio) will follow the following *asymptotic* shape as the frequency is increased from zero:

- a constant of 20 dB at low frequency up to 20 rad/s,
- a slope of −20 dB/decade from 20 rad/s up to 100 rad/s,
- a slope of −40 dB/decade from 100 rad/s onward.

The Bode diagram is shown in Fig. 5.19.

It will be seen from Fig. 5.19 that the asymptotic approximation is actually a good first estimate of the gain (amplitude ratio) variation with frequency and has historically been used extensively to speed up the design process in the absence of computer technology now in common use. This has shown its value when, for example, additional compensating circuits are to be considered and the dynamic performance assessed.

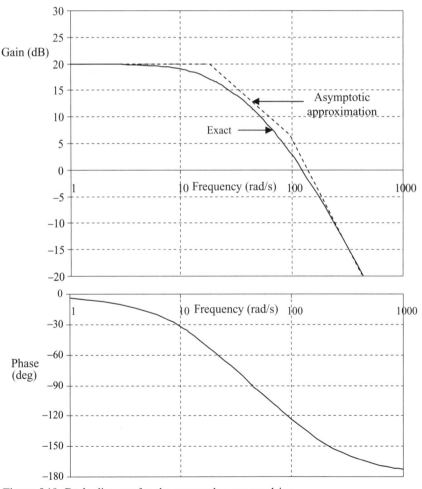

Figure 5.19. Bode diagram for the servovalve–motor drive.

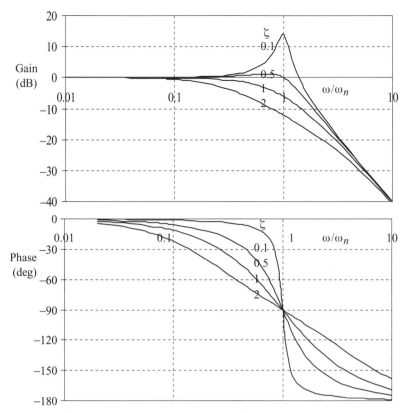

Figure 5.20. Bode diagram for a second-order transfer function.

Second-order transfer functions are commonly met in fluid power systems and it is important to understand their properties for both the overdamped case and the underdamped case. Consider such a transfer function:

$$G(s) = \frac{1}{\left(1 + \dfrac{2\zeta}{\omega_n}s + \dfrac{s^2}{\omega_n^2}\right)},$$

$$G(j\omega) = \frac{1}{\left(1 - \dfrac{\omega^2}{\omega_n^2} + j\dfrac{2\zeta\omega}{\omega_n}\right)}, \tag{5.87}$$

$$|G(j\omega)| = \frac{1}{\sqrt{\left(1 - \dfrac{\omega^2}{\omega_n^2}\right)^2 + \left(\dfrac{2\zeta\omega}{\omega_n}\right)^2}}, \quad \angle G(j\omega) = -\tan^{-1}\frac{\dfrac{2\zeta\omega}{\omega_n}}{1 - \dfrac{\omega^2}{\omega_n^2}}.$$

Figure 5.20 shows the Bode diagram for this second-order transfer function. This transfer function has properties that are useful for identifying it by using measured data:

- When $\omega = \omega_n$, then the phase angle is $\angle G(j\omega) = -90°$.
- When $\omega = \omega_n$, then the magnitude is $|G(j\omega)| = \dfrac{1}{2\zeta}$.

- The peak magnitude occurs when:

$$\omega = \omega_n\sqrt{1 - 2\zeta^2}, \qquad |G(j\omega)| = \frac{1}{2\zeta\sqrt{1 - \zeta^2}}. \qquad (5.88)$$

Some practical issues are as follows:

- The frequency-response method described is applicable to linear systems. It may also be applied to nonlinear systems for small variations about a steady-state operating point, the transfer function to be verified being deduced by a linearized analysis.
- The practical measurement of frequency response will become difficult as the frequency is increased because of the reducing output amplitude, which will also be masked to some extent by noise.
- It is particularly difficult to obtain practical data for small amplitudes of oscillation – for example, as required by a linearized analysis – because of both noise and transducer sensitivity for small fluctuations about the operating point.

Worked Example 5.4

The following Bode diagram has been experimentally determined for an open-loop servovalve–cylinder drive with the servovalve current as the input and cylinder position as the output. Identify the most probable transfer function.

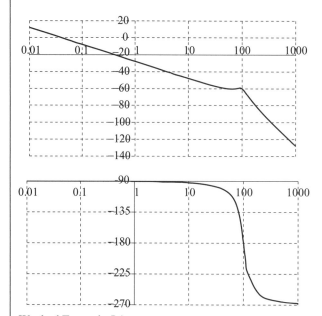

Worked Example 5.4

It is seen that:

(i)   at low frequency, the magnitude slope is -20 dB/decade,
(ii) at high frequency, the magnitude slope is -60 dB/decade,
(iii) at low frequency, the phase is asymptotic to $-90°$, and
(iv) at high frequency, the phase is asymptotic to $-270°$

This suggests a transfer function of the following form:

$$G(s) \approx \frac{K}{s\left(1 + \dfrac{2\zeta s}{\omega_n} + \dfrac{s^2}{\omega_n^2}\right)}.$$

The gain K may be evaluated by considering the low-frequency part of the magnitude plot, say at $\omega = 1\,\text{rad/s}$. The transfer function at such a frequency may be approximated with sufficient accuracy by:

$$G(s) \approx \frac{K}{s} \rightarrow G(j\omega) = \frac{K}{j\omega}.$$

So, the magnitude is read from the data as $-28\,\text{dB}$ at $\omega = 1\,\text{rad/s}$:

$$|G(j\omega)| = \frac{K}{\omega} \rightarrow 20\,\log\frac{K}{1} = -28,$$

$$K = 0.04.$$

The second-order transfer function has a peak at $\omega \approx 100\,\text{rad/s}$, and this has an associated phase angle of $-90°$ from the second-order contribution alone. Because the damping is low, this frequency must be very near to the undamped natural frequency ω_n. At the undamped natural frequency,

$$G(j\omega) = \frac{K}{j\omega(2\zeta)} \rightarrow |G(j\omega)| = \frac{K}{\omega(2\zeta)}.$$

It can be seen from the data that because the magnitude is $-60\,\text{dB}$, then:

$$20\,\log\frac{0.04}{100(2\zeta)} = -60,$$

$$\zeta = 0.2.$$

This low damping ratio validates the assumption that the undamped natural frequency $\omega_n \approx 100\,\text{rad/s}$, and the deduction of the damping ratio is probably a good estimate.

5.8 Optimum Transfer Functions, the ITAE Criterion

In many cases in which there are system parameters in a transfer function that can be varied, it is possible to select the best parameters such that the response to a step input to the system is at a defined optimum. This is done by solving the step response and selecting the parameters in the transfer function such that the response satisfies a particular criterion by minimizing the error in some way between the demand and the actual output. The most popular error criteria are the integral of error squared (IES) and the integral of time multiplied by absolute error (ITAE):

$$\text{IES min} \int_{t=0}^{\infty} e^2(t)\,dt, \tag{5.89}$$

$$\text{ITAE min} \int_{t=0}^{\infty} t|e(t)|\,dt. \tag{5.90}$$

Table 5.3. *Some transfer functions based on the*
ITAE criterion

$n = 2$	$\dfrac{1}{1 + \dfrac{1.4s}{\omega_o} + \dfrac{s^2}{\omega_o^2}}$
$n = 3$	$\dfrac{1}{1 + \dfrac{2.15s}{\omega_o} + \dfrac{1.75s^2}{\omega_o^2} + \dfrac{s^3}{\omega_o^3}}$
$n = 4$	$\dfrac{1}{1 + \dfrac{2.7s}{\omega_o} + \dfrac{3.4s^2}{\omega_o^2} + \dfrac{2.1s^3}{\omega_o^3} + \dfrac{s^4}{\omega_o^4}}$
$n = 5$	$\dfrac{1}{1 + \dfrac{3.4s}{\omega_o} + \dfrac{5.5s^2}{\omega_o^2} + \dfrac{5.0s^3}{\omega_o^3} + \dfrac{2.8s^4}{\omega_o^4} + \dfrac{s^5}{\omega_o^5}}$

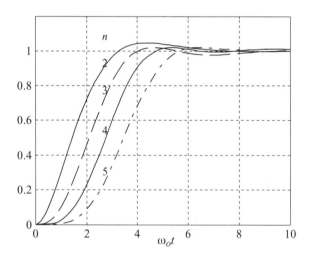

Figure 5.21. Responses to a unit step input for a range of transfer functions satisfying the minimum ITAE.

The solution for the ITAE criterion gives a maximum overshoot of typically less than 5% in response to a step input for a range of transfer functions. Table 5.3 shows the required structure for transfer functions up to fifth order and using a standard transfer function notation with respect to a defining frequency ω_0.

For example, it will be deduced that the solution for a second-order transfer function has a damping ratio of $\zeta = 0.7$. A comparison of transient responses to a step input is shown in Fig. 5.21 for $n = 2, 3, 4, 5$.

Worked Example 5.5

Consider Worked Example 5.3, which analyzed a line with a linear terminal resistance, using a single π lumped approximation for line dynamics.

Determine an expression for the load resistance such that the flow-rate transfer function satisfies the appropriate ITAE criterion, recalling that the transfer function was developed as:

$$\frac{Q_o(s)}{Q_i(s)} = \frac{1}{1 + s\left(CR_L + \dfrac{CR}{2}\right) + s^2\left(\dfrac{LC}{2} + \dfrac{C^2RR_L}{4}\right) + s^3\dfrac{LC^2R_L}{4}}.$$

It is seen that the transfer function is third order and, from Table 5.3:

$$\omega_0^3 = \frac{4}{LC^2R_L}, \quad \frac{1.75}{\omega_0^2} = \frac{LC}{2} + \frac{C^2RR_L}{4}, \quad \frac{2.15}{\omega_0} = CR_L + \frac{CR}{2}.$$

Rearranging gives the equation:

$$2y^2 + y(1 - 3.525X) + 2X = 0,$$

$$\text{where } y = \frac{R_L}{R} \quad \text{and} \quad X = \frac{L}{CR^2}.$$

The solution is easily obtained from this quadratic equation once X is known. A graph showing the solution is shown in the figure:

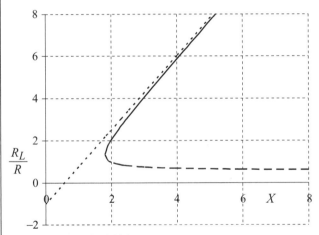

Worked Example 5.5(a)

The graph has a minimum value of $X \approx 1.81$ below which the optimum ITAE criterion cannot be achieved. Also shown is a good approximation given by:

$$\frac{R_L}{R} = 1.763X - 1 \quad \text{for } X > 4.$$

In fact, for larger values of X, the -1 term is negligible and results in:

$$R_L = \frac{0.07\rho\beta}{\mu\ell}.$$

Thus, the load resistance necessary is independent of the line diameter. However, a large value of X is associated with the dynamic behavior's being dominated by the load resistance. In other words, the line dynamics becomes less significant. For this example, the achievement of the optimum ITAE response is associated with very small diameter lines.

For example, consider a 2-mm-diameter line, 4 m long, used for remotely sampling the fluid. Assume also the following data:

fluid density, $\rho = 860 \text{ kg/m}^3$, fluid viscosity $\mu = 0.025 \text{ N s/m}^2$
effective bulk modulus $\beta = 1.4 \times 10^9 \text{ N/m}^2$

$$\text{line inductance,} \quad L = \frac{\rho \ell}{a} = 1.1 \times 10^9$$

$$\text{line capacitance,} \quad C = \frac{a\ell}{\beta} = 0.9 \times 10^{-14}$$

$$\text{line resistance,} \quad R = \frac{128 \mu \ell}{\pi d^4} = 0.26 \times 10^{12}$$

$$X = \frac{L}{CR^2} = 1.81$$

This value of X just happens to be the minimum possible, and the two equal solutions for the load resistance are:

$$R_L = 1.35 R.$$

Selecting this load resistance gives the following behavior for the flow at the end of the line, given a step input flow rate of 2 L/min.

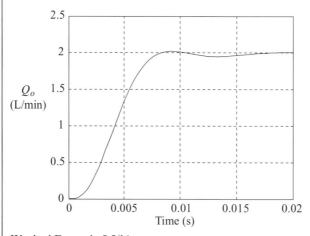

Worked Example 5.5(b)

The ITAE criterion is not usually applied to passive systems such as Worked Example 5.5 and becomes particularly useful when active dynamical components and systems are to be designed, particularly closed-loop control systems.

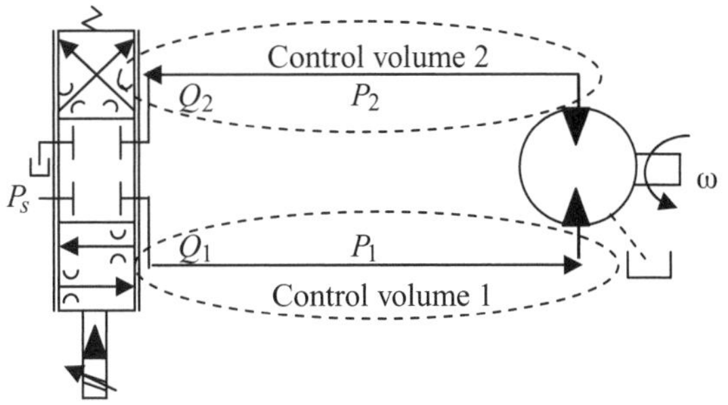

Figure 5.22. A servovalve–motor drive.

5.9 Application to a Servovalve–Motor Open-Loop Drive

5.9.1 Forming the Equations

It is instructive to first consider a motor drive because of the likelihood of fluid volume symmetry on either side. The flow-continuity and torque equations are brought together, but because motor leakage cannot usually be avoided, the various leakages must be taken into account, as discussed in Chapters 3 and 4. Consider the basic drive concept shown in Fig 5.22 with a critically lapped servovalve.

Servovalve flow rates:

$$Q_1 = k_f i \sqrt{P_s - P_1} \quad \text{when } i > 0, \qquad = k_f i \sqrt{P_1} \quad \text{when } i < 0,$$
$$Q_2 = k_f i \sqrt{P_2} \quad \text{when } i > 0, \qquad = k_f i \sqrt{P_s - P_2} \quad \text{when } i < 0. \tag{5.91}$$

Motor flow continuity:

$$\text{control volume 1,} \quad Q_1 - \left[\frac{(P_1 - P_2)}{R_i} + \frac{P_1}{R_e} \right] = D_m \omega + \frac{V_1}{\beta} \frac{dP_1}{dt},$$

$$\text{control volume 2,} \quad \left[\frac{(P_1 - P_2)}{R_i} - \frac{P_2}{R_e} \right] - Q_2 = -D_m \omega + \frac{V_2}{\beta} \frac{dP_2}{dt},$$

$$Q_1 = D_m \omega + \frac{(P_1 - P_2)}{R_i} + \frac{P_1}{R_e} + \frac{V_1}{\beta} \frac{dP_1}{dt}, \tag{5.92}$$

$$Q_2 = D_m \omega + \frac{(P_1 - P_2)}{R_i} - \frac{P_2}{R_e} - \frac{V_2}{\beta} \frac{dP_2}{dt}.$$

Motor torque:

$$D_m(P_1 - P_2) = T_{cf} + B_v \omega + T_{load} + J \frac{d\omega}{dt}. \tag{5.93}$$

5.9.2 An Estimate of Dynamic Behavior by a Linearized Analysis

Linearizing the flow-continuity and torque equations about a steady-state operating condition gives:

$$k_{i1} \delta i - k_{p1} \delta P_1 = D_m \delta \omega + \frac{\delta(P_1 - P_2)}{R_i} + \frac{\delta P_1}{R_e} + \frac{V_1(0)}{\beta} \frac{d\delta P_1}{dt},$$

$$k_{i2} \delta i + k_{p2} \delta P_2 = D_m \delta \omega + \frac{\delta(P_1 - P_2)}{R_i} - \frac{\delta P_2}{R_e} - \frac{V_2(0)}{\beta} \frac{d\delta P_2}{dt}, \tag{5.94}$$

$$D_m \delta(P_1 - P_2) = B_v \delta \omega + \delta T_{load} + J \frac{d\delta \omega}{dt}. \tag{5.95}$$

The linearized coefficients cannot be determined explicitly, but it is possible if the approximation for flow rates is made as discussed in Chapters 3 and 4 by assuming that $P_1 + P_2 \approx P_s$. The operating conditions for pressure differential $P_1(0) - P_2(0)$ and speed $\omega(0)$ are then as follows:

$$k_f i(0) \sqrt{\frac{P_s - [P_1(0) - P_2(0)]}{2}} \approx D_m \omega(0) + \frac{[P_1(0) - P_2(0)]}{R_m},$$

$$\frac{1}{R_m} = \frac{1}{R_i} + \frac{1}{2R_e}, \qquad (5.96)$$

$$[P_1(0) - P_2(0)] = \frac{T_{cf}(0) + B_v \omega(0) + T_{load}(0)}{D_m}. \qquad (5.97)$$

If the two flow gains and the two pressure gains are now equal because of the pressure assumptions made, then:

$$k_i \delta i - k_p \delta P_1 = D_m \delta \omega + \frac{\delta(P_1 - P_2)}{R_i} + \frac{\delta P_1}{R_e} + \frac{V_1(0)}{\beta} \frac{d\delta P_1}{dt},$$

$$k_i \delta i + k_p \delta P_2 = D_m \delta \omega + \frac{\delta(P_1 - P_2)}{R_i} - \frac{\delta P_2}{R_e} - \frac{V_2(0)}{\beta} \frac{d\delta P_2}{dt}, \qquad (5.98)$$

$$D_m \delta(P_1 - P_2) = B_v \delta \omega + \delta T_{load} + J \frac{d\delta \omega}{dt}, \qquad (5.99)$$

$$\text{flow gain,} \ k_i = k_f \sqrt{\frac{P_s - [P_1(0) - P_2(0)]}{2}}, \qquad (5.100)$$

$$\text{pressure gain,} \ k_p = \frac{k_f i(0)}{2\sqrt{2}\sqrt{P_s - [P_1(0) - P_2(0)]}}. \qquad (5.101)$$

Taking Laplace transforms and assuming equal volumes on either side then gives the transfer function:

$$\delta\omega(s) = \frac{\dfrac{k_i \delta i(s)}{D_m} - \left(\dfrac{1}{2R} + \dfrac{1}{R_m} + s\dfrac{C}{2} \right) \dfrac{\delta T_{load}(s)}{D_m^2}}{1 + \dfrac{R_v}{2R} + \dfrac{R_v}{R_m} + s\left(\dfrac{L}{2R} + \dfrac{L}{R_m} + \dfrac{CR_v}{2} \right) + s^2 \dfrac{LC}{2}},$$

$$R = \frac{1}{k_p} = \frac{2(P_s - [P_1(0) - P_2(0)])}{D_m \omega(0) + \dfrac{[P_1(0) - P_2(0)]}{R_m}}, \qquad (5.102)$$

$$R_v = \frac{B_v}{D_m^2}, \quad C = \frac{V}{\beta}, \quad L = \frac{J}{D_m^2}.$$

Considering changes in speed with reference to changes in servovalve current then gives:

$$\delta\omega(s) = \frac{\dfrac{k_i \delta i(s)}{D_m}}{1 + \dfrac{R_v}{2R} + \dfrac{R_v}{R_m} + s\left(\dfrac{L}{2R} + \dfrac{L}{R_m} + \dfrac{CR_v}{2} \right) + s^2 \dfrac{LC}{2}}. \qquad (5.103)$$

Clearly, this is a second-order open-loop system and will have a damping ratio that is heavily dependent on the servovalve resistance R. This resistance changes with steady-state pressure differential because of the steady-state load torque and the steady-state speed.

Worked Example 5.6

Consider the practical system discussed in Chapters 3 and 4 for three steady-state speed conditions of 0, 234 rpm, and 903 rpm. Data are as follows:

Shell Tellus ISO 32 mineral oil at 40°C
Fluid effective bulk modules $\beta = 1.4 \times 10^9$ N/m^2
Supply pressure $P_s = 100$ bar
$D_m = 1.68 \times 10^{-6}$ m^3/rad
$R_m = 1.59 \times 10^{12}$ N m^{-2}/m^3 s^{-1}
Motor torque losses $= 2$ N m at 234 rpm and 3 N m at 903 rpm
Motor viscous torque loss coefficient $B_v = 0.02$ N m/rad s^{-1}
Volume on one side $V = 9.2 \times 10^{-6}$ m^3
Motor and load inertia $J = 0.014$ kg m^2

Evaluating the various constant terms of the transfer function gives:

$$R_v = \frac{B_v}{D_m^2} = \frac{0.02}{1.68^2 \times 10^{-12}} = 7.09 \times 10^9 \text{ N m}^{-2}/\text{m}^3\text{ s}^{-1},$$

$$C = \frac{V}{\beta} = \frac{9.2 \times 10^{-6}}{1.4 \times 10^9} = 6.57 \times 10^{-15}, \quad L = \frac{J}{D_m^2} = \frac{0.014}{1.68^2 \times 10^{-12}} = 5 \times 10^9.$$

Now consider the three steady-state speed conditions of 0, 234 rpm, and 903 rpm. An estimate of the most poorly damped case can be made by considering the limiting zero speed with a zero steady-state pressure differential. This gives, at zero speed:

$$R = \frac{2\{P_s - [P_1(0) - P_2(0)]\}}{D_m\omega(0) + \dfrac{[P_1(0) - P_2(0)]}{R_m}} = \infty \text{ N m}^{-2}/\text{m}^3\text{ s}^{-1},$$

$$\frac{R_v}{2R} = 0, \quad \frac{R_v}{R_m} = 0.00446, \quad 1 + \frac{R_v}{2R} + \frac{R_v}{R_m} = 1.0045,$$

$$\frac{L}{2R} = 0 \text{ s}, \quad \frac{L}{R_m} = 0.0031 \text{ s}, \quad \frac{CR_v}{2} = 0.23 \times 10^{-4} \text{ s},$$

undamped natural frequency $\omega_n = 247$ rad s^{-1},   damping ratio $\zeta = 0.38$.

At 234 rpm $= 24.5$ rad s^{-1}:

$$R = \frac{2\{P_s - [P_1(0) - P_2(0)]\}}{D_m\omega(0) + \dfrac{[P_1(0) - P_2(0)]}{R_m}} = \frac{176.2 \times 10^5}{(41.2 \times 10^{-6} + 0.7 \times 10^6)},$$

$$R = 4.2 \times 10^{11} \text{ N m}^{-2}/\text{m}^3\text{ s}^{-1} \quad \frac{R_v}{2R} = 0.00844, \quad \frac{R_v}{R_m} = 0.00446,$$

$$\frac{L}{2R} = 0.006 \text{ s}, \quad \frac{L}{R_m} = 0.0031 \text{ s}, \quad \frac{CR_v}{2} = 0.23 \times 10^{-4} \text{ s},$$

undamped natural frequency $\omega_n = 247$ rad s^{-1}, damping ratio $\zeta = 1.12$.

At 903 rpm $= 94.6$ rad s^{-1}:

$$R = \frac{2\{P_s - [P_1(0) - P_2(0)]\}}{D_m\omega(0) + \frac{[P_1(0) - P_2(0)]}{R_m}} = \frac{164.2 \times 10^5}{(158.9 \times 10^{-6} + 1.1 \times 10^6)}$$

$$R = 1.0 \times 10^{11} \text{ N m}^{-2}/\text{m}^3 \text{ s}^{-1}, \quad \frac{R_v}{2R} = 0.0355, \quad \frac{R_v}{R_m} = 0.00446,$$

$$\frac{L}{2R} = 0.025 \text{ s}, \quad \frac{L}{R_m} = 0.0031 \text{s}, \quad \frac{CR_v}{2} = 0.23 \times 10^{-4} \text{ s},$$

undamped natural frequency $\omega_n = 252$ rad s^{-1}, damping ratio $\zeta = 3.39$.

Worked Example 5.6 leads to the following observations:

- Viscous friction effects contribute at very low speeds when servovalve resistance becomes less significant.
- Motor inherent leakage may also be negligible in terms of providing adequate damping.
- Around zero speed, at which viscous and friction damping becomes dominant, the system damping ratio may be very low with the distinct possibility of damaging oscillation if the motor is suddenly stopped. The motor stiction characteristic will provide additional damping but oscillations will probably still occur.
- It can be seen from these two steady-state speed conditions that the small signal dynamic response about the steady-state conditions considered are heavily damped.
- The damping ratio increases as the motor steady-state speed increases, as is well known.

Considering transfer function equation (5.103) further, it will be seen that if the servovalve resistance dominates, then the damping ratio may be written as:

$$\zeta = \frac{L}{2R}\sqrt{\frac{2}{LC}} \quad \text{and aim for a damping ratio of } \frac{\sqrt{2}}{2},$$

$$X = \frac{L}{CR^2} = 1. \tag{5.104}$$

Expanding further and considering the no-load condition gives an approximate solution for motor speed as:

$$R \rightarrow \frac{2P_s}{D_m\omega(0)}, \quad \omega(0) = \sqrt{\frac{4V(0)P_s^2}{J\beta}}. \tag{5.105}$$

For the current example, this gives a motor speed of 13.7 rad s^{-1} (130 rpm).

Determining the small-signal transient response about an operating point is difficult in practice because of both hydraulic noise and speed transducer accuracy and response over such a small speed change necessary to validate a linearized analysis.

Considering the large-signal response does tend to reveal system characteristics similar to those of the small-signal response. For the present example, it was shown

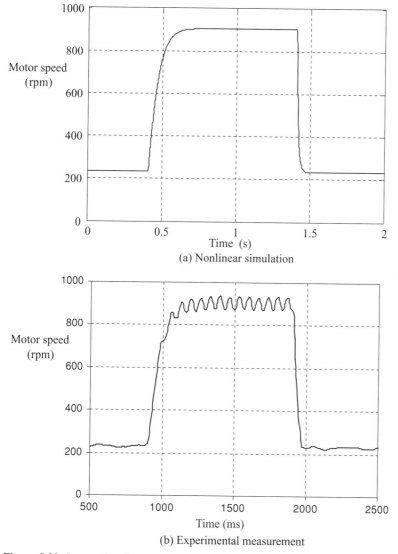

Figure 5.23. Large-signal response of a servovalve–motor open-loop drive.

that the linearized damping ratio around 234 rpm was $\zeta = 1.12$ and around 903 rpm was $\zeta = 3.39$. The exact nonlinear simulation result for speed changes between these two speed values is shown in Fig. 5.23 and is compared with the experimental measurement for the same speed range.

The large-signal response shows that the system is more heavily damped as the speed is increased. In general, the damping of a nonlinear hydraulic system response is greater than predicted by the linearized solution. The measurement was taken with a torque–speed transducer mounted between the test motor shaft and its load pump and illustrates some practical realities. The speed sensor uses an optical technique that has its own dynamic response characteristic. For large speed increases, the response will be further yet slightly damped and, for rapid speed decreases with much reduced system damping, the transducer has a limiting speed rate of change-tracking characteristics. In addition, the transducer is mounted by connecting between the motor and pump shafts with flexible couplings, and this reflects an out-of-balance speed effect and motor–pump-ripple effects at motor speed, and more evident at higher speeds. Given the transducer and mounting limitations, the measurement shows a good correlation with the simulation result, validating system dynamics damping predicted by a linearized analysis.

5.9.3 A Comparison of Nonlinear and Linearized Equations Using the Phase-Plane Method

Phase-plane methods are graphical methods that can handle differential equations up to second order with the possibility of the inclusion of first-order controllers if followed by a restricted class of nonlinear element. As such, the phase-plane method is restricted to a narrow range of dynamical system modeling, but it does have the advantage of being able to handle a range of nonlinearities without the need to utilize computer analysis.

    The system differential equations are rearranged into a form that is suitable for the construction of trajectories in the phase plane, the phase plane being a plot of the time differential of the parameter being studied against the parameter. For a speed control system, the phase-plane plot would be motor acceleration against motor speed, and for a position control system, the phase-plane plot would be velocity against position. The phase-plane plot therefore gives no indication of time, although it can be computed, but the general shape of the plot gives a good indication of the type of time response that would exist, and peak magnitudes may be determined. Consider then the motor open-loop speed control system previously discussed by means of a linearized analysis. Leakage and friction will be neglected to understand the basic dynamics when fluid compressibility and load inertia are dominant. Therefore, it follows that the two differential equations become:

$$k_f i \sqrt{\frac{P_s - P_\ell}{2}} = D_m \omega + \frac{V}{\beta} \frac{dP_\ell}{dt}, \tag{5.106}$$

$$D_m P_\ell = J \frac{d\omega}{dt}, \tag{5.107}$$

$$\text{where } P_\ell = P_1 - P_2.$$

These two equations may then be combined and written in nondimensional form and nondimensional time as follows:

$$\sqrt{1 - X \frac{d\overline{\omega}}{d\overline{t}}} = \overline{\omega} + X \frac{d^2 \overline{\omega}}{d\overline{t}^2},$$

$$\overline{\omega} = \frac{\omega}{\omega(0)}, \quad \overline{t} = \frac{t}{\tau}, \quad \omega(0) = \frac{k_f i(0)}{D_m} \sqrt{\frac{P_s}{2}}, \tag{5.108}$$

$$\tau = \frac{V P_s}{\beta D_m \omega(0)}, \quad X = \frac{J \beta \omega^2(0)}{V P_S^2}.$$

In this example, the phase plane requires a plot of acceleration versus speed a/ω; in this case, $\overline{a}/\overline{\omega}$. To aid this graphical approach, an isocline m is defined as the slope of the phase-plane trajectory and, therefore, it follows that for this example,

$$\frac{d^2 \overline{\omega}}{d\overline{t}^2} = \frac{d\overline{a}}{d\overline{t}} = \frac{d\overline{a}}{d\overline{\omega}} \frac{d\overline{\omega}}{d\overline{t}} = m\overline{a}. \tag{5.109}$$

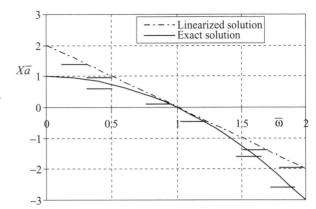

Figure 5.24. Phase-plane isoclines for $m = 0$ for an open-loop motor drive.

Consequently, the first of Eq. (5.108) becomes:

$$\sqrt{1 - X\bar{a}} = \bar{\omega} + Xm\bar{a}.$$ (5.110)

The linearized form of Eq. (5.108), by means of Eq. (5.109), is given by:

$$1 - \frac{X\bar{a}}{2} = \bar{\omega} + Xm\bar{a}.$$ (5.111)

Therefore, by selecting m, the isocline line of constant slope m may be constructed with either the exact equation or the linearized equation. Hence, the response trajectory may be graphically constructed. The *particular isocline, $m = 0$*, is useful because this gives the upper limit of acceleration and, from Eqs. (5.110) and (5.111):

$$\text{exact solution, } \bar{a} = \frac{(1 - \bar{\omega}^2)}{X},$$ (5.112)

$$\text{linearized solution } \bar{a} = \frac{2(1 - \bar{\omega})}{X}.$$ (5.113)

Equations (5.112) and (5.113) are shown in Fig. 5.24.

It is seen from this plot that the maximum positive acceleration predicted by the linearized solution is greater than that predicted by the exact solution and for any steady-state motor speed. Consequently, it is deduced that *the linearized solution is always more oscillatory that the exact solution*. A comparison of trajectories is given in Fig. 5.25 for the dynamic parameter $X = 1$.

It can be seen from Fig. 5.25 that the linearized approximation still gives a good feel for the dynamic behavior of the drive and allows an element of predesign before a more complex computer simulation is pursued.

5.10 Application to a Servovalve–Linear Actuator Open-Loop Drive

5.10.1 Forming the Equations

The equations for the servovalve flow rates, actuator flow rates, fluid compressibility, load mass, and load force equation may now be combined to characterize the connected system shown in Fig. 5.26 and in a similar manner as that of the previous section, in which a motor actuator was used.

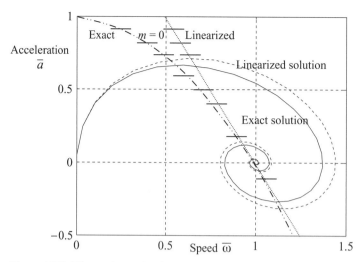

Figure 5.25. Phase-plane plot for a motor open-loop speed drive $X = 1$.

Considering the critically lapped servovalve equations, with flow continuity on both sides, and the load force equation then gives the *servovalve flow equations:*

$$Q_1 = k_f i \sqrt{P_s - P_1} \quad \text{where } i > 0, \quad = k_f i \sqrt{P_1} \quad \text{where } i < 0,$$
$$Q_2 = k_f i \sqrt{P_2} \quad \text{where } i > 0, \quad = k_f i \sqrt{P_s - P_2} \quad \text{where } i < 0. \tag{5.114}$$

When considering the flow-continuity equations for a linear actuator, any small leakage across the piston is usually neglected. Considering, therefore, the generalized flow-continuity equation gives the *actuator flow-continuity equations:*

$$Q_i - Q_o = \frac{dV}{dt} + \frac{V}{\beta}\frac{dP}{dt}, \tag{5.115}$$

$$\text{control volume } 1, Q_1 - 0 = A_1 U + \frac{V_1}{\beta}\frac{dP_1}{dt},$$

$$\text{control volume } 2, \ 0 - Q_2 = -A_2 U + \frac{V_2}{\beta}\frac{dP_2}{dt}, \tag{5.116}$$

$$Q_1 = A_1 U + \frac{V_1}{\beta}\frac{dP_1}{dt},$$

$$Q_2 = A_2 U - \frac{V_2}{\beta}\frac{dP_2}{dt}.$$

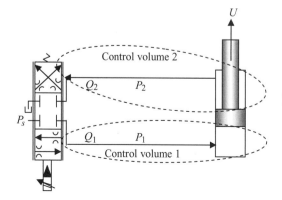

Figure 5.26. A servovalve–linear actuator drive.

Note that both volumes V_1 and V_2 change with time because of actuator motion and also include any initial volumes and connecting line volumes, both lumped together as $V_1(0)$ and $V_2(0)$:

$$V_1 = V_1(0) + A_1 y,$$
$$V_2 = V_2(0) - A_2 y. \tag{5.117}$$

In reality, the change in volume does not often influence the transient response at starting and stopping conditions during the transient stage, although the responses will be different because of the different volumes at each condition. Bringing the flow equations together – for example, for positive current – gives:

$$k_f i \sqrt{P_s - P_1} = A_1 U + \frac{V_1}{\beta} \frac{dP_1}{dt},$$
$$k_f i \sqrt{P_2} = A_2 U - \frac{V_2}{\beta} \frac{dP_2}{dt}. \tag{5.118}$$

Actuator force equation:

$$P_1 A_1 - P_2 A_2 = F_{cf} + B_v U + F_{\text{load}} + M\frac{dU}{dt}. \tag{5.119}$$

A simplification made with an average flow rate cannot be done here, as was the case for a motor drive, because of area asymmetry for the general case. If the areas are the same, $A_1 = A_2 = A$, and because actuator leakage is considered negligible, then it is clear from Eqs. (5.118) and (5.119) that dynamically, the sum of pressure is equal to supply pressure for equal volumes, $P_1 + P_2 = P_s$.

5.10.2 An Estimate of Dynamic Behavior by a Linearized Analysis

The linearization process in this case is more complicated for the general case because of the different actuator areas. The servovalve linearized flow equations combined with the linearized actuator flow equations are written as follows:

$$\delta Q_1 = k_{i1}\delta i - k_{p1}\delta P_1 = A_1 \delta U + \frac{V_1(0)}{\beta} \frac{d\delta P_1}{dt},$$
$$\delta Q_2 = k_{i2}\delta i + k_{p2}\delta P_2 = A_2 \delta U - \frac{V_2(0)}{\beta} \frac{d\delta P_2}{dt}. \tag{5.120}$$

Considering the steady-state pressures at the steady-state operating condition given in Chapter 4, the flow and pressure gains are given in Table 5.4.

The linearized force equation becomes:

$$\delta P_1 A_1 - \delta P_2 A_2 = B_v \delta U + \delta F_{\text{load}} + M\frac{d\delta U}{dt}. \tag{5.121}$$

The flow and torque linearized equations developed are applicable to both extending and retracting cases, although, of course, the current sign is changed and the sign of velocity is changed. We then may write the generalized transfer function by combining Eqs. (5.120) and (5.121) when Laplace transformed:

$$\delta U(s) = \frac{[k_{i1} A_1 n_2(s) + k_{i2} A_2 n_1(s)]\delta i(s) - n_1(s)n_2(s)\delta F(s)}{b_0 + b_1 s + b_2 s^2 + b_3 s^3},$$

Table 5.4. *Linearized coefficients for a servovalve–linear actuator*

Extending	Retracting
$k_{i1} = k_f\sqrt{P_s}\gamma\sqrt{\dfrac{\gamma - \overline{F}}{(1+\gamma^3)}}$	$k_{i1} = k_f\sqrt{P_s}\gamma\sqrt{\dfrac{1 + \overline{F}}{(1+\gamma^3)}}$
$k_{i2} = k_f\sqrt{P_s}\sqrt{\dfrac{\gamma - \overline{F}}{(1+\gamma^3)}}$	$k_{i2} = k_f\sqrt{P_s}\sqrt{\dfrac{1 + \overline{F}}{(1+\gamma^3)}}$
$k_{p1} = \dfrac{k_f i(0)}{2\gamma\sqrt{P_s}}\sqrt{\dfrac{1+\gamma^3}{\gamma - \overline{F}}}$	$k_{p1} = \dfrac{k_f i(0)}{2\gamma\sqrt{P_s}}\sqrt{\dfrac{1+\gamma^3}{1 + \overline{F}}}$
$k_{p2} = \dfrac{k_f i(0)}{2\sqrt{P_s}}\sqrt{\dfrac{1+\gamma^3}{\gamma - \overline{F}}}$	$k_{p2} = \dfrac{k_f i(0)}{2\sqrt{P_s}}\sqrt{\dfrac{1+\gamma^3}{1 + \overline{F}}}$

$$\overline{F} = \frac{F}{P_s A_2}$$

$$n_1(s) = \left(k_{p1} + s\frac{V_1}{\beta}\right), \qquad n_2(s) = \left(k_{p2} + s\frac{V_2}{\beta}\right),$$

$$b_0 = A_1^2 k_{p2} + A_2^2 k_{p1} + B_v k_{p1} k_{p2}, \tag{5.122}$$

$$b_1 = A_1^2\frac{V_2}{\beta} + A_2^2\frac{V_1}{\beta} + B_v\left(k_{p1}\frac{V_2}{\beta} + k_{p2}\frac{V_1}{\beta}\right) + M k_{p1} k_{p2},$$

$$b_2 = M\left(k_{p1}\frac{V_2}{\beta} + k_{p2}\frac{V_1}{\beta}\right) + B_v\frac{V_2}{\beta}\frac{V_1}{\beta}, \qquad b_3 = M\frac{V_2}{\beta}\frac{V_1}{\beta}.$$

5.10.3 Transfer Function Simplification for a Double-Rod Actuator

Transfer function equation (5.122) is difficult to graphically represent because of the many variables. However, for servovalve–actuator systems, it is common and often dynamically preferable to use a double-rod actuator. In addition, it has been established that the actuator undamped natural frequency occurs with equal volumes on either side of the actuator. With the assumptions that $A_1 = A_2 = A$, $V_1 = V_2 = V$, $k_{p1} = k_{p2} = k_p$, and $k_{i1} = k_{i2} = k_i$, it follows that:

$$\delta U(s) = \frac{\dfrac{k_i}{A}\delta i(s) - \dfrac{(1+sCR)}{2A^2 R}\delta F(s)}{1 + \dfrac{R_v}{2R} + s\left(\dfrac{L}{2R} + \dfrac{CR_v}{2}\right) + s^2\dfrac{LC}{2}}, \tag{5.123}$$

$$R = \frac{P_s(1 - \overline{F})}{AU_e(0)}, \quad \text{extending}, \quad R = \frac{P_s(1 + \overline{F})}{AU_r(0)}, \quad \text{retracting},$$

$$k_i = k_f\sqrt{P_s}\sqrt{\frac{1 - \overline{F}}{2}}, \quad \text{extending}, \quad k_i = k_f\sqrt{P_s}\sqrt{\frac{1 + \overline{F}}{2}}, \quad \text{retracting},$$

$$\overline{F} = \frac{F}{P_s A}, \quad R_v = \frac{B_v}{A^2}, \quad C = \frac{V}{\beta}, \quad L = \frac{M}{A^2}. \tag{5.124}$$

This transfer function is similar to the transfer function developed for the servovalve–motor drive and given by Eq. (5.102). If viscous effects are negligible, then a good design in terms of the damping ratio is the same as that defined earlier for a servovalve–motor drive:

$$\zeta = \frac{L}{2R}\sqrt{\frac{2}{LC}} \quad \text{and aim for } \frac{\sqrt{2}}{2}; \quad \text{then } X = \frac{L}{CR^2} = 1. \tag{5.125}$$

This gives the solution for the no-load condition as:

$$R \to \frac{P_s}{AU(0)}, \quad U(0) = \sqrt{\frac{V(0)P_s^2}{J\beta}}. \tag{5.126}$$

The appropriate speed may be achieved in both directions for the no-load condition. Note that the definition of servovalve resistance is different from that defined for the motor actuator.

5.11 Further Considerations of the Nonlinear Flow-Continuity Equations of a Servovalve Connected to a Motor or a Double-Rod Linear Actuator

Bringing the flow equations together for a motor and a double-rod linear actuator gives the equations developed earlier and repeated here, with the assumption that $V_1 = V_2 = V$:

Motor:

$$k_f i \sqrt{P_s - P_1} = D_m \omega + \frac{(P_1 - P_2)}{R_i} + \frac{P_1}{R_e} + \frac{V}{\beta}\frac{dP_1}{dt}, \tag{5.127}$$

$$k_f i \sqrt{P_2} = D_m \omega + \frac{(P_1 - P_2)}{R_i} - \frac{P_2}{R_e} - \frac{V}{\beta}\frac{dP_2}{dt}.$$

Double-rod linear actuator:

$$k_{fi} \sqrt{P_s - P_1} = AU + \frac{V}{\beta}\frac{dP_1}{dt}, \tag{5.128}$$

$$k_{fi} \sqrt{P_2} = AU - \frac{V}{\beta}\frac{dP_2}{dt}.$$

Subtracting the two flow rates then gives the following:

Motor:

$$k_f i \sqrt{P_s - P_1} - k_f i \sqrt{P_2} = \frac{(P_1 + P_2)}{R_e} + \frac{V}{\beta}\frac{d(P_1 + P_2)}{dt}. \tag{5.129}$$

Double-rod linear actuator:

$$k_{fi} \sqrt{P_s - P_1} - k_f \, i \sqrt{P_2} = \frac{V}{\beta}\frac{d(P_1 + P_2)}{dt}. \tag{5.130}$$

- If the two servovalve flow rates are postulated to be dynamically equal, then the left-hand side of both Eqs. (5.129) and (5.130) are zero and can be true only if dynamically $P_1 + P_2 = P_s$.
- If $P_1 + P_2 \approx P_s$ is assumed, then the second term on the right-hand sides of Eqs. (5.129) and (5.130) must also be zero. However, the first term on the right-hand side is nonzero for a motor because of leakage and exactly zero for the linear actuator because leakage is usually negligible.

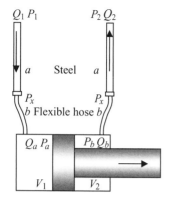

Figure 5.27. Short lines connecting a servovalve to an actuator.

- It is therefore concluded that if the external leakage term is small for a motor, then it is probably true that dynamically $P_1 + P_2 \approx P_s$ for both a motor and a double-rod linear actuator.

Following the previous discussion, the flow-continuity equations are combined into a simplified flow-continuity equation as follows:

Motor:

$$k_f i \sqrt{\frac{P_s - P_{\text{load}}}{2}} = D_m \omega + \frac{V}{\beta} \frac{dP_{\text{load}}}{dt}. \tag{5.131}$$

Double-rod linear actuator:

$$k_f i \sqrt{\frac{P_s - P_{\text{load}}}{2}} = AU + \frac{V}{\beta} \frac{dP_{\text{load}}}{dt}; \tag{5.132}$$

$$P_{\text{load}} = P_1 - P_2,$$

$$P_1 = \frac{P_s + P_{\text{load}}}{2}, \quad P_2 = \frac{P_s - P_{\text{load}}}{2}. \tag{5.133}$$

These similar flow-continuity equations are often used in the analysis of servovalve-controlled systems. It will be deduced that the dynamic increase in one line pressure will correspond to an equal dynamic decrease in the other line pressure.

5.12 The Importance of Short Connecting Lines When the Load Mass Is Small

Consider an example in which the servovalve is connected to the actuator by a pair of similar lines, each line being a combination of steel pipe and flexible hose. This is typical of what happens in practice: A short hose is often used to minimize vibration coupling between components and a steel pipe used to minimize airborne noise. Details of each line are as shown in Fig. 5.27.

Additional data are as follows:

Fluid density $\rho = 860 \, \text{kg/m}^3$
Fluid viscosity $\mu = 0.032 \, \text{N s/m}^2$
Velocity damping $B_v = 2 \times 10^4 \, \text{N/m s}^{-1}$
$\beta_{\text{hose}} = 0.7 \times 10^9 \, \text{N m}^{-2}$
$\beta_{\text{oil}} = 1.4 \times 10^9 \, \text{N m}^{-2}$
Moving mass $m = 25 \, \text{kg}$

Cylinder bore diameter $= 76.2$ mm
Cylinder rod diameter $= 38.1$ mm
Total stroke $= 0.254$ mm
Piston initially centralized
Steel lines 0.9 m long, 13-mm diameter
Hose 0.6 m long, 7-mm diameter

The actuator is horizontal, hence no load force F_{load}, and is not connected to its load mechanism, hence having its minimum mass m. The cylinder is a high-quality product (supplied by Eland Engineering UK) designed for precision control applications and has low-friction seals. Consequently, the stiction–friction force was found to be negligible. Calculating the relevant volumes then gives:

$$A_1 = 4.56 \times 10^{-3}\,\text{m}^2, \quad A_2 = 3.41 \times 10^{-3}\,\text{m}^2,$$
$$a_a = 1.33 \times 10^{-4}\,\text{m}^2, \quad a_b = 0.38 \times 10^{-4}\,\text{m}^2,$$
$$V_1 = 0.58 \times 10^{-3}\,\text{m}^3, \quad V_2 = 0.433 \times 10^{-3}\,\text{m}^3,$$
$$V_a = 0.12 \times 10^{-3}\,\text{m}^3, \quad V_b = 0.023 \times 10^{-3}\,\text{m}^3.$$

Total volume on side 1, $V_{t1} = 0.72 \times 10^{-4}\,\text{m}^3$
Total volume on side 2, $V_{t2} = 0.56 \times 10^{-4}\,\text{m}^3$

From Section 5.5, the undamped natural frequency, neglecting line dynamics and damping, is given by:

$$\omega_n = \sqrt{\frac{\dfrac{A_1^2 \beta_{e1}}{V_{t1}} + \dfrac{A_2^2 \beta_{e2}}{V_{t2}}}{m}} = 1544\,\text{rad/s}\,(246\ \text{Hz}). \tag{5.134}$$

The measured frequency is assessed as 120 Hz, almost half that calculated assuming line dynamics could be neglected. To get a feel for line dynamics, consider inertia effects. This is conveniently done by comparing the inductance for each side. Assuming an estimate for load inductance by combining both A_1 and A_2 then gives:

$$\text{Line } a, \quad L_a = \frac{\rho \ell_a}{a_a} = 5.83 \times 10^6,$$

$$\text{Line } b, \quad L_b = \frac{\rho \ell_b}{a_b} = 13.41 \times 10^6, \tag{5.135}$$

$$\text{Load}, \quad L = \frac{m}{A_1 A_2} = 1.61 \times 10^6.$$

It would therefore seem that for the cylinder with minimum mass, the lines are more dominant than the load in terms of the pressure difference to cause acceleration. In fact, the mass m would have to be increased by a factor of 4 if Eq. (5.134) is appropriate. One way of defining a new mass is to add all the inductances together and evaluate an equivalent inductance L_{eq} and, hence, an equivalent load mass m_{eq}. This is not theoretically correct from a modeling point of view, but it does give a feel for the effect:

$$\frac{m_{eq}}{A_1 A_2} = \frac{m}{A_1 A_2} + \frac{\rho \ell_a}{a_a} + \frac{\rho \ell_b}{a_b},$$

$$m_{eq} = m + \frac{A_1 A_2 \rho \ell_a}{a_a} + \frac{A_1 A_2 \rho \ell_b}{a_b}. \tag{5.136}$$

This gives $m_{eq} = 349.2$ kg, an increase by a factor of 14 and not the factor of 4 required. The conclusion is that line dynamics are significant in this example and must be included in more detail than considering additive inductance effects only; a good dynamic model of each line is required.

As a starting point, consider therefore a simple lumped approximation, as discussed in Section 5.6, in which a single π network is assumed for each steel line and each flexible hose. The servovalve dynamic response is represented by its manufacturer's recommended second-order transfer function with an undamped natural frequency $\omega_{ns} = 140$ Hz and a damping ratio $\zeta_s = 1$. Note that this has no effect on the system's natural frequency. Servovalve spool-position dynamics are represented symbolically by a suitable differential equation transformation to current as follows:

$$i + \frac{2\zeta_s}{\omega_{ns}}\frac{di}{dt} + \frac{1}{\omega_n^2}\frac{d^2 i}{dt^2} = i_d, \tag{5.137}$$

$$Q_1 = k_f i \sqrt{P_s - P_1}, \quad Q_2 = k_f i \sqrt{P_2} \ i > 0,$$

where i_d is the input current.

$$
\begin{array}{ll}
\text{Line-in} & \text{Line-out} \\[6pt]
Q_1 - Q_a = \dfrac{C_a}{2}\dfrac{dP_1}{dt} & A_2 U - Q_b = \left(\dfrac{C_b}{2} + C_2\right)\dfrac{dP_b}{dt} \\[12pt]
P_1 - P_x = R_a Q_a + L_a \dfrac{dQ_a}{dt} & P_b - P_x = R_b Q_b + L_b \dfrac{dQ_b}{dt} \\[12pt]
Q_a - Q_b = \dfrac{(C_a + C_b)}{2}\dfrac{dP_x}{dt} & Q_b - Q_a = \dfrac{(C_a + C_b)}{2}\dfrac{dP_x}{dt} \\[12pt]
P_x - P_a = R_b Q_b + L_b \dfrac{dQ_b}{dt} & P_x - P_2 = R_a Q_a + L_a \dfrac{dQ_a}{dt} \\[12pt]
Q_b - A_1 U = \left(\dfrac{C_b}{2} + C_1\right)\dfrac{dP_a}{dt} & Q_a - Q_2 = \dfrac{C_a}{2}\dfrac{dP_2}{dt}
\end{array}
\tag{5.138}
$$

The pressure at each steel–hose junction is P_x, the internal flow through each steel line is Q_a, and the internal flow through each hose is Q_b:

$$\text{load force,} \quad P_a A_1 - P_b A_2 = F_{cf} + B_v U + F_{load} + M\frac{dU}{dt}, \tag{5.139}$$

$$F_{cf} + F_{load} \approx 0.$$

Computed and measured results for the cylinder extending are shown in Fig. 5.28.

The effect of even a simple model for line dynamics is clear from Fig. 5.28 because the natural frequency has now been reduced from 246 Hz without lines to 127 Hz with short lines and close to the measured frequency of 120 Hz. This frequency reduction is dominated by line inertia effects because line volumes are small compared with the actuator volumes, and line resistance is small compared with the servovalve at steady-state conditions. The initial condition for the measured case, with the actuator initially at rest, has pressures of $P_a = 52.5$ bar and $P_b = 70$ bar as a result of retracting the actuator to its midposition prior to the extending test. These pressures match those predicted by the steady-state theory discussed in Chapter 4. The ratio of pressures at the initial condition is 1.33; the area ratio A_1/A_2. During

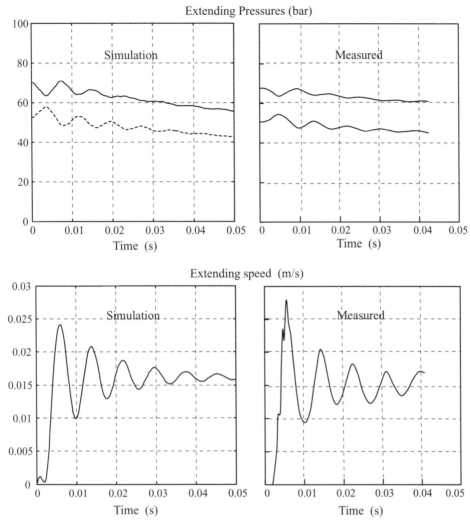

Figure 5.28. The response of a servovalve–cylinder open-loop drive, unloaded and with short lines.

constant-speed motion when extending, the speed $U = 0.016$ m/s, the pressure then will settle to values of $P_a = 30$ bar and $P_b = 40$ bar.

It is therefore concluded that short lines must be considered, particularly if small-diameter sections are used, when the system dynamics is analyzed under low load conditions.

5.13 A Single-Stage PRV with Directional Damping

5.13.1 Introduction

Consider the single-stage valve shown schematically in Fig. 5.29. Damper unit a is preloaded by the main spring b, which is used to set the system pressure, and the spring–damper assembly rests on top of valve spindle c. During operation, the spindle rests in a position to maintain the system pressure required, the surplus flow passing through radial circular exit ports d.

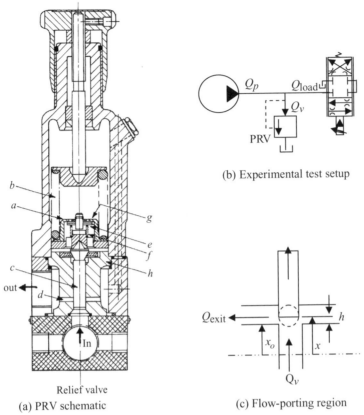

(b) Experimental test setup

(a) PRV schematic   (c) Flow-porting region

Figure 5.29. A single-stage PRV with directional damping (manufactured by The Oilgear Company, USA).

A basic single-stage configuration would generally be dynamically undesirable because of the relatively low viscous damping that would naturally occur in practice between the moving parts. The manufacturer has neatly overcome this problem by introducing directional damping at damper element a. Directional damping is achieved by spring pad e, which is restrained in the upward direction by damper body g, but may move downward against weak restraining spring f. Hence, if the valve assembly moves upward (which is defined as positive), fluid behind the damper unit is allowed to pass through the holes in g because they will be opened by movement of spring pad e. This creates a negligible pressure drop across the damper unit and offers minimum resistance to motion. However, when the damper unit moves downward, the spring pad now closes the flow path through the damper unit, which means that fluid trapped between the damper and the top of the spindle is compressed. A small radial clearance (typically 0.12 mm) exists between the damper unit and the valve main body, and this allows the compressed fluid to flow behind the damper unit. Consequently, a damping force is generated in the negative-velocity direction and can be controlled to some extent at the manufacturing stage.

The significant advantage of this relief-valve design is that it will open rapidly in response to load flow-rate changes and will consequently maintain system pressure with minimum fluctuation. In practice, this result in an extremely stable valve that has widespread applications, particularly in high-flow-rate systems.

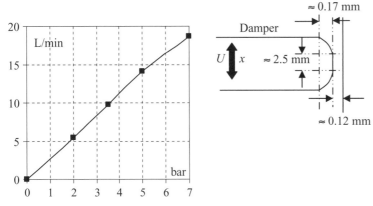

Figure 5.30. Measured pressure–flow data for the PRV damper.

5.13.2 Forming the Equations, Transient Response

Control-Volume Flow Continuity

$$Q_p - Q_{\text{load}} = Q_v + \frac{V}{\beta}\frac{dP}{dt}. \tag{5.140}$$

PRV Flow

Assuming small displacements of the spindle allows the exposed flow area to be expressed by the following approximation:

$$Q_v = 1.7 C_q a_{\text{rad}} \left(\frac{h}{d}\right)^{1.5} \sqrt{\frac{2P}{\rho}} + a_s \frac{dx}{dt}. \tag{5.141}$$

Force Balance at the Spindle

Equating the hydraulic force to the spring force, viscous damper force, flow-reaction force, and acceleration force gives:

$$Pa_s = F(0) + k_x x + B^* \frac{dx}{dt} + 3.4\cos 69° \, C_q a_{\text{rad}} P \left(\frac{h}{d}\right)^{1.5} + m\frac{d^2 x}{dt^2}, \tag{5.142}$$

$$B^* = B^+ \quad \text{for} \quad \frac{dx}{dt} > 0, \quad \text{and} \quad B^- \text{ for } \frac{dx}{dt} < 0.$$

The total cross-sectional area of the radial holes when fully opened by the spindle is a_{rad} and the main spindle cross-sectional area is a_s. A crucial aspect of the PRV design is the damping coefficient for negative-velocity B^-. Therefore, consider Fig. 5.30.

Consider the equation for annular laminar flow resistance (Chapter 3):

$$R_a = \frac{6\mu\ell}{\pi r_o h^3} = \frac{6(0.028)(0.0025)}{\pi(0.0268)(0.00012^3)},$$

$$R_a = 2.89 \times 10^9 \,\text{N m}^{-2}/\text{m}^3\,\text{s}^{-1}.$$

The damper diameter $D = 53.8\,\text{mm}$, its width $\ell = 6.5\,\text{mm}$, and its clearance $h \approx 0.12\,\text{mm}$. The outer edge has a crown design with only typically 2.5 mm at the center causing the flow restriction, and the application of the pressure-drop equation for a uniform annular gap is not strictly applicable. However, it is useful to apply the pressure-drop equation developed in Chapter 3 to get a feel for the resistance of the annular gap.

The measured resistance from the inverse slope of Fig. 5.30 gives a value of $R_a = 2.33 \times 10^9 \, \text{N m}^{-2}/\text{m}^3 \, \text{s}^{-1}$ and close to that estimated from laminar flow theory.

When the damper has negative velocity, it compresses the fluid between the damper and the spindle bush, and the process of velocity damping is actually a dynamic process. The pressure buildup is due to the combination of annular gap restriction and fluid compressibility, leading to the following equation for the force generation F_d:

$$F_d + C R_a \frac{dF_d}{dt} = R_a a_d^2 \frac{dx}{dt},$$

$$C = \frac{V_d}{\beta},$$

(5.143)

where a_d is the damper cross-sectional area. Inserting data $V_d = 3.2 \times 10^{-5} \, \text{m}^3$, $\beta = 1.4 \times 10^9 \, \text{N/m}^2$, shows that the time constant $C R_a = 74 \, \mu\text{s}$ and, hence, the transient force may be neglected. The negative-velocity damping coefficient is given by $B^- = R_a a_d^2 = 1.2 \times 10^4 \, \text{N/m s}^{-1}$. In the computer simulation, a value of $B^- = 1.4 \times 10^4 \, \text{N/m s}^{-1}$ was used.

Other data applicable to the PRV are as follows:

Spindle area	$a_s = 0.635 \times 10^{-4} \, \text{m}^2$ (9-mm diameter)
Flow coefficient	$C_q = 0.55$ (see Chapter 3)
Radial ports diameter	$d = 4 \, \text{mm}$
Radial ports total area	$a_{\text{rad}} = 0.76 \times 10^{-4} \, \text{m}^3$ (six ports)
Main spring stiffness	$k_x = 34.3 \times 10^3 \, \text{N/m}$
Spring preload	$F(0) = 408.3 \, \text{N}$
Damper coefficient	$B^+ = 0.05 \times 10^4 \, \text{N/m s}^{-1}$
Damper coefficient	$B^- = 1.4 \times 10^4 \, \text{N/m s}^{-1}$
Enclosed length	$x_0 = 6.1 \, \text{mm}$
Spindle mass	$m = 0.33 \, \text{kg}$
Test volume	$V = 2.3 \times 10^{-3} \, \text{m}^3$
Fluid density	$\rho = 860 \, \text{kg/m}^3$
Fluid bulk modulus	$\beta = 1.4 \times 10^9 \, \text{N/m}^2$

A comparison between measurement and simulation is shown in Fig. 5.31.

The PRV is tested by use of a load servovalve with the load lines connected, and is rapidly switched on and off to create a load-resistive path that is dynamically much faster than the PRV–fluid volume combination being tested. In reality, when measurements are compared with theory, the servovalve dynamics do have a slight effect and should be included. For this study, the manufacturer's data indicate a second-order lag transfer function $G_{\text{sv}}(s)$ with a damping ratio $\zeta \approx 1$ and an undamped natural frequency $f_n \approx 100 \, \text{Hz}$:

$$(Q_p - Q_{\text{load}})(s) = \frac{k_i i(s)}{\left(1 + \dfrac{2\zeta}{\omega_n} s + \dfrac{s^2}{\omega_n^2}\right)}.$$

(5.144)

The test condition is that the initial pressure is $P(0) = 96.5$ bar with the PRV inoperative and the spindle in a position to just open the exit ports; that is, $x = 6.1$ mm. The pump flow, 15 L/min, and initially passing through the servovalve is then rapidly

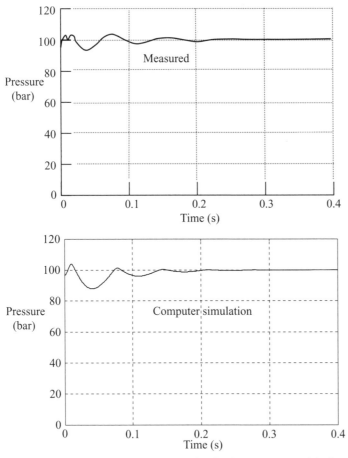

Figure 5.31. The transient response of a single-stage PRV with directional damping.

switched to the PRV by switching the servovalve current to zero. These conditions then result in the pressure's increasing to $P(0) = 100$ bar, with a valve lift $h(0) = 0.325$ mm.

5.13.3 Frequency Response from a Linearized Transfer Function Analysis

The comparison between measurement and simulation is encouraging in the sense that the predicted trend and damped frequency are similar. It is therefore useful to next consider the linearized transfer function because this will give an indication of which PRV properties are dominant. The linearized equations, when Laplace transformed neglecting initial conditions, are:

$$\delta Q_v(s) = k_q \delta h(s) + \frac{\delta P(s)}{R_v} + a_s s h(s) + sC\delta P(s), \tag{5.145}$$

$$k_q = \frac{1.5 Q_v(0)}{h(0)}, \quad R_v = \frac{2P(0)}{Q_v(0)}, \quad C = \frac{V}{\beta}, \tag{5.146}$$

$$\delta P(s)a_s = k_x \delta h(s) + B^* s \delta h(s) + k_f \delta h(s) + k_r a_{\mathrm{rad}} \delta P(s) + ms^2 \delta h(s), \tag{5.147}$$

$$k_f = \frac{a_{\mathrm{rad}} P(0)\sqrt{\dfrac{h(0)}{d}}}{d}, \quad k_r = 0.67 \left(\frac{h(0)}{d}\right)^{1.5}. \tag{5.148}$$

Combining these equations then gives the transfer function:

$$\delta P(s) = \frac{R_v[\delta Q_p - \delta Q_{\text{load}}](s)[k_x + k_f + B^*s + ms^2]}{[R_v(a_s - k_r a_{\text{rad}})(k_q + a_s s) + (1 + sCR_v)(k_x + k_f + B^*s + ms^2)]}. \quad (5.149)$$

The first condition to be satisfied is that all coefficients in the denominator of the transfer function should be positive; that is:

$$a_s - k_r a_{\text{rad}} > 0 \rightarrow 1 - 0.67 \left(\frac{h(0)}{d}\right)^{1.5} \frac{a_{\text{rad}}}{a_s} > 0. \quad (5.150)$$

This is satisfied and, for the valve being considered, the second term may be neglected. Also, by considering the total linearized spring stiffness, we have:

$$R_v k_q (a_s - k_r a_{\text{rad}}) + k_x + k_f \approx R_v k_q a_s + k_x + k_f$$

$$= \frac{3P(0)a_s}{h(0)} + k_x + \frac{a_{\text{rad}}P(0)}{d}\left(\frac{h(0)}{d}\right)^{0.5} \quad (5.151)$$

$$= 5.91 \times 10^6 + 0.034 \times 10^6 + 0.054 \times 10^6.$$

It can be seen that the first term in Eq. (5.151) significantly dominates, and the transfer is then simplified to:

$$\frac{\delta P(s)}{K[\delta Q_p - \delta Q_{\text{load}}](s)} = \frac{[1 + a_1 s + a_2 s^2]}{[1 + b_1 s + b_2 s^2 + b_3 s^3]} \quad (5.152)$$

$$K = \frac{(k_x + k_f)h(0)}{1.5a_s Q_v(0)}, \quad a_1 = \frac{B^*}{(k_x + k_f)}, \quad a_2 = \frac{m}{(k_x + k_f)}, \quad (5.153)$$

$$b_1 = \frac{a_s h(0)}{1.5 Q_v(0)} + \frac{B^* h(0)}{3P(0)a_s} + \frac{(k_x + k_f)Ch(0)}{1.5a_s Q_v(0)},$$

$$b_2 = \frac{mh(0)}{3P(0)a_s} + \frac{CB^* h(0)}{1.5a_s Q_v(0)}, \quad b_3 = \frac{mCh(0)}{1.5a_s Q_v(0)}. \quad (5.154)$$

This transfer function can be evaluated once a decision on an appropriate value for the nonlinear velocity damping coefficient B^* has been made. It seems reasonable to plot the transfer function using only the average of the two velocity damping coefficients, $B^* = 0.73 \times 10^4$ N/m s^{-1}.

The experimental data were obtained by applying a test signal to the servovalve and measuring the pressure response. A two-channel transfer function analyzer was used that generated a pseudorandom binary sequence signal (prbs) output to excite the servovalve, the pressure transducer signal being then used to determine the transfer function by the cross-spectrum-analysis technique. A comparison between the linearized transfer function and the measurement is shown in Fig. 5.32.

The theoretical transfer function should be modified by adding the servovalve transfer function as described earlier. It is evident from the data that this will make only a minor modification to the theoretical transfer function. However, it is clear that the linearized transfer function does indicate a good general trend with the experimental data, given the highly nonlinear aspect of the unique velocity damping unit.

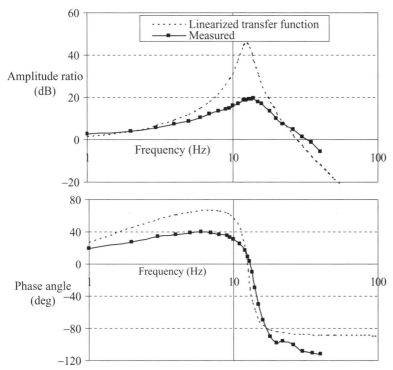

Figure 5.32. A comparison of computed and measured transfer functions for a single-stage PRV with directional damping.

5.14 Servovalve Dynamics

Considering the design of a servovalve – for example, of the force-feedback type shown in Fig. 5.33, will make it clear what components contribute toward the overall dynamic performance.

The use of a current-feedback servoamplifier means that the current-buildup characteristic is extremely fast when compared with other elements of the servo-valve. For example, Fig. 5.34 shows some measured characteristics.

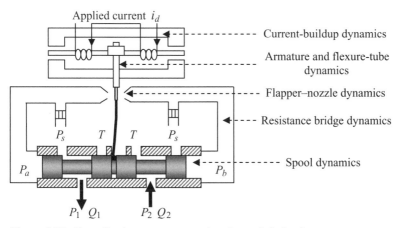

Figure 5.33. Contributions to a servovalve dynamic behavior.

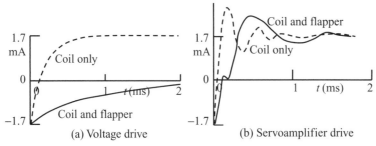

Figure 5.34. Current buildup for a servovalve.

If the current is derived from a direct-voltage source, then the current buildup will be relatively sluggish, as shown in Fig. 5.34(a); although, of course, the servovalve will still be functional. With a servoamplifier current drive, as shown in Fig. 5.34(b), the natural frequency ω_{nc} is of the order of 1 kHz.

A feeling for the complexity of the remaining dynamic characteristic can be obtained by considering some aspects of each internal element. The following analysis is therefore intended only as an indication of the way forward, and a particular servovalve will have its unique set of defining equations.

First-Stage, Armature, and Flapper–Nozzle

Considering the steady-state servovalve first-stage and flapper–nozzle stage theory and the notation developed previously in Chapter 3, the current-buildup and the dynamic torque equations will be of the following types:

$$i + \frac{2\zeta}{\omega_n}\frac{di}{dt} + \frac{1}{\omega_{nc}}\frac{d^2i}{dt^2} = i_d, \tag{5.155}$$

$$k_t i = (k_a - k_m)\theta + (P_a - P_b)a_n r + k[x_s + (r+b)\theta](r+b) + B_v\frac{d\theta}{dt} + J\frac{d^2\theta}{dt^2}. \tag{5.156}$$

Here, the flapper viscous velocity rotational damping torque, coefficient B_v, and rotational acceleration torque, inertia J, have been added to the steady-state torque balance. The spool displacement is x_s. The applied current is i_d.

Flapper–Nozzle and Resistance Bridge Flow Characteristic

Considering flow continuity on each side gives:

$$C_{qo}a_o\sqrt{\frac{2(P_s - P_a)}{\rho}} - C_{qn}a_{nx}\sqrt{\frac{2P_a}{\rho}} = +a_s\frac{dx_s}{dt} + \frac{V_a}{\beta}\frac{dP_a}{dt},$$

$$\tag{5.157}$$

$$C_{qo}a_o\sqrt{\frac{2(P_s - P_b)}{\rho}} - C_{qn}a_{ny}\sqrt{\frac{2P_b}{\rho}} = -a_s\frac{dx_s}{dt} + \frac{V_b}{\beta}\frac{dP_b}{dt},$$

$$a_{nx} = \pi d_n(x_{nm} - x), \quad a_{ny} = \pi d_n(x_{nm} + x), \quad x = r\theta, \tag{5.158}$$

where V_a and V_b are the internal, small volumes on either side of and within the resistance bridge. The flapper displacement at the nozzle is x, and its maximum displacement is x_{nm}. The flapper rotation θ and displacement x are obtained from the previous torque balance equation and the spool displacement x_s and is obtained from the spool force balance equation.

Force Balance at the Spool

The static force balance, including the flow-reaction force, is now modified to include the dynamic flow-reaction force, the spool viscous damping, and acceleration effects:

$$(P_a - P_b)a_s = k[x_s + (r + b)\theta] + 2C_q^2 w x_s \cos\theta[P_s - P_{\text{load}}],$$

$$+ \rho\ell\left(\frac{dQ_1}{dt} - \frac{dQ_2}{dt}\right) + B_s\frac{dx_s}{dt} + m\frac{d^2x_s}{dt^2}, \qquad (5.159)$$

$$Q_1 = C_q w x_s\sqrt{\frac{2(P_s - P_1)}{\rho}}, \quad Q_2 = C_q w x_s\sqrt{\frac{2P_2}{\rho}},$$

$$P_{\text{load}} = P_1 - P_2. \qquad (5.160)$$

Here, the spool viscous velocity rotational damping force, coefficient B_s, and acceleration force, mass m, have been added to the steady-state torque balance. For a direct-drive servovalve, the spool-force equation requires a knowledge of the solenoid force variation with applied current and position.

Clearly, the defining equations are nonlinear, and the solution also requires the load specification so that the load pressure differential $P_1 - P_2$ can be derived. Considering the equations presented, it will be seen that a servovalve dynamic performance depends on not only electrical–electromagnetic–geometry parameters but also on the load it is supplying $P_1 - P_2$ and, hence, the load flow rate, the supply pressure P_s, and the magnitude of the input current.

In practice, the dynamic characteristic is often specified by the manufacturer as a frequency-response diagram and is intended to indicate a typical performance range. The frequency response is obtained experimentally and usually with the output ports connected; that is, for the no-load condition. It therefore represents the best performance that can be expected for that particular servovalve. Figure 5.35 shows a typical frequency-response diagram for a servovalve with force feedback. Note, however, that the frequency response could be significantly worse than that shown or even better for high-performance servovalves at a higher cost. The dynamic performance also depends on the servovalve type and is usually slightly worse for proportional valves of the same flow rating but with integrated electrical feedback. In essence, a servovalve is probably available to meet a particular system dynamics requirement.

For the example shown in Fig. 5.35, the small-signal frequency response may be represented by a second-order transfer function approximation with an undamped natural frequency of $\omega_n \approx 170$ Hz and a damping ratio $\zeta \approx 0.8$. This will be valid certainly up to a frequency equal to the undamped natural frequency. For the entire frequency range measured, 500 Hz, a third-order transfer function will have to be fitted to the data.

5.15 An Open-Loop Servovalve–Motor Drive with Line Dynamics Modeled by Lumped Approximations

Servovalve, Dynamics Included, Underlapped Spool

Consider Fig. 5.36. Because this example considers the open-loop behavior, the following approximation for the spool-flow characteristic is sufficient to account for

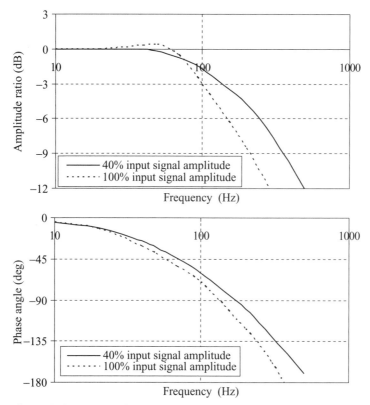

Figure 5.35. A servovalve measured frequency response, 38 L/min rated flow rate.

the flow gain change away from the zero-current condition:

$$Q_1 = k_f(0.071 + 0.212i)\sqrt{P_s - P_1}, \quad i > 1.5\,\text{mA},$$
$$Q_2 = k_f(0.071 + 0.212i)\sqrt{P_2}, \qquad i > 1.5\,\text{mA}.$$

$$(5.161)$$

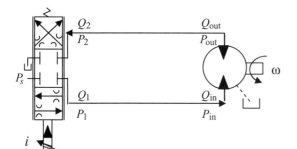

Figure 5.36. A servovalve–motor open-loop drive with interconnecting lines.

Lines, Laminar Mean Flow, Two π Lump Approximations per Line, Negligible Motor Internal Volume

For both lines, the internal pressure node at the center of each line is P_x with flow rate Q_a into it and Q_b after it:

Line 1   Line 2

$$
\begin{aligned}
Q_1 - Q_a &= \frac{C}{4}\frac{dP_1}{dt} & Q_{\text{out}} - Q_a &= \frac{C}{4}\frac{dP_{\text{out}}}{dt} \\
P_1 - P_x &= \frac{R}{2}Q_a + \frac{L}{2}\frac{dQ_a}{dt} & P_{\text{out}} - P_x &= \frac{R}{2}Q_a + \frac{L}{2}\frac{dQ_a}{dt} \\
Q_a - Q_b &= \frac{C}{2}\frac{dP_x}{dt} & Q_a - Q_b &= \frac{C}{2}\frac{dP_x}{dt} \\
P_x - P_{\text{in}} &= \frac{R}{2}Q_b + \frac{L}{2}\frac{dQ_b}{dt} & P_x - P_2 &= \frac{R}{2}Q_b + \frac{L}{2}\frac{dQ_b}{dt} \\
Q_b - Q_{\text{in}} &= \frac{C}{4}\frac{dP_{\text{in}}}{dt} & Q_b - Q_2 &= \frac{C}{4}\frac{dP_2}{dt}
\end{aligned}
\tag{5.162}
$$

If the motor has a significant volume on either side, then the capacitance at the end of line 1 and at the beginning of line 2 must be increased to include the extra volume to give:

$$
\frac{C}{4} = \frac{\text{line volume}/4}{\beta} \rightarrow \frac{\text{line volume}/4 + \text{motor volume}}{\beta}.
\tag{5.163}
$$

Motor Flow and Torque Equations

Flow rates:

$$
Q_{\text{in}} = D_m\omega + \frac{(P_{\text{in}} - P_{\text{out}})}{R_i} + \frac{P_{\text{in}}}{R_e},
\tag{5.164}
$$

$$
Q_{\text{out}} = D_m\omega + \frac{(P_{\text{in}} - P_{\text{out}})}{R_i} - \frac{P_{\text{out}}}{R_e}.
\tag{5.165}
$$

Torque:

$$
D_m(P_1 - P_2) = T_{\text{load}} + T_{\text{cf}} + B_v\omega + J\frac{d\omega}{dt}.
\tag{5.166}
$$

For this study, there is no load torque applied, $T_{\text{load}} = 0$, and the stiction–friction characteristic has a stiction equivalent pressure of 24 bar and a Coulomb friction equivalent pressure of 12 bar, the latter becoming constant for servovalve currents beyond 2 mA. This friction characteristic is therefore important for motor speeds around zero. Further data are as follows:

Servovalve flow constant $k_f = 5.27 \times 10^{-8}$ (current mA)
Both lines 13 mm diameter, 4.63 m long
Mineral oil density $\rho = 860\,\text{kg/m}^3$, viscosity $\mu = 0.033\,\text{N s/m}^2$
Motor leakage dominated by external losses, $R_e = 3 \times 10^{12}\,\text{N m}^{-2}/\text{m}^3\,\text{s}^{-1}$
Motor displacement $D_m = 2.61 \times 10^{-6}\,\text{m}^3/\text{rad}$
Motor and load inertia $J = 0.0069\,\text{kg m}^2$, viscous coefficient $B_v = 0.01\,\text{N m/rad s}^{-1}$
Fluid bulk modulus $\beta = 1.4 \times 10^9\,\text{N m}^{-2}/\text{m}^3\,\text{s}^{-1}$
Velocity of sound in the fluid $C_o = 1276\,\text{m/s}$

Line delay $T = \ell/C_o = 3.63\,\text{ms}$
For one line, $C = 4.39 \times 10^{-13}\,\text{m}^3/\text{N}\,\text{m}^{-2}$, $L = 3 \times 10^7\,\text{kg/m}^4$, $R = 2.18 \times 10^8\,\text{N}\,\text{m}^{-2}/\text{m}^3\,\text{s}^{-1}$.

The servovalve dynamics are represented by a second-order transfer function with an undamped natural frequency of $\omega_{ns} = 110\,\text{Hz}$ and a damping ratio $\zeta = 1$, obtained from experimental work.

A comparison between computer simulation and measurement of pressures at the motor and motor speed is shown in Fig. 5.37. The line effect is theoretically evident from motor pressures but is heavily filtered experimentally. Trend comparisons are good, the results suggesting that the dynamics are actually dominated by the motor rather than by the lines. However, the small pure delay time $\ell/C_o = 3.63\,\text{ms}$ is evident both experimentally and theoretically.

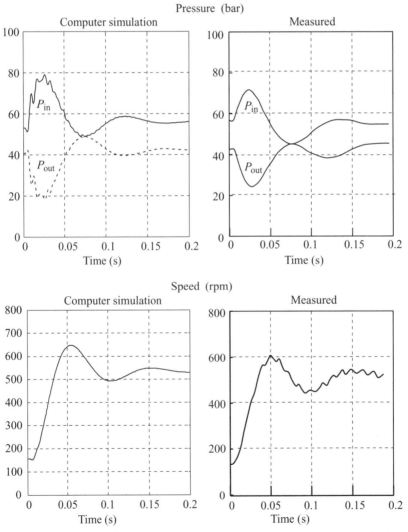

Figure 5.37. The transient response of a motor coupled to a servovalve by long lines and a comparison with a lumped-parameter approximation.

Considering the results and the low-frequency oscillation remote from the line frequency, it is of interest to determine approximate system damping and natural frequency from a linearized analysis at the steady-state operating point. Considering also the previous example on the linearized transfer function for line volume effects only, and neglecting viscous damping effects, then gives:

$$\delta\omega(s) \approx \frac{\dfrac{k_i \delta i(s)}{D_m}}{1 + s\left(\dfrac{L}{2R} + \dfrac{L}{R_m}\right) + s^2 \dfrac{LC}{2}}, \tag{5.167}$$

where L now represents motor inductance J/D_m^2, not line inductance, and R represents the servovalve resistance at the steady-state condition, not line resistance. Inserting the data, $\omega(0) = 55.9\,\text{rad/s}$ and $P_1(0) - P_2(0) = 14.4\,\text{bar}$ gives $R = 1.17 \times 10^{11}\,\text{N m}^{-2}/\text{m}^3\,\text{s}^{-1}$. This results in the transfer function damping ratio $\zeta = 0.15$ and a damped frequency of oscillation of 10.6 Hz. The simulation and measured results both indicate a similar highly oscillatory behavior with the measurement showing a damped frequency just slightly in excess of 10 Hz.

5.16 Transmission Line Dynamics

5.16.1 Introduction

Earlier in this chapter, lumped system components were discussed and, in particular, an interconnecting line between components was considered to be a combination of resistive, inductive, and capacitive elements. Of course, transmission line dynamics are always present in hydraulic systems, but the frequencies of interest are often so high for a good system design that pressure ripples, for example, may appear as only a very small-amplitude oscillation on the large-scale dynamic variation. An approximate guide to determining the significance of the transmission line frequency was established earlier by simply considering the time for a pressure wave to propagate at the speed of sound, C_o, down the line and back again to produce the frequency:

$$f = \frac{C_o}{2\ell}, \tag{5.168}$$

where ℓ is the length of the line. Considering the previous analyses on a servovalve–cylinder and a servovalve–motor, it was found that line dynamics had different influences, depending on not only the line length but also on the servovalve resistance and actuator dynamic characteristics. Considering the appropriate servovalve resistance R, the average capacitance of both sides C, and the load inductance L, then the ratio of mechanical time constant L/R divided by the fluid time constant CR gives the following characteristics for the two examples.

Servovalve–Cylinder with Short Lines and Significant Actuator Volumes

$$X = \frac{L}{CR^2} = \frac{1.61 \times 10^6}{(0.49 \times 10^{-13})(1.56 \times 10^{11})^2} = 0.00135.$$

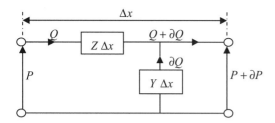

Figure 5.38. Element of a fluid transmission line.

Servovalve–Motor with Long Lines and Negligible Actuator Volumes

$$X = \frac{L}{CR^2} = \frac{10^9}{(4.4 \times 10^{-13})(1.2 \times 10^{11})^2} = 0.16.$$

The study in Section 5.13 showed that the lines had only a secondary effect on the open-loop transient response. A useful rule of thumb seems to be that transmission-line effects probably need to be taken into account when $X \ll 1$.

The analysis of transmission lines is often referred to as *distributed-parameter analysis* because the fluid momentum, state, and mass flow state continuity equation must be applied to an infinitesimal length of line and then "integrated" to produce the required solution when boundary conditions are included. The approach is therefore to consider a small length of line Δx, as shown in Fig. 5.38.

The transmission-line element is uniquely defined by two components referred to as the *series impedance per unit length Z* and the *shunt admittance per unit length Y*. The change in pressure and volumetric flow rate along the element, assuming laminar flow, is then given by:

$$\frac{\partial P}{\partial x} = -ZQ, \quad \frac{\partial Q}{\partial x} = -YP. \tag{5.169}$$

These two equations are then combined to form the wave equation:

$$\frac{\partial^2 P}{\partial x^2} = ZYP, \quad \text{or} \quad \frac{\partial^2 Q}{\partial x^2} = ZYQ. \tag{5.170}$$

The solution of the wave equation may then be obtained by use of the boundary conditions:

$$\begin{aligned} x = 0, \quad P = P_1, \quad \text{and} \quad Q = Q_1; \\ x = \ell, \quad P = P_2, \quad \text{and} \quad Q = Q_2. \end{aligned} \tag{5.171}$$

The solution to the wave equation is then:

$$\begin{aligned} P &= \frac{(P_1 + Z_c Q_1)e^{-\Gamma x}}{2} + \frac{(P_1 - Z_c Q_1)e^{+\Gamma x}}{2}, \\ Q &= \frac{(Q_1 + P_1/Z_c)e^{-\Gamma x}}{2} + \frac{(Q_1 - P_1/Z_c)e^{+\Gamma x}}{2}, \end{aligned} \tag{5.172}$$

where the parameters Z_c and Γ are defined as:

$$\text{characteristic impedance } Z_c = \sqrt{\frac{Z}{Y}}, \tag{5.173}$$

$$\text{propagation factor } \Gamma = \sqrt{ZY}. \tag{5.174}$$

A specific solution using Eq. (5.172) depends on the model chosen to determine the characteristic impedance Z_c and the propagation factor Γ. However, the first terms in Eq. (5.172) suggest a contribution from the forward-traveling wave, whereas the second terms suggest a contribution from the backward or reflected wave. Clearly, the pressure and flow variation down a line will change, depending on the characteristic impedance Z_c and propagation factor Γ. If, for example, the input impedance is chosen to be equal to the characteristic impedance, then:

$$\text{if } \frac{P_1}{Q_1} = Z_{\text{input}} = Z_c,$$

$$P = P_1 e^{-\Gamma x}, \quad Q = Q_1 e^{-\Gamma x}. \tag{5.175}$$

The pressure distribution down the line therefore monotonically decreases with distance x and there are no high- or low-pressure points. This would not be the case for any other input impedance.

Development of the basic transmission-line approach to fluid systems really developed from about 1950, resulting in the foundation on which further studies were based. Initially, work was carried out on pneumatic and water systems, and a variety of analytical techniques in both the frequency domain and the time domain have been used. The frequency domain is perhaps the easiest to handle, particularly for the case in which complicated average friction or distributed friction series impedance Z and shunt admittance Y models are used. For time-domain analysis, perhaps the most efficient method uses modal approximations to the transmission-line equations. For both frequency-domain and time-domain studies, the transmission-line equations are therefore best handled when they are in hyperbolic form. By rearranging Eq. (5.172), the conditions at the end of the line may be expressed in terms of the conditions at the inlet to the line:

$$\begin{bmatrix} P_2 \\ Q_2 \end{bmatrix} = \begin{bmatrix} \cosh \Gamma \ell & -Z_c \sinh \Gamma \ell \\ -\dfrac{\sinh \Gamma \ell}{Z_c} & \cosh \Gamma \ell \end{bmatrix} \begin{bmatrix} P_1 \\ Q_1 \end{bmatrix}. \tag{5.176}$$

To apply the developed solutions to a hydraulic control circuit, it is now necessary to consider the three forms for series impedance Z and shunt admittance Y; that is, lossless line model, average friction model, and distributed friction model.

5.16.2 Lossless Line Model for Z and Y

Recalling the definitions of pressure drop and compressibility flow, per unit length, gives:

$$dP = R'dQ + L'\frac{dQ}{dt}, \tag{5.177}$$

$$dQ = C'\frac{dP}{dt}. \tag{5.178}$$

So, recalling Laplace transforms then gives the following terms for the general case:

$$\text{series impedance,} \quad Z = \frac{dP(s)}{dQ(s)} = R' + sL', \tag{5.179}$$

$$\text{shunt admittance,} \quad Y = \frac{dQ(s)}{dP(s)} = sC', \tag{5.180}$$

where s is the Laplace operator symbolic of differentiation. The per-unit-length parameters are given by:

$$R' = \frac{R}{\ell} = \frac{128\mu}{\pi d^4}, \quad L' = \frac{L}{\ell} = \frac{\rho}{a}, \quad C' = \frac{C}{\ell} = \frac{a}{\beta}, \tag{5.181}$$

where a is the line cross-sectional area and d is its diameter. Hence, for a lossless line, $R' = 0$, and consequently the characteristic impedance and propagation factor are in their simplest form and become:

$$\text{characteristic impedance,} \ Z_c = Z_{ca} = \sqrt{\frac{Z}{Y}} = \sqrt{\frac{L'}{C'}} = \sqrt{\frac{\rho\beta}{a^2}}, \tag{5.182}$$

$$\text{propagation factor,} \quad \Gamma_{ca} = \sqrt{ZY} = s\sqrt{L'C'} = \frac{s}{C_o}, \tag{5.183}$$

$$C_o = \sqrt{\frac{\beta}{\rho}} = \text{velocity of sound in the fluid.} \tag{5.184}$$

The solution for pressure and flow rate at the end of the line is then obtained from Eq. (5.172) to give:

$$P_2 = \frac{(P_1 + Z_{ca}Q_1)e^{-sT}}{2} + \frac{(P_1 - Z_{ca}Q_1)e^{+sT}}{2},$$

$$Q_2 = \frac{(Q_1 + P_1/Z_{ca})e^{-sT}}{2} + \frac{(Q_1 - P_1/Z_{ca})e^{+sT}}{2}, \tag{5.185}$$

$$T = \frac{\ell}{C_o} \ (s).$$

The first terms in Eq. (5.185) contain a pure delay for a forward-propagating wave, e^{-sT}, and the second terms contains a pure delay for a backward-propagating wave, e^{sT}. This means that the lossless line can easily be modeled as a series of pure delays and therefore used, given its limitations, for time-domain simulation of a system. The form of the equations used depends on the system being considered, particularly whether or not actuator volumes are significant. This is illustrated in Fig. 5.39 for a servovalve–motor with insignificant motor volumes and a servovalve–cylinder with significant cylinder volumes.

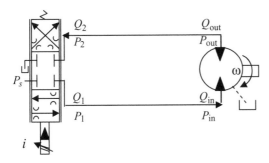

$$P_1 = Z_{ca} Q_1 + (P_{in} - Z_{ca} Q_{in})e^{-sT},$$
$$P_{in} = (P_1 + Z_{ca} Q_1) e^{-sT} - Z_{ca} Q_{in},$$
$$Q_1 = k_{fi} \sqrt{P_s - P_1}, \quad Q_{in} \approx D_m \omega + \frac{P_{in}}{R_m},$$

$$P_{out} = Z_{ca} Q_{out} + (P_2 - Z_{ca} Q_2)e^{-sT},$$
$$P_2 = (P_{out} + Z_{ca} Q_{out})e^{-sT} - Z_{ca}Q_2,$$
$$Q_2 = k_{fi} \sqrt{P_2}, \quad Q_{out} \approx D_m \omega \frac{P_{out}}{R_m},$$

$$D_m(P_{in} - P_{out}) = T_{load} + T_{cf} + B_v \omega + J \frac{d\omega}{dt}.$$

(a) A servovalve–motor open-loop drive with negligible motor volumes

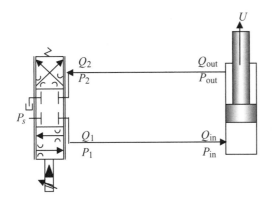

$$P_1 = Z_{ca} Q_1 + (P_{in} - Z_{ca} Q_{in})e^{-sT},$$
$$Q_{in} = \left(\frac{P_1}{Z_{ca}} + Q_1\right)e^{-sT} - \frac{P_{in}}{Z_{ca}},$$
$$Q_1 = k_{fi} \sqrt{P_s - P_1},$$

$$P_2 = (P_{out} + Z_{ca} Q_{out})e^{-sT} - Z_{ca}Q_2,$$
$$Q_{out} = \frac{P_{out}}{Z_{ca}} - \left(\frac{P_2}{Z_{ca}} - Q_2\right)e^{-sT},$$
$$Q_2 = k_{fi} \sqrt{P_2},$$

$$Q_{out} = A_2 U - \frac{V_2}{\beta} \frac{dP_{out}}{dt},$$
$$P_{in}A_1 - P_{out} A_2 = F_{load} + F_{cf} + B_v U + M \frac{dU}{dt},$$
$$Q_{in} = A_1 U + \frac{V_1}{\beta} \frac{dP_{in}}{dt}.$$

(b) A servovalve–cylinder open-loop drive actuator with significant actuator volumes

Figure 5.39. Modeling systems with lossless transmission line models.

Consider therefore *the servovalve–motor drive*, Fig. 5.39(a), and also studied in Section 5.13 with lumped-parameter models used for each line. Further data are as follows:

Motor leakage dominated by external losses $R_e = 3 \times 10^{12}$ N m^{-2}/m^3 s^{-1}
Motor displacement $D_m = 2.61 \times 10^{-6}$ m^3/rad
Motor and load inertia $J = 0.0069$ kg m^2

Motor viscous coefficient $B_v = 0.01 \, \text{N m/rad s}^{-1}$

The stiction–friction characteristic has a stiction equivalent pressure of 24 bar

The Coulomb friction equivalent pressure is 12 bar and constant for $|i| > 2 \, \text{mA}$

For this study, there is no load torque applied, $T_{\text{load}} = 0$

Lines $d = 13 \, \text{mm}$, $\ell = 4.63 \, \text{m}$

Line cross-sectional area $a = 1.33 \times 10^{-4} \, \text{m}^2$

Mineral oil density $\rho = 860 \, \text{kg/m}^3$, viscosity $\mu = 0.033 \, \text{N s/m}^2$

Fluid bulk modulus $\beta = 1.4 \times 10^9 \, \text{N m}^{-2}/\text{m}^3 \, \text{s}^{-1}$

Velocity of sound in the fluid $C_o = 1276 \, \text{m/s}$

Line delay $T = \ell/C_o = 3.63 \, \text{ms}$

For one line, $C = 4.39 \times 10^{-13} \, \text{m}^3/\text{N m}^{-2}$, $L = 3 \times 10^7 \, \text{kg/m}^4$, $R = 2.18 \times 10^8 \, \text{N m}^{-2}/\text{m}^3 \, \text{s}^{-1}$

Characteristic impedance $Z_{ca} = \sqrt{\dfrac{\rho\beta}{a^2}} = 0.83 \times 10^{10} \, \text{N m}^{-2}/\text{m}^3 \, \text{s}^{-1}$

Servovalve flow constant $k_f = 5.27 \times 10^{-8}$ (current mA)

The servovalve dynamics are represented by a second-order transfer function with an undamped natural frequency of $\omega_{ns} = 110 \, \text{Hz}$ and a damping ratio $\zeta = 1$, and obtained from experimental measurements.

A comparison between a computer simulation, using MATLAB Simulink, and practical measurement is shown in Fig. 5.40. Because each line identifying mathematical equations contains no integration blocks, it is important to set the correct initial conditions on each pure delay block. The simulation results are similar to those shown in Fig. 5.37 using lumped-parameter π approximations for each line, but slightly more oscillatory, as might be expected with line resistance neglected for the lossless line model.

5.16.3 Average and Distributed Line Friction Models for Z and Y

For laminar flow, the reintroduction of line friction gives rise to two approaches to determine the series impedance, one based on the lossless line case with friction added, the other based on a more complex solution including heat transfer effects:

(i) *Average friction:*

$$\text{series impedance,} \quad Z = R' + sL', \tag{5.186}$$

$$\text{shunt admittance,} \quad Y = sC', \tag{5.187}$$

$$\text{characteristic impedance,} \quad Z_c = \sqrt{\frac{Z}{Y}} = \sqrt{\frac{R' + sL'}{sC'}}, \tag{5.188}$$

$$\text{propagation factor,} \quad \Gamma = \sqrt{ZY} = \sqrt{(R' + sL')sC'}. \tag{5.189}$$

(ii) *Distributed friction:*

$$\text{series impedance,} \quad Z = \frac{sL'}{1 - \dfrac{2J_1(j\sqrt{sr^2/v})}{j\sqrt{sr^2/v}J_0(j\sqrt{sr^2/v})}}, \tag{5.190}$$

$$\text{shunt admittance,} \quad Y = sC', \tag{5.191}$$

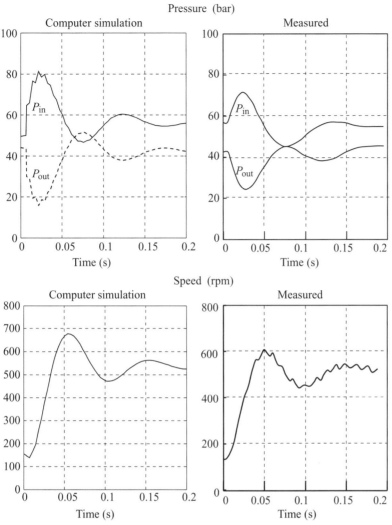

Figure 5.40. The transient response of a motor coupled to a servovalve by long lines and a comparison with a lossless line approximation.

where J_0 is the Bessel function of the first kind of zero order and J_1 is the Bessel function of the first kind of first order. The shunt admittance is unchanged but not if a gas is considered.

The series impedances for average and distributed frictions for a fluid are therefore complex functions and cannot be expressed in a usable explicit form for simulation purposes by using software packages such as MATLAB Simulink. However, it is possible to numerically evaluate the propagation factor for average friction in the frequency domain. Various approximations to these functions have understandingly been considered over many years, with transfer function approximations perhaps offering the best way forward for simulation purposes.

5.16.4 Frequency-Domain Analysis

Letting $s = j\omega$ allows evaluation of the various transmission-line functions for the three cases of zero friction, average friction, and distributed friction. Table 5.5 gives

Table 5.5. *Transmission line functions for various models*

Velocity of sound

Lossless
$$\frac{C}{C_o} = 1$$

Average friction
$$\frac{C}{C_o} = \frac{1}{\left[\frac{1}{2}\left(1 + \frac{64}{F^4}\right)^{1/2} + \frac{1}{2} \right]^{1/2}}$$

Distributed friction
$$\frac{C}{C_o} = \frac{1}{\mathrm{Re}\left(\dfrac{\Gamma}{\Gamma_{ca}}\right)} \qquad \mathrm{Re}\left(\frac{\Gamma}{\Gamma_{ca}}\right) = \text{real component of } \frac{\Gamma}{\Gamma_{ca}}$$

Propagation factor and series impedance

Lossless
$$\frac{\Gamma}{\Gamma_{ca}} = \frac{Z_c}{Z_{ca}} = 1$$

Average friction
$$\frac{\Gamma}{\Gamma_{ca}} = \frac{Z_c}{Z_{ca}} = \left[1 - j\frac{8}{F^2}\right]^{1/2}$$

Distributed friction
$$\frac{\Gamma}{\Gamma_{ca}} = \frac{Z_c}{Z_{ca}} = \frac{1}{\left[1 - \dfrac{2J_1(j^{3/2}F)}{j^{3/2}FJ_0(j^{3/2}F)}\right]^{1/2}}$$

$$C_o = \sqrt{\frac{\beta}{\rho}}, \quad Z_{ca} = \sqrt{\frac{\rho\beta}{a^2}}, \quad \Gamma_{ca} = \frac{j\omega}{C_o}, \quad F = \sqrt{\frac{\omega r^2}{v}}$$

the expressions for the variation of the velocity of sound propagation C_o and the propagation factor Γ with frequency.

The average friction and distributed friction models in the frequency domain are shown in Figs. 5.41 and 5.42.

The distributed friction theory predicts a lower wave propagation velocity, and the propagation factor and series impedance show increased magnitudes as the frequency is increased.

As an example of frequency-domain analysis, consider a simple test set up with a servovalve to generate a sinusoidal input flow rate to a line that has a restrictor termination, as shown in Fig. 5.43. This is, of course, a nonlinear system and, therefore, a transfer function analysis is applicable only for small-signal variations about a steady-state operating point. From the transmission line matrix

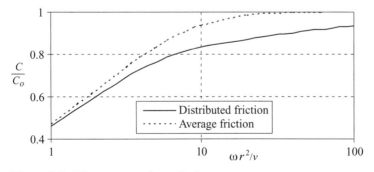

Figure 5.41. Wave propagation velocity.

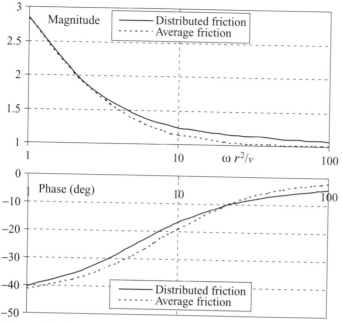

Figure 5.42. Propagation factor and series impedance $\Gamma/\Gamma_{ca} = Z/Z_{ca}$.

equation (5.176), the relationship between the line inlet conditions and outlet conditions may be expressed in a linearized form as follows:

$$\begin{bmatrix} \delta P_2 \\ \delta Q_2 \end{bmatrix} = \begin{bmatrix} \cosh \Gamma\ell & -Z_c \sinh \Gamma\ell \\ -\dfrac{\sinh \Gamma\ell}{Z_c} & \cosh \Gamma\ell \end{bmatrix} \begin{bmatrix} \delta P_1 \\ \delta Q_1 \end{bmatrix}. \tag{5.192}$$

Assuming that the load restrictor is purely resistive and higher-frequency inductive effects can be neglected for a preliminary analysis, the load linearized equation is written as:

$$\delta P_2 = R_L \delta Q_2. \tag{5.193}$$

Rearranging these two equations then gives the following linearized transfer function relating the two pressures:

$$\frac{\delta P_2(j\omega)}{\delta P_1(j\omega)} = \frac{1}{\cosh \Gamma\ell + \dfrac{Z_c}{R_L} \sinh \Gamma\ell}. \tag{5.194}$$

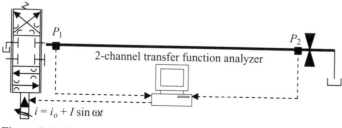

Figure 5.43. Frequency-response testing of a long line with a restrictor termination.

Pressure transducers are positioned at the ends of the line, a variable-frequency current about a mean is applied to the servovalve, and the frequency response is measured with a two-channel frequency-response analyzer. Data for the system are as follows:

> Line length 6 m, 8.5-mm diameter, mean flow rate $= 3.5$ L/min
> Dynamic viscosity $\mu = 0.033$ N s/m^2, density $\rho = 860$ kg/m^3
> Bulk modulus $\beta = 1.4 \times 10^9$ N/m^2, $C_o = 1276$ m/s
> Load orifice linearized resistance $R_L = 1.08 \times 10^{10}$ N m^{-2}/m^3 s^{-1}
> Line resistance $R = 1.54 \times 10^9$ N m^{-2}/m^3 s^{-1}
> Line characteristic impedance $Z_{ca} = 1.93 \times 10^{10}$ N m^{-2}/m^3 s^{-1}

If the simplest, lossless line case is first considered, then Eq. (5.194) becomes

$$\frac{\delta P_2(j\omega)}{\delta P_1(j\omega)} = \frac{1}{\cos\dfrac{\omega\ell}{C_o} + j\dfrac{Z_{ca}}{R_L}\sin\dfrac{\omega\ell}{C_o}}. \tag{5.195}$$

The pressure at the end of the line will therefore not experience any change with frequency if the load is matched to the line characteristic impedance, $R_L = Z_{ca}$. It is evident that, in general, the magnitude ratio has repeated maximum and minimum magnitudes given by:

$$Maximum \text{ when } \quad \sin\frac{\omega\ell}{C_o} = 0 \rightarrow \frac{\omega\ell}{C_o} = 0, \pi, 2\pi, \ldots,$$

$$\left|\frac{\delta P_2(j\omega)}{\delta P_1(j\omega)}\right| = 1, \quad \text{phase angle} = 0°. \tag{5.196}$$

For this example, the frequencies at the maximum magnitude of 0 dB are:

$$f = 106 \text{ Hz}, 213 \text{ Hz}, 319 \text{ Hz}, \ldots,$$

$$Minimum \text{ when } \quad \cos\frac{\omega\ell}{C_o} = 0 \rightarrow \frac{\omega\ell}{C_o} = \frac{\pi}{2}, \frac{3\pi}{2}, \frac{5\pi}{2}, \ldots,$$

$$\left|\frac{\delta P_2(j\omega)}{\delta P_1(j\omega)}\right| = \frac{R_L}{Z_{ca}}, \quad \text{phase angle} = -90°. \tag{5.197}$$

For this example, the frequencies at a minimum magnitude of −5.04 dB are:

$$f = 53 \text{ Hz}, 160 \text{ Hz}, 266 \text{ Hz}, \ldots. \tag{5.198}$$

Line friction in practical hydraulic lines has only a small effect on these frequencies, and a comparison between measurement and theory made with the average friction model for the present example is shown in Fig. 5.44.

The experimental approach is restricted to a frequency upper limit of around 500 Hz when signals become difficult to generate by the servovalve and also to record and analyze by the transfer function analyzer. Note that the average friction solution is little different to the lossless line solution, and comparisons with measurement are reasonable up to a frequency of 200 Hz. For higher frequencies, the distributed friction model should be considered, particularly for noise analysis.

Any transmission line analysis should also take into account the fact that for long lines the temperature will drop along the line and, in the experience of the

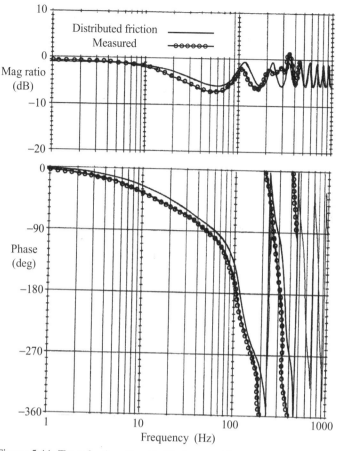

Figure 5.44. Transfer function P_2/P_1 for an orifice-terminated line.

author, can be up to $10°C$. This is difficult to accommodate analytically using the theory presented apart from the lumping-by-length approach, in which viscosity changes with temperature, and can be included for each lump. Of course, the use of other numerical analysis techniques makes it possible to include temperature effects, but these methods are difficult to embrace within a general systems analysis block diagram approach.

5.16.5 Servovalve-Reflected Linearized Coefficients

Now consider the case in which the lines between the servovalve and the actuator may introduce a significant dynamic effect, and consider Fig. 5.45. It is now useful to express the servovalve linearized flow equations in terms of the pressures and flow rates at the actuator rather than pressures and flow rates at the servovalve. By use of the servovalve linearized coefficients and the transmission line equations developed earlier, it follows that:

$$\delta Q_1 = k_i \delta i - k_{p1} \delta P_1 \rightarrow \delta Q_a(s) = \frac{1}{R_1(s)} k_i \delta i(s) - \frac{R_2(s)}{R_1(s)} k_p \delta P_a(s),$$

$$\delta Q_2 = k_i \delta i + k_p \delta P_2 \rightarrow \delta Q_b(s) = \frac{1}{R_1(s)} k_i \delta i(s) + \frac{R_2(s)}{R_1(s)} k_p \delta P_b(s).$$

(5.199)

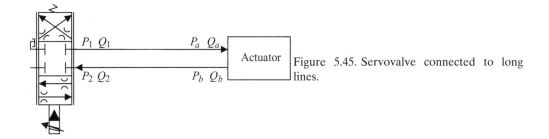

Figure 5.45. Servovalve connected to long lines.

The original servovalve linearized static equations have now been reflected into their dynamic equivalents at the load. The functions $R_1(s)$ and $R_2(s)$ are given by:

$$R_1(s) = \cosh \Gamma\ell + k_p Z_c \sinh \Gamma\ell,$$

$$R_2(s) = \cosh \Gamma\ell + \frac{1}{k_p Z_c} \sinh \Gamma\ell. \tag{5.200}$$

The linearized coefficients are selected appropriately for each line. For the inlet line, then $k_i = k_{i1}$ and $k_p = \mathrm{k_{p1}}$. For the return line, then $k_i = k_{i2}$ and $k_p = \mathrm{k_{p2}}$.

Worked Example 5.7

Determine the linearized transfer function relating motor speed to servovalve current for a servovalve–motor open-loop drive with equal-length transmission lines and assuming that motor load inertia exists and leakages may be neglected.

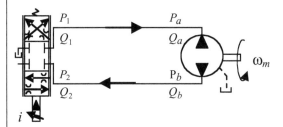

Worked Example 5.7(a)

The transmission line equations are expressed in linearized form for the two lines:

$$\begin{bmatrix} \delta P_a \\ \delta Q_a \end{bmatrix} = \begin{bmatrix} \cosh \Gamma\ell & -Z_c \sinh \Gamma\ell \\ -\dfrac{\sinh \Gamma\ell}{Z_c} & \cosh \Gamma\ell \end{bmatrix} \begin{bmatrix} \delta P_1 \\ \delta Q_1 \end{bmatrix},$$

$$\begin{bmatrix} \delta P_2 \\ \delta Q_2 \end{bmatrix} = \begin{bmatrix} \cosh \Gamma\ell & -Z_c \sinh \Gamma\ell \\ -\dfrac{\sinh \Gamma\ell}{Z_c} & \cosh \Gamma\ell \end{bmatrix} \begin{bmatrix} \delta P_b \\ \delta Q_b \end{bmatrix}.$$

Reflected linearized coefficients are:

$$\delta Q_1 = \frac{1}{R_1(s)} k_i \delta i - \frac{R_2(s)}{R_1(s)} k_p \delta P_a,$$

$$\delta Q_2 = \frac{1}{R_1(s)} k_i \delta i + \frac{R_2(s)}{R_1(s)} k_p \delta P_b.$$

Motor leakages are neglected; hence:

$$\delta Q_a = \delta Q_b = D_m \delta \omega_m.$$

The load is defined as inertia only so that:

$$\delta P_a - \delta P_b = \frac{J}{D_m} \frac{d \delta \omega_m}{dt}.$$

Combining the various equations developed then gives the required transfer function:

$$\frac{D_m \delta \omega_m(s)}{k_i \delta i(s)} = \frac{1}{\cosh \Gamma \ell + k_p Z_c \sinh \Gamma \ell + \dfrac{s L_m k_p}{2} \left(\cosh \Gamma \ell + \dfrac{1}{k_p Z_c} \sinh \Gamma \ell \right)}.$$

The first observation is that if the servovalve linearized resistance is matched to the line characteristic impedance, $1/k_p = R = Z_c$, then:

$$\frac{D_m \delta \omega_m(s)}{k_i \delta i(s)} = \frac{1}{\left(1 + \dfrac{s L_m}{2R} \right) (\cosh \Gamma \ell + \sinh \Gamma \ell)}.$$

This can be achieved only when frequency effects on Z_c are neglected. The lines are then dynamically isolated from the servovalve–motor interaction between resistance and load inertia. In fact, the magnitude ratio is not affected by line dynamics if a lossless model is used, although a monotonically decreasing phase exists because of the line delay effect:

$$\text{magnitude} = \frac{1}{\sqrt{1 + \left(\dfrac{\omega L_m}{2R} \right)^2}}, \quad \text{phase} = -\tan^{-1} \frac{\omega L_m}{2R} - \frac{\omega \ell}{C_o}.$$

A feel for resonant frequencies can be gained to a good accuracy by considering the lossless line case:

$$\frac{D_m \delta \omega_m(j\omega)}{k_i \delta i(j\omega)} = \frac{1}{\underbrace{\left(\cos \dfrac{\omega \ell}{C_o} - \dfrac{\omega L_m}{2 Z_{ca}} \sin \dfrac{\omega \ell}{C_o} \right)}_{A} + j \dfrac{Z_{ca}}{R} \underbrace{\left(\sin \dfrac{\omega \ell}{C_o} + \dfrac{\omega L_m}{2 Z_{ca}} \cos \dfrac{\omega \ell}{C_o} \right)}_{B}}.$$

Assume the following parameters:

$$\frac{L_m}{2 Z_{ca}} = 0.002 \, \text{s}, \quad \frac{\ell}{C_o} = 0.005 \, \text{s}, \quad \frac{Z_{ca}}{R} = 2.$$

The transfer function frequency response may then be constructed.

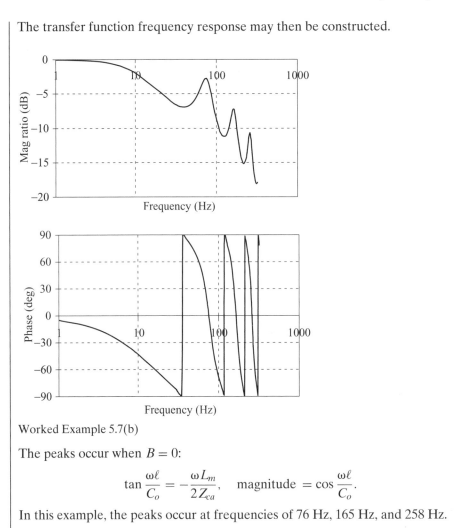

Worked Example 5.7(b)

The peaks occur when $B = 0$:

$$\tan \frac{\omega \ell}{C_o} = -\frac{\omega L_m}{2 Z_{ca}}, \quad \text{magnitude} = \cos \frac{\omega \ell}{C_o}.$$

In this example, the peaks occur at frequencies of 76 Hz, 165 Hz, and 258 Hz.

5.16.6 Modeling Systems with Nonlossless Transmission Lines, the Modal Analysis Method

The problem here is that a method is required that allows the distributed friction hyperbolic functions to be integrated within a block diagram approach such as MATLAB Simulink. Hydraulic transmission lines have received a great deal of attention with regard to the understanding and prediction of dynamic signal transmissions in a range of applications with air, water, and oil as the working fluid. With respect to oil-hydraulic transmission lines, consideration has been given to both frequency-domain and time-domain analysis with a variety of approximations and interpretations of the fundamental distributed-parameter equations. Analysis of typical fluid lines is perhaps most easily achieved in the frequency domain with the usual limitations of small-signal linearization techniques. This has been proved to be highly effective in problems involving pressure-ripple propagation, the evaluation of component impedance, and the allied topic of vibration and noise emission. When large signal fluctuations in pressure and flow rate are experienced, as in conventional hydraulic control systems, together with inherent nonlinear effects, time-domain analysis is inevitably required. A number of approaches are possible such as transmission line modeling (TLM), the method of characteristics, modal analysis, and finite-element techniques.

The TLM technique fundamentally utilizes the lossless, dispersionless transmission line model and requires detailed consideration of other dynamic components within the system and the mathematical linking to the appropriate transmission line termination equations. Such lines that can be characterized by a transmission delay also allow component models to be decoupled for the current numerical time step, enabling a parallel solution technique to be utilized.

The method of characteristics utilizes forward-traveling and backward-traveling wave fronts developed from consideration of the transmission line total differential equation and resulting in two finite-difference equations that may be integrated into a system numerical analysis routine. This method again requires boundary conditions for linked components to be expressed in finite-difference form, and the simplest of circuits requires careful thought and some skill in producing a soluble set of equations. The numerical solution technique is also significantly complicated if the distributed friction transmission line model is to be considered, and the feel of the hydraulic system as a set of real components is rapidly lost.

The technique of modal analysis offers a promising way forward in that the various transmission line transfer functions are developed by frequency-response analysis to produce appropriate first-order and second-order modal approximations, the combination of modes being determined by the accuracy of solution required. The method may be integrated into a modern fluid power simulation package in block diagram form in a relatively easy manner and in a form that is in keeping with conventional simulation techniques. It was proposed that the finite-element technique may alleviate some of the computational problems, such as numerical instability and variable time steps, and the resulting equations are expressed in state-space form. Solutions were obtained for a blocked line, the mathematical modeling requiring consideration of the range of undamped natural frequencies in advance. Results for the lossless line case have been compared with a further theoretical solution using the method of characteristics.

In addition to these various approaches outlined, many approximation techniques have been used to simplify the transmission line equations into a usable form for computer simulation. The various methods have advantages and disadvantages, but the solution technique in most cases is based on distributed-parameter theory with its restriction to laminar flow and uniform fluid properties. In reality, this is unlikely to be the case for lines with large pressure and flow-rate fluctuations and with significant temperature variations between the input and the output of the line.

A method was developed using the modal analysis technique as the foundation theory to establish the form of a set of discrete equations relating pressures and flow rates at both ends of the line. The unknown coefficients of each time-domain equation may then be determined for the practical line by use of measured transient pressure and flow-rate data with the least-squares estimation technique; this is referred to as *data-based modeling*. The modal analysis approach has become popular because it uses modal approximations to the hyperbolic functions defined in the distributed-parameter analysis. The number of modes used is decided by the user; often, four modes are adequate, and the form of the system block diagram determines which hyperbolic functions are to be used. As an example of a practical situation, consider a pair of lines coupled to a component, as shown in Fig. 5.46.

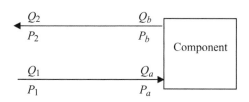

Figure 5.46. A system with a pair of long lines.

Considering Example 5.7, the transmission line matrix equation was manipulated to produce the appropriate pressures and flow rates for both lines to generate the required solution. In terms of a block diagram approach, the way forward becomes obvious when it is recalled that the integration of flow-rate differences is used to generate pressure. Reworking the transmission line equations then gives:

$$\text{inlet line, } P_1 = \frac{Z_c \cosh \Gamma \ell}{\sinh \Gamma \ell}\left[Q_1 - \frac{Q_a}{\cosh \Gamma \ell}\right],$$

$$P_a = \frac{Z_c \cosh \Gamma \ell}{\sinh \Gamma \ell}\left[\frac{Q_1}{\cosh \Gamma \ell} - Q_a\right],$$

$$\text{outlet line, } P_b = \frac{Z_c \cosh \Gamma \ell}{\sinh \Gamma \ell}\left[Q_b - \frac{Q_2}{\cosh \Gamma \ell}\right],$$

$$P_2 = \frac{Z_c \cosh \Gamma \ell}{\sinh \Gamma \ell}\left[\frac{Q_b}{\cosh \Gamma \ell} - Q_2\right]. \tag{5.201}$$

The modal analysis technique now requires the two hyperbolic functions in Eq. (5.201) to be replaced with a series of rational polynomial approximations to the distributed-parameter solution. These approximations were extensively developed (Hsue and Hullender, 1983) for a range of hyperbolic function combinations. In this study, the first function required for n modes is:

$$\frac{1}{\cosh \Gamma \ell} = T_1(s) = \sum_{i=1}^{n} \frac{a_i \bar{s} + b_i}{\bar{s}^2 + 2\zeta_i \omega_{ni} \bar{s} + \omega_{ni}^2}, \quad \bar{s} = \frac{sr^2}{v}. \tag{5.202}$$

The constants for each mode are shown in Table 5.6, and each mode is defined by its root index given by:

$$\lambda_i = \frac{(2i-1)}{2D_n} \quad \text{where the dissipation number } D_n = \frac{v\ell}{C_o r^2}. \tag{5.203}$$

Once the number of modes n has been decided, then each b_i in the summation given in Eq. (5.202) must be multiplied by a suitable gain K such that in the steady-state condition:

$$K \sum_{i=1}^{n} \frac{b_i}{\omega_{ni}^2} = 1. \tag{5.204}$$

The functions are defined as $(-1)^{i+1}(1-2i)a_i$, and $(-1)^{i+1}(1-2i)a_i$ and therefore the sign of each modal transfer function alternates in the pattern $+ - + -$ as the mode number i is incremented. It is therefore common sense to choose an even number of modes, $n = 2, 4, 6$, and so on. The second function required for n modes is:

$$\frac{Z_c \cosh \Gamma \ell}{\sinh \Gamma \ell} = \frac{Z_{ca}}{D_n \bar{s}} + \sum_{i=1}^{n} \frac{a_i \bar{s} + b_i}{\bar{s}^2 + 2\zeta_i \omega_{ni} \bar{s} + \omega_{ni}^2}. \tag{5.205}$$

Table 5.6. *Modal constants for* $\dfrac{1}{\cosh \Gamma \ell}$ *(Hsue and Hullender, 1983)*

λ_i	ω_{ni}	ζ_i	$(-1)^{i+1}(1-2i)a_i$	$(-1)^{i+1}(1-2i)b_i$
0.1	0.2670	10.832	-6.7404×10^{-4}	-9.0758×10^{-2}
0.2	0.5340	5.4172	-2.6952×10^{-3}	-3.6306×10^{-1}
0.4	1.0681	2.7113	-1.0765×10^{-2}	-1.4528
0.6	1.6023	1.8106	-2.4165×10^{-2}	-3.2707
0.8	2.1369	1.3611	-4.2817×10^{-2}	-5.8194
1.0	2.6718	1.0921	-6.6615×10^{-2}	-9.1026
1.4	3.7432	0.7862	-1.2908×10^{-1}	-17.891
2.00	5.3553	0.5593	-2.5710×10^{-1}	-36.725
3.00	8.0609	0.3868	-5.4497×10^{-1}	-83.734
4.00	10.796	0.3036	-8.9161×10^{-1}	-151.34
5.00	13.566	0.2554	-1.2545	-240.78
10.0	27.887	0.1613	-2.4968	-1037.6
15.0	42.640	0.1262	-3.0243	-2419.6
20.0	57.558	0.1061	-3.4562	-4388.6
25.0	72.570	0.0930	-3.8496	-6952.0
30.0	87.647	0.0836	-4.2000	-10113
35.0	102.77	0.0764	-4.5144	-15875
40.0	117.94	0.0707	-4.8011	-18238
45.0	133.14	0.0661	-5.0651	-23204
50.0	148.36	0.0622	-5.3101	-28774
55.0	163.60	0.0589	-5.5387	-34950
60.0	178.86	0.0560	-5.7528	-41731
70.0	209.43	0.0513	-6.1438	-57113
80.0	240.05	0.0475	-6.4932	-74925
90.0	270.71	0.0444	-6.8079	-95171
100	301.41	0.0418	-7.0929	-1.1785×10^5
150	455.23	0.0330	-8.1857	-2.6786×10^5
200	609.42	0.0278	-8.8940	-4.7895×10^5
250	763.82	0.0243	-9.3460	-7.5117×10^5
300	918.36	0.0217	-9.6132	-1.0846×10^6
350	1073.0	0.0197	-9.7418	-1.4791×10^6
400	1227.7	0.0180	-9.7642	-1.9348×10^6
450	1382.4	0.0166	-9.7046	-2.4515×10^6
500	1537.2	0.0155	-9.5817	-3.0294×10^6
550	1692.0	0.0145	-9.4101	-3.6683×10^6
600	1846.8	0.0136	-9.2018	-4.3683×10^6
650	2001.6	0.0129	-8.9662	-5.1293×10^6
700	2156.4	0.0122	-8.7111	-5.9513×10^6
800	2466.1	0.0110	-8.1668	-7.7783×10^6
900	2775.7	0.0100	-7.6061	-9.8490×10^6
1000	3085.4	0.0092	-7.0529	-1.2164×10^7

This approximation is rewritten in the following form for simulation purposes:

$$
\frac{Z_c \cosh \Gamma \ell}{\sinh \Gamma \ell} = \frac{Z_{ca}}{D_n \bar{s}} \left[1 + \sum_{i=1}^{n} \frac{\dfrac{D_n a_i}{Z_{ca}} \bar{s}^2 + \dfrac{D_n b_i}{Z_{ca}} \bar{s}}{\bar{s}^2 + 2\zeta_i \omega_{ni} \bar{s} + \omega_{ni}^2} \right]. \tag{5.206}
$$

The final integration to obtain pressure is therefore done when all the modes have been added together, including the unity gain. Simplifying the nondimensional Laplace operator $\bar{s} = sr^2/v$ then gives:

$$\frac{Z_{ca}}{D_n\bar{s}} = \frac{1}{sC},\qquad(5.207)$$

where $C = (\text{line volume}/\beta)$. Transfer function Eq. (5.206) is then modified to the following form:

$$\frac{Z_c\cosh\Gamma\ell}{\sinh\Gamma\ell} = \frac{T_2(s)}{sC};\quad\text{where}\quad T_2(s) = \left[1 + \sum_{i=1}^{n}\frac{\dfrac{D_na_i}{Z_{ca}}\bar{s}^2 + \dfrac{D_nb_i}{Z_{ca}}\bar{s}}{\bar{s}^2 + 2\zeta_i\omega_{ni}\bar{s} + \omega_{ni}^2}\right],\qquad(5.208)$$

where, in this case, the root index is:

$$\lambda_i = \frac{i}{D_n}.\qquad(5.209)$$

The block diagram operation given by Eq. (5.208) is equivalent to obtaining pressure by integration of flow-rate differences when used with transmission line equation (5.201). The modal constants D_na_i/Z_{ca} and D_nb_i/Z_{ca} may be read directly from Table 5.7.

5.16.7 Modal Analysis Applied to a Servovalve–Motor Open-Loop Drive

Again consider this system that was analyzed experimentally and analytically by a lumped-parameter analysis in Section 5.13 and with the lossless-line theory earlier in this section. The collection of equations is shown in Fig. 5.47.

$$P_{\text{out}} = \frac{T_2(s)}{sC}[Q_{\text{out}} - T_1(s)Q_2],$$

$$P_2 = \frac{T_2(s)}{sC}[T_1(s)Q_{\text{out}} - Q_2],$$

$$Q_2 = k_{fi}\sqrt{P_2},\quad Q_{\text{out}} \approx D_m\omega - \frac{P_{\text{out}}}{R_m},$$

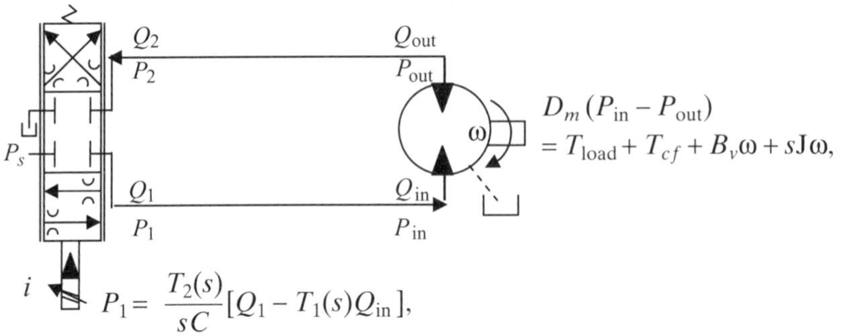

$$D_m(P_{\text{in}} - P_{\text{out}}) = T_{\text{load}} + T_{cf} + B_v\omega + sJ\omega,$$

$$P_1 = \frac{T_2(s)}{sC}[Q_1 - T_1(s)Q_{\text{in}}],$$

$$P_{\text{in}} = \frac{T_2(s)}{sC}[T_1(s)Q_1 - Q_{\text{in}}],$$

$$Q_1 = k_{fi}\sqrt{P_s - P_1},\quad Q_{\text{in}} \approx D_m\omega + \frac{P_{\text{in}}}{R_m}.$$

Figure 5.47. A servovalve–motor open-loop drive with long lines.

Table 5.7. *Modal constants for* $\dfrac{Z_c \cosh \Gamma \ell}{\sinh \Gamma \ell}$ *(Hsue and Hullender, 1983)*

λ_i	ω_{ni}	ζ_i	$D_n a_i / Z_{ca}$	$D_n b_i / Z_{ca}$
0.10	0.2670	10.832	2.0000	11.567
0.20	0.5340	5.4172	2.0000	11.567
0.40	1.0681	2.7113	2.0005	11.569
0.60	1.6023	1.8106	2.0011	11.573
0.80	2.1369	1.3611	2.0019	11.578
1.00	2.6718	1.0921	2.0030	11.584
1.40	3.7432	0.7862	2.0058	11.603
2.00	5.3553	0.5593	2.0114	11.645
3.00	8.0609	0.3866	2.0242	11.765
4.00	10.796	0.3036	2.0394	11.967
5.00	13.566	0.2554	2.0550	12.267
10.0	27.887	0.1613	2.0957	14.935
15.0	42.640	0.1262	2.0904	17.737
20.0	57.558	0.1061	2.0808	19.996
25.0	72.570	0.0930	2.0736	21.941
30.0	87.647	0.0836	2.0680	23.698
35.0	102.77	0.0764	2.0634	25.313
40.0	117.94	0.0707	2.0596	26.815
45.0	133.14	0.0661	2.0564	28.223
50.0	148.36	0.0622	2.0536	29.552
55.0	163.60	0.0589	2.0511	30.813
60.0	178.86	0.0560	2.0490	32.017
70.0	209.43	0.0513	2.0454	34.275
80.0	240.05	0.0475	2.0424	36.369
90.0	270.71	0.0444	2.0399	38.327
100	301.41	0.0418	2.0377	40.172
150	455.23	0.0330	2.0303	48.150
200	609.42	0.0278	2.0257	54.724
250	763.82	0.0243	2.0224	60.376
300	918.36	0.0217	2.0200	65.354
350	1073.0	0.0197	2.0180	69.807
400	1227.7	0.0180	2.0164	73.831
450	1382.4	0.0166	2.0150	77.496
500	1537.2	0.0155	2.0138	80.850
550	1692.0	0.0145	2.0128	83.933
600	1846.8	0.0136	2.0119	86.776
650	2001.6	0.0129	2.0111	89.402
700	2156.4	0.0122	2.0103	91.834
800	2466.1	0.0110	2.0090	96.182
900	2775.7	0.0100	2.0080	99.937
1000	3085.4	0.0092	2.0071	103.19

For solution continuity, the system data are repeated here as follows:

Motor leakage dominated by external losses $R_e = 3 \times 10^{12} \, \text{N m}^{-2}/\text{m}^3 \, \text{s}^{-1}$
Motor displacement $D_m = 2.61 \times 10^{-6} \, \text{m}^3/\text{rad}$
Motor and load inertia $J = 0.0069 \, \text{kg m}^2$
Motor viscous coefficient $B_v = 0.01 \, \text{N m/rad s}^{-1}$

The stiction–friction characteristic has a stiction equivalent pressure of 24 bar
The Coulomb friction equivalent pressure is 12 bar and constant for $|i| > 2\,\text{mA}$
For this study, there is no load torque applied, $T_{\text{load}} = 0$
Lines $d = 13\,\text{mm}$, $\ell = 4.63\,\text{m}$
Line cross sectional area $a = 1.33 \times 10^{-4}\,\text{m}^2$
Mineral oil density $\rho = 860\,\text{kg/m}^3$, viscosity $\mu = 0.033\,\text{N s/m}^2$
Fluid bulk modulus $\beta = 1.4 \times 10^9\,\text{N m}^{-2}/\text{m}^3\,\text{s}^{-1}$
Velocity of sound in the fluid $C_o = 1276\,\text{m/s}$
Line delay $\text{T} = \ell/C_o = 3.63\,\text{ms}$

For one line, $C = 4.39 \times 10^{-13}\,\text{m}^3/\text{N m}^{-2}$, $L = 3 \times 10^7\,\text{kg/m}^4$, $R = 2.18 \times 10^8\,\text{N m}^{-2}/\text{m}^3\,\text{s}^{-1}$:

$$\frac{r^2}{v} = 1.1, \quad D_n = \frac{v\ell}{C_o r^2} = 0.0033. \tag{5.210}$$

The servovalve flow constant $k_f = 5.27 \times 10^{-8}$ (current mA). The servovalve dynamics are represented by a second-order transfer function with an undamped natural frequency of $\omega_{ns} = 110\,\text{Hz}$ and a damping ratio $\zeta = 1$ and are obtained from experimental measurements.

The modal constants are then evaluated for $T_1(s)$ and $T_2(s)$ as follows:

$$\frac{1}{\cosh \Gamma \ell} = T_1(s) = \sum_{i=1}^{n} \frac{b_i}{\omega_{ni}^2} \frac{\left(\dfrac{a_i r^2}{b_i v}s + 1\right)}{\left(\dfrac{r^4}{v^2\omega_{ni}^2}s^2 + \dfrac{2r^2\zeta_i}{v\omega_{ni}}s + 1\right)}, \tag{5.211}$$

$$\lambda_i = 303(i - 1/2), \tag{5.212}$$

$$\frac{Z_c \cosh \Gamma \ell}{\sinh \Gamma \ell} = \frac{T_2(s)}{sC},$$

$$T_2(s) = 1 + \sum_{i=1}^{n} \frac{1}{\omega_{ni}^2} \frac{\left(\dfrac{D_n a_i r^4}{Z_{ca} v^2}s^2 + \dfrac{D_n b_i r^2}{Z_{ca} v}s\right)}{\left(\dfrac{r^4}{v^2\omega_{ni}^2}s^2 + \dfrac{2\zeta_i r^2}{\omega_{ni} v}s + 1\right)}, \tag{5.213}$$

$$\lambda_i = 303\,i. \tag{5.214}$$

Again, this model was simulated using MATLAB Simulink, and a comparison between simulation and measurements is shown in Fig. 5.48. The simulated pressures and speed are slightly better than those obtained with lossless line theory, shown in Fig. 5.40, in terms of peak values. However, the higher modal frequencies are evident analytically on the pressures but not on the speed. Note from Tables 5.8 and 5.9 that the fourth-mode true undamped natural frequency is 479 Hz in both cases, and high for a hydraulic control system. Adding more modes therefore takes the undamped natural frequencies much higher and well beyond the operating dynamic range of the system, particularly if closed-loop control is added.

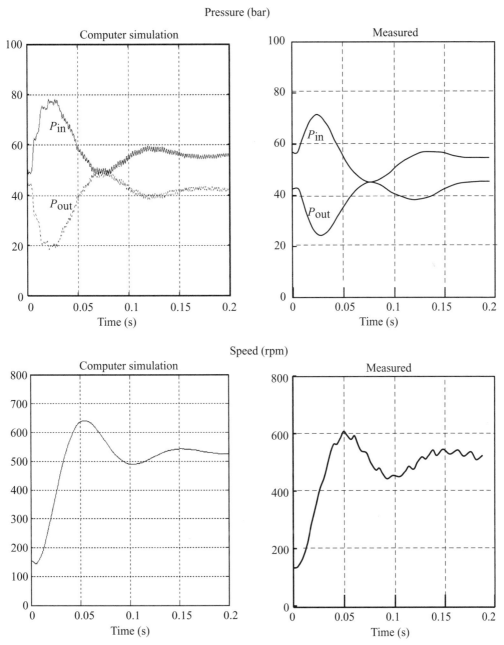

Figure 5.48. The transient response of a motor coupled to a servovalve by long lines and a comparison with a modal analysis approximation.

5.17 The State-Space Method for Linear Systems Modeling

5.17.1 Modeling Principles

The state-space approach is essentially a matrix analysis technique whereby an nth-order system set of linear differential equations is broken down into n first-order linear differential equations. It is therefore a generalized approach and is ideally suited

Table 5.8. *Modal constants for* $\dfrac{1}{\cosh \Gamma \ell}$

i	λ_i	ω_{ni}	ζ_i	a_i	b_i	$b_i/\omega_{ni}{}^2$
1	152	464	0.0327	8.23	2.81×10^5	1.305
2	455	1407	0.0165	-3.23	-8.3×10^5	-0.419
3	758	2351	0.0114	1.67	14.2×10^5	0.257
4	1061	3295	0.0087	-0.95	-19.6×10^5	-0.181
						0.962

i	$a_i r^2/b_i\, \nu$	$r^4/\nu^2\omega_{ni}{}^2$	$2r^2\zeta_i/\nu\omega_{ni}$	corrected $b_i/\omega_{ni}{}^2$
1	3.220×10^{-5}	0.562×10^{-5}	1.550×10^{-4}	1.357
2	0.428×10^{-5}	0.061×10^{-5}	0.258×10^{-4}	-0.436
3	0.129×10^{-5}	0.022×10^{-5}	0.107×10^{-4}	0.267
4	0.053×10^{-5}	0.011×10^{-5}	0.058×10^{-4}	-0.188

to computer analysis and system design. The approach may be extended to multi-variable systems and is a first step toward the application of modern control theory and feedback–adaptive control. To illustrate the approach, consider a servovalve–motor open-loop system. The linear differential equation relating motor speed ω to servovalve input current i, and including fluid compressibility and load inertia, is given for example purposes by:

$$\frac{1}{\omega_n^2}\frac{d^2\omega}{dt^2} + \frac{2\zeta}{\omega_n}\frac{d\omega}{dt} + \omega = Ki. \tag{5.215}$$

Now define the two state variables working from right to left:

$$x_1 = \omega = \text{speed},$$
$$x_2 = \frac{dx_1}{dt} = \frac{d\omega}{dt} = \text{angular acceleration}. \tag{5.216}$$

And, from the original differential equation:

$$x_1 + \frac{2\zeta}{\omega_n}x_2 + \frac{1}{\omega_n^2}\frac{dx_2}{dt} = Ki. \tag{5.217}$$

Table 5.9. *Modal constants for* $\dfrac{Z_{ca}\cosh \Gamma \ell}{\sinh \Gamma \ell}$

i	λ_i	ω_{ni}	ζ_i	$D_n a_i/Z_{ca}$	$D_n b_i/Z_{ca}$
1	303	927	0.0215	2.02	66
2	606	1878	0.0135	2.01	87
3	909	2822	0.0099	2.01	100
4	1212	3300	0.0074	2.00	110

i	$D_n a_i r^4/Z_{ca}\nu^2$	$D_n b_i r^2/Z_{ca}\nu$	$r^4/\nu^2\omega_{ni}{}^2$	$2\zeta_i r^2/\omega_{ni}\nu$	$1/\omega_{ni}{}^2$
1	2.444	72.6	1.408×10^{-6}	5.102×10^{-5}	1.164×10^{-6}
2	2.432	95.7	0.343×10^{-6}	1.581×10^{-5}	0.284×10^{-6}
3	2.432	110	0.152×10^{-6}	0.772×10^{-5}	0.126×10^{-6}
4	2.420	121	0.111×10^{-6}	0.493×10^{-5}	0.009×10^{-6}

The second-order differential equation has now been broken down into two first-order differential equations, as follows:

$$\frac{dx_1}{dt} = x_2, \tag{5.218}$$

$$\frac{dx_2}{dt} = \omega_n^2 \left(Ki - \frac{2\zeta}{\omega_n} x_2 - x_1 \right), \tag{5.219}$$

$$\begin{bmatrix} \dfrac{dx_1}{dt} \\ \dfrac{dx_2}{dt} \end{bmatrix} = \begin{bmatrix} 0 & 1 \\ -\omega_n^2 & -2\zeta\omega_n \end{bmatrix} \begin{bmatrix} x_1 \\ x_2 \end{bmatrix} + \omega_n^2 Ki. \tag{5.220}$$

Using the "overdot" notation to represent differentiation, for both convenience and historical correctness, then the matrix form is written in the following general state-space notation:

$$\dot{x} = Ax + Bu, \tag{5.221}$$

where u is the system input; in this case, the servovalve current i. Taking Laplace transforms then gives:

$$sx(s) - x(0) = Ax(s) + Bu(s),$$
$$x(s) = [sI - A]^{-1}[x(0) + Bu(s)];$$
$$x(s) = [sI - A]^{-1}Bu(s) \quad + \quad [sI - A]^{-1}x(0) \tag{5.222}$$
$$\qquad\qquad \downarrow \qquad\qquad\qquad\qquad \downarrow$$

response to the input   response to the initial conditions.

For control systems studies, the main concerns are the response to the input and closed-loop stability. For linear systems, the conditions for instability of a closed-loop control system do not depend on the system initial conditions. Hence, for conventional transfer function analysis, the first part of Eq. (5.222) is appropriate. Both parts of Eq. (5.222) may be used if the complete transient response is required. However, it will be seen that both parts require the evaluation of $[sI - A]^{-1}$, and this inverse may be evaluated as follows:

$$[sI - A]^{-1} = \frac{\text{adjoint}\,[sI - A]}{\text{determinant}\,[sI - A]}. \tag{5.223}$$

The denominator of Eq. (5.223), that is, *the determinant of* $[sI - A]$, *is therefore the system characteristic equation when equated to zero*:

$$\text{characteristic equation det}\,[sI - A] = 0,$$
$$\text{and often written as}\ |sI - A| = 0. \tag{5.224}$$

This is an important equation in control because, as previously discussed, the roots of the characteristic equation must contain negative real parts if the response is to be stable. This is a necessary but insufficient requirement, as will be further discussed in Chapter 6.

For the example being used here, it will be seen that:

$$[sI - A] = \begin{bmatrix} s & 0 \\ 0 & s \end{bmatrix} - \begin{bmatrix} 0 & 1 \\ -\omega_n^2 & -2\zeta\omega_n \end{bmatrix} = \begin{bmatrix} s & -1 \\ \omega_n^2 & (s + 2\zeta\omega_n) \end{bmatrix}. \tag{5.225}$$

The characteristic equation is the determinant of $[sI - A]$, resulting in:

$$s(s + 2\zeta\omega_n) + \omega_n^2 = 0,$$
$$s^2 + 2\zeta\omega_n s + \omega_n^2 = 0. \tag{5.226}$$

This, of course, could have been easily deduced directly from the left-hand side of the original differential equation (5.215) when Laplace transformed.

Worked Example 5.8

A second-order system is defined by the following state-space equation:

$$\begin{bmatrix} \dot{x}_1 \\ \dot{x}_2 \end{bmatrix} = \begin{bmatrix} 0 & 1 \\ -2 & -3 \end{bmatrix} \begin{bmatrix} x_1 \\ x_2 \end{bmatrix} + \begin{bmatrix} 0 \\ 1 \end{bmatrix} u,$$

with initial conditions $\begin{bmatrix} x_1(0) \\ x_2(0) \end{bmatrix} = \begin{bmatrix} 0 \\ 0 \end{bmatrix}$.

If the input is a unit step $u = 1$, determine the transient response for x_1 and x_2.

The state-space solution is $x(s) = [sI - A]^{-1}[Bu(s) + x(0)]$.

The Laplace transform $u(s)$ of the unit step input u from Table 5.2 is $(1/s)$:

$$[sI - A] = \begin{bmatrix} s & 0 \\ 0 & s \end{bmatrix} - \begin{bmatrix} 0 & 1 \\ -2 & -3 \end{bmatrix} = \begin{bmatrix} s & -1 \\ 2 & (s+3) \end{bmatrix},$$

$$[sI - A]^{-1} = \frac{\text{adjoint}}{\text{determinant}} = \frac{1}{[s(s+3)+2]} \begin{bmatrix} (s+3) & 1 \\ -2 & s \end{bmatrix},$$

$$x(s) = \frac{1}{[s(s+3)+2]} \begin{bmatrix} (s+3) & 1 \\ -2 & s \end{bmatrix} \left(\begin{bmatrix} 0 \\ \frac{1}{s} \end{bmatrix} + \begin{bmatrix} 0 \\ 0 \end{bmatrix} \right),$$

$$x(s) = \begin{bmatrix} x_1(s) \\ x_2(s) \end{bmatrix} = \frac{1}{(s+2)(s+1)} \begin{bmatrix} \frac{1}{s} \\ 1 \end{bmatrix},$$

$$x_1(s) = \frac{1}{s(s+1)(s+2)} \rightarrow \frac{1}{2s} - \frac{1}{(s+1)} + \frac{1}{2(s+2)},$$

$$x_2(s) = \frac{1}{(s+1)(s+2)} \rightarrow \frac{1}{(s+1)} - \frac{1}{(s+2)}.$$

Taking inverse Laplace transforms then gives the solution:

$$x_1(t) = \frac{1}{2} - e^{-t} + \frac{e^{-2t}}{2},$$
$$x_2(t) = e^{-t} - e^{-2t}.$$

Note the steady-state conditions as $t \rightarrow \infty$; then $x_1 \rightarrow 0.5$ and $x_2 \rightarrow 0$.

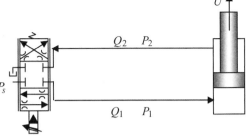

Figure 5.49. A servovalve–linear actuator drive.

In practice, the state-space approach is used to formulate the system model directly from the individual equations, thereby avoiding any further mathematical processing.

EXAMPLE 5.1. Consider the servovalve–linear actuator open-loop system previously analyzed in Section 5.10 and shown in Fig. 5.49.
The linearized equations were shown to be:

$$\delta Q_1 = k_{i1}\delta i - k_{p1}\delta P_1 = A_1\delta U + \frac{V_1(0)}{\beta}\frac{d\delta P_1}{dt}, \tag{5.227}$$

$$\delta Q_2 = k_{i2}\delta i + k_{p2}\delta P_2 = A_2\delta U - \frac{V_2(0)}{\beta}\frac{d\delta P_2}{dt}, \tag{5.228}$$

$$\delta P_1 A_1 - \delta P_2 A_2 = B_v\delta U + \delta F_{\text{load}} + M\frac{d\delta U}{dt}. \tag{5.229}$$

Now define the state variables as follows, recalling that speed is the "output" required:

$$x_1 = \delta U, \quad x_2 = \delta P_1, \quad x_3 = \delta P_2. \tag{5.230}$$

The three system equations then become:

$$k_{i1}\delta i - k_{p1}x_2 = A_1 x_1 + \frac{V_1(0)}{\beta}\dot{x}_2, \tag{5.231}$$

$$k_{i2}\delta i + k_{p2}x_3 = A_2 x_1 - \frac{V_2(0)}{\beta}\dot{x}_3, \tag{5.232}$$

$$x_2 A_1 - x_3 A_2 = B_v x_1 + \delta F_{\text{load}} + M\dot{x}_1. \tag{5.233}$$

Three first-order linear differential equations have now been established for this third-order system. Rewriting these equations in state-space notation then gives:

$$\begin{bmatrix} \dot{x}_1 \\ \dot{x}_2 \\ \dot{x}_3 \end{bmatrix} = \begin{bmatrix} -B_v/M & A_1/M & -A_2/M \\ -A_1/C_1 & -1/R_1C_1 & 0 \\ A_2C_2 & 0 & -1/R_2C_2 \end{bmatrix} \begin{bmatrix} x_1 \\ x_2 \\ x_3 \end{bmatrix} + \begin{bmatrix} -\delta F_{\text{load}}/M \\ k_{i1}\delta i/C_1 \\ -k_{i2}\delta i/C_2 \end{bmatrix}, \tag{5.234}$$

where $C_1 = V_1(0)/\beta$, $C_2 = V_2(0)/\beta$, $R_1 = 1/k_{p1}$, $R_2 = 1/k_{p2}$;

$$[sI - A] = \begin{bmatrix} (s + B_v/M) & -A_1/M & A_2/M \\ A_1/C_1 & (s + 1/R_1C_1) & 0 \\ -A_2C_2 & 0 & (s + 1/R_2C_2) \end{bmatrix}. \tag{5.235}$$

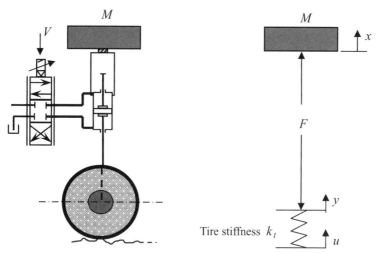

Figure 5.50. A vehicle active-suspension concept.

The system characteristic equation is then obtained by equating the determinant $|sI - A| = 0$ to give:

$$(s + B_v/M)(s + 1/R_1C_1)(s + 1/R_2C_2) + A_1^2(s + 1/R_2C_2)/MC_1$$
$$+ A_2^2(s + 1/R_1C_1)/MC_2 = 0. \tag{5.236}$$

In this example, the characteristic equation is obtained with probably less effort than with conventional manipulation of the Laplace transformed equations, a process prone to manipulation errors. This third-order characteristic equation will always have roots with negative real parts; that is, the open-loop speed response is inherently dynamically stable.

EXAMPLE 5.2. Consider an active-suspension open-loop system shown in Fig. 5.50.

The modeling and control of a practical 1/4 car test rig is discussed in more detail in Chapter 7, where it is deduced that the system model is either idealized or quite complicated. For the purpose of this introductory example, it is sufficient to realize that under closed-loop control, the steady-state servovalve current is zero. If actuator piston leakage resistance R_i exists, the fluid capacitance C on both sides are equal and the time constant CR_i is sufficiently small to be negligible; then, the highly simplified equations are reduced to:

$$\text{force generated,} \quad F = B_i G_a k_i V - B_i \left(\frac{dx}{dt} - \frac{dy}{dt} \right), \tag{5.237}$$

$$\text{force generated,} \quad F = M\frac{d^2x}{dt^2}, \tag{5.238}$$

$$\text{tire force, with mass neglected,} \quad F = k_t(u - y). \tag{5.239}$$

Thus, the piston leakage acts like a viscous damper with a damping coefficient given by $B_i = R_iA^2$ N/m s^{-1}, where A is the double-rod actuator annulus cross-sectional area. The servovalve flow coefficient k_i is assumed to be the mean value of the extending condition and the retracting condition, as discussed earlier.

Now define the state variables:

$$x_1 = x, \quad x_2 = \dot{x}_1, \quad x_3 = y. \tag{5.240}$$

Rearranging the system equations then gives:

$$\begin{bmatrix} \dot{x}_1 \\ \dot{x}_2 \\ \dot{x}_3 \end{bmatrix} = \begin{bmatrix} 0 & 1 & 0 \\ 0 & 0 & -k_t/M \\ 0 & 1 & -k_t/R_i A^2 \end{bmatrix} \begin{bmatrix} x_1 \\ x_2 \\ x_3 \end{bmatrix} + \begin{bmatrix} 0 \\ k_t/M \\ k_t/R_i A^2 \end{bmatrix} u + \begin{bmatrix} 0 \\ 0 \\ -G_a k_i \end{bmatrix} V, \tag{5.241}$$

$$\dot{x} = Ax + Bu + GV.$$

The system characteristic equation is then obtained by equating $\det |sI - A| = 0$ to give:

$$\begin{vmatrix} s & -1 & 0 \\ 0 & s & k_t/M \\ 0 & -1 & (s + k_t/R_i A^2) \end{vmatrix} = 0, \tag{5.242}$$

$$s^2 + s\frac{k_t}{R_i A^2} + \frac{k_t}{M} = 0.$$

It is then deduced that the system modeled has the following second-order properties:

$$\text{undamped natural frequency,} \quad \omega_n = \sqrt{\frac{k_t}{M}}, \tag{5.243}$$

$$\text{damping ratio,} \quad \zeta = \frac{\sqrt{k_t M}}{2 R_i A^2}. \tag{5.244}$$

In other words, the simple active-suspension open-loop system has the dynamic characteristics of a mass–spring–viscous damper equivalent system. However, damping relies on piston seal leakage for this simple model and is not a realistic option in practice for an active suspension with the normal viscous damper unit removed. Note that to complete the active control, a feedback link is required that relates the servovalve voltage V to the chosen state variable transducer signals. This is pursued as a further example in Chapter 6.

5.17.2 Some Further Aspects of the Time-Domain Solution

It will be recalled that the solution to the state-space equation is given as:

$$\dot{x} = Ax + Bu,$$

$$e^{-At}[\dot{x} - Ax] = e^{-At} Bu,$$

$$\frac{d}{dt}[e^{-At}x(t)] = e^{-At} Bu(t).$$

Integrating both sides between the limits of 0 and t gives:

$$e^{-At}x(t) - x(0) = \int_0^t e^{-A\tau} Bu(\tau)d\tau,$$

$$x(t) = e^{At}x(0) + e^{At} \int_0^t e^{-A\tau} Bu(\tau)d\tau. \tag{5.245}$$

The right-hand term is known as the *convolution integral* and Eq. (5.237) finally becomes:

$$x(t) = \mathcal{L}^{-1}x(s) = e^{At}x(0) + \int_0^t e^{A(t-\tau)} Bu(\tau)d\tau, \tag{5.246}$$

where:

$$e^{At} = \mathcal{L}^{-1}[sI - A]^{-1} \tag{5.247}$$

and often has the symbol $\Phi(t)$.

Worked Example 5.9

From Worked Example 5.8, it was shown that:

$$[sI - A]^{-1} = \begin{bmatrix} \dfrac{(s+3)}{(s+1)(s+2)} & \dfrac{1}{(s+1)(s+2)} \\ -\dfrac{2}{(s+1)(s+2)} & \dfrac{s}{(s+1)(s+2)} \end{bmatrix},$$

$$\Phi(t) = e^{At} = \mathcal{L}^{-1}[sI - A]^{-1} = \begin{bmatrix} (2e^{-t} - e^{-2t}) & (e^{-t} - e^{-2t}) \\ (-2e^{-t} + 2e^{-2t}) & (-e^{-t} + 2e^{-2t}) \end{bmatrix}.$$

The time response is then given by:

$$x(t) = \mathcal{L}^{-1}x(s) = e^{At}x(0) + \int_0^t e^{A(t-\tau)}Bu(\tau)d\tau,$$

$$x(t) = \begin{bmatrix} (2e^{-t} - e^{-2t}) & (e^{-t} - e^{-2t}) \\ (-2e^{-t} + 2e^{-2t}) & (-e^{-t} + 2e^{-2t}) \end{bmatrix}\begin{bmatrix} 0 \\ 0 \end{bmatrix},$$

$$+ \int_0^t \begin{bmatrix} [2e^{-(t-\tau)} - e^{-2(t-\tau)}] & [e^{-(t-\tau)} - e^{-2(t-\tau)}] \\ [-2e^{-(t-\tau)} + 2e^{-2(t-\tau)}] & [-e^{-(t-\tau)} + 2e^{-2(t-\tau)}] \end{bmatrix}\begin{bmatrix} 0 \\ 1 \end{bmatrix}d\tau,$$

$$x(t) = \int_0^t \begin{bmatrix} [e^{-(t-\tau)} - e^{-2(t-\tau)}] \\ [-e^{-(t-\tau)} + 2e^{-2(t-\tau)}] \end{bmatrix}d\tau,$$

$$x(t) = \begin{bmatrix} e^{-(t-\tau)} - \dfrac{e^{-2(t-\tau)}}{2} \\ -e^{-(t-\tau)} + e^{-2(t-\tau)} \end{bmatrix}_0^t = \begin{bmatrix} \left(1 - \dfrac{1}{2}\right) - \left(e^{-t} - \dfrac{e^{-2t}}{2}\right) \\ (-1 + 1) - (-e^{-t} + e^{-2t}) \end{bmatrix},$$

$$\begin{bmatrix} x_1(t) \\ x_2(t) \end{bmatrix} = \begin{bmatrix} \dfrac{1}{2} - e^{-t} + \dfrac{e^{-2t}}{2} \\ e^{-t} - e^{-2t} \end{bmatrix}.$$

This is as shown in Worked Example 5.8.

5.17.3 The Transfer Function Concept in State Space

Recall the solution to the state-space equation:

$$x(s) = [sI - A]^{-1}Bu(s) + [sI - A]^{-1}x(0). \tag{5.248}$$

The output of concern is usually just one of the state variables, and this is handled by defining the system output as:

$$y(s) = Cx(s), \tag{5.249}$$

where C is a row vector having the dimension of the system order. Consider just the relationship between the output $y(s)$ and the input $x(s)$ then from these two equations:

$$y(s) = C[sI - A]^{-1} Bu(s). \tag{5.250}$$

The system transfer function $G(s)$, in matrix notation, is then:

$$G(s) = C[sI - A]^{-1} B. \tag{5.251}$$

Worked Example 5.10

Considering Worked Examples 5.8 and 5.9, it was shown that:

$$[sI - A]^{-1} = \begin{bmatrix} \dfrac{(s+3)}{(s+1)(s+2)} & \dfrac{1}{(s+1)(s+2)} \\ -\dfrac{2}{(s+1)(s+2)} & \dfrac{s}{(s+1)(s+2)} \end{bmatrix}, \quad B = \begin{bmatrix} 0 \\ 1 \end{bmatrix}.$$

If the output of the system is the state variable x_1, then:

$$C = [1 \quad 0],$$

$$G(s) = C[sI - A]^{-1} B = [1 \quad 0] \begin{bmatrix} \dfrac{(s+3)}{(s+1)(s+2)} & \dfrac{1}{(s+1)(s+2)} \\ -\dfrac{2}{(s+1)(s+2)} & \dfrac{s}{(s+1)(s+2)} \end{bmatrix} \begin{bmatrix} 0 \\ 1 \end{bmatrix},$$

$$G(s) = \tfrac{1}{(s+1)(s+2)}.$$

5.18 Data-Based Dynamic Modeling

5.18.1 Introduction

In many situations, it is not possible to specify an adequate mathematical model for a number of reasons:

- A subcomponent may simply not be sufficiently understood from a dynamic viewpoint.
- A subcomponent may be highly nonlinear and difficult to express mathematically.
- Performance data from the manufacturer may simply not be available.

What is usually attainable is a measurement of the input–output characteristic; for example, current into a servovalve and output position of a linear actuator. In practice, a sensible estimate of the overall dynamic behavior may be deduced from previous experience, and this allows a dynamic model of the system to be estimated, providing the appropriate operating bounds of the system are experimentally covered. The dynamic model estimated will then be applicable only for other operating conditions that fall within the test boundary, although in practice a modest encroachment beyond the test boundary can sometimes be tolerated.

From a testing viewpoint, one important issue is the determination of the type of dynamic input signal that should be applied. Considering a servovalve–linear actuator, the application of a step input, with varying magnitudes or a varying frequency input with varying magnitudes, may not be sufficient because of the nonlinear flow characteristic of the servovalve. Hence, a deterministic method of determining the dynamic model may be limited in its use. This can be overcome to some extent by use of artificial neural networks (ANNs), although this approach also needs some thought regarding the probable dynamic model.

Dynamic test data are clearly needed, and these will be acquired by a data-acquisition system linked to a computer. These data-acquisition boards can be relatively low cost, yet sufficiently accurate – for example, 8-bit resolution (1 in 256 or better than 0.4%) – having two input channels and sometimes an output drive channel. More expensive boards may have, for example, 12 input channels with a host of other advanced features and at least 12-bit resolution (1 in 4096 or better than 0.025%). The acquired test signals will inevitably be noisy, and some prefiltering may be required before analytical processing to determine the dynamic model. If this is necessary, then a filtering frequency minimum must be selected that does not fall within the dynamic range expected. This almost always presents a restriction resulting in the removal of some noise but probably not sufficient from a visual viewpoint. An additional problem occurs when motors form part of the system because the multiple-piston effect results in an inherent superimposed oscillation on the measured signals, and the frequency, of course, is proportional to the rotational speed; as the motor speed changes in a transient test, then so does the superimposed oscillation.

5.18.2 Time-Series Modeling

This method requires an estimate of the input–output relationship in a sampled data form. If the input is u and the output is y, then this will be of the general form:

$$y(t) = a_1 y(t - T) + a_2 y(t - 2T) + \cdots + b_0 u(t) + b_1 u(t - T) + \cdots +, \qquad (5.252)$$

where T is the sampling interval. Various terms of this time-series could be tried, or the time-series could be derived from a typical transfer function that could be expected from experience that is then transformed into a time-series by a linear approximation for the delay function $z^{-1} = e^{-sT}$. Two common transformations are the backward-difference approximation and the more accurate bilinear approximation, which are both obtained by taking just the first two terms of the Taylor expansion for the appropriate exponential function.

$$
\begin{array}{cc}
\textit{backward difference} & \textit{Bilinear} \\[4pt]
z^{-1} = e^{-sT} \approx 1 - sT & z^{-1} = e^{-sT} = \dfrac{e^{-sT/2}}{e^{sT/2}} \approx \dfrac{1 - sT}{1 + sT} \qquad (5.253) \\[10pt]
s \approx \dfrac{1 - z^{-1}}{T} & s \approx \dfrac{2}{T}\dfrac{(1 - z^{-1})}{(1 + z^{-1})}
\end{array}
$$

For example, a system with an expected first-order time response could be developed into a times series, with a sampling interval T, by an appropriate transformation, as follows.

Using the backward-difference approximation gives:

$$\frac{y(s)}{u(s)} = \frac{1}{(1+s\tau)} \rightarrow \frac{y(z^{-1})}{u(z^{-1})} = \frac{1}{1 + \frac{\tau}{T}(1 - z^{-1})},$$

$$y(z^{-1}) = \frac{\frac{\tau}{t}}{\left(1 + \frac{\tau}{T}\right)} z^{-1} y(z^{-1}) + \frac{1}{\left(1 + \frac{\tau}{T}\right)} u(z^{-1}), \tag{5.254}$$

$$y(t) = a_1 y(t - T) + b_o u(t).$$

Using the bilinear approximation gives:

$$\frac{y(s)}{u(s)} = \frac{1}{(1+s\tau)} \rightarrow \frac{y(z^{-1})}{u(z^{-1})} = \frac{1}{1 + \frac{2\tau}{T} \frac{(1 - z^{-1})}{(1 + z^{-1})}},$$

$$y(z^{-1}) = -\frac{\left(1 - \frac{2\tau}{t}\right)}{\left(1 + \frac{2\tau}{T}\right)} z^{-1} y(z^{-1}) + \frac{1}{\left(1 + \frac{2\tau}{T}\right)} u(z^{-1}) + \frac{1}{\left(1 + \frac{2\tau}{T}\right)} z^{-1} u(z^{-1}), \tag{5.255}$$

$$y(t) = a_1 y(t - T) + b_o u(t) + b_1 u(t - T).$$

The input u and output y are now in measured sampled data form, which for corresponding sample times may be expressed in matrix form for the bilinear approximation example:

$$\begin{bmatrix} y(1) \\ y(2) \\ y(3) \\ \vdots \\ y(n) \end{bmatrix} = \begin{bmatrix} y(0) & u(1) & u(0) \\ y(1) & u(2) & u(1) \\ y(2) & u(3) & u(2) \\ \vdots & & \vdots \\ y(n-1) & u(n) & u(n-1) \end{bmatrix} \begin{bmatrix} a_1 \\ b_o \\ b_1 \end{bmatrix}. \tag{5.256}$$

In general terms, a time-series is represented in matrix notation as follows:

$$Y = A\beta. \tag{5.257}$$

The least-squares solution for the unknown coefficients is:

$$\beta = (A^T A)^{-1} A^T Y. \tag{5.258}$$

The error vector e is defined as:

$$e = Y - A\beta. \tag{5.259}$$

The variance σ^2 is given by:

$$\sigma^2 = e^T e/(N - n). \tag{5.260}$$

$(N - n)$ is referred to as the number of degrees of freedom. The variance of estimates is given by:

$$V(\beta) = \sigma^2 (A^T A)^{-1}. \tag{5.261}$$

5.18.3 The Group Method of Data Handling (GMDH) Algorithm

Recall the previous section in which a transfer function was evaluated in the z domain such that a time-series could then be generated. The resulting time-series, Eq. (5.252), is repeated here:

$$y(t) = a_1 y(t - T) + a_2 y(t - 2T) + \cdots + b_0 u(t) + b_1 u(t - T) + \cdots + . \quad (5.262)$$

In general GMDH terminology,

$$y = w_1 x_1 + w_2 x_2 + w_3 x_3 + w_4 x_4 + w_5 x_5 + w_6 x_6, \quad (5.263)$$

where w_i are now called the weights – that is, the unknown coefficients – and x_i are the measured states. To determine the current output, previous samples of output together with the current sample and previous samples of input are added in a linear combination once the coefficients have been determined from the least-squares training process.

The GMDH starting point, and particularly useful for nonlinear systems, is to consider polynomial relationships between combinations of samples. For example, Eq. (5.263) could be modified to:

$$\begin{aligned} y = {} & w_1 x_1 + w_2 x_2 + w_3 x_3 + w_4 x_4 + w_5 x_5 + w_6 x_6 \\ & + w_7 x_1^2 + w_8 x_2^2 + w_9 x_3^2 + w_{10} x_4^2 + w_{11} x_5^2 + w_{12} x_6^2. \end{aligned} \quad (5.264)$$

Products of all the measured states could also be used and, clearly, a large combination of parameters could be chosen. It does not seem that there is a rule for which combinations should be chosen, and experience is inevitably required in practice. The unknown parameters are calculated with the matrix method of least squares, as used in the time-series analysis.

The process of performing calculations and transmitting information has a similarity with ANN concepts, by means of a series of connected neurons. However, the GMDH approach is based on a principle postulated by Ivakhnenko, a Ukrainian cyberneticist who realized that highly structured models were inadequate for complex time-domain processes (Ivakhnenko, 1971). The concept of neurons is a useful precursor to ANNs and will be used in the GMDH training approach. Some important issues of the GMDH approach are as follows:

- All states considered of importance must be used.
- The approach usually considers pairs or triplets of inputs.
- A hidden layer is used to perform the specified activation function that is a polynomial function of a suitable order operating on the inputs.
- The polynomial function coefficients are evaluated with the least-squares method applied to the sampled data string.
- The number of neurons in the hidden layer increases significantly as the number of parameter combinations is increased; for example, from pairs to triplets.
- If the number of neurons becomes too large, then a limit may have to be set. This is particularly true if a second hidden layer is considered necessary.
- The rms error between the test data and the computed data is then calculated.
- If the error is not acceptable, then another hidden layer may be added.
- The process is repeated for all pairs or triplets of inputs.
- The training process is stopped when a suitable rms error has been achieved.

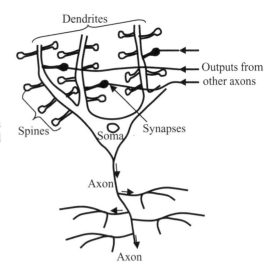

Figure 5.51. Schematic of the human neuron concept (for more information, see Cichocki and Unbehauen, 1993).

- All connections are removed other than the output neuron route that has been selected as having the lowest rms error.
- The selection of the fittest will simplify the training topology and possibly remove some of the assumed input states.

5.18.4 Artificial Neural Networks

An ANN is a collection of neurons coupled together and capable of mapping input data to output data for the fluid power component or system in a similar way to the GMDH mapping approach. However, each neuron calculation is nonlinear and the overall computation concept is quite different and has evolved from a consideration of the possible way that the human brain treats information by means of its billions of human neurons. From neurophysiology, it is estimated that the human brain contains a complex interconnection net of 10^{10}–10^{11} neurons or nerve cells. There are typically 10^3–10^4 dendrites (inputs) per neuron and these dendrites connect the neuron to other neurons. They either receive inputs from other neurons via specialized contacts called synapses or connect other dendrites to the synaptic outputs. The synapses are specialized contacts on a neuron that are the termination points for the axons from other neurons. They are capable of changing a dendrite's local potential in a positive or negative direction. Because of their function, the synapses can be either excitatory or inhibitory in accordance with the ability to strengthen or damp the neuron excitation. A simplified schematic of a biological neuron is shown in Fig. 5.51.

Information storage in a neuron is thought to be concentrated at the synaptic connections or, more precisely, in the pattern of these connections and strengths (weights) of the synaptic connections. Human synapses are of a complex chemical nature, whereas synapses in nervous systems of primitive animals, such as insects, are predominantly based on electrical signal transmission.

According to the simplified model of the neuron, the cell body (soma) receives inputs from other neurons through adjustable or adaptive synaptic connections to the dendrites. The output signal (consisting of nerve impulses) from a cell is transmitted along a branching axon to the synapses of other neurons. When a neuron is

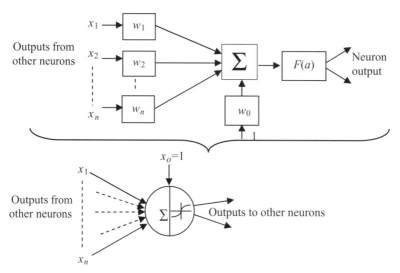

Figure 5.52. The artificial neuron.

excited, it produces nerve impulses (a train of pulses) that are transmitted along an axon to the synaptic connections of other neurons. The output pulse rate depends on both the strength of the input signals and the strengths, or weights, of the corresponding synaptic connections. The maximum firing rate is proposed as 1000 pulses per second. The input signals at the excitatory synapses increase the pulse rate, whereas the input signals at the inhibitory synapses reduce the pulse rate or even block the output signal.

The artificial neuron concept is shown in Fig. 5.52 and brings together the inputs x_i, the weights w_i, the summation, and the activation function F(a):

$$F(a) = w_0 + w_1 x_1 + w_2 x_2 + \cdots + w_n x_n. \tag{5.265}$$

Although the neuron's response function is in general nonlinear, neurophysiologists have discovered that for many biological neurons, a linear summation approximation is appropriate. Therefore, the neuron's output is proportional, in some range, to a linear combination of the neuron's input signal values. This introduces the concept of a neuron "firing" when only particular conditions of information occur at the input, and the mechanism that determines how much information is fired is called the *activation function*.

A suitable topology of neurons is selected, usually by iteration because there is no prescriptive way of guaranteeing what the topology should be for a particular application. However, there are some common construction rules, as illustrated in Fig. 5.53, that have two inputs (e.g., the servovalve current and load pressure differential), and one output (e.g., the servovalve flow rate). This ANN has one hidden layer, and it is common to have $(2n + 1)$ neurons in the first hidden layer, where n is the number of inputs. In some cases, an ANN may contain more hidden layers to map the input–output relationship with sufficient accuracy.

If a network has two hidden layers, then the second hidden layer should have a smaller number of neurons than the first hidden layer; for example, two if needed for the problem of Fig. 5.53.

The next issue is the selection of a suitable activation function $F(a)$. This must be compatible with the numerical algorithm used to calculate the weights from the

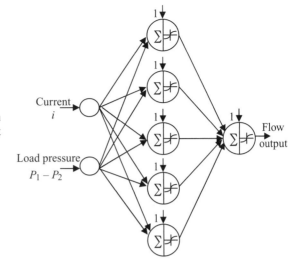

Figure 5.53. A neural network construction to predict servovalve flow with known input current and load pressure differential.

training input–output data. It usually has a smooth characteristic that is analytically differentiable, and the unipolar and bipolar exponential functions of the type shown in Fig. 5.54 are quite common and often called sigmoid activation functions. The derivative of each activation function is given by:

$$\begin{array}{cc} \textit{Bipolar} & \textit{Unipolar} \\ F(a) = \dfrac{1 - e^{-2a}}{1 + e^{-2a}} & F(a) = \dfrac{1}{1 + e^{-2a}} \\ \dfrac{dF(a)}{da} = 1 - F^2(a) & \dfrac{dF(a)}{da} = 2F(a)[1 - F(a)] \end{array} \qquad (5.266)$$

A common training method is the backpropagation method. Each column of the input data matrix is used to calculate each output of the network. Each weight is then modified by considering the (sum of error^2) and the backpropagation algorithm is used to reduce the error. This is repeated with new data until a satisfactory convergence has been achieved. In its simplest form, consider the output error function E:

$$E = \frac{1}{2}\sum e^2, \quad e = d - y, \qquad (5.267)$$

where d is the desired output and y is the actual output. The change of a weight at the output layer is then considered to be proportional to the change of E with the weight value and is given by:

$$\Delta w_i = -\eta \frac{\partial E}{\partial w_i} = \eta[e]\frac{dF}{da}x_i, \qquad (5.268)$$

where $0 < \eta < 1$ is known as the learning rate.

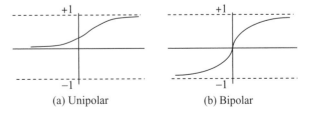

Figure 5.54. Sigmoid activation functions.

(a) Unipolar　　　(b) Bipolar

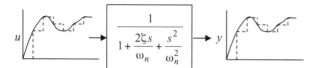

Figure 5.55. Sampled measurements of a second-order system.

All these terms can be determined once the activation function and its corresponding differential have been selected. The sigmoid activation functions in Fig. 5.54 are therefore convenient because their derivatives may be expressed in terms of the actual value at that point. Equation (5.268) is valid at the output layer, but the local error cannot be evaluated for the hidden layers. Therefore, the local error for a hidden layer is evaluated with the knowledge of the local error at the output layer; hence, the term *backpropagation*. The delta method is applied as follows:

$$\Delta w_{ij} = \eta[e]F(a),$$

$$e_j^{output} = y_j^{desired} - y_j^{actual}, \tag{5.269}$$

$$e_i^{hidden} = \frac{dF_i^{hidden}(a)}{da} \sum_{j=1}^{n} w_{ij} e_j^{output}.$$

Momentum, $0 < \alpha < 1$, is usually added to speed the convergence process:

$$\Delta w_{ij} = \eta[e]F(a) + \alpha \Delta w_{ij}^{previous}. \tag{5.270}$$

A large number of training sets and corresponding iterations are usually needed to obtain convergence, often because small values of learning rate and momentum are needed. In addition, values may have to be changed during the iteration process to aid convergence.

5.18.5 A Comparison of Time-Series, GMDH, and ANN Modeling of a Second-Order Dynamic System

An example, undertaken by colleague Y. Xue at Cardiff University during 1994, is now considered whereby the dynamics of a closed-loop position control system is to be modeled with the assumption that the system is probably second order. Therefore, consider the system in Fig. 5.55 where the input–output data are sampled every T seconds.

The training input data must be selected to cover the dynamic range over which the model is to be used, and Fig. 5.56 shows the input data for this example. The output data follow from the second-order transfer function output. The transfer function selected has an undamped natural frequency $\omega_n = 20\,\text{Hz}$ and a damping ratio $\zeta = 0.5$.

To arrive at an appropriate discrete model, the bilinear transformation is used:

$$s \to \frac{2}{T} \frac{(1 - z^{-1})}{(1 + z^{-1})}. \tag{5.271}$$

Substituting Eq. (5.271) into the transfer function shown in Fig. 5.55 gives:

$$\frac{y(z)}{u(z)} = \frac{b_o + b_1 z^{-1} + b_2 z^{-2}}{1 - a_1 z^{-1} - a_2 z^{-2}}. \tag{5.272}$$

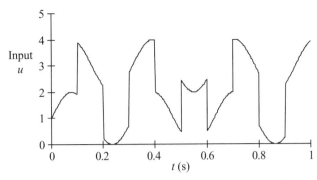

Figure 5.56. Input training data for the position control system.

The discrete time model is then determined by considering the transformation of Eq. (5.272) into the time domain as follows:

$$y(t) = a_1 y(t - T) + a_2 y(t - 2T) + b_o u(t) + b_1 u(t - T) + b_2 u(t - 2T). \quad (5.273)$$

Beginning with the *time-series model* results in the following coefficients using the least-squares analysis outlined earlier:

$$a_1 = 1.716, \qquad\qquad b_o = 2.239 \times 10^{-2},$$
$$a_2 = -7.722e \times 10^{-1}, \qquad b_1 = 3.420 \times 10^{-2}, \qquad\qquad (5.274)$$
$$b_2 = -6.844 \times 10^{-4}.$$

A check on the accuracy of the coefficients can be made by noting from z transform theory that the steady-state gain of the transfer function is determined from the final-value theorem when considered in the z domain:

$$\left.\frac{y(z)}{u(z)}\right|_{\text{steady state}} = \left.\frac{b_0 + b_1 z^{-1} + b_2 z^{-2}}{1 - a_1 z^{-1} - a_2 z^{-2}}\right|_{z=1} = 1,$$
$$\frac{b_0 + b_1 + b_2}{1 - a_1 - a_2} = 0.995 \approx 1. \qquad\qquad (5.275)$$

Converting the z domain model back to the s domain using the computed coefficients gives $\omega_n = 20.13$ Hz and $\zeta = 0.517$ compared with the actual undamped natural frequency $\omega_n = 20$ Hz and damping ratio $\zeta = 0.5$.

Now considering the *GMDH algorithm*, the equation selected is as follows:

$$y = \sum_{1}^{4} w_i x_i + \sum_{1}^{4} g_i x_i^2 + \sum_{1}^{4} h_i x_i^3 + \sum_{i=1}^{3} \sum_{j=i+1}^{4} w_{ij} x_i x_j + C. \qquad (5.276)$$

The training network and the trained network are shown in Fig. 5.57, with z^{-1} representing one sample delay.

Considering that here there are $n = 5$ inputs and $r = 4$ parameters for function Eq. (5.276), then the number of combinations is given by:

$$C_r^n = \frac{n!}{(n-r)!r!} = \frac{5!}{1!4!} = 5. \qquad\qquad (5.277)$$

Hence, there are five output neurons for the training topology. From the least-squares analysis of all the input–output data, the rms errors for the output layer were calculated to be 0.0120, 0.0160, 0.0381, 0.194, and 0.198. Therefore, the first

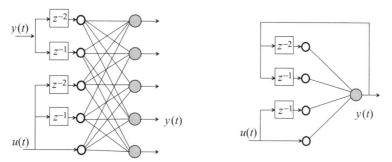

Figure 5.57. GMDH training network and trained network for the second-order system.

output neuron was selected as having the lowest and acceptable rms error, and only the connections from this output neuron are retained. This resulted in four inputs only from the original five selected, as shown in the trained network in Fig. 5.57.

Finally, consider the ANN approach for this example. It is now postulated that the ANN should emulate the discrete time equation as a nonlinear recurrent mapping as follows:

$$y(t) = f[y(t - T), y(t - 2T), u(t), u(t - T), u(t - 2T)]. \qquad (5.278)$$

So, the inputs used for training this recurrent network are $y(t - T)$, $y(t - 2T)$, $u(t)$, $u(t - T)$, and $u(t - 2T)$, as used in the GMDH algorithm, and the training network and the trained network for prediction are shown in Fig. 5.58. Of course, this approach retains all the inputs originally selected.

Two hidden layers were necessary for this apparently straightforward system, with five neurons in the first hidden layer and three neurons in the second hidden layer, and determined by consideration of many different topologies. This example showed a rapid drop in rms error during the early stages, the learning rate and momentum being set to 0.01. It can be fruitful if convergence is checked and the momentum adjusted during the training process. The training results for the three approaches discussed are shown in Fig. 5.59.

Training requires a large amount of training data and iterations to approach a reasonable convergence, and the convergence profile is shown in Fig. 5.60.

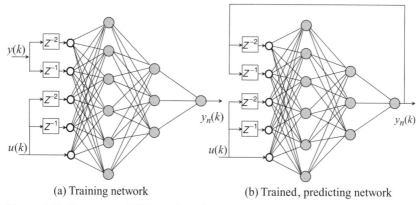

(a) Training network   (b) Trained, predicting network

Figure 5.58. Recurrent ANN topology for a second-order system.

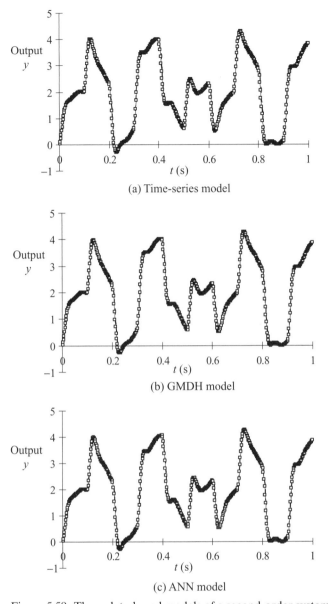

(a) Time-series model

(b) GMDH model

(c) ANN model

Figure 5.59. Three data-based models of a second-order system.

Observing the three results of Fig. 5.59 suggests little difference between each approach. The rms errors are 0.011 for the time-series approach, 0.012 for the GMDH approach, and 0.023 for the ANN approach. However, the time-series approach is less time-consuming and good for a linear system. The GMDH method requires that many functions be tried, but there is a tendency for the trained model to require fewer inputs than originally assumed. The ANN method is very time-consuming and requires a more complex topology than would be imagined for a linear second-order system. It does seem from this work and other work that the GMDH approach offers more promise than the ANN approach, particularly for nonlinear systems.

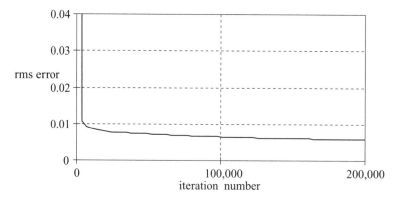

Figure 5.60. The rms error convergence with increasing iteration number.

5.18.6 Time-Series Modeling of a Position Control System

Consider a closed-loop double-rod cylinder position control system with a horizontally moving mass. Such a system has a reasonably linear transient response characteristic such as that shown in Fig. 5.61 for a demanded change of 10 mm. The position transducer signal is sampled at a frequency of 100 Hz for a sample length of 0.5 s, and it will be seen from Fig. 5.60 that it has a noise component, probably because of pump-ripple effects. However, the transient response is clearly overdamped, and it will be assumed that a first-order dynamic approximation is adequate. A first approach is to assume that the response shown to a step input is first order and therefore given by:

$$y = y_d(1 - e^{-t/\tau}). \qquad (5.279)$$

Evaluating the time constant τ when $t = 63\%$ of the steady-state value gives a value of $\tau \approx 0.1$ s. Alternatively, rearranging Eq. (5.238) gives:

$$-\ln\left(1 - \frac{y}{y_d}\right) = \frac{t}{\tau}. \qquad (5.280)$$

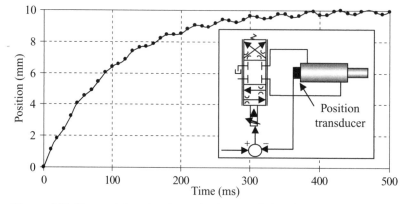

Figure 5.61. Determining the dynamic characteristic of a closed-loop servovalve–actuator position control system.

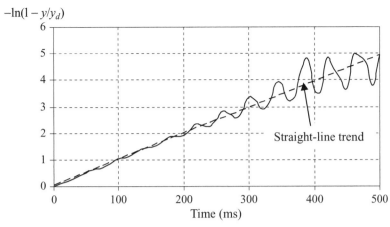

$-\ln(1 - y/y_d)$

Figure 5.62. Modified plot of the servocylinder position response.

Clearly, the slope relating the right-hand side of Eq. (5.280) to time is $1/\tau$. These rearranged data are plotted in Fig. 5.62 for $y_d = 10$ mm.

The best straight line, by means of a least-squares analysis, through all the data is also shown, and the slope is calculated to be $4.7/0.5 = 9.4$, giving $\tau = 0.106$ s. Note the data corruption as time increases and the transient response is approaching the steady-state condition with little change, and this is typical of a fluid power system measurement.

Consider now a time-series analysis; in this example, the differential equation equivalent of Eq. (5.279) is:

$$y + \tau \frac{dy}{dt} = y_d. \tag{5.281}$$

Considering the simplest backward-difference transformation then gives:

$$y(t) = \beta_1 y(t - T) + \beta_2 y_d(t),$$

$$\beta_1 = \frac{\tau/T}{(1 + \tau/T)}, \quad \beta_2 = \frac{1}{(1 + \tau/T)}. \tag{5.282}$$

Equation (5.282) may then be written in the following matrix form for N data points and n unknown coefficients:

$$\begin{bmatrix} y(2) \\ y(3) \\ y(4) \\ \vdots \\ y(N) \end{bmatrix} = \begin{bmatrix} y(1) & y_d(2) \\ y(2) & y_d(3) \\ y(3) & y_d(4) \\ \vdots & \vdots \\ y(N-1) & y_d(N) \end{bmatrix} \begin{bmatrix} \beta_1 \\ \beta_2 \end{bmatrix}. \tag{5.283}$$

In matrix form:

$$Y = A\beta. \tag{5.284}$$

For the current example, the first 49 sampled data points for the column vector $y(t)$ are measured and selected as follows:

$$y(t) = \begin{bmatrix} 0 & 1.101 & 1.832 & 2.445 & 3.259 & 4.077 & 4.568 \\ 4.899 & 5.434 & 6.060 & 6.410 & 6.557 & 6.884 & 7.376 \\ 7.651 & 7.682 & 7.853 & 8.244 & 8.484 & 8.451 & 8.503 \\ 8.810 & 9.040 & 8.981 & 8.943 & 9.176 & 9.407 & 9.350 \\ 9.245 & 9.409 & 9.644 & 9.608 & 9.458 & 9.556 & 9.791 \\ 9.789 & 9.614 & 9.647 & 9.876 & 9.916 & 9.732 & 9.705 \\ 9.918 & 10.002 & 9.827 & 9.745 & 9.932 & 10.057 & 9.904 \end{bmatrix}^{T}. \tag{5.285}$$

For all the samples taken, the demand value is constant at $y_d(t) = 10\,\text{mm}$. Therefore, $N = 48$ data points are then used to satisfy $y(t)$ and $y(t - T)$ in Eq. (5.283), and the $n = 2$ unknown coefficients are evaluated by the MATLAB command window. This gives:

$$\beta = \begin{bmatrix} 0.9030 \\ 0.0966 \end{bmatrix}, \sigma^2 = 0.02,$$

$$V(\beta) = 0.02 \begin{bmatrix} 0.0031 & -0.0024 \\ -0.0024 & 0.0021 \end{bmatrix} \rightarrow \beta = \begin{bmatrix} 0.9030 \pm 0.0079 \\ 0.0966 \pm 0.0065 \end{bmatrix}. \tag{5.286}$$

The accuracy of β_1 is better than that of β_2, although both are acceptable. Evaluating each coefficient then gives:

$$\beta_1 = \frac{\tau/T}{(1 + \tau/T)} = 0.9030, \quad \beta_2 = \frac{1}{(1 + \tau/T)} = 0.0966 \tag{5.287}$$

$$\tau = 0.093\,\text{s}, \quad \tau = 0.094\,\text{s}.$$

These values of the time constant τ are close, although smaller, to the value of 0.106 s calculated earlier. Had Fig. 5.62 first been considered more closely, then it might have been more sensible to choose the first 25 data points. Repeating the analysis gives $\beta_1 = 0.9026$ and $\beta_2 = 0.0965$, barely different from using 48 data points.

5.18.7 Time-Series Modeling for Fault Diagnosis

A similar approach, using a least-squares analysis, may be used for fault diagnostic purposes by considering a time-series analysis on a suitable signal, a deteriorating condition sometimes leading to changes in the signal dynamic shape. In this case, consider the pressure ripple at the output of an axial piston pump.

Chapter 3 discussed the flow source, in which it was shown that the piston-pumping effect produced an ideal flow ripple that is remarkably close to a rectified sine wave at the pumping frequency. This ideal flow source will be considered the sampled "input." If it is assumed for this application that the load impedance does not change, then the pressure ripple is related to the ideal flow source ripple by its impedance transfer function. This transfer function is likely to be of the following form:

$$\frac{p(s)}{q(s)} = \frac{R(1 + c_1 s + c_2 s^2)}{(1 + d_1 s + d_2 s^2 + d_3 s^3)}, \tag{5.288}$$

where R is the low-frequency pump resistance. This transfer function is transformed into the time domain again by a finite-difference representation. The backward-difference transformation, as used in the previous examples, may be used to achieve this, and Eq. (5.288) becomes:

$$s \rightarrow \frac{(1 - z^{-1})}{T},$$

$$\frac{p(z)}{q(z)} = \frac{b_0 + b_1 z^{-1} + b_2 z^{-2}}{1 + a_1 z^{-1} + a_2 z^{-2} + a_3 z^{-3}}, \qquad (5.289)$$

$$(1 + a_1 z^{-1} + a_2 z^{-2} + a_3 z^{-3})p(z) = (b_0 + b_1 z^{-1} + b_2 z^{-2})q(z).$$

In the time domain, this becomes:

$$p(t) = b_0 q(t) + b_1 q(t - T) + b_2 q(t - 2T)$$
$$- a_1 p(t - T) - a_2 p(t - 2T) - a_3 p(t - 3T). \qquad (5.290)$$

This means that when the data are acquired, then the least-squares method requires knowledge of the previous two samples for flow ripple and the previous three samples for pressure ripple. The amplitude of the assumed sine wave acting over the measured time interval is not important if changes from only a reference are sought. In addition, the phase shift between pressure ripple and flow source ripple is not important for the same reason and also because it is constant between comparable samples. In this example, a flow source ripple amplitude of unity is used and the flow source ripple and measured pressure ripple are divided into $N = 25$ samples. The number of unknown β coefficients to be determined is $n = 6$ and the matrix form used is as follows:

$$\begin{bmatrix} p(3) \\ p(4) \\ p(5) \\ \vdots \\ p(n) \end{bmatrix} = \begin{bmatrix} q(3) & q(2) & q(1) & p(2) & p(1) & p(0) \\ q(4) & q(3) & q(2) & p(3) & p(2) & p(1) \\ q(5) & q(4) & q(3) & p(4) & p(3) & p(2) \\ \vdots & \vdots & \vdots & \vdots & \vdots & \vdots \\ q(n) & q(n-1) & q(n-2) & p(n-1) & p(n-2) & p(n-3) \end{bmatrix} \begin{bmatrix} b_0 \\ b_1 \\ b_2 \\ a_1 \\ a_2 \\ a_3 \end{bmatrix}. \qquad (5.291)$$

Data used are as follows:

NEW CONDITION

$$p(N) = [0.00 \quad 0.88 \quad 1.75 \quad 2.50 \quad 2.95 \quad 3.08 \quad 3.10 \quad 3.08 \quad 3.00$$
$$2.93 \quad 2.88 \quad 2.88 \quad 2.88 \quad 2.88 \quad 2.90 \quad 2.90 \quad 2.88 \quad 2.85$$
$$2.78 \quad 2.68 \quad 2.58 \quad 2.48 \quad 1.80 \quad 0.88 \quad 0.00]^T.$$

WORN CONDITION

$$p(N) = [0.00 \quad 0.88 \quad 1.75 \quad 2.50 \quad 3.30 \quad 3.60 \quad 3.55 \quad 3.30 \quad 3.05$$
$$2.70 \quad 2.63 \quad 2.85 \quad 3.00 \quad 3.03 \quad 3.00 \quad 2.90 \quad 2.88 \quad 2.85$$
$$2.78 \quad 2.68 \quad 2.58 \quad 2.48 \quad 1.80 \quad 0.88 \quad 0.00]^T.$$

Table 5.10. *Time-series coefficients for pump ripple*

Coefficient	New condition	Worn condition
b_0	14.811	−24.540
b_1	−28.395	49.850
b_2	14.311	−25.170
a_1	1.834	1.758
a_2	−1.123	−1.000
a_3	0.140	0.072
σ^2	0.0141	0.028

For all conditions, the assumed ideal flow source sinusoidal ripple is:

$$q(N) = [0.00 \quad 0.131 \quad 0.259 \quad 0.383 \quad 0.500 \quad 0.609 \quad 0.707 \quad 0.793 \quad 0.866$$
$$0.924 \quad 0.966 \quad 0.991 \quad 1.000 \quad 0.991 \quad 0.966 \quad 0.924 \quad 0.866 \quad 0.793 \quad (5.292)$$
$$0.707 \quad 0.609 \quad 0.500 \quad 0.383 \quad 0.259 \quad 0.131 \quad 0.000]^T.$$

Table 5.10 shows the computed coefficients and variances for these test conditions and with $N - n = 16$ degrees of freedom (DOF). The determination of a changing condition is then decided by observation of the change in coefficients. The dominant effects are the changes in the b coefficients, all of which significantly change in magnitude and sign as the fault develops.

One restriction of any diagnostic approach is that several different faults may affect pump ripple, and it is a difficult task to identify which particular fault is responsible for coefficient changes. In practice, known single-fault conditions and multiple-fault conditions greatly help the fault identification process. Figure 5.63 shows two pressure ripples measured in the new condition and in a condition in which a fault is suspected of changing the ripple shape. Also shown are the ripples predicted by least-squares analysis.

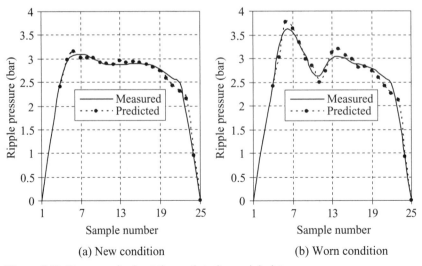

(a) New condition   (b) Worn condition

Figure 5.63. Pressure ripple at the outlet of an axial piston pump.

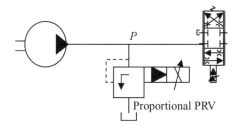

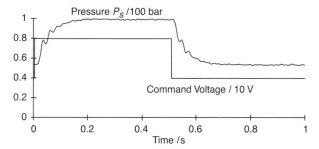

Figure 5.64. The transient pressure response between a pump and servovalve supply pressure port for step demand changes in supply pressure using a proportional PRV.

The signal reconstruction is acceptable although, of course, it can be determined only with knowledge of the three previous pressure samples, as indicated by time-series equation (5.290).

5.18.8 Time-Series Modeling of a Proportional PRV

Now consider an electrically controlled proportional relief valve located at a pump outlet. The analysis is again based on measured data as the PRV is demanded to change pressure levels; in this case, between values of 53 and 100 bar. The intention is to develop a transfer function that can be used in the simulation of a system in which pump supply pressure is varied to track a varying load torque on a motor, thus improving system performance; such a PRV was discussed in Chapter 1. The dynamics of pressure change, following a sudden change in the applied voltage, would not be expected to be significantly fast because of the combination of the sluggish electromagnetic first stage and the dynamics of spool motion within the valve. The pressure transient response will also vary if the connecting volume between the pump and its load servovalve, in this example, is changed. Figure 5.64 shows the practical system with a load servovalve together with a measured pressure response to step changes in demand voltage.

With a pragmatic view to system dynamic behavior, experience suggests that an approximate transfer function would be third order, and this is feasible from Fig. 5.64. There appears to be a dominant first-order component together with an oscillatory second-order component, and this is assumed in the following analysis. The relationship between pressure and applied voltage is therefore assumed to be of the following form:

$$\frac{P(s)}{V(s)} = \frac{G}{(s+a)(s^2 + 2bs + b^2 + c^2)}. \tag{5.293}$$

It is now necessary to make the suitable transformation from the s domain to the z domain. In this example, the technique of *matched transforms* is used rather than the backward-difference substitution used in the previous example. This now gives:

$$(s + a) \rightarrow (1 - z^{-1}e^{-aT}), \tag{5.294}$$

$$(s^2 + 2bs + b^2 + c^2) \rightarrow (1 - z^{-1}e^{-bT}\cos cT + z^{-2}e^{-2bT}); \tag{5.295}$$

z transforms are discussed in more detail in Chapter 6. Substituting these transforms into Eq. (5.293) then gives:

$$\frac{P(z)}{V(z)} = \frac{b_0}{(1 - a_1 z^{-1} - a_2 z^{-2} - a_3 z^{-3})},$$
$$a_1 = 2e^{-bT}\cos cT + e^{-aT}, \tag{5.296}$$
$$a_2 = -(e^{-2bT} + 2e^{-aT}e^{-bT}\cos cT),$$
$$a_3 = e^{-aT}e^{-2bT}.$$

The time-domain equation then becomes:

$$P(t) = a_1 P(t - T) + a_2 P(t - 2T) + a_3 P(t - 3T) + b_0 V(t). \tag{5.297}$$

By use of the method of least squares as before, a sampling frequency of 500 Hz was selected to give a sampling period of $T = 2$ ms. A very good convergence of parameters was achieved with only 25 samples, in which the variation in transient data is usable. The result is:

$$a_1 = 2.7113, \quad a_2 = -2.6278, \quad a_3 = 0.9071, \quad b_0 = 0.1061. \tag{5.298}$$

From Eq. (5.296), it can be shown that:

$$x^3 - a_1 x^2 - a_2 x - a_3 = 0, \quad x = e^{-aT}. \tag{5.299}$$

Hence, it follows that:

$$a = 25.12, \quad b = 11.82, \quad c = 225.5. \tag{5.300}$$

Rearranging transfer function Eq. (5.293) into the more usual standard form gives:

$$\frac{P(s)}{V(s)} = \frac{K}{(s + 1/\tau)(s^2 + 2\zeta\omega_n s + \omega_n^2)}. \tag{5.301}$$

From Eq. (5.294) and (5.295), then:

$$\tau \approx 1/a = 0.04\,\text{s},$$

$$\text{undamped natural frequency, } \omega_n = \sqrt{b^2 + c^2} = 225.8\,\text{rad/s}, \tag{5.302}$$

$$\text{damping ratio, } \zeta\omega_n = b \rightarrow \zeta = 0.052.$$

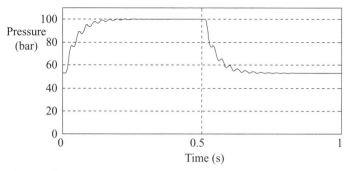

Figure 5.65. Reconstructed transient response of a proportional PRV using data-based modeling of the measured response.

The waveform reconstructed with transfer function Eq. (5.301) by computer simulation is shown in Fig. 5.65 with the addition of a measured pure delay of 0.01 s and for step changes in demand pressure at 0 and 0.5 s.

Comparing this with the measured data, Fig. 5.64, indicates a sufficiently accurate model for design purposes.

5.18.9 GMDH Modeling of a Nitrogen-Filled Accumulator

Consider identifying the dynamic behavior of a nitrogen-filled accumulator. Gas dynamics is analytically challenging because it is well known that the index of expansion–compression varies with the rate of change of pressure. A data-based approach to modeling therefore offers an attractive way forward for inclusion in circuit-analysis problems. A pressure-fluctuation source is connected to the accumulator, a pressure transducer is used to measure the pressure fluctuation, and a fast-acting flow meter is placed at the entrance to the accumulator, as shown in Fig. 5.66. The excellent Parker Hannifin fast-acting flow meter has a response time quoted as better than 4 ms, and several applications by the author have been successful in identifying frequencies of the order of 250 Hz. For this application, the flow meter response is more than adequate, as will be deduced from the transient data shown later. A 1-L hydracushion accumulator was tested and data were sampled at a frequency of 500 Hz and used directly to train the GMDH network.

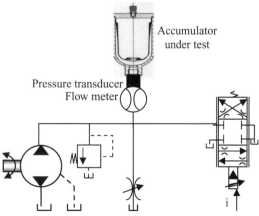

Figure 5.66. Measuring the dynamic behavior of an accumulator.

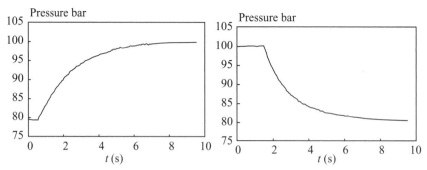

Figure 5.67. Transient pressure used for training.

A three-parameter function was used to train each neuron in a single hidden layer, the combination being every (1 flow rate x_i + 2 pressures x_j and x_k). The three-parameter function chosen is as follows:

$$y = [w_0 x_i + w_1 x_j + w_2 x_k] + [w_3 x_i^2 + w_4 x_j^2 + w_5 x_k^2]$$
$$+ [w_6 x_i x_j + w_7 x_i x_k + w_8 x_j x_k] + [w_9 x_i x_j x_k + w_{10}]. \tag{5.303}$$

Hence, 11 weights have to be estimated by the matrix least-squares method. The load servovalve is controlled with a series of step on–off current signals to provide the training data shown in Fig. 5.67.

The dynamic states selected were 2 previous samples of flow rate and 10 previous samples of pressure plus the current pressure; that is, 13 inputs. The large number of pressure samples is necessary to cover at least 5 s of testing because of the sluggish charging and discharging behavior of the accumulator. The large number of pressure states necessary resulted in a hidden layer of 110 neurons; in this study, 5985 data points were used for training. The training network topology is shown in Fig. 5.68 and indicates just a few of the 110 neuron connections.

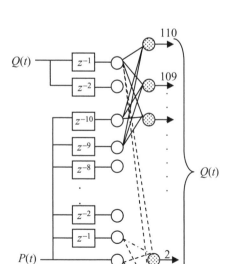

Figure 5.68. GMDH topology for training.

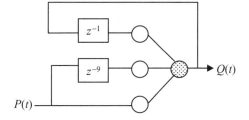

Figure 5.69. GMDH trained network.

When the least-squares calculation has been done on each of the hidden neurons, then the neuron with the lowest rms error is selected, the rest being discarded. For this example, this resulted in just two pressures and one flow-rate input for the trained network, as shown by the highly reduced topology shown in Fig. 5.69.

The states required for training and subsequent prediction are therefore $Q(t - T)$, $P(t - 9T)$, $P(t)$, and the calculated weights are:

$$w_0 = 1, \quad w_3 = 6.81 \times 10^{-3}, \quad w_6 = -9.71 \times 10^{-2}, \quad w_9 = -9.84 \times 10^{-8},$$

$$w_1 = 6.48 \times 10^{-2}, \quad w_4 = 7.12 \times 10^{-4}, \quad w_7 = 9.69 \times 10^{-2}, \quad w_{10} = -4.75 \times 10^{-2},$$

$$w_2 = -6.40 \times 10^{-2}, \quad w_5 = -3.02 \times 10^{-4}, \quad w_8 = -4.13 \times 10^{-4}. \tag{5.304}$$

The flow rates measured and identified by this trained GMDH network are shown in Fig. 5.70 for similar, although not the same, pressure transients as used for training. This is often called *unseen data network validation*.

It is interesting to see from the trained network that only the ninth previous pressure sample is retained, a total time of $9T = 0.45$ s, and this reflects on the first-order-type pressure response of the pressure training data. It is important to test the trained network for other transient pressure conditions to verify the validity of the GMDH model. Figure 5.71 shows some predicted and measured flow rates for cyclic variations in system pressure. Note that the pressure levels are different from those used for the original pressure training data.

The GMDH approach can be a powerful aid to system modeling and is easily integrated within a simulation package. Perhaps one drawback is the selection of a suitable polynomial, and there does not seem to be a structured approach to this task.

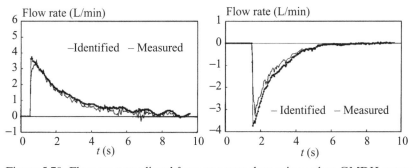

Figure 5.70. Flow rates predicted from pressure data using only a GMDH network (Watton and Xue, 1995).

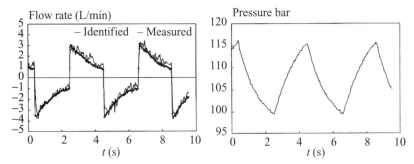

Figure 5.71. Flow rates predicted and measured by pressure data only and for different test conditions, GMDH network (Watton and Xue, 1995).

5.19 Some Comments on the Effect of Coulomb Friction

The three types of friction – speed, stiction, and Coulomb – were introduced in Section 5.3. Removing the speed term that is due to viscosity leaves only the combined stiction–friction term that is difficult to determine at very low speeds. It is well known that the initial stiction value at zero speed rapidly falls to its Coulomb value, which is usually considered constant with speed, but changes in sign as the speed changes sign. The total friction effect can be measured with pressure transducers placed in each line, or a torque meter for a motor, and then slowly increasing the control valve signal to just create motion. A few carefully controlled tests usually give a good picture of the overall friction effect. Some results for a motor and a cylinder, undertaken by the author, are shown in Fig. 5.72.

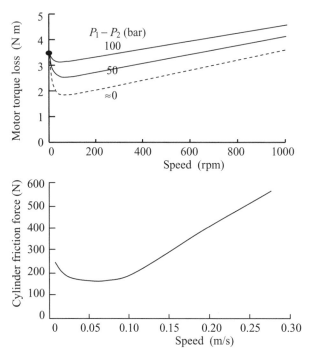

Figure 5.72. Some measured total friction loss characteristics for a motor and a cyclinder.

Clearly, the total friction loss for an axial piston motor is a complicated function of pressure differential and speed. However, in both cases, the stiction, Coulomb friction, and speed components are evident. In addition, there will be a stiction–Coulomb friction contribution from load bearings, sliding surfaces, or both. Traditionally, the Coulomb friction effect on transient behavior has been explained in terms of an equivalent second-order system load such as a mass–spring–damper system, and often a phase-plane analysis is used to describe the transient characteristic in terms of a plot of velocity against displacement. An alternative approach to understanding system dynamics is to assume that Coulomb friction is the dominant term and then to represent its nonlinear characteristic by an equivalent linear viscous damper characteristic:

$$F_c \, \text{sign}(U) \approx B_v U. \tag{5.305}$$

The damping coefficient B_v is determined by considering the energy absorbed per cycle for the two friction models, assuming that the motion is lightly damped, and may be considered as a sine wave $y = y_o \sin(\omega_n t)$ at the undamped natural frequency ω_n and with an amplitude y_o. Equating the energy absorbed gives:

$$4 \int_0^{y_o} F_c dy = 4 \int_0^{y_o} B_v \frac{dy}{dt} dy,$$

$$4 F_c y_o = \pi B_v y_o^2 \omega_n, \tag{5.306}$$

$$B_v = \frac{4 F_c}{\pi \omega_n y_o}.$$

Therefore, the linear damping coefficient B depends also on the amplitude of the oscillation assumed, the Coulomb friction level, and the undamped natural frequency of oscillation. However, the analysis is valid only for highly oscillatory conditions, and the use of an equivalent linear viscous damper is often of limited application in practice, particularly for closed-loop systems under stable control.

Consider therefore a simplified open-loop system example whereby the force F generated by a linear actuator is immediately available as required for moving the load that has a mass m, Coulomb friction F_c, and a resisting spring of stiffness k. This system for hydraulic opening and closing of a valve poppet to control the flow of fuel is shown in Fig. 5.73.

The equation of motion is given by:

$$F = ky + F_c \, \text{sign}\left(\frac{dy}{dt}\right) + m\frac{d^2 y}{dt^2}. \tag{5.307}$$

Coulomb friction only has been used as the dominant component of friction, and an explicit solution would not be possible in the presence of a small stiction component around zero velocity. Rearranging Eq. (5.307) into a nondimensional form for the poppet-open requirement then gives:

$$1 = \overline{y} + \alpha \, \text{sign}(\overline{U}) + \frac{d\overline{U}}{d(\omega_n t)}, \tag{5.308}$$

$$\overline{y} = \frac{y}{y_d}, \quad \overline{U} = \frac{U}{\omega_n y_d}, \quad y_d = \frac{F}{k}, \quad \alpha = \frac{F_c}{k y_d} \quad \omega_n = \sqrt{\frac{k}{m}}.$$

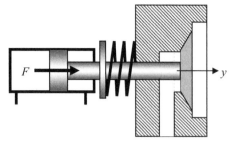

Figure 5.73. Open-loop control of a valve poppet by hydraulic actuation.

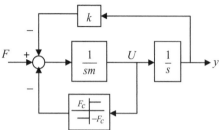

This second-order differential equation may then be integrated and placed into phase-plane form to give:

$$\left\{\overline{y} - [1 - \alpha\,\text{sign}(\overline{U})]\right\}^2 + \overline{U}^2 = \overline{R}^2. \tag{5.309}$$

This represents a set of circles centered at $\overline{y} = (1 - \alpha)$, $\overline{U} = 0$ for positive velocities and at $\overline{y} = (1 + \alpha)$, $\overline{U} = 0$ for negative velocities. The appropriate radius $\overline{R}$ is set by the initial condition and subsequent crossings of the zero-velocity axis. Results for switch-open with a small value of $\alpha = 0.04$ are shown in Fig. 5.74, and with a zero initial position condition.

The effect of Coulomb friction damping is to produce a linear decay in position response and with a possible steady-state error depending on the value of α. The position dead-band is given by $(1 + \alpha) < \overline{y} < (1 + \alpha)$. The transient response decay is different from the exponential decay characteristic evident when viscous damping is dominant. Considering an equivalent viscous damper as given by Eq. (5.306), and assuming that $y_o = y_d/2$ as the mean value, it is an easy matter to show that the damping ratio ζ for the approximate second-order system is given by:

$$\zeta = \frac{4\alpha}{\pi}. \tag{5.310}$$

For the example used, this gives $\zeta = 0.051$. A comparison with the exact position response and the approximate second-order system position response is shown in Fig. 5.75.

An important feature of Coulomb friction damping is, therefore, that it does not change the undamped natural frequency.

From this simple example, it is understandable that it is expected that Coulomb friction damping will produce a linear decay characteristic for position response in a practical control system. However, for fluid power control circuits, such an assumption cannot be made because fluid compressibility cannot usually be neglected and

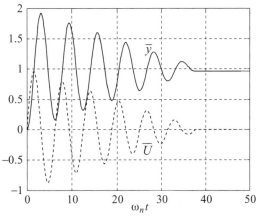

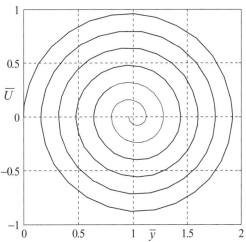

Figure 5.74. Transient response and phase-plane plot for the valve poppet control example.

the actuator dynamics cannot be isolated from the load. In addition, the system equations are nonlinear. The resulting linearized differential equation for position control is inevitably third order, as indicated earlier, and Chapter 6 illustrates the practical reality for a more realistic example.

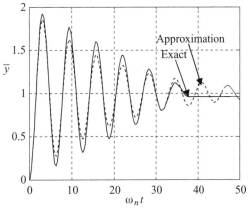

Figure 5.75. A comparison of transient responses.

5.20 References and Further Reading

Achten PAJ and Fu Z [2000]. Valving land phenomena of the Innas hydraulic transformer. *Int. J. Fluid Power* 1(1), 39–48.

Achten PAJ, Van Den Brink TL, Potma JW [2004]. Movement of the cups on the barrel plate of a floating cup, axial piston machine. *Int. J. Fluid Power* 5(2), 25–34.

Almondo A and Sorli M [2006]. Time domain fluid transmission line modelling using a passivity preserving rational approximation of the frequency dependent transfer matrix. *Int. J. Fluid Power* 7(1), 41–50.

Ansari JS and Oldenburger R [1967]. Propagation of disturbance in fluid lines. *Trans ASME J. Basic Eng.* 415–452.

Baum H and Murrenhoff H [2001]. Use of neural networks for the simulation of hydraulic systems including fluid temperature-dependent component efficiencies. In *Power Transmission and Motion Control 2001*, Professional Engineering Publications Ltd., 57–71.

Brown FT [1962]. Transient response of fluid lines. *Trans. ASME J. Basic Eng.* 84, 547–553.

Burton JD, Edge KA, Burrows CR [1964]. Modelling requirements for the parallel simulation of hydraulic systems. *Trans. ASME J. Dyn. Syst. Meas. Control* 116, 137–145.

Cichocki A and Unbehauen R [1993]. *Neural Networks for Optimisation and Signal Processing*, Wiley.

Dagupta K, Watton J, Pan S [2006]. Open-loop dynamic performance of a servo-valve controlled motor transmission system with pump loading using steady-state characteristics. *Mechanism Machines Theory* 41, 262–282.

Dahlen L, Carlsson P [2003]. Numerical optimization of a distributor valve. *Int. J. Fluid Power* 4(3), 17–26.

Davies AM and Davies RM [1969]. Non-linear behaviour including jump resonance of hydraulic servomechanisms. *Inst. Mech. Eng. J. Mech. Eng. Sci.* 11, 837–846.

Del Vescovo G and Lippolis A [2006]. A review analysis of unsteady forces in hydraulic valves. *Int. J. Fluid Power* 7(3), 29–40.

Edge KA and Johnston DN [1986]. A new method for evaluating the fluid-borne noise characteristics of positive displacement pumps. In *Proceedings of Seventh BHRA Fluid Power Symposium*, British Hydromechanics Research Association, 253–260.

Fales R [2006]. Stability and performance analysis of a metering poppet valve. *Int. J. Fluid Power* 7(2), 11–18.

Giuffrida A and Laforgia D [2005]. Modelling and simulation of a hydraulic breaker. *Int. J. Fluid Power* 6(2), 47–56.

Glaze SG [1960]. Analogue technique and the non-linear jack servomechanism. In *Proceedings of the IMechE Automatic Control Conference*, Institute of Mechanical Engineering, pp. 178–188.

Gordic D, Babic M, Jovicic N [2004]. The modelling of a spool position feedback servovalve. *Int. J. Fluid Power* 5(1), 37–50.

Grabbel J and Ivantysynova M [2005]. An investigation of swash-plate control concepts for displacement controlled actuators. *Int. J. Fluid Power* 6(2), 19–36.

Guillon M and Blondel JP [1971]. Non-symmetrical cylinders and valves under non-symmetrical loading. In *Proceedings of the 2nd BHRA Fluid Power Symposium*, British Hydromechanics Research Association, B5, 85–111.

Habibi SR and Singh G [2000]. Derivation of design requirements for optimization of a high performance hydrostatic actuation system. *Int. J. Fluid Power* 1(2), 11–28.

Harper NF [1953]. Some considerations of hydraulic servos of jack type. In *Proceedings of the IMechE Conference on Hydraulic Servos*, Institute of Mechanical Engineers, 41–50.

Hilton DJ [1978]. Interactions between a pressure reducing valve and the upstream pipe. In *Proceedings of the 5th BHRA Fluid Power Symposium*, British Hydromechanics Research Association, G2-23–44.

Hsue CY and Hullender DA [1983]. Modal approximations for the fluid dynamics of hydraulic and pneumatic transmission lines. In *Fluid Transmission Line Dynamics*, ASME, 51–77.

Hullender DA and Healey AJ [1981]. Rational polynomial approximation for fluid transmission line models. In *Fluid Transmission Line Dynamics*, ASME, pp. 33–56.

Iberall AS [1950]. Attenuation of oscillatory pressures in instrument lines. *J. Res. Nat. Bur. Stand.* 45, 2115.

Ivakhnenko A [1971]. Polynomial theory of complex systems. *IEEE Trans. Syst. Man Cybern.* SMC-1, 364–378.

Ivantysynova M and Lasaar R [2004]. An investigation into micro- and macrogeometric design of piston/cylinder assembly of swash-plate machines. *Int. J. Fluid Power* 5(1), 23–36.

Johnston DN [1991]. Numerical modelling of reciprocating pumps with self-acting valves. *Proc. Inst. Mech. Eng. J. Syst. Control Eng.* 205, 87–95.

Johnston DN and Drew JE [1996]. Measurement of positive displacement pump flow ripple and impedance. *Proc. IMechE J. Syst. Control Eng.* 210, 65–74.

Johnston DN and Edge KA [1989]. Simulation of the pressure ripple characteristics of hydraulic circuits. *Proc. IMechE Part C*, 203(C4), 275–282.

Kannisto S and Virvalo T [2002]. Hydraulic pressure in long hose. In *Power Transmission and Motion Control 2002*, Professional Engineering Publications Ltd., 165–176.

Karam JT and Franke ME [1967]. The frequency response of pneumatic lines. *Trans. ASME J. Basic Eng.* 89(3), 371–377.

Karam JT and Leonard RG [1972]. A simple but complete solution for the stop response of semi-infinite circular fluid transmission line systems. *ASME J. Basic Eng.* 94(2).

Katz S [1977]. Transient response of fluid lines by frequency response conversion. *Trans. ASME J. Dyn. Syst. Meas. Control*, 311–313.

Khrapak AV [2001]. Controlled distributive valve plate in axial piston hydraulic motors. *Int. J. Fluid Power* 2(2), 65–74.

Kitsios EE and Boucher RF [1986]. Transmission line modelling of a hydraulic position control system. *Proc. IMechE Part B* 200(B4), 229–236.

Kojima E [2003]. Development of a quieter variable-displacement vane pump for automotive hydraulic power steering system. *Int. J. Fluid Power* 4(2), 5–14.

Kojima E and Shinada M [2002]. Development of accurate and practical simulation technique based on the modal approximations for fluid transients in compound fluid line systems. 1st report: Establishment of fundamental calculation algorithm and basic considerations for verification of its availability. *Int. J. Fluid Power* 4(2), 5–15.

Kojima E and Shinada M [2003]. Development of accurate and practical simulation technique based on the modal approximations for fluid transients in compound fluid line systems. 2nd report: Enhancement of analytic functions for generalization. *Int. J. Fluid Power* 4(3), 35–45.

Kontz ME and Book WJ [2007]. Electronic control of pump pressure for a small haptic backhoe. *Int. J. Fluid Power* 8(2), 5–16.

Koskinen KT and Vilenius MJ [2000]. Steady-state and dynamic characteristics of water hydraulic proportional ceramic spool valve. *Int. J. Fluid Power* 1(1), 5–16.

Krus P, Weddfelt K, Palmberg JO [1994]. Fast pipeline models for simulation of hydraulic systems. *Trans. ASME J. Dyn. Syst. Meas. Control* 115, 132–136.

Lanzetta F, Desevaux P, Bailly Y [2002]. Optimization performance of a microfluid flow power converter. *Int. J. Fluid Power* 3(3), 5–12.

Leino T, Linjama M, Koskinen K, Vilenius M [2001]. Applicability of a laminar flow-based model in pipe flow modelling of water hydraulic systems. *Int. J. Fluid Power* 2(2), 37–46.

Lim J, Jackson PR, Yang Q, Jones BE [2001]. Optically powered hydraulic pilot valve using piezo-electric multilayer actuator. *Int. J. Fluid Power* 2(3), 15–22.

Longmore DK and Schlesinger A [1991]. Transmission of vibration and pressure fluctuations through hydraulic hoses. *Proc. IMechE Part I*, 205(12), 97–104.

Macor A and Tramontan M [2007]. Hydrostatic hybrid system: System definition and application. *Int. J. Fluid Power* 8(2), 47–62.

Manco G, Manco S, Rundo M, Nervegna N [2000]. Computerized generation of novel gearings for internal combustion engines lubricating pumps. *Int. J. Fluid Power* 1(1), 49–58.

Manco S, Nervegna N, Rundo M [2002]. A contribution to the design of hydraulic lube pumps. *Int. J. Fluid Power* 3(1), 31–32.

Manhartsgruber B [2004]. Passivity of fluid transmission line models. In *Proceedings of the Power Transmission and Motion Control Workshop, PTMC 2004*, Professional Engineering Publications Ltd., 99–108.

Martin KF [1970]. Stability and step response of a hydraulic servo with special reference to unsymmetrical oil volume conditions. *Proc. Inst. Mech. Eng. J Mech. Eng. Sci.* 12, 331–338.

Martin KF [1974]. Flow saturated response of a hydraulic servo. *ASME J. Dyn. Syst. Meas. Control* 341–346.

Mookherjee S, Acharyya S, Majumdar K, Sanyal D [2001]. Static-performance based computer-aided design of a DDV and its sensitivity analysis. *Int. J. Fluid Power* 2(2), 47–64.

Murin J [2005]. A controlled diesel drive with hydrostatic transmission: Part 1—Mathematical model. *Int. J. Fluid Power* 38(2/3), 105–120.

Murrenhoff H and Scharf S [2006]. Wear and friction of ZRCG-coated pistons of axial piston pumps. *Int. J. Fluid Power* 7(3), 13–20.

Muto T and Kanei T [1980]. Resonance and transient response of pressurised complex systems. *Bulletin JSME* 23, 1610–1617.

Nichols NB [1962]. The linear properties of pneumatic transmission lines. *Trans. Instrum. Soc. Am.* (1), 15–14.

Nikiforuk PN and Westland BE [1965]. The large signal response of a loaded high-pressure hydraulic servomechanism. *Proc. Inst. Mech. Eng.* 180, 757–786.

Olems L [2000]. Investigations of the temperature behaviour of the piston cylinder assembly in axial piston pumps. *Int. J. Fluid Power* 1(1), 27–38.

Piche R and Ellman A [1996]. A standard hydraulic fluid transmission line model for use with ODE simulators. In *Proceedings of the 8th Bath International Fluid Power Workshop*, Research Studies Press, 221–236.

Qian Y and Xiang MG [2007]. Reducing influence of eccentric load on dynamic characteristics of rotary actuator. *Int. J. Fluid Power* 8(2), 17–24.

Rohmann CP and Grogan EC [1957]. On the dynamics of pneumatic transmission lines. *Trans ASME*, 79, 853–874.

Royle JK [1959]. Inherent non-linear effects of hydraulic control systems with inertia loading. *Proc. Inst. Mech. Eng.* 173, 257–269.

Ruan J, Ukrainetz PR, Burton R [2000]. Frequency domain modelling and identification of 2d digital servo valve. *Int. J. Fluid Power* 1(2) 49–58.

Sanada K, Richards CW, Longmore DK, Johnston DN [1993]. A finite element model of hydraulic pipelines using an optimized interlacing grid system. *Proc. Inst. Mech. Eng. J. Syst. Control Eng.* 207, 213–222.

Scharfand S and Murrenhoff H [2005]. Measurement of friction forces between piston and bushing of an axial piston displacement unit. *Int. J. Fluid Power* 6(1), 7–18.

Shinada M and Kojima E [2002]. Development of a practical and high accuracy simulation technique based on numerical modal approximation for fluid transients in compound fluid line systems. In *Proceedings of the 5th JFPS International Symposium on Fluid Power*, Japan Fluid Power Society, 871–876.

Silberberg MY [1956]. A note on the describing function of an element with Coulomb, static and viscous friction. *Trans. AIEE* 75, Part 2, 423–425.

Stecki JS and Davis DC [1986]. Fluid transmission lines-distributed parameter models. Part 1: A review of the state of the art. *Proc. Inst. Mech. Eng.* 200, Part A, 215–228.

Stecki JS and Davis DC [1986]. Fluid transmission lines-distributed parameter models. Part 2: Comparison of models. *Proc. Inst. Mech. Eng.* 200, Part A, 229–236.

Suzuki K and Urata E [2005]. Dynamic characteristics of a direct-pressure sensing water hydraulic relief valve. In *Proceedings of the 6th JHPS International Symposium on Fluid Power*, 461–466.

Suzuki K, Taketomi T, Sato S [1991]. Improving Zielke's method of simulating frequency-dependent friction in laminar liquid pipe flow. *Trans. ASME. J. Fluids Eng.* 113, 569–573.

Tahmeen M, Yamada H, Muto T [2001]. The dynamic characteristics of tapered fluid lines with viscoelastic walls (transfer matrix and frequency response). *Int. J. Fluid Power* 2(2), 33–40.

Takahashi K and Takahashi Y [1980]. Dynamic characteristics of a spool valve controlled servomotor with a non-symmetrical cylinder. *Bull. JSME* 23, 1155–1162.

Tanahashi T [1982]. Distorted pressure histories due to the step response in a linear tapered line. *Bull. JSME* 25, 1521–1528.

Taylor SEM, Johnston DN, Longmore DK [1997]. Modelling of transient flow in hydraulic pipelines. *Proc. Inst. Mech. Eng. J. Syst. Control Eng.* 211, 447–456.

Tou J and Sculthesis PM [1958]. Static and sliding friction in feedback systems. *J. Appl. Phys.* 21, 1210–1217.

Trikha AK [1975]. An efficient method for simulating frequency-dependent friction in transient liquid flow. *ASME J. Fluids Eng.* 97–104.

Turnbull DE [1959]. The response of a loaded hydraulic servomechanism. *Proc. Inst. Mech. Eng.* 173, 270–284.

Urata E [2004]. One-degree-of-freedom model for torque-motor dynamics. *Int. J. Fluid Power* 5(2), 35–42.

Watton J [1984]. The generalised response of servovalve-controlled, single rod, linear actuators and the influence of transmission line dynamics. *ASME J. Dyn. Syst. Meas. Control* 106, 157–162.

Watton J [1988]. Modelling of electrohydraulic systems with transmission lines using modal approximations. *Proc. Inst. Mech. Eng. Part B*, 202(83), 153–163.

Watton J [1990]. Optimum response design guides for electrohydraulic cylinder control systems. *J. Appl. Math. Model.* 14, 598–604.

Watton J and Hawkley CJ [1996]. An approach for the synthesis of oil hydraulic transmission line dynamics utilising in situ measurements. *Proc. Inst. Mech. Eng. J. Syst. Control Eng.* 210, 77–93.

Watton J and Kwon K-S [1996]. Neural network modelling of fluid power control systems using internal state variables. *Mechatronics* 6, 817–827.

Watton J and Tadmori MJ [1988]. A comparison of techniques for the analysis of transmission line dynamics in electrohydraulic control systems. *J. Appl. Math. Model.* 12, 457–466.

Watton J and Xue Y [1995]. Identification of fluid power component behaviour using dynamic flow rate measurement. *Proc. Inst. Mech. Eng. J. Syst. Control Eng.* 209, 179–191.

Watton J and Xue Y [1997]. Simulation of fluid power circuits using artificial network models, Part 1: Selection of component models. *Proc. Inst. Mech. Eng. J. Syst. Control Eng.* 211, 111–122.

Wieczorek U and Ivantysynova M [2002]. Computer-aided optimization of bearing and sealing gaps in hydrostatic machines: The simulation tool CASPAR. *Int. J. Fluid Power* 3(1), 7–20.

Wiens T, Burton R, Schoenau G, Ruan J [2005]. Optimization and experimental verification of a variable ratio flow divider valve. *Int. J. Fluid Power* 6(3), 45–54.

Wu D, Burton R, Schoenau G, Bitner D [2002]. Establishing operating points for a linearised model of a load sensing system. *Int. J. Fluid Power* 3, 47–54.

Wu K, Zhang Q, Hansen A [2004]. Modelling and identification of a hydrostatic transmission hardware-in-the-loop simulator. *Int. J. Vehicle Design* 34(1), 63–75.

Xu L, Schueller JK, Harrell R [1996]. Dynamic response of a servovalve controlled hydraulic motor driven centrifugal pump. *Trans. ASME J. Dyn. Syst. Meas. Control* 118, 253–258.

Xue Y and Watton J [1995]. A self-organising neural network approach to data-based modelling of fluid power systems dynamics using the GMDH algorithm. *Proc. Inst. Mech. Eng. J. Syst. Control Eng.* 209, 229–240.

Xue Y and Watton J [1997]. Simulation of fluid power circuits using artificial network models, Part 2: Circuit simulation. *Proc. Inst. Mech. Eng. J. Syst. Control Eng.* 211, 429–438.

Yamada H, Wennmacher G, Muto T, Suematsu Y [2000]. Development of a high-speed on/off digital valve for hydraulic control systems using a multilayered pzt actuator. *Int. J. Fluid Power* 1(2), 5–10.

Yang WC and Tobler WE [1991]. Dissipative modal approximation of fluid transmission lines using linear friction models. *Trans. ASME J. Dyn. Syst. Meas. Control* 113, 152–162.

Yuge A, Tomioka K, Tanaka K, Nagayama K, Tokuda K [2005]. Dynamic characteristics of a spool valve coupled with electromagnetic and mechanical effects. In *Proceedings of the 6th JHPS International Symposium on Fluid Power*, Japan Fluid Power Society, 340–345.

Zhang R, Alleyne AG, Prasetiawan EA [2002]. Performance limitations of a class of two-stage electro-hydraulic flow valves. *Int. J. Fluid Power* 3(1), 47–54.

Zielke W [1968]. Frequency-dependent friction in transient pipe flow. *Trans. ASME J. Basic Eng.* 90(1), 109–115.

6 Control Systems

6.1 Introduction to Basic Concepts, the Hydromechanical Actuator

The interconnection of components to form a closed-loop control system introduces new features, primarily the consideration of steady-state error and the speed of control. Steady-state error is a function of component design, such as the existence of a servovalve spool underlap, and the speed of control is a function of the way control is dynamically achieved combined with the dynamic characteristics of the system. It has been shown in Chapter 5 that fluid compressibility and load mass–inertia play the dominant part when characterizing system dynamics, and these aspects cannot be neglected when the dynamic stability of a closed-loop system is considered. Increasing system gain – for example, by increasing a servovalve servoamplifier current gain in a closed-loop servovalve–actuator position controller – will eventually lead to closed-loop instability. This will result in severe oscillations that could rapidly lead to component damage.

The hydromechanical actuator is one of the simplest forms of closed-loop control with applications reaching back to the very earliest days of industrial manufacturing using cast-iron components and low-pressure water as the working fluid. It incorporates a spool valve and an actuator – for example as shown in Fig. 6.1.

It will be seen that as the handle is moved to the right, the spool valve opens, allowing pressurized fluid to move the actuator in the same direction as the handle movement. In its basic form, it therefore acts in a manner similar to that of a servovalve–actuator system. The body of the actuator dynamically follows the position of the handle until the error sensed by the spool eventually becomes zero; the servocontrol is often referred to as a "follow-up controller." Figure 6.1 shows an application in which the body is connected to the lever of a pump swash plate, thus allowing the pump displacement to be manually adjusted, which is particularly useful where a servovalve is not feasible, such as low-cost marine applications. Given that the absolute movement of the handle is x, the absolute movement of the body is y, and the absolute movement of the spool is z, then the spool relative opening, or error, $e = z - y$. The spool-valve ports are assumed to have a linear area variation with displacement, similar to a servovalve spool, and the flow rates through the spool valve are defined as follows:

$$Q_1 = k_f e \sqrt{P_s - P_1}, \qquad Q_2 = k_f e \sqrt{P_2}. \tag{6.1}$$

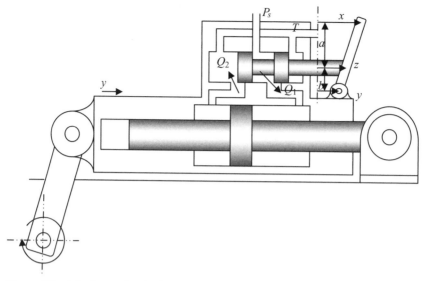

Figure 6.1. A hydromechanical actuator.

It will be assumed that:

- The load force F on the actuator is negligible.
- Viscous friction exists.
- The actuator is double rod $A_1 = A_2 = A$.
- The moving mass is M.
- Fluid compressibility exists.
- The actuator is centralized, having its lowest undamped natural frequency.

Linearizing these equations and incorporating the open-loop dynamic analysis discussed in Chapter 5 then gives the following transfer function for actuator velocity:

$$\delta U(s) = \frac{\dfrac{k_i}{A} \delta e(s)}{1 + \dfrac{R_v}{2R} + s\left(\dfrac{L}{2R} + \dfrac{C R_v}{2}\right) + s^2 \dfrac{LC}{2}}, \tag{6.2}$$

$$R = \frac{P_s}{A U_e(0)}, \quad k_i = k_f \sqrt{\frac{P_s}{2}}, \quad R_v = \frac{B_v}{A^2}, \quad C = \frac{V}{\beta}, \quad L = \frac{M}{A^2}. \tag{6.3}$$

Under closed-loop position control, the steady-state velocity $U(0) = 0$ and therefore $R = \infty$. This introduces the conclusion that *a critically lapped spool does not contribute to system damping under closed-loop position control.*
    Considering also the mechanical linkage for small rotations gives:

$$\frac{(z - y)}{b} = \frac{e}{b} = \frac{(x - y)}{(a + b)}, \tag{6.4}$$

$$e = \lambda(x - y), \quad \lambda = \frac{b}{(a + b)}.$$

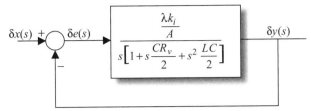

Figure 6.2. Block diagram for the hydromechanical position actuator.

The transfer function relating actuator position to spool error is then given by integrating velocity transfer function (6.2) to give:

$$\delta y(s) = \frac{\dfrac{\lambda\, k_i}{A}\delta e(s)}{s\left(1 + s\dfrac{C R_v}{2} + s^2 \dfrac{LC}{2}\right)}.$$ (6.5)

This transfer function is referred to as the *open-loop transfer function*. Combining Eq. (6.4) and open-loop transfer function Eq. (6.5) then allows the concept of the block diagram with negative feedback to be developed as shown in Fig. 6.2.

When block diagrams are considered for closed-loop control systems, a common unifying notation is used, as shown in Fig. 6.3.

In general, there will be a feedback transfer function $H(s)$ that may well contain dynamic components. The forward transfer function $G(s)$ contains the hydraulic system dynamics and any other element; for example, a compensating network or spool dynamics that have been neglected so far. The block diagram terminology is as follows:

$$\text{the open-loop transfer function,} \quad \text{OLTF} = G(s)H(s);$$ (6.6)

$$\text{the closed-loop transfer function, CLTF} \quad \frac{y(s)}{y_d(s)} = \frac{G(s)}{1 + G(s)H(s)}.$$ (6.7)

For the hydromechanical actuator, the CLTF is:

$$\frac{\delta y(s)}{\delta x(s)} = \frac{\dfrac{\lambda\, k_i}{A}}{s^3 \dfrac{LC}{2} + s^2 \dfrac{C R_y}{2} + s + \dfrac{\lambda\, k_i}{A}}.$$ (6.8)

The CLTF is therefore third order and the closed-loop transient response to any input demand depends on the roots of its denominator. For example, if the input is a demanded step change in position, then the response can be sluggish, oscillatory, or even unstable as the gain term $\lambda k_i/A$ is increased. The potential for instability can occur only for a linear system of third order or higher for fluid power control systems met in practice.

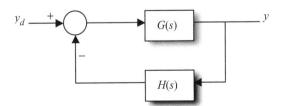

Figure 6.3. Standard block diagram notation for a feedback control system.

Worked Example 6.1

Determine the conditions that satisfy the optimum closed-loop ITAE criterion for the hydromechanical position actuator.

Write the two transfer functions as follows:

$$\frac{\delta y(s)}{\delta x(s)} = \frac{1}{\dfrac{LCs^3}{2K} + \dfrac{CR_v s^2}{2K} + \dfrac{s}{K} + 1}, \quad K = \frac{\lambda k_i}{A}.$$

Seek the ITAE form:

$$\frac{\delta y(s)}{\delta x(s)} \rightarrow \frac{1}{\dfrac{s^3}{\omega_o^3} + \dfrac{1.75s^2}{\omega_o^2} + \dfrac{2.15s}{\omega_o} + 1}.$$

Equating coefficients gives:

$$\frac{2.15}{\omega_o} = \frac{1}{K}, \quad \frac{1.75}{\omega_o^2} = \frac{CR_v}{2K}, \quad \frac{1}{\omega_o^3} = \frac{LC}{2K}.$$

Rearranging then gives the solution:

$$K = \frac{\lambda k_i}{A} = 0.266\frac{R_v}{L}, \quad \frac{L}{CR_v^2} = 0.35,$$

$$\text{giving } \omega_o = 0.571\frac{R_v}{L}.$$

6.2 Stability of Closed-Loop Linear Systems

6.2.1 Nyquist's Stability Criterion

For closed-loop systems that are third-order or above, it is possible for the system to be unstable if parameters, such as system gain, are not within a specific range of values. In hydraulic systems, it is unusual to have unstable open-loop components, and therefore all the coefficients in the open-loop transfer function will have positive coefficients. Recall the closed-loop transfer function:

$$\text{CLTF} = \frac{G(s)}{1 + G(s)H(s)}. \tag{6.9}$$

In general, $G(s)H(s)$ may contain polynomials of s in both the numerator $N(s)$ and denominator $D(s)$. It then follows that:

$$1 + G(s)H(s) = 1 + \frac{N(s)}{D(s)} = \frac{D(s) + N(s)}{D(s)},$$

$$1 + G(s)H(s) = \frac{(s - z_1)(s - z_2)\dots(s - z_m)}{(s - p_1)(s - p_2)\dots(s - p_n)}. \tag{6.10}$$

Definitions are as follows:

- z_i are the zeros and p_i are the poles of $1 + G(s)H(s)$ and can be real, complex, or a combination of both.

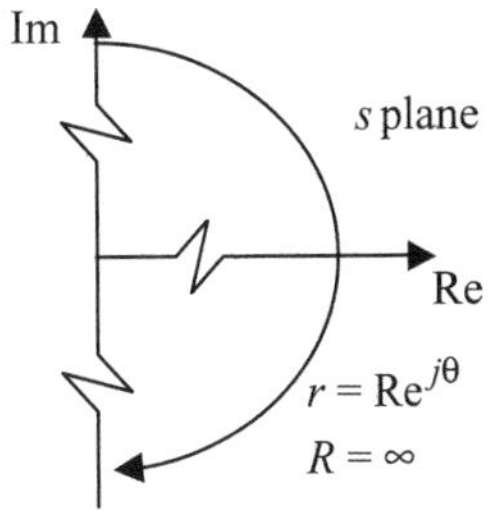

Figure 6.4. A suitable closed path for s.

- The zeros are the roots of the numerator polynomial.
- The poles are the roots of the denominator polynomial.
- Note that the poles of $G(s)H(s)$ are also the poles of $1 + G(s)H(s)$.
- Stability depends on the properties $1 + G(s)H(s)$.

Considering the transient response of the closed-loop system, its dynamic characteristic is determined by the roots of $1 + G(s)H(s)$, the denominator of Eq. (6.9), or the zeros of $1 + G(s)H(s)$, which should not contain positive real parts. Now consider what happens to a plot of $1 + G(s)H(s)$ as s varies through its infinite range of values around a closed path, embracing the entire right-hand half of the s plane as shown in Fig. 6.4.

It can be shown that:

$$N = Z - P. \tag{6.11}$$

N is the net number of encirclements of $1 + G(s)H(s)$ about the origin, Z is the number of zeros of $1 + G(s)H(s)$ within the closed path chosen, and P is the number of poles of $1 + G(s)H(s)$ within the closed path chosen. The Nyquist stability criterion can then be stated as follows:

If a closed path for s *encloses the right-hand half of the* s *plane and* $1 + G(s)H(s)$ *makes a net clockwise encirclement of the origin, then there is an excess of zero over poles. Because these lie in the right-hand half of the* s *plane, then the system must be unstable.*

The procedure is as follows:

- The poles of $1 + G(s)H(s)$ are also the poles of $G(s)H(s)$ and are usually readily observed from the OLTF. Hence, P is readily deduced.
- Select s to follow a path that encloses the right-hand half of the s plane. This is ensured by choosing the entire imaginary axis and an infinite semicircle. Selecting the clockwise path for s shown in Fig. 6.4 results in three parts:

$$\begin{aligned}
&\text{(i)} \quad s = j\omega, \omega = 0 \to \infty; \\
&\text{(ii)} \quad s = R^{j\theta}, \quad R = \infty; \\
&\text{(iii)} \quad s = -j\omega, \quad \omega = -\infty \to 0.
\end{aligned} \tag{6.12}$$

Part (iii) is the reflection of Part (i), and Part (ii) usually reduces $G(s)H(s)$ to a single point at the origin because of the infinite radius of the semicircle path for s.

- It is necessary to plot only $G(s)H(s)$. The number of encirclements of the -1 point is now determined to give N. From Eq. (6.12), it follows that:

$$Z = N + P. \tag{6.13}$$

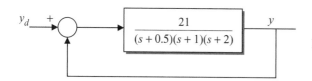

Figure 6.5. A position control system.

- If Z is positive, then there is an excess of zeros over poles, and the closed-loop system is unstable.

In hydraulic control systems, P is usually zero and, therefore, it is necessary to ensure only that $G(j\omega)H(j\omega)$, the conventional frequency response, does not encircle the -1 point at least once. The roots of the characteristic equation are obtained simply from:

$$\text{characteristic equation, } 1 + G(s)H(s) = 0, \tag{6.14}$$

$$\text{or, alternatively, } G(s)H(s) = -1. \tag{6.15}$$

To illustrate these principles, consider the following closed-loop position control system shown in Fig. 6.5.

The system OLTF is:

$$G(s)H(s) = \frac{21}{(s + 0.5)(s + 1)(s + 2)}. \tag{6.16}$$

It is immediately deduced that the three poles of $1 + G(s)H(s)$ are the three poles of $G(s)H(s)$ and are determined directly from Eq. (6.16) as $p_i = -0.5, -1, -2$, and therefore there are no poles of $G(s)H(s)$ with positive real parts, $P = 0$. Figure 6.6 shows the Nyquist plot as s travels around its closed path shown in Fig. 6.4. $G(s)H(s)$ collapses to a single point at the origin as s traverses the semicircle of infinite radius. It can be seen that there are *two encirclements of the -1 point* and, therefore:

$$Z = N - P,$$
$$Z = 2 - 0 = 2. \tag{6.17}$$

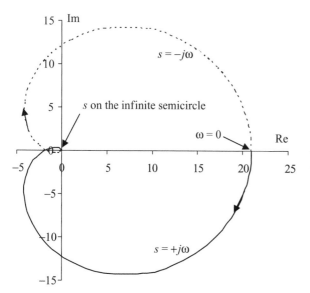

Figure 6.6. Nyquist plot for the OLTF as s travels around the right-hand half of the s plane in a clockwise direction.

Therefore, *there are two zeros that lie in the right-hand half of the* s *plane*. They must have positive real parts and, therefore, the closed-loop system is unstable. Considering characteristic equation (6.14) gives:

$$1 + G(s)H(s) = 0,$$
$$s^3 + 3.5s^2 + 3.5s + 22 = 0,$$
$$(s + 4)(s^2 - 0.5s + 5.5) = 0,$$
$$s = -4, \ +0.25 + j2.33, \ +0.25 - j2.33.$$

(6.18)

The two complex zeros with positive real parts are evident in Eq. (6.18).

Two roots of Eq. (6.18) clearly have positive real parts, validating that the closed-loop system is unstable. It would seem, therefore, that it is necessary to plot only the frequency response $G(j\omega)H(j\omega)$ and then ensure for stability that it does not cross the real axis beyond the -1 point. However, this can be misleading in the general sense if the OLTF contains poles with positive real parts.

Worked Example 6.2

Consider the following OLTF:

$$G(s)H(s) = \frac{2}{(s - 0.5)(s + 1)(s + 2)}.$$

Determine whether the closed-loop system will be stable.

Clearly, $P = 1$. Plot $G(s)H(s)$ as s traverses around the right-hand half of the s plane. For s on the positive imaginary axis $s = j\omega$, then:

$$G(j\omega)H(j\omega) = \frac{2}{(j\omega - 0.5)(j\omega + 1)(j\omega + 2)},$$

$$G(j\omega)H(j\omega) = 2\frac{[-(1 + 2.5\omega^2) - j(0.5\omega - \omega^3)]}{(0.25 + \omega^2)(1 + \omega^2)(4 + \omega^2)}.$$

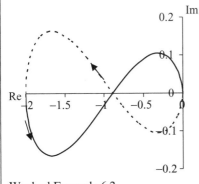

Worked Example 6.2.

$P = 1$ and $N = -1$ because there is one encirclement of the -1 point but in an anticlockwise direction. Hence, it follows that:

$$N = Z - P,$$
$$Z = N + P,$$
$$Z = -1 + 1 = 0.$$

There are no zeros with positive real parts; the closed-loop system is stable. This can be validated by considering the characteristic equation:

$$1 + G(s)H(s) = 0,$$
$$(s + 2.46)(s^2 + 0.04s + 0.412) = 0.$$

Clearly, there cannot be any roots with positive real parts, and the closed-loop system is stable even though the open loop is apparently unstable because of its pole with a positive real part. This example also shows that just a plot of the OLTF frequency response $G(j\omega)H(j\omega)$ would not have revealed closed-loop instability in the sense of its crossing the real axis beyond the -1 point.

6.2.2 Root Locus Method

The main problem in practice is the determination of the roots of a high-order polynomial representing the characteristic equation $1 + G(s)H(s) = 0$. There is a graphical approach known as the *root locus plot* that has received a comprehensive treatment prior to the availability of standard graphics and mathematical analysis software packages. The root locus method combines a number of rules to aid graphical construction of $G(s)H(s)$ and can be pursued through standard control theory textbooks.

For example, if the earlier example is considered but now with a variable system gain, then the OLTF becomes:

$$G(s)H(s) = \frac{K}{(s + 0.5)(s + 1)(s + 2)}. \tag{6.19}$$

It was shown earlier that the closed-loop system was unstable with $K = 21$. The characteristic equation now becomes for a general gain K:

$$1 + G(s)H(s) = 0,$$
$$(s + 0.5)(s + 1)(s + 2) + K = 0, \tag{6.20}$$
$$s^3 + 3.5s^2 + 3.5s + 1 + K = 0.$$

To determine the value of the gain K to cause closed-loop instability, simply vary K and determine the roots of Eq. (6.20). Notice that when $K = 0$, the zeros of the characteristic equation are the poles of the OLTF: -0.5, -1, -2. The root locus diagram is shown in Fig. 6.7 and indicates instability when $K = 11.25$.

The following points arise:

- The number of loci is equal to the order of the OLTF, three in this example.
- As the gain K is increased from $K = 0$, then all the roots move away from the three real poles of the OLTF. One root increases in the negative real direction. The two other roots move toward a breakaway point that may be determined from:

$$\frac{dK}{ds} = 0. \tag{6.21}$$

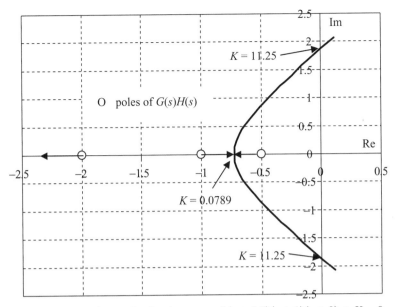

Figure 6.7. Root locus plot for the roots of $(s + 0.5)(s + 1)(s + 2) + K = 0$.

In this example:

$$K = -(s^3 + 3.5s^2 + 3.5s + 1),$$

$$\frac{dK}{ds} = -(3s^2 + 7s + 3.5) = 0,$$

$$s = -0.726, \quad K = 0.0789,$$

$$s = -1.608, \quad K = -0.264. \tag{6.22}$$

Clearly, the appropriate solution is $s = -0.726$ when $K = 0.0789$.

• To determine the asymptotes of the root loci, the angle condition must be satisfied as $s \to \infty$. It is then noted that for this condition:

$$1 + G(s)H(s) = 0 \quad \to \quad G(s)H(s) = -1,$$

$$\angle G(s)H(s) = \pm 180°(2k + 1), \quad k = 0, 1, 2, \ldots, \tag{6.23}$$

$$\text{asymptote angle} = \frac{\pm 180°(2k + 1)}{r},$$

where r is the excess of poles over zeros in $G(s)H(s)$.

• For asymptotes extending to infinity, they intersect the real axis at the value:

$$s = \frac{1}{r} \left(\sum_{i=1}^{n} p_i - \sum_{i=1}^{m} z_i \right), \tag{6.24}$$

where n are the poles and m are the zeros of $G(s)H(s)$.

For the current example,

$$G(s)H(s) = \frac{K}{(s + 0.5)(s + 1)(s + 2)} \to r = 3, \tag{6.25}$$

$$\text{angle of asymptotes} = \pm 60°(2k + 1).$$

The angles are therefore $+60°$, $-60°$, $180°$, and $-180°$ and are repeated. Thus, there are three asymptotes, one being the negative real axis. The asymptotes intersect the real axis at the point:

$$s = \frac{1}{r}\left(\sum_{i=1}^{n} p_i - \sum_{i=1}^{m} z_i\right) = \frac{1}{3}[(-0.5 - 1 - 2) - (0)] = -1.17. \qquad (6.26)$$

- As the gain is increased, two roots occur in conjugate pairs until the right-hand half of the s plane is reached.

6.2.3 Routh Stability Criterion

An alternative approach is to use the *Routh array method* that determines how many roots of the characteristic equation have positive real parts and is particularly useful when the order of the system is high. It may also be used to determine limits of system parameters to ensure closed-loop stability and is therefore a powerful design tool. It is not strictly necessary to know how many exist because just one will render the closed-loop unstable. It also follows that the numerical value of any root having a real part is also not necessary. To apply the Routh array method, the characteristic equation is set down in a specific order, as follows:

$$1 + G(s)H(s)$$

$$b_n s^n + b_{n-1} s^{n-1} + b_{n-2} s^{n-2} + \cdots + b_3 s^3 + b_2 s^2 + b_1 s + b_0. \qquad (6.27)$$

Write the first two rows of the Routh array:

$$\text{odd-order row } 1 \quad b_n \quad b_{n-2} \quad b_{n-4} \ldots 0,$$
$$\text{even-order row } 2 \quad b_{n-1} \quad b_{n-3} \quad b_{n-5} \ldots 0. \qquad (6.28)$$

Now arrange new rows, one at a time, using the pair of rows above the new row being constructed. Each new row utilizes the first element of the row immediately above, known as the pivotal element, and a multiplication sequence is undertaken. Rows are added until the last row, m, contains all zeros.

row							
1	b_n	b_{n-2}	b_{n-4}	.	.	.	0
2	b_{n-1}	b_{n-3}	b_{n-5}	.	.	.	0
3	c_1	c_2	c_3	.	.	.	0
4	d_1	d_2	d_3	.	.	.	0
.	.	.	.		.	.	0
.	.	.	.	.	.	.	
.	.	.	.	.	.	.	0
m	0	0	.	.	.	.	0

$$c_1 = \frac{b_{n-1}b_{n-2} - b_{n-3}b_n}{b_{n-1}}, \quad c_2 = \frac{b_{n-1}b_{n-4} - b_{n-5}b_n}{b_{n-1}}, \ldots;$$

$$d_1 = \frac{c_1 b_{n-3} - c_2 b_{n-1}}{c_1}, \quad d_2 = \frac{c_1 b_{n-5} - c_3 b_{n-1}}{c_1}, \ldots, \text{ etc.} \qquad (6.29)$$

The *Routh stability criterion* states that:

- For a stable closed-loop system, there should be no sign changes of all the elements of the first column of the array.
- Furthermore, the number of sign changes is equal to the number of zeros of the characteristic equation with positive real parts.
- Because the first elements in the first two rows should normally be positive, then it is just necessary that to ensure closed-loop stability all the coefficients in the first column of the array should also be positive.

So, for the present example, the characteristic equation is:

$$s^3 + 3.5s^2 + 3.5s + 1 + K = 0. \tag{6.30}$$

Construct the Routh array beginning with the first two rows:

row			
1	1	3.5	0
2	3.5	$(1+K)$	0
3	$\dfrac{11.25 - K}{3.5}$	0	0
4	$(1+K)$	0	0
5	0	0	0

$$\tag{6.31}$$

To ensure closed-loop stability, it is necessary that all the signs of the first column be positive. Noting that the gain K is inherently positive, then the condition for stability requires, from row 3, that $K < 11.25$, as deduced from the previous root locus analysis. The *auxiliary equation* is obtained from the row above that which contains the first possible row of zeros; that is, row 3 in Eq. (6.32) when $K = 11.25$. This may be used to determine the frequency of oscillation at the point of instability. In this example:

$$\text{Auxiliary equation} \quad 3.5s^2 + (1+K) = 0,$$
$$3.5s^2 + 12.25 = 0, \tag{6.32}$$
$$s = \pm j1.87,$$

that is, a frequency of oscillation of 1.87 rad/s as also deduced from the previous root locus analysis.

A *much simpler method* to determine the condition for closed-loop instability for practically realizable systems is to note that:

$$1 + G(s)H(s) = \text{Re} + j\text{Im} = 0. \tag{6.33}$$

So, $\text{Re} = 0$ and $\text{Im} = 0$ with $s = j\omega$.

For the present example, this condition results in:

$$1 + G(j\omega)H(j\omega) = 0,$$
$$[K + 1 - 3.5\omega^2] + j[3.5\omega - \omega^3] = 0. \tag{6.34}$$

The imaginary part is zero when $\omega^2 = 3.5$ rad/s, giving a frequency of oscillation $\omega = 1.87$rad/s. The real part is zero when $K = 3.5\omega^2 - 1$ giving $K = 11.25$, as previously determined from the root locus plot calculations.

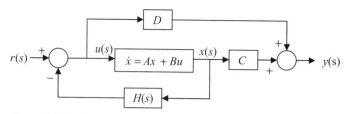

Figure 6.8. A closed-loop system in state-space notation.

For the general third-order characteristic equation:

$$b_3 s^3 + b_2 s^2 + b_1 s + b_0 = 0. \tag{6.35}$$

- The closed-loop system should be stable, providing:

$$b_1 b_2 > b_0 b_3. \tag{6.36}$$

- At the point of instability:

$$\omega = \sqrt{\frac{b_1}{b_3}}, \quad b_1 b_2 = b_0 b_3. \tag{6.37}$$

6.2.4 The State-Space Approach

Chapter 5 introduced the state-space modeling concept for linear systems, in which it was shown that the open-loop system may be described in the following general form:

$$\dot{x} = Ax + Bu. \tag{6.38}$$

Having the open-loop solution in Laplace transform notation then gives:

$$x(s) = [sI - A]^{-1}[x(0) + Bu(s)]. \tag{6.39}$$

A feedback control system is now developed in which the state feedback gains are embodied in the row vector $H(s)$. In addition, it is common in state-space analysis to write the output of the closed-loop system in the general form:

$$y(s) = Cx(s) + Du(s). \tag{6.40}$$

In hydraulic control systems in which a mechanical output is being considered, the output does not contain elements of the input signal $u(s)$ and therefore $D = 0$. However, if the output of the system is, for example, a chemical or combustion process, then Eq. (6.40) may well apply. Figure 6.8 shows the diagram for the closed-loop system.

The closed-loop state-space equation and solution now become:

$$\dot{x} = (A - BH)x + Br, \tag{6.41}$$

$$x(s) = [sI - A + BH]^{-1}[x(0) + Br(s)]. \tag{6.42}$$

Therefore, the characteristic equation for the closed-loop system is now given by:

$$\text{characteristic equation} \rightarrow \det[sI - A + BH] = 0. \qquad (6.43)$$

Worked Example 6.3

The dynamics of an open-loop position control system are defined in the following state-space form:

$$\begin{bmatrix} \dot{x}_1 \\ \dot{x}_2 \\ \dot{x}_3 \end{bmatrix} = \begin{bmatrix} 0 & 1 & 0 \\ 0 & 0 & 1 \\ 0 & -1 & -2 \end{bmatrix} \begin{bmatrix} x_1 \\ x_2 \\ x_3 \end{bmatrix} = \begin{bmatrix} 0 \\ 0 \\ K \end{bmatrix} V,$$

where V is the servovalve input voltage, x_1 is position, x_2 is velocity, and x_3 is acceleration of the load. All the states are used for feedback control. Determine the condition(s) such that closed-loop stability is ensured.

Define the three feedback gains as k_p for position, k_u for velocity, and k_p for acceleration:

$$sI - A + BH = \begin{bmatrix} s & 0 & 0 \\ 0 & s & 0 \\ 0 & 0 & s \end{bmatrix} - \begin{bmatrix} 0 & 1 & 0 \\ 0 & 0 & 1 \\ 0 & -1 & -2 \end{bmatrix} + \begin{bmatrix} 0 \\ 0 \\ K \end{bmatrix} \begin{bmatrix} k_p & k_u & k_a \end{bmatrix},$$

$$sI - A + BH = \begin{bmatrix} s & -1 & 0 \\ 0 & s & -1 \\ Kk_p & (1 + Kk_u) & (s + 2 + Kk_a) \end{bmatrix}.$$

The characteristic equation is then given by:

$$s[s(s + 2 + Kk_a) + (1 + Kk_u)] + 1[0 + Kk_p] = 0,$$
$$s^3 + s^2(2 + Kk_a) + s(1 + Kk_u) + Kk_p = 0.$$

Considering Routh's method for this third-order system, it is then necessary that for stability:

$$(2 + Kk_a)(1 + Kk_u) > Kk_p.$$

Therefore, the sensors for position, speed, and acceleration must be chosen with the correct individual gains as given in the preceding equation.

6.2.5 Servovalve–Motor Closed-Loop Speed Control

It was established in the previous chapters that hydraulic elements of a control system have inherent nonlinear characteristics. To aid design and, to some extent, computer-simulation-aided design, it has been shown that a small-signal, or linearized, analysis can provide valuable information on system dynamics. From a closed-loop stability point of view, it is argued that if the system is predicted as being unstable for small perturbations about an operating condition, then it is probably unstable in general about that operating condition. Thus, a linearized analysis of servovalve-controlled systems is inevitably used to get a feel for the conditions that must be satisfied to ensure closed-loop stability. The closed-loop system is shown in Fig. 6.9.

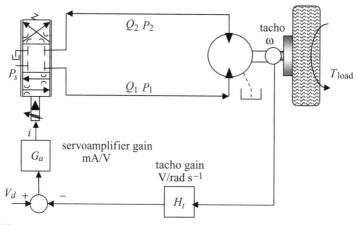

Figure 6.9. A servovalve–motor closed-loop system.

Consider the linearized OLTF that was developed in Chapter 5 and includes motor leakage, fluid compressibility, and load inertia. The open-loop linearized transfer function is as follows:

$$\delta\omega(s) = \frac{\dfrac{k_i\,\delta i(s)}{D_m} - \left(\dfrac{1}{2R} + \dfrac{1}{R_m} + s\dfrac{C}{2}\right)\dfrac{\delta T_{\text{load}}(s)}{D_m^2}}{1 + \dfrac{R_v}{2R} + \dfrac{R_v}{R_m} + s\left(\dfrac{L}{2R} + \dfrac{L}{R_m} + \dfrac{CR_v}{2}\right) + s^2\dfrac{LC}{2}}, \tag{6.44}$$

$$k_i = k_f\sqrt{\frac{P_s - P_{\text{load}}}{2}}, \quad R = \frac{1}{k_p} = \frac{2(P_s - P_{\text{load}})}{D_m\omega(0) + \dfrac{P_{\text{load}}}{R_m}},$$

$$R_v = \frac{B_v}{D_m^2}, \quad C = \frac{V(0)}{\beta}, \quad L = \frac{J}{D_m^2}. \tag{6.45}$$

Considering the response of motor speed to input current then gives:

$$\delta\omega(s) = \frac{\dfrac{k_i\,\delta i(s)}{D_m}}{1 + \dfrac{R_v}{2R} + \dfrac{R_v}{R_m} + s\left(\dfrac{L}{2R} + \dfrac{L}{R_m} + \dfrac{CR_v}{2}\right) + s^2\dfrac{LC}{2}}. \tag{6.46}$$

The control system block diagram is shown in Fig. 6.10, neglecting load torque variations.

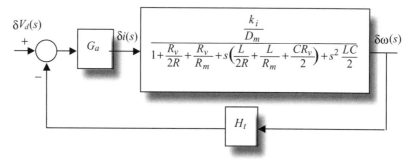

Figure 6.10. Closed-loop block diagram for a servovalve–motor drive.

This closed-loop system as modeled here cannot be unstable because the OLTF is second order. However, the addition of servovalve dynamics in practice will create conditions for instability; for example, if the servoamplifier gain is increased.

The OLTF is given by:

$$G(s)H(s) = \frac{\dfrac{G_a H_t k_i}{D_m}}{1 + \dfrac{R_v}{2R} + \dfrac{R_v}{R_m} + s\left(\dfrac{L}{2R} + \dfrac{L}{R_m} + \dfrac{CR_v}{2}\right) + s^2 \dfrac{LC}{2}}. \qquad (6.47)$$

The CLTF is given by:

$$\frac{H_t\,\omega(s)}{V_d(s)} = \frac{K}{1 + \dfrac{R_v}{2R} + \dfrac{R_v}{R_m} + K + s\left(\dfrac{L}{2R} + \dfrac{L}{R_m} + \dfrac{CR_v}{2}\right) + s^2 \dfrac{LC}{2}},$$

$$\text{open-loop system gain } K = \frac{G_a H_t k_i}{D_m}. \qquad (6.48)$$

The CLTF given as Eq. (6.48) can be placed in optimum second-order form, $\zeta \approx 0.7$, once each term in the transfer function has been evaluated:

$$\frac{H_t\omega(s)}{V_d(s)} = \frac{K_{CL}}{1 + \dfrac{2\zeta s}{\omega_n} + \dfrac{s^2}{\omega_n^2}}, \qquad K_{CL} = \frac{K}{\left(1 + \dfrac{R_v}{2R} + \dfrac{R_v}{R_m} + K\right)},$$

$$\frac{2\zeta}{\omega_n} = \frac{\left(\dfrac{L}{2R} + \dfrac{L}{R_m} + \dfrac{CR_v}{2}\right)}{\left(1 + \dfrac{R_v}{2R} + \dfrac{R_v}{R_m} + K\right)}, \qquad \omega_n^2 = \frac{2\left(1 + \dfrac{R_v}{2R} + \dfrac{R_v}{R_m} + K\right)}{LC}. \qquad (6.49)$$

Worked Example 6.4

Consider the open-loop control system studied in Section 5.9. Determine the closed-loop response characteristic.

Considering open-loop steady-state speeds of 0 rpm, 234 rpm, and 903 rpm, it was shown that for the three cases:

$$1 + \frac{R_v}{2R} + \frac{R_v}{R_m} \cong 1 \text{ to within an accuracy of better than } 4\%$$

speed	$\dfrac{L}{2R} + \dfrac{L}{R_m} + \dfrac{CR_v}{2}$
0 rpm	0.0031
234 rpm	0.0091
903 rpm	0.0281

$$\frac{2}{LC} = 6.1 \times 10^4.$$

The CLTF parameters are now simplified to:

$$\frac{H_t\omega(s)}{V_d(s)} = \frac{K_{CL}}{1 + \frac{2\zeta s}{\omega_n} + \frac{s^2}{\omega_n^2}}, \qquad K_{CL} \approx \frac{K}{(1+K)}, \qquad \omega_n^2 = 6.1 \times 10^4(1+K)$$

speed	$\dfrac{2\zeta}{\omega_n}$	ζ
0 rpm	$\dfrac{0.0031}{(1+K)}$	$\dfrac{0.383}{\sqrt{1+K}}$
234 rpm	$\dfrac{0.0091}{(1+K)}$	$\dfrac{1.124}{\sqrt{1+K}}$
903 rpm	$\dfrac{0.0281}{(1+K)}$	$\dfrac{3.470}{\sqrt{1+K}}$

The gain K may now be selected to give the desired closed-loop damping, the input voltage being changed to produce the appropriate steady-state speed applicable:

Speed	ζ	K	ω_n rad/s
0 rpm	0.383 max	0 min	247 min
234 rpm	0.700	1.58	397
903 rpm	0.700	23.57	1224

Note the improved steady-state speed characteristic discussed in Section 4.7. The motor–load pump has a significant speed ripple, which is evident from the open-loop transient measurement shown in Section 5.9. Gains of $K > 2.6$ were not achievable in practice because servovalve and tachometer dynamics, not included in the design, become important at higher gains.

The corrupting influence of a large motor–pump speed ripple under closed-loop control meant that it was not possible to record transient behavior around a steady-state condition. Large-signal response measurements were also practically very difficult to obtain with clarity because, as deduced from the preceding discussion, the damping ratio is very small at lower speeds, even with a gain of $K = 1.3$; this particular system and its loading do not work well under closed-loop control.

6.2.6 Servovalve–Linear Actuator Position Control

The closed-loop system is shown in Fig. 6.11. The open-loop linearized transfer function for speed was developed in Chapter 5 for the general single-rod case. For position control, the steady-state linearized pressure coefficients k_{p1} and k_{p2} are zero for a critically lapped servovalve; that is, a critically lapped servovalve provides no

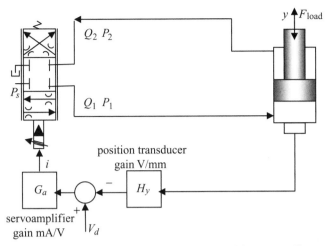

Figure 6.11. A servovalve–linear actuator position controller.

damping at the steady-state rest position. The general transfer function may then be rearranged to give:

$$\delta y(s) = \frac{\dfrac{\alpha k_f \sqrt{P_s}}{A_2}\delta i(s) - \dfrac{sC_1}{A_2^2(\varepsilon + \gamma^2)}\delta F(s)}{s\left[1 + \dfrac{C_1 R_v}{(\varepsilon + \gamma^2)}s + \dfrac{LC_1}{(\varepsilon + \gamma^2)}s^2\right]},$$

$$\alpha = \sqrt{\frac{\gamma - \overline{F}}{1 + \gamma^3}} \text{ when extending,} \qquad \sqrt{\frac{1 + \overline{F}}{1 + \gamma^3}} \text{ when retracting;} \qquad (6.50)$$

$$\varepsilon = \frac{V_1(0)}{V_2(0)}, \qquad \gamma = \frac{A_1}{A_2}, \qquad C_1 = \frac{V_1(0)}{\beta}, \qquad R_v = \frac{B_v}{A_2^2}, \qquad L = \frac{M}{A_2^2}.$$

Note that damping here is provided solely by viscous friction. The control system block diagram is shown in Fig. 6.12, neglecting load-force variations.

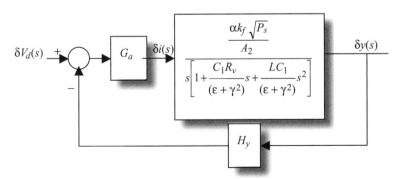

Figure 6.12. Closed-loop block diagram for a servocylinder drive.

The OLTF is therefore given by:

$$G(s)H(s) = \frac{\dfrac{G_a H_y \alpha k_f \sqrt{P_s}}{A_2}}{s\left[1 + \dfrac{C_1 R_v}{(\varepsilon + \gamma^2)}s + \dfrac{LC_1}{(\varepsilon + \gamma^2)}s^2\right]}. \tag{6.51}$$

The ratio of open-loop gains is given by:

$$\frac{\text{gain extending}}{\text{gain retracting}} = \sqrt{\frac{\gamma - \overline{F}}{1 + \overline{F}}}. \tag{6.52}$$

The extending gain is greater than the retracting gain when:

$$\text{gain extending} > \text{gain retracting when } \overline{F} < \frac{(\gamma - 1)}{2}. \tag{6.53}$$

The CLTF is given by:

$$\frac{\delta y(s)}{\delta V_d(s)} = \frac{\dfrac{G_a H_y \alpha k_f \sqrt{P_s}}{A_2}}{\dfrac{LC_1}{(\varepsilon + \gamma^2)}s^3 + \dfrac{C_1 R_v}{(\varepsilon + \gamma^2)}s^2 + s + \dfrac{G_a H_y \alpha k_f \sqrt{P_s}}{A_2}}. \tag{6.54}$$

It will be deduced that the response to a step demand change in position results in a zero steady-state position error for an ideal critically lapped servovalve controller. However, the system is third-order, and closed-loop instability is possible if incorrect gains are chosen. The characteristic equation for this third-order position control system is then given by:

$$\frac{LC_1}{(\varepsilon + \gamma^2)}s^3 + \frac{C_1 R_v}{(\varepsilon + \gamma^2)}s^2 + s + \frac{G_a H_y \alpha k_f \sqrt{P_s}}{A_2} = 0. \tag{6.55}$$

Using the Routh criterion, the closed-loop is stable, providing:

$$\frac{B_v A_2}{G_a H_y k_f \sqrt{P_s}} > \alpha, \tag{6.56}$$

$$\alpha = \sqrt{\frac{\gamma - \overline{F}}{1 + \gamma^3}} \text{ when extending,} \qquad \sqrt{\frac{1 + \overline{F}}{1 + \gamma^3}} \text{ when retracting.} \tag{6.57}$$

The boundaries for both extending and retracting cases are shown in Fig. 6.13 as the load force and area ratio are changed.

Assuming a fixed area A_2, these results show that the following:

- When extending, the stability limit is determined at minimum load force.
- When retracting, the stability limit is determined at maximum load force.
- The largest setting from the left-hand side of Eq. (6.56) occurs for a double-rod actuator when $\gamma = 1$.
- The system should be designed for the largest load and for retracting conditions.
- The speed of position response will usually be different in the two directions.

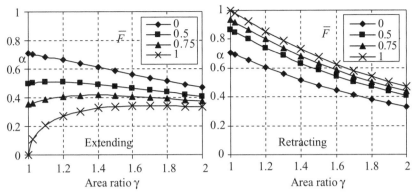

Figure 6.13. Stability boundaries for a single-rod servoactuator position control system.

Consider, therefore, the application that was introduced Chapter 5 and reproduced here as Fig. 6.14. The actuator is double rod, and the actual position response will be compared with the linearized transfer function prediction.

In practice, the servoactuator is controlled by computer and its associated data-acquisition card. Data for this example are as follows:

Load mass, $M = 80\,\text{kg}$; double-rod actuator cross-sectional area, $A = 2.22 \times 10^{-4}\,\text{m}^2$

Viscous damping effective coefficient, $B_v = 4800\,\text{N/m s}^{-1}$

For the actuator in midposition, $V_1(0) = V_2(0) = V(0) = 1.2 \times 10^{-5}\,\text{m}^3$

Servovalve flow constant, $k_f = 2.26 \times 10^{-8}$; supply pressure, $P_s = 210\,\text{bar}$

ISO 32 mineral oil bulk modulus, $\beta = 1.4 \times 10^9\,\text{N/m}^2$

Servoamplifier gain, $G_a = 1\,\text{mA/V}$; position transducer gain, $H_y = 100\,\text{V/m}$

The servovalve linearized flow gain was shown previously to depend on the factor α that varies with the direction of motion and the load. For position control, the linearized transfer function must be evaluated at one of these conditions and, clearly, both are not possible unless the load force is zero. An approximation is

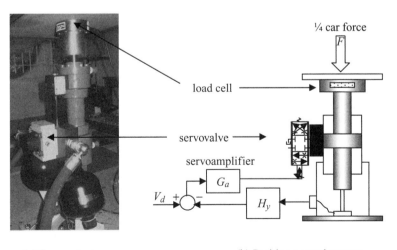

     (a) Servoactuator   (b) Position control system

Figure 6.14. A servoactuator, one of four forming part of a four-poster vehicle test rig (Cardiff University, School of Engineering).

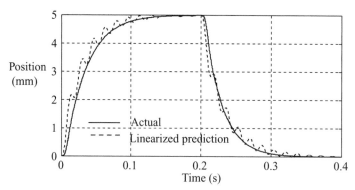

Figure 6.15. A comparison of step responses for position control of a servoactuator.

therefore made with the average value given by:

$$\alpha \approx \frac{\sqrt{1 - \overline{F}} + \sqrt{1 + \overline{F}}}{2\sqrt{2}}, \quad \overline{F} = \frac{F}{P_s A}. \tag{6.58}$$

Given that $F/P_s A = 0.17$, then $\alpha \approx 1/\sqrt{2}$ and is little different than for the no-load condition. The linearized OLTF from Eq. (6.51) becomes:

$$G(s)H(s) = \frac{\dfrac{H_y G_a k_f \sqrt{P_s/2}}{A}}{s\left(1 + \dfrac{CR_v}{2}s + \dfrac{LC}{2}s^2\right)}. \tag{6.59}$$

Placing this transfer function because of standard second-order notation then gives:

$$G(s)H(s) = \frac{KH_y}{s\left(1 + \dfrac{2\zeta}{\omega_n}s + \dfrac{s^2}{\omega_n^2}\right)}.$$

$$K = \frac{G_a k_f \sqrt{P_s/2}}{A}, \quad \frac{2\zeta}{\omega_n} = \frac{CR_v}{2}, \quad \omega_n^2 = \frac{2}{LC}. \tag{6.60}$$

The characteristic equation is therefore:

$$\frac{s^3}{\omega_n^2} + \frac{2\zeta s^2}{\omega_n} + s + KH_y = 0. \tag{6.61}$$

Closed-loop stability is therefore ensured, providing:

$$\overline{K} = \frac{KH_y}{\omega_n} < 2\zeta. \tag{6.62}$$

Substituting the system data then gives the open-loop undamped natural frequency $\omega_n = 379$ rad/s (60.3 Hz), the damping ratio $\zeta = 0.079$, and the forward gain $K = 0.33$ ms^{-1}/V. It then follows that $\overline{K} = 0.087$, $2\zeta = 0.158$, and from Eq. (6.62) closed-loop stability is predicted. A comparison between the actual system response and that predicted from the approximate linearized transfer function is shown in Fig. 6.15.

It can be seen that the linearized solution predicts the general position trend with time, but there is a superimposed oscillation that does not appear to exist on

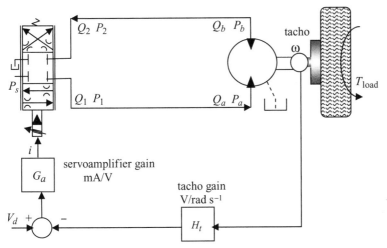

Figure 6.16. A servovalve–motor closed-loop system with long lines.

the actual response. Note, however, that the linearized solution neglects servovalve dynamics. The linearized transfer function approach is therefore conservative in the sense that it tends to overestimate the gain for closed-loop stability; this is a virtue from a design point of view.

6.2.7 The Effect of Long Lines on Closed-Loop Stability, Speed Control of a Motor

Consider a closed-loop motor speed control system. The relationship between motor speed and servovalve current was developed in Chapter 5 for a motor with no losses. The closed-loop system is shown in Fig. 6.16.

Now considering leakage and torque losses and using the servovalve reflected linearized coefficients discussed in Chapter 5.16.5 to embrace line dynamics gives servovalve and lines:

$$
\begin{aligned}
\delta Q_a(s) &= \frac{1}{R_1(s)} k_i \delta i(s) - \frac{R_2(s)}{R_1(s)} k_p \delta P_a(s), \\
\delta Q_b(s) &= \frac{1}{R_1(s)} k_i \delta i(s) + \frac{R_2(s)}{R_1(s)} k_p \delta P_b(s),
\end{aligned}
\tag{6.63}
$$

line functions:

$$
\begin{aligned}
R_1(s) &= \cosh \Gamma \ell + k_p Z_c \sinh \Gamma \ell, \\
R_2(s) &= \cosh \Gamma \ell + \frac{1}{k_p Z_c} \sinh \Gamma \ell.
\end{aligned}
\tag{6.64}
$$

Considering motor leakage and viscous friction then also gives motor flow continuity:

$$
\begin{aligned}
\delta Q_a &= D_m \delta \omega + \frac{(\delta P_a - \delta P_b)}{R_i} + \frac{\delta P_a}{R_e}, \\
\delta Q_b &= D_m \delta \omega + \frac{(\delta P_a - \delta P_b)}{R_i} - \frac{\delta P_b}{R_e},
\end{aligned}
\tag{6.65}
$$

Motor torque:

$$D_m(\delta P_a - \delta P_b) = B_v \delta\omega + J \frac{d\delta\omega}{dt}. \tag{6.66}$$

Collecting these equations together then gives the following OLTF:

$$G(s)H(s) = \frac{K}{R_1(s) + (R_v + s L_m)\left[\dfrac{R_1(s)}{R_m} + \dfrac{R_2(s)}{2R}\right]},$$

$$\frac{1}{R_m} = \frac{1}{R_i} + \frac{1}{2R_e}, \quad R = \frac{1}{k_p}, \quad L_m = \frac{J}{D_m^2}, \quad K = \frac{G_a H_t k_i}{D_m}. \tag{6.67}$$

The closed-loop characteristic equation then becomes:

$$R_1(s) + (R_v + s L_m)\left[\frac{R_1(s)}{R_m} + \frac{R_2(s)}{2R}\right] + K = 0. \tag{6.68}$$

This cannot be solved explicitly, but an approximate solution can be obtained by assuming that each line resistance is negligible compared with servovalve and motor resistance. Recall that for a lossless line:

$$\cosh \Gamma\ell \to \cos\frac{\omega\ell}{C_o}, \quad \sinh \Gamma\ell \to j\sin\frac{\omega\ell}{C_o}, \quad Z_c \to Z_{ca} = \frac{\rho C_o}{a}. \tag{6.69}$$

Then, the characteristic equation solution is obtained by equating the real and imaginary parts to zero, which then gives:

$$\text{frequency,} \quad \tan\theta = -\theta\alpha,$$

$$\theta = \frac{\omega\ell}{C_o}, \quad \alpha = \frac{\left(\dfrac{L_m}{2L}\right)\left(1 + \dfrac{2R}{R_m}\right)}{\left(1 + \dfrac{R_v}{R_m} + \dfrac{R R_v}{2 Z_{ca}^2}\right)},$$

$$\text{gain,} \quad K = \theta\left(\frac{L_m}{2L}\right)\left(1 + \frac{2 Z_{ca}^2}{R R_m}\right)\sin\theta - \left(1 + \frac{R_v}{R_m} + \frac{R_v}{2R}\right)\cos\theta, \tag{6.70}$$

where L is the line total inductance $L = \rho\ell/a$. It can be seen that the solution for frequency embraces all the system resistances and is shown in Fig. 6.17 for a value of $\alpha = 1$.

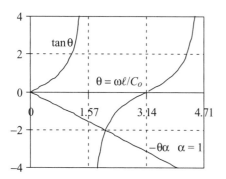

Figure 6.17. Instability condition for closed-loop motor control, no losses.

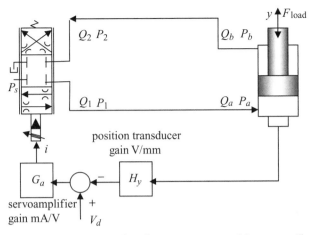

Figure 6.18. A servovalve–linear actuator position controller with long lines.

It is clear that the closed-loop frequency of oscillation lies between the boundaries defined by:

$$\frac{\pi}{2} < \frac{\omega\ell}{C_o} < \pi. \tag{6.71}$$

A value of $\omega\ell/C_o \to \pi/2$ is approached for a relatively low line inductance and a value of $\omega\ell/C_o \to \pi$ is approached for a relatively high value of line inductance. Given a typical value of $C_o = 1276$ m/s for mineral oil, then the frequency of oscillation will be typically $319/\ell < f\,\text{Hz} < 638/\ell$. Care has to be taken in this interpretation of this frequency if short lines lengths are such that the frequency response of the servovalve then becomes important.

It may also be deduced from Worked Example 5.7 that the frequency-condition for closed-loop instability represents the resonant frequency range on the open-loop frequency-response magnitude diagram if losses are neglected.

6.2.8 The Effect of Long Lines on Closed-Loop Stability, Position Control of a Linear Actuator

Because a critically lapped servovalve is being considered, then from earlier work, it has been established that the pressure coefficients are zero at the steady-state position. Consider therefore Fig. 6.18 and the established system equations.

The following conditions apply for this analysis:

- zero position error, the servovalve steady-state current is zero, $k_{p1} = k_{p2} = 0$
- actuator volumes are not neglected as in the case for a load motor
- actuator friction B_v provides only system damping

The simplified equations then become servovalve and lines:

$$\delta Q_a(s) = \frac{k_{i1}}{\cosh \Gamma\ell}\delta i(s) - \frac{\tanh \Gamma\ell}{Z_c}\delta P_a(s),$$

$$\delta Q_b(s) = \frac{k_{i2}}{\cosh \Gamma\ell}\delta i(s) + \frac{\tanh \Gamma\ell}{Z_c}\delta P_b(s), \tag{6.72}$$

actuator flow continuity:

$$\cdot \delta Q_a = A_1 \delta U + \frac{V_1(0)}{\beta} \frac{d \, \delta P_a}{dt},$$

$$\delta Q_b = A_2 \delta U - \frac{V_2(0)}{\beta} \frac{d \, \delta P_b}{dt}, \tag{6.73}$$

load force equation:

$$\delta P_a A_1 - \delta P_b A_2 = B_v \delta U + M \frac{d \, \delta U}{dt}. \tag{6.74}$$

The OLTF then becomes:

$$G(s)H(s) = \frac{\dfrac{G_a H_y}{A_2 G_3(s)} \left[\gamma k_{i1} G_1(s) + k_{i2} G_2(s) \right]}{s \left[\gamma^2 G_2(s) + G_1(s) + (R_v + s L_m) G_1(s) G_2(s) \right]},$$

$$G_1(s) = \frac{\tanh \Gamma \ell}{Z_c} + s C_1, \quad G_2(s) = \frac{\tanh \Gamma \ell}{Z_c} + s C_2, \quad G_3(s) = \cosh \Gamma \ell,$$

$$C_1 = \frac{V_1(0)}{\beta}, \quad C_2 = \frac{V_2(0)}{\beta}. \tag{6.75}$$

This transfer function is difficult to interpret in general terms from a stability point of view. Some progress can be made if it is assumed that the volumes on either side of the actuator are equal such that $C_1 = C_2 = C$. This would be exact for a double-rod actuator, $\gamma = 1$, with the piston at the central position. The OLTF then becomes:

$$G(s)H(s) = \frac{K \alpha}{s \left[\cosh \Gamma \ell + \dfrac{(R_v + s L_m)(\sinh \Gamma \ell + s C Z_c \cosh \Gamma \ell)}{(\gamma^2 + 1) Z_c} \right]},$$

$$K = \frac{G_a H_y k_f \sqrt{P_s}}{A_2}, \quad R_v = \frac{B_v}{A_2^2}, \quad L_m = \frac{M}{A_2^2}, \tag{6.76}$$

$$\alpha = \sqrt{\frac{\gamma - \overline{F}}{(1 + \gamma^3)}} \text{ extending}, \quad \alpha = \sqrt{\frac{1 + \overline{F}}{(1 + \gamma^3)}} \text{ retracting}.$$

The closed-loop characteristic equation is given by:

$$s \left[(\gamma^2 + 1) \cosh \Gamma \ell + \frac{R_v \sinh \Gamma \ell}{Z_c} \right] + s^2 \left[\frac{L_m \sinh \Gamma \ell}{Z_c} + C R_v \cosh \Gamma \ell \right]$$

$$+ s^3 L_m C \cosh \Gamma \ell + (\gamma^2 + 1) K \alpha = 0. \tag{6.77}$$

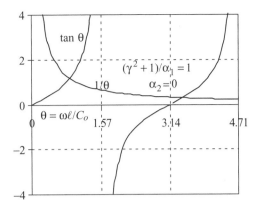

Figure 6.19. Instability condition for closed-loop cylinder position control.

A good feel for the conditions for instability can be obtained by considering the lines to be lossless, and the transmission line functions are again simplified to:

$$\cosh \Gamma\ell \to \cos\frac{\omega\ell}{C_o}, \quad \sinh \Gamma\ell \to j\,\sin\frac{\omega\ell}{C_o}, \quad Z_c \to Z_{ca} = \frac{\rho C_o}{a}. \tag{6.78}$$

Then, the characteristic equation solution is again obtained by equating the real and imaginary parts to zero, which then gives:

$$\text{frequency,} \quad \theta^2\alpha_1\alpha_2\cos\theta + \theta\sin\theta\alpha_1 - (\gamma^2+1)\cos\theta = 0,$$

$$\text{gain,} \quad K\alpha = \frac{\theta^2\alpha_2\alpha_3\cos\theta}{(\gamma^2+1)} + \frac{\theta\sin\theta\alpha_3}{(\gamma^2+1)},$$

$$\alpha = \sqrt{\frac{\gamma - \overline{F}}{1+\gamma^3}}, \text{ extending,} \quad \alpha = \sqrt{\frac{1+\overline{F}}{1+\gamma^3}}, \text{ retracting}$$

$$\alpha_1 = \frac{L_m}{L}, \quad \alpha_2 = \frac{V(0)}{V_{\text{line}}}, \quad \alpha_3 = \frac{R_v}{L},$$

(6.79)

where L is the line inductance, $V(0)$ is the volume on each side of the actuator, and V_{line} is the volume of one line.

For this system model, there is no solution for gain if the viscous damping B_v is neglected. The frequency solution may be written as:

$$\tan\theta = \frac{(\gamma^2+1)}{\theta\,\alpha_1} - \theta\alpha_2. \tag{6.80}$$

A graphical representation of Eq. (6.80) is shown in Fig. 6.19.

It will be seen that whatever value of γ, α_2 and α_1 chosen, then the frequency will always satisfy:

$$0 < \frac{\omega\ell}{C_o} < \frac{\pi}{2}. \tag{6.81}$$

Given a typical value of $C_o = 1276$ m/s for mineral oil, then the frequency of oscillation will be typically f Hz $< 319/\ell$. Again, care has to be taken when interpreting this frequency if short line lengths are such that the frequency response of the servo-valve then becomes important.

Worked Example 6.5

Consider a practical example undertaken by the author for a position control system with the following data:

> ISO 32 mineral oil @ 50°C, density $\rho = 860$ kg/m³, bulk modulus $\beta = 1.4 \times 10^9$ N/m², servovalve supply pressure $P_s = 100$ bar
> Lines $\ell = 10.73$ m long, 7-mm internal diameter
> Actuator bore, 50.8-mm diameter, rod 28.58-mm diameter, stroke 254-mm
> Load mass $M = 156$ kg
> Position transducer gain, $H_y = 41.3$ V/m
> $\overline{F} = F/P_s A_2 = 0.11$
> extending $\alpha = 0.57$, retracting $\alpha = 0.53$, average 0.55
> Line $a = 0.385 \times 10^{-4}$ m², $A_1 = 0.00203$ m², $A_2 = 0.00139$ m², area ratio $\gamma = 1.46$
> Line inductance, $L = \rho\ell/a = 2.4 \times 10^8$ kg/m⁴
> $C_o = \sqrt{\beta/\rho} = 1276$ m/s
> Load inductance, $L_m = M/A_2^2 = 0.8 \times 10^8$ kg/m⁴
> Actuator equal volumes on either side, $V(0) = 0.21 \times 10^{-3}$ m³
> Single line volume, $V_{\text{line}} = 0.413 \times 10^{-3}$ m³
> $\alpha_1 = L_m/L = 0.333$, $\alpha_2 = V(0)/V_{\text{line}} = 0.508$

At the point of instability, using Eq. (6.79) gives:

$$\theta^2 \alpha_1 \alpha_2 \cos\theta + \theta \sin\theta\,\alpha_1 - (\gamma^2 + 1)\cos\theta = 0,$$

$$0.169\theta^2\cos\theta + 0.333\theta \sin\theta - 3.13\cos\theta = 0.$$

The solution is $\theta = \omega\ell/C_o = 1.4$. Further calculations and comparisons are then as shown in the following list:

> predicted frequency of oscillation $\theta = 1.4$, $f = 26.5$ Hz;
> neglecting actuator volumes $\alpha_2 = 0$, $\theta = 1.42$, $f = 26.9$ Hz;
> measured value, $f = 23.8$ Hz.

Therefore, the predicted frequency of oscillation is within 11.3% of the measured value. Neglecting actuator volumes produced a prediction accuracy of 13%. The actual frequency of oscillation justified neglecting servovalve dynamics in this example. Assuming a mean value of $\alpha = 0.55$ with a negligible loss in accuracy between extending and retracting conditions, then the gain at the point of instability is given by:

$$\text{gain} \quad K\alpha = \frac{\theta^2\alpha_2\alpha_3\cos\theta}{(\gamma^2 + 1)} + \frac{\theta\sin\theta\,\alpha_3}{(\gamma^2 + 1)},$$

$$\frac{G_a H_y k_f \sqrt{P_s}}{A_2} = \frac{1.27 R_v}{L}.$$

The gain prediction at the point of instability was inconclusive because of the difficulty in determining an accurate estimate of the low actuator viscous friction coefficient B_v. Some closed-loop position and velocity responses to a step position demand are shown in the following figure and illustrates that the servo-amplifier gain $G_a \approx 40$ mA/V at the point of instability.

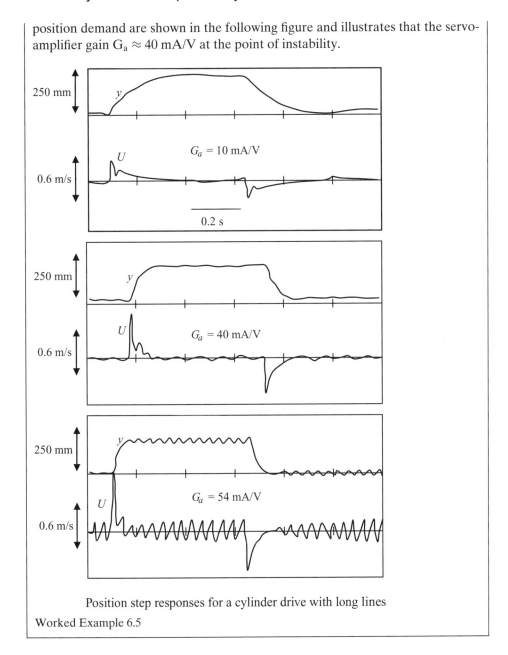

Position step responses for a cylinder drive with long lines

Worked Example 6.5

6.2.9 The Effect of Coulomb Friction Damping on the Response and Stability of a Servovalve–Linear Actuator Position Control System

Now consider the case in which actuator and load damping is dominated by Coulomb friction. This was introduced in Chapter 5, in which it was shown that for a simple and idealized open-loop example, the position response had a linear decay characteristic. The system diagram is shown in Fig. 6.20.

For a conventional servovalve–linear actuator position control system, the non-linear flow characteristic of the servovalve must be considered when Coulomb friction is present. However, a realistic simplification can be made for a double-rod

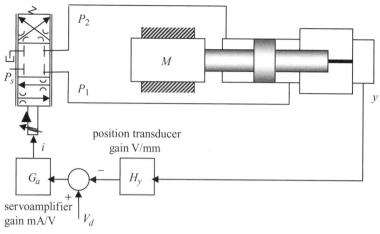

Figure 6.20. A servovalve–linear actuator position controller.

actuator as outlined in Chapter 5, in which it was shown that in the absence of piston seal leakage, the sum of line pressures is dynamically constant and equal to supply pressure. Hence, the system equations are given by:

$$k_f i \sqrt{\frac{P_s - P_{\text{load}}}{2}} = AU + \frac{V}{\beta} \frac{dP_{\text{load}}}{dt},$$

$$P_{\text{load}} A = F_c \, \text{sign}(U) + M \frac{d^2 y}{dt^2}, \quad (6.82)$$

$$i = G_a(V_d - H_p y) \quad P_{\text{load}} = P_1 - P_2.$$

The system equations may then be placed in nondimensional form to significantly simplify the choice of parameters, as follows:

$$(1 - \bar{y})\sqrt{1 - \overline{P}_{\text{load}}} = \overline{U} + \frac{d\overline{P}_{\text{load}}}{dt/\tau},$$

$$\overline{P}_{\text{load}} = \alpha \, \text{sign}(\overline{U}) + X \frac{d\overline{U}}{dt/\tau},$$

$$\frac{d\bar{y}}{dt/\tau} = Y\overline{U},$$

$$\bar{y} = \frac{y}{y_d}, \quad \overline{U} = \frac{U}{U_{\text{ref}}}, \quad \overline{P}_{\text{load}} = \frac{P_{\text{load}}}{P_s}, \quad U_{\text{ref}} = \frac{Q_{\text{ref}}}{A}, \quad (6.83)$$

$$X = \frac{L}{CR^2}, \quad Y = \frac{V}{V_{\text{disp}}} \frac{P_s}{\beta}, \quad L = \frac{M}{A^2}, \quad C = \frac{V}{\beta}, \quad R = \frac{P_s}{Q_{\text{ref}}},$$

$$y_d = \frac{V_d}{H_p}, \quad \alpha = \frac{F_c}{P_s A}, \quad \tau = CR, \quad Q_{\text{ref}} = k_f G_a V_d \sqrt{\frac{P_s}{2}}.$$

V is the actuator volume, and initially equal on each side, and V_{disp} is the demanded volume change during position control; for example, 10% of the half-stroke. In practice for servoactuator applications, it is likely that the dynamic parameter X will be small, possibly around the unity value. In this situation, the closed-loop position response will be dominated by the actuator integrator characteristic combined with the servovalve flow gain. This means that the transient response to a step input will look like a first-order response with perhaps a small superimposed oscillation that

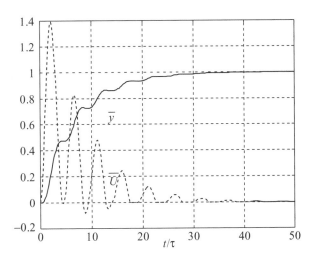

Figure 6.21. Transient response and phase-plane plot for the servoactuator position control system $X = 0.5$, $Y = 0.15$, $\alpha = 0.05$.

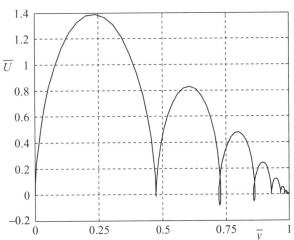

is due to fluid compressibility, load inertia, and damping effects. The time constant of the dominant first-order response is $\tau = V_{\text{disp}}/Q_{\text{ref}}$. In such a case, it is difficult to assess whether viscous, Coulomb, or a combination of both frictions exists from practical dynamic tests. Servovalve dynamics will usually reduce this superimposed oscillation because of system dynamics.

For example, consider the solution of Eq. (6.83) for $X = 0.5$ and $Y = 0.15$. The value of Y is assessed from a demanded displaced volume of $1/10$ of the actuator volume on each side, with the piston initially centralized. Therefore, the change in volume on the transient behavior is negligible over the transient period. A typical servopressure of 210 bar is assumed with an effective fluid bulk modulus of 14,000 bar. A friction factor of $\alpha = 0.05$ has also been selected that would result in a pressure difference of 10.5 bar because of Coulomb friction. Figure 6.21 shows the simulation result. It will be seen from system equation (6.83) that in the absence of leakage, there will not be a steady-state position error that is due to Coulomb friction.

Different values of the system parameters X, Y, and α produce different transient responses and phase-plane plots. For a transient response similar to that shown in Fig. 6.21, the velocity is positive dominant. The effect of Coulomb friction on closed-loop position control in practice is quite different from that traditionally used

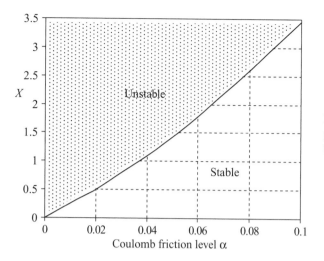

Figure 6.22. Stability boundary for the position control example with Coulomb friction damping, $Y = 0.15$.

to explain its effect by use of a simple mass–spring–damper system. At the condition for closed-loop instability, severe oscillation in pressure differential will usually occur rather than severe oscillations in position. Considering the transient behavior shown in Fig. 6.21, it is unwise to rely on a linearized analysis to understand the effect of Coulomb friction in a position control system. The stability boundary variation with friction level is not predicted with sufficient accuracy with a linearized analysis. When the closed-loop response is dominated by the integration characteristic of the actuator, the amplitude of the position oscillation at the point of instability is small compared with the steady-state demanded position; the oscillation amplitude is not half the step change demanded. Therefore, the use of an equivalent linear viscous damping coefficient B_v, as outlined in Section 5.19, is difficult to interpret because of the requirement that the amplitude of oscillation y_o must be specified; it varies with the choice of α and X. In addition, the frequency of oscillation ω_n varies with α and X. For the example being used here, Fig. 6.22 shows the stability boundary as the friction level is increased and determined from the exact simulation.

The stability boundary is given by:

$$X = 23\alpha + 115\alpha^2. \tag{6.84}$$

Considering the definition of the dynamic parameter X then, for a fixed friction level, instability caused by an increase in X would occur by:

- increasing mass M
- increasing the servoamplifier gain G_a
- decreasing the servovalve supply pressure P_s
- increasing the servovalve rated flow (increasing k_f)

6.3 Digital Control

6.3.1 Introduction

Digital control is usually concerned with using a microcomputer or equivalent system to control a hydraulic system and with some sort of sampling of the parameter to be controlled followed by a decision-making process and then control actuation.

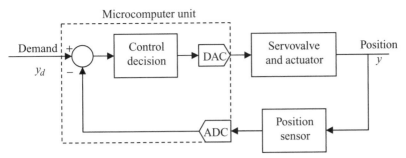

Figure 6.23. Elements of a digital control system.

In a perfect world, the digital control process would not appear any different from conventional analog control, and the ever-increasing processing speed of microprocessors might suggest that this should be the case. However, issues of bandwidth and noise mean that, in reality, the process of digital control is not an idealized process, and this section introduces some basic concepts of digital control theory. Consider therefore the basic control system shown in Fig. 6.23; in this example, a cylinder position control system, with the digital processing unit placed in the error part of the closed-loop control system.

It is assumed in Fig. 6.23 that the position sensor is an analog device and, therefore, its voltage must be converted into a form that can be processed by the computer. Once the control decision with its associated computations has then been carried out, the signal to be sent to the servovalve must be converted to an analog voltage that is passed to the servoamplifier.

In practice, the digital controller can be a data-acquisition card plugged into the back of a PC with analog-to-digital converters (ADCs), digital-to-analog converters (DACs), logic gates, and so forth, and with a range of sampling speeds and bit accuracy. The actual sampling frequency of an ADC can be set by the user and can be many kilohertz. The actual sampling frequency, when computations are complete together with DAC, is usually between 500 Hz and 2 kHz for hydraulic systems and with 12-bit data resolution. A commercial programming and software library may then be added to produce a flexible input–control decision–output control tool. Alternatively, a dedicated unit may be available from the servovalve manufacturer, usually having two channels and with dedicated programming code and comprehensive software functions. This is less flexible than the former but it is designed for industrial applications while still containing all the main programming functions needed to significantly improve and often optimize closed-loop performance in the most complicated of process requirements. Figure 6.24 shows both an industrial two-channel servocontroller and a data-acquisition card.

6.3.2 The Process of Sampling

Consider a continuous signal $f(t)$ that is sampled by an ideal sampler to produce $f^*(t)$, as shown in Fig. 6.25.

The continuous signal is sampled every T seconds; the sampling interval, and the time required for acquiring the signal, is shown but in reality is considered negligible in comparison with the sampling interval. The output of the ideal sampler therefore may be represented by the input multiplied by the sampling process; in

(a) Moog 2-channel industrial programmable servocontroller

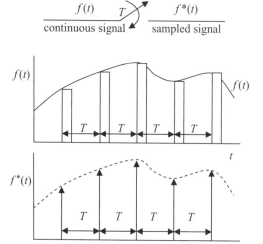

(b) Multichannel data-acquisition input–output card

Figure 6.24. An industrial controller and a data-acquisition card.

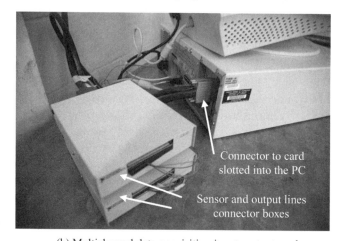

Figure 6.25. Sampling of a continuous signal.

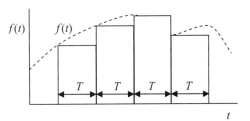

Figure 6.26. The practical approach to sampling with zero-order hold.

this instance, the unit impulse (Dirac delta function). Each contribution from previous samples is included by choosing the appropriate delay period, which will be a multiple of the sampling period, as follows:

$$f^*(t) = f(0)\delta(t) + f(T)\delta(t - T) + f(2T)\delta(t - 2T) + f(3T)\delta(t - 3T) + \cdots +,$$

$$f^*(t) = \sum_{n=0}^{\infty} f(nT)\delta(t - nT), \tag{6.85}$$

where $\delta(t - nT)$ is the unit impulse at $t = nT$.

The Laplace transform of the sampled signal is therefore:

$$F^*(s) = f(0) + f(T)e^{-sT} + f(2T)e^{-2sT} + f(3T)e^{-3sT} + \cdots +,$$

$$F^*(s) = \sum_{n=0}^{\infty} f(nT)e^{-nsT}. \tag{6.86}$$

The Laplace transform of a sampled signal is therefore an infinite series. Now, to use the sampled information, it must be reconstructed in some practical manner other than the idealized form. The most common form of signal reconstruction is referred to as a *zero-order hold* (ZOH), whereby the sampled signal is simply held constant over the whole of the sampling period T. This is shown in Fig. 6.26.

Recalling the Laplace transform for a step function held constant for T seconds then allows Eq. (6.86) to be modified as follows:

$$F(s) = f(0)\frac{(1 - e^{-sT})}{s} + f(T)\frac{(e^{-sT} - e^{-2sT})}{s} + f(2T)\frac{(e^{-2sT} - e^{-3sT})}{s} + \cdots +,$$

$$= \frac{(1 - e^{-sT})}{s}[f(0) + f(T)e^{-sT} + f(2T)e^{-2sT}] + \cdots +,$$

$$= \frac{(1 - e^{-sT})}{s}F^*(s). \tag{6.87}$$

Therefore, zero-order hold sampling may be considered to have a transfer function $G_o(s)$ in conjunction with the ideal sampler and given by:

$$G_o(s) = \frac{(1 - e^{-sT})}{s}. \tag{6.88}$$

Therefore, to analyze a control system with digital control, it is now only necessary to determine the Laplace transform of common signals and functions, and this is greatly helped by the use of z-transform theory.

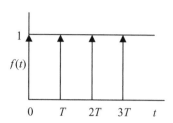

Figure 6.27. Ideal sampling of a unit step function.

6.3.3 The z Transform

This is simply defined as:

$$z = e^{sT}. \tag{6.89}$$

Consequently, the z transform of an ideally sampled signal, from Eq. (6.86), is given by:

$$F(z) = \mathbf{Z}F^*(s) = \sum_{n=0}^{\infty} f(nT)z^{-n}. \tag{6.90}$$

EXAMPLE 1: A UNIT STEP FUNCTION. Consider ideal sampling of the unit step function as shown in Fig. 6.27:

$$F(z) = \sum_{n=0}^{\infty} f(nT)z^{-n} = \sum_{n=0}^{\infty} (1)z^{-n},$$
$$= 1 + \frac{1}{z} + \frac{1}{z^2} + \frac{1}{z^3} + \cdots + = \frac{z}{(z-1)}. \tag{6.91}$$

EXAMPLE 2: AN EXPONENTIALLY DECAYING SIGNAL. Consider ideal sampling of the unit step function as shown in Fig. 6.28:

$$F(z) = \sum_{n=0}^{\infty} f(nT)z^{-n} = \sum_{n=0}^{\infty} (e^{-anT})z^{-n},$$
$$= 1 + \left(\frac{e^{aT}}{z}\right) + \left(\frac{e^{aT}}{z}\right)^2 + \left(\frac{e^{aT}}{z}\right)^3 + \cdots + = \frac{z}{(z - e^{-aT})}. \tag{6.92}$$

Some common z transforms are shown in Table 6.1.

6.3.4 Closed-Loop Analysis with Zero-Order-Hold Sampling

Consider a cylinder position control system as shown in Fig. 6.29 together with the block diagram. The digital processor is sampling the error signal with no modification to it, but it is a ZOH sampler.

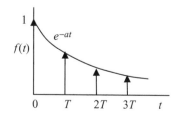

Figure 6.28. Ideal sampling of an exponentially decaying function.

Table 6.1. *Some common z transforms*

	$f(t)$	$F(s)$	$F(z)$
1	Unit step	$\dfrac{1}{s}$	$\dfrac{z}{(z-1)}$
2	t	$\dfrac{1}{s^2}$	$\dfrac{Tz}{(z-1)^2}$
3	$t^2/2$	$\dfrac{1}{s^3}$	$\dfrac{T^2z(z+1)}{2(z-1)^3}$
4	e^{-at}	$\dfrac{1}{(s+a)}$	$\dfrac{z}{(z-e^{-aT})}$
5	$(1-e^{-at})$	$\dfrac{a}{s(s+a)}$	$\dfrac{z(1-e^{-aT})}{(z-1)(z-e^{-aT})}$
6	$t-\dfrac{(1-e^{-aT})}{a}$	$\dfrac{a}{s^2(s+a)}$	$\dfrac{Tz}{(z-1)^2}-\dfrac{z(1-e^{-aT})}{a(z-1)(z-e^{-aT})}$
7	$e^{-at}\sin bt$	$\dfrac{b}{(s+a)^2+b^2}$	$\dfrac{ze^{-aT}\sin bT}{(z^2-2ze^{-aT}\cos bT+e^{-2aT})}$
8	$e^{-at}\cos bt$	$\dfrac{(s+a)}{(s+a)^2+b^2}$	$\dfrac{z^2-ze^{-aT}\cos bT}{(z^2-2ze^{-aT}\cos bT+e^{-2aT})}$
9	Final-value theorem	$\lim\limits_{t\to\infty} f(t) = (z-1)F(z)]_{z=1}$	
10	Initial-value theorem	$\lim\limits_{t\to 0} f(t) = F(z)]_{z=\infty}$	

The CLTF in the z domain is obtained in the usual way:

$$\frac{y(z)}{V_d(z)} = \frac{\mathbf{Z}G_o(s)G(s)}{1+\mathbf{Z}G_o(s)GH(s)}. \tag{6.93}$$

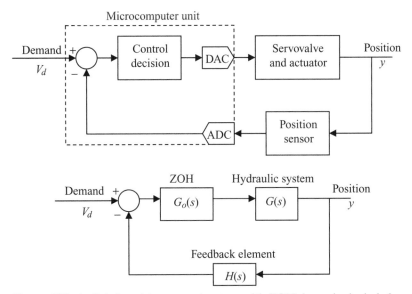

Figure 6.29. A digital position control system with ZOH dynamics included.

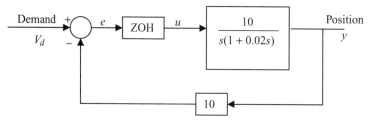

Figure 6.30. A digital position control system with ZOH dynamics included.

It will be recalled that the ZOH has a transfer function $G_o = \dfrac{(1 - e^{-sT})}{s}$ and this is easily handled once it is recognized that:

$$ZG_o(s)G(s) = (1 - z^{-1})Z\frac{G(s)}{s}, \tag{6.94}$$

$$ZG_o(s)GH(s) = (1 - z^{-1})Z\frac{GH(s)}{s}. \tag{6.95}$$

Consider the position control system shown in Fig. 6.30 with a sampling frequency of 200 Hz, giving $T = 5$ ms. The computer is simply a data-transfer device with a ZOH sampler:

$$ZG_o(s)G(s) = (1 - z^{-1})Z\frac{100}{s^2(1 + 0.02s)}, \tag{6.96}$$

$$ZG_o(s)GH(s) = (1 - z^{-1})Z\frac{100}{s^2(1 + 0.02s)}. \tag{6.97}$$

From the z transform, Table 6.1 Entry 6, and noting that $a = 50$ and $aT = 0.25$, then Eq. (6.97) becomes:

$$ZG_o(s)GH(s) = \frac{(0.058z + 0.0526)}{(z - 1)(z - 0.779)}. \tag{6.98}$$

Notice that a single pole s, shown in block diagram Fig. 6.28, in the OLTF is mapped to $(z - 1)$ in the z domain. The CLTF is given by:

$$\frac{10y(z)}{V_d(z)} = \frac{y(z)}{y_d(z)} = \frac{(0.058z + 0.0526)}{(z^2 - 1.721z + 0.832)}. \tag{6.99}$$

The remaining issue is now how to determine the closed-loop transient response for a particular applied input, and there are three ways of doing this:

 (i) Apply the input z transform and find the inverse using z-transform Table 6.1.
(ii) Apply the input z transform and use long division to find the inverse.
(iii) Use a recursive algorithm, applying the input step by step.

This will now be done for the example using a unit step input signal:

$$y_d(s) = \frac{1}{s} \rightarrow y_d(z) = \frac{z}{(z - 1)}. \tag{6.100}$$

(i) Transient response method 1, apply the input and use z-transform Table 6.1:

$$\frac{y(z)}{y_d(z)} = \frac{(0.058z + 0.0526)}{(z^2 - 1.721z + 0.832)},$$

$$y(z) = \frac{z(0.058z + 0.0526)}{(z-1)(z^2 - 1.721z + 0.832)}. \tag{6.101}$$

Using partial fraction expansion gives:

$$y(z) = \frac{z}{(z-1)} - \frac{(z^2 - 0.779z)}{(z^2 - 1.721z + 0.832)},$$

$$= \frac{z}{(z-1)} - \frac{(z^2 - 0.86z)}{(z^2 - 1.721z + 0.832)} - \frac{0.081z}{(z^2 - 1.721z + 0.832)}. \tag{6.102}$$

Using Table 6.1, Entries 1, 7, and 8, then gives the inverse:

$$y(t) = 1 - e^{-at}(\cos bt + 0.268 \sin bt), \tag{6.103}$$

where $a = 18.4$, $b = 67.54$ rad/s, and Eq. (6.103) should be evaluated every $T = 0.005$ s. This gives the following sequence:

$$y(t) = 0.058z^{-1} + 0.211z^{-2} + 0.426z^{-3} + 0.668z^{-4} + \cdots +. \tag{6.104}$$

The values shown are valid at only the appropriate sampling interval.

(ii) *Transient response method 2, apply the input and use long division.*

Recall the transfer function with the step input applied:

$$y(z) = \frac{z(0.058z + 0.0526)}{(z-1)(z^2 - 1.721z + 0.832)}. \tag{6.105}$$

Now, from long division:

$$\begin{array}{r}
0.058z^{-1} + 0.21z^{-2} + 0.423z^{-3}, \ldots, \\
(z^3 - 2.721z^2 + 2.553z - 0.832) \overline{) \ 0.058z^2 + 0.0526z} \\
0.058z^2 - 0.1578z + 0.1481 - 0.0483z^{-1}, \\
0.2104z - 0.1481 + 0.0483z^{-1}.
\end{array} \tag{6.106}$$

(iii) *Transient response method 3, recursive algorithm applying the input at each step.*

Recall the transfer function without the step input applied:

$$\frac{y(z)}{y_d(z)} = \frac{(0.058z + 0.0526)}{(z^2 - 1.721z + 0.832)}. \tag{6.107}$$

Now rearrange this by dividing the numerator and denominator by the highest power of z in the denominator and rearrange:

$$\frac{y(z)}{y_d(z)} = \frac{(0.058z^{-1} + 0.0526z^{-2})}{(1 - 1.721z^{-1} + 0.832z^{-2})},$$

$$(1 - 1.721z^{-1} + 0.832z^{-2})y(z) = (0.058z^{-1} + 0.0526z^{-2})y_d(z),$$

$$y(z) = (1.721z^{-1} - 0.832z^{-2})y(z) + (0.058z^{-1} + 0.0526z^{-2})y_d(z). \tag{6.108}$$

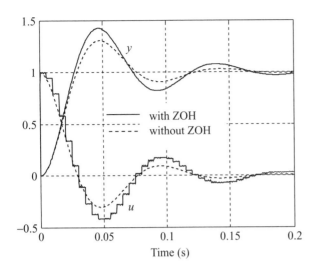

Figure 6.31. Position and ZOH responses for a unit step input.

This means that the current sampled value of y may be computed from previous samples of y and y_d. The calculation proceeds as follows:

$$y(z) = (1.721z^{-1} - 0.832z^{-2})y(z) + (0.058z^{-1} + 0.0526z^{-2})y_d(z),$$
$$y(1) = 0 - 0 + 0.058 + 0 = 0.058,$$
$$y(2) = 1.721(0.058) - 0 + 0.058 + 0.0526 = 0.211,$$
$$y(3) = 1.721(0.211) - 0.832(0.058) + 0.111 = 0.426,$$
$$y(4) = 1.721(0.426) - 0.832(0.211) + 0.111 = 0.668. \tag{6.109}$$

This method is perhaps the better of the three methods and is also useful in that a nonlinear, sampled input may be considered. The closed-loop behavior is, of course, easily achieved by computer simulation, and the result is shown in Fig. 6.31 and matches the earlier calculations.

It will be seen that the effect of ZOH sampling alone is to make the position response more oscillatory. In addition, it will also be noticed that the ZOH sampling effect is not evident at the output because of the filtering effect of the hydraulic system dynamics. The output of the ZOH, shown as u, illustrates the importance of the sampling interval selection T such that the system natural frequency is captured with sufficient accuracy.

6.3.5 Closed-Loop Stability

The determination of the condition for closed-loop instability in the z plane is again concerned with the evaluation of the closed-loop system characteristic equation, which in this case is given by:
$$1 + ZG_o(s)GH(s) = 0. \tag{6.110}$$

This cannot be handled in the same way as previously done for a linear system using s domain theory. It will be recalled that the condition for instability in the s plane occurs when the roots of the characteristic equation just lie on the imaginary axis of the s plane. Considering the z plane equivalence gives:

$$s = j\omega,$$
$$z = e^{sT} = e^{j\omega T} = \cos \omega T + j \sin \omega T. \tag{6.111}$$

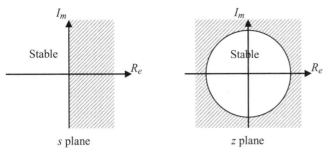

Figure 6.32. Stability boundaries for roots of the closed-loop characteristic equation.

Therefore, the *magnitude* of z is unity; that is, Eq. (6.111) defines a unit circle in the z plane. A comparison between the conditions for instability in the s and z planes is shown in Fig. 6.32.

Alternatively, when Eq. (6.110) is expressed in z notation, it can be transformed into s notation by making a suitable linear approximation for $z = e^{sT}$. The reasoning is that if the approximation is sufficiently accurate, then the characteristic equation in the approximate s domain can be treated by conventional s plane theory previously discussed. The issue then is one of selecting a suitable approximation for $z = e^{sT}$, and the most common approach is to use a bilinear approximation. This considers just the first two terms of the Taylor series expansion to the following approximation:

$$z = e^{sT} = \frac{e^{sT/2}}{e^{-sT/2}} \approx \frac{1 + sT/2}{1 - sT/2}. \tag{6.112}$$

Because s is only an approximation to its true value, then it is given another symbol that is defined as $w' \approx s$ and, therefore:

$$z = \frac{1 + w'T/2}{1 - w'T/2}. \tag{6.113}$$

It is then an easy matter to show that the substitution of $z = e^{j\omega T}$ gives:

$$w' = j\frac{2}{T}\tan\frac{\omega T}{2}. \tag{6.114}$$

When this is compared with $s = j\omega$, it can be seen that w' has exactly the same phase of $+90°$ but a magnitude that is a function of both frequency and sampling interval. However, it follows from Eq. (6.114) that:

$$|w'| \to \omega \quad \text{when} \quad \frac{\omega T}{2} \ll 1. \tag{6.115}$$

Given that the sampling frequency $f_s = 1/T$ and the system frequency $f = \omega/2\pi$, then the inequality in Eq. (6.115) is satisfied when:

$$f \ll \frac{f_s}{\pi}. \tag{6.116}$$

The importance of this background work is that the bilinear transformation is a good representation from z to s because the phase is always correct and the magnitude will be sufficiently accurate if the correct sampling frequency is selected. Therefore, when the characteristic equation in z is developed, then the bilinear approximation may be used to convert it to a polynomial in s. Furthermore, if the

Routh criterion is used, then the factor $T/2$ in Eq. (6.113) becomes redundant and it is simpler to use a modified approximation as follows:

$$z = \frac{1 + w'T/2}{1 - w'T/2} \rightarrow \frac{1+w}{1-w}. \tag{6.117}$$

Worked Example 6.6

Consider the previous control system but with the hydraulic gain of 10 now a variable gain K. Note that without the ZOH, the closed-loop system is second-order and cannot be unstable. The figure shows the block diagram again.

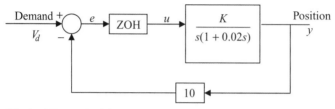

Worked Example 6.6.

A digital position control system with ZOH dynamics included:

$$\text{OLTF} \quad ZG_o(s)GH(s) = \frac{K(0.0058z + 0.00526)}{(z-1)(z-0.779)}.$$

The characteristic equation then becomes:

$$z^2 - z(1.779 - 0.0058K) + 0.00526K + 0.779 = 0.$$

The roots may now be evaluated as K is varied until a unity-magnitude root occurs, the root then just lying on the unit circle. In this example:

$K \approx 42.1$ with $z = 0.767 \pm j0.641$

Now use the bilinear transformation to give:

$$z^2 - z(1.779 - 0.0058K) + 0.00526K + 0.779 = 0,$$

$$z = \frac{1+w}{1-w},$$

$$w^2(0.011K) + w(0.442 - 0.0105K) + (3.558 - 0.00054K) = 0.$$

Now apply the Routh array method, treating the polynomial in w as though it is a polynomial in s. However, in this case, because the polynomial is quadratic, it is necessary only that all coefficients be positive for a stable closed-loop system. There are two conditions for this given by $(0.442 - 0.0105K) > 0 \rightarrow K < 42.1$ and $(3.558 - 0.00054K) > 0 \rightarrow K < 6589$. Clearly, as K is increased, then instability occurs when $K = 42.1$.

6.4 Improving the Closed-Loop Response

6.4.1 Servovalve Spool Underlap for Actuator Position Control, a Linearized Transfer Function Approach

Consider position control of an actuator shown in Fig. 6.33.

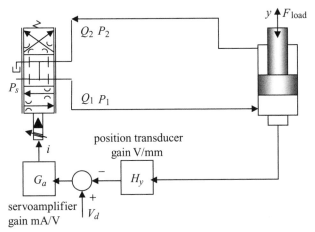

Figure 6.33. An underlapped servovalve–linear actuator position controller.

The steady-state flow characteristic was discussed in Section 3.5, the two flow equations within the underlap region of a symmetrically underlapped servovalve being given by:

$$Q_1 = k_f(i_u + i)\sqrt{P_s - P_1} - k_f(i_u - i)\sqrt{P_1}, \qquad (6.118)$$

$$Q_2 = k_f(i_u + i)\sqrt{P_2} - k_f(i_u - i)\sqrt{P_s - P_2}. \qquad (6.119)$$

Section 4.9 also showed that a single-rod actuator under closed-loop control with a load force can result in a steady-state position error that can be removed for a unique load force–supply pressure relationship.

Earlier in Section 6.2, position control was considered for a critically lapped servovalve spool in which it was evident that the servovalve provides no damping at the zero steady-state error condition. In general, this is not the case with an underlapped spool, and the effect of underlap will be pursued by means of a linearized analysis for small variations around the steady-state condition.

To aid understanding of underlap damping, the following reasonable restrictions will be set:

- It will be assumed that the supply pressure and rod cross-sectional area are matched to the load force to give zero position error at the steady-state condition:

$$F = \frac{P_s A_{\text{rod}}}{2} \quad \text{where } A_{\text{rod}} = A_1 - A_2. \qquad (6.120)$$

- The volumes on either side of the actuator are equal because of connecting volumes, $V_1(0) = V_2(0) = V(0)$.
- Dynamics are dominated by load mass, fluid compressibility, and viscous friction.

The linearized equations may then be written:

$$\delta Q_1 = k_{i1}\delta i - k_{p1}\delta P_1 = A_1\delta U + \frac{V_1(0)}{\beta}\frac{d\delta P_1}{dt}, \qquad (6.121)$$

$$\delta Q_2 = k_{i2}\delta i + k_{p2}\delta P_2 = A_2\delta U - \frac{V_2(0)}{\beta}\frac{d\delta P_2}{dt}, \qquad (6.122)$$

$$\delta P_1 A_1 - \delta P_2 A_2 = B_v\delta U + \delta F_{\text{load}} + M\frac{d\delta U}{dt}. \qquad (6.123)$$

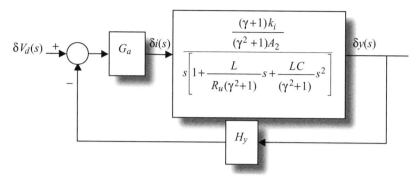

Figure 6.34. Closed-loop block diagram for a servocylinder drive.

Given load force condition equation (6.120), then within the underlap region and for zero steady-state error:

$$k_{i1} = k_{i2} = k_i = k_f\sqrt{2P_s}, \tag{6.124}$$

$$k_{p1} = k_{p2} = k_p = k_f i_u\sqrt{\frac{2}{P_s}}. \tag{6.125}$$

The transfer function relating actuator speed to servovalve current and load force is then given by:

$$\delta U(s) = \frac{\dfrac{(\gamma+1)k_i}{(\gamma^2+1)A_2}\delta i(s) - \dfrac{(1+sCR_u)}{R_u(\gamma^2+1)A_2^2}\delta F_{\text{load}}(s)}{\left[1 + \dfrac{R_v}{(\gamma^2+1)R_u} + s\dfrac{\left(CR_v + \dfrac{L}{R_u}\right)}{(\gamma^2+1)} + s^2\dfrac{LC}{(\gamma^2+1)}\right]},$$

$$R_u = \frac{1}{k_p}, \quad R_v = \frac{B_v}{A_2^2}, \quad C = \frac{V(0)}{\beta}, \quad L = \frac{M}{A_2^2}. \tag{6.126}$$

The significant damping is to be provided by underlapping the servovalve spool, viscous friction effects being negligible in comparison. The block diagram then becomes as shown in Fig. 6.34, in the absence of load force changes.

The OLTF may then be written:

$$G(s)H(s) = \frac{\dfrac{(\gamma+1)G_a H_y k_i}{(\gamma^2+1)A_2}}{s\left[1 + +s\dfrac{L}{(\gamma^2+1)R_u} + s^2\dfrac{LC}{(\gamma^2+1)}\right]}. \tag{6.127}$$

The closed-loop characteristic equation then becomes:

$$s^3\frac{LC}{(\gamma^2+1)} + s^2\frac{L}{(\gamma^2+1)R_u} + s + \frac{(\gamma+1)G_a H_y k_i}{(\gamma^2+1)A_2} = 0. \tag{6.128}$$

Applying the Routh array method, it is deduced that the closed-loop system will be stable, providing:

$$i_u > G_a\frac{P_s(\gamma+1)}{\beta(\gamma^2+1)}H_y\frac{V(0)}{A_2}. \tag{6.129}$$

Given that $V(0)/A_2$ has the unit of length ℓ and a measure of half-stroke, then $G_a H_y \ell$ represents the current generated. A working rule of thumb for stability is therefore:

$$i_u > i_s \frac{P_s}{\beta}, \tag{6.130}$$

where i_s is approximately the current developed at half-stroke and is less than the servovalve rating. It is common to start the design approach with a gain approximately a factor of 4 less than that at the condition for instability. Using the ITAE criterion to match characteristic equation (6.128) gives a factor of 3.76 using the assumptions leading to Eq. (6.130). Therefore, it follows that the closed-loop design should consider the following starting point:

$$i_u \approx 4 i_s \frac{P_s}{\beta}, \quad \frac{(\gamma + 1) G_a H_y k_i}{(\gamma^2 + 1) A_2} \approx \frac{0.3}{C R_u}. \tag{6.131}$$

6.4.2 Phase Compensation, Gain and Phase Margins

When the system damping is low, then simply changing the open-loop gain will not result in a preferred dynamic behavior; for example, the transient response to a demanded step change in position. One way of improving the dynamic behavior is to use a phase-compensating network, often an electrical phase advance network in the servovalve electrical drive circuit. It will be recalled that *the condition for instability* for a linear system occurs, in fluid power reality, when:

$$G(s)H(s) = -1. \tag{6.132}$$

Therefore, considering the frequency domain, the plot of the OLTF $G(j\omega)H(j\omega)$ will have unity magnitude, $|G(j\omega)H(j\omega)| = 1$, at a phase angle $\angle G(j\omega)H(j\omega) = -180°$ at the point of instability. The function of a phase lead network is therefore to shape the OLTF around the -1 point such that when the magnitude is unity, the phase angle is greater then $-180°$. Another way of stating this is that when the phase angle is $-180°$, then the magnitude is less than unity. The *phase margin* and *gain margin* are defined in Fig. 6.35. Needless to say, the greater the gain–phase margin, the more stable the system will be.

One electrical circuit that will provide phase lead is shown in Fig. 6.36. Considering impedances, it is a simple matter to show that the voltage ratio is given by:

$$\frac{V_2(s)}{V_1(s)} = \frac{\alpha(1 + s\tau)}{(1 + \alpha s\tau)}, \quad \alpha = \frac{R_1}{R_1 + R_2}, \quad \tau = C R_2. \tag{6.133}$$

This transfer function has a low-frequency attenuation of α that can be removed by the inclusion of a series gain $1/\alpha$. When this is done, the phase lead network then becomes:

$$G_c(s) = \frac{(1 + s\tau)}{(1 + \alpha s\tau)}. \tag{6.134}$$

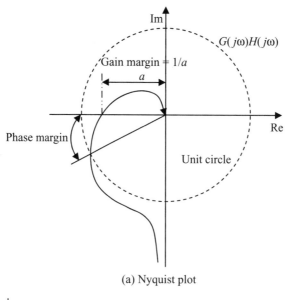

(a) Nyquist plot

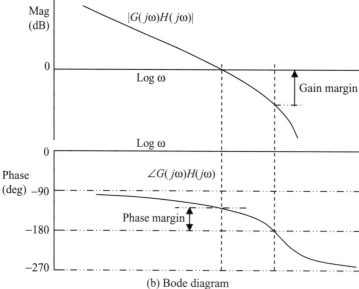

(b) Bode diagram

Figure 6.35. Gain and phase margins in the frequency domain.

This transfer function is plotted in the frequency domain and in Bode diagram form, $s = j\omega$, as shown in Fig. 6.37.

It will be seen that for any selected value of α, then a *maximum phase lead* occurs and is defined by the following properties:

$$\angle G_c(j\omega) = \sin^{-1}\frac{(1 - \alpha)}{(1 + \alpha)}, \qquad |G_c(j\omega)| = \frac{1}{\sqrt{\alpha}}, \qquad \omega\tau = \frac{1}{\sqrt{\alpha}}. \qquad (6.135)$$

Graphs representing this condition are shown in Fig. 6.38.

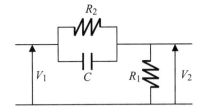

Figure 6.36. An electrical phase lead circuit.

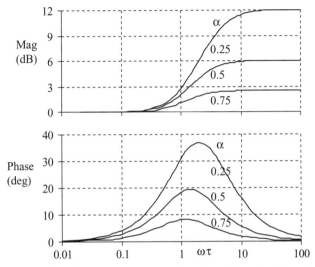

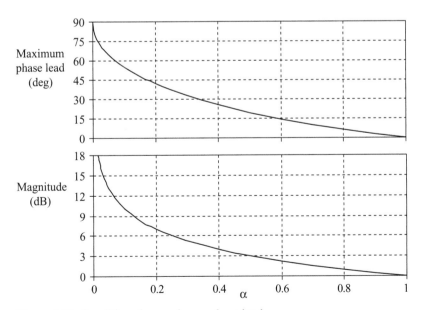

Figure 6.37. Frequency response plot for a phase lead circuit.

Figure 6.38. Conditions for maximum phase lead.

Worked Example 6.7

The block diagram of a position control system is shown.

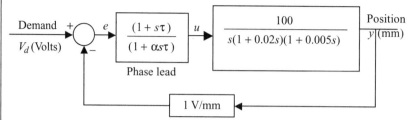

Worked Example 6.7a.

The closed-loop response is highly oscillatory and undesirable, as will be deduced from the Bode diagram of the OLTF shown.

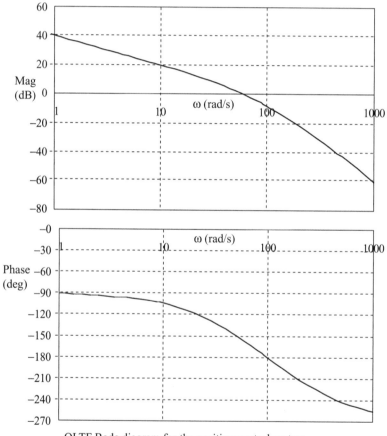

OLTF Bode diagram for the position control system

Worked Example 6.7b.

The 0-db frequency is approximately at a frequency of 61 rad/s and the $-180°$ phase occurs at a frequency of approximately 100 rad/s. The gain margin is therefore approximately 8 dB, and the phase margin is approximately 22°, and too small. The objective is therefore to design the phase lead network to improve the closed-loop response by increasing the phase margin. It therefore seems sensible to select a maximum phase lead at a frequency of $\omega = 100$ rad/s, hence defining the phase margin providing the modified magnitude is 0 dB. The steps therefore are as follows:

Step 1. At $\omega = 100$ rad/s, the magnitude is ≈ -8dB:

$$\text{select } |G_c(j\omega)| = \frac{1}{\sqrt{\alpha}} = +8\,\text{dB}$$

to give a = 0.16.

Step 2.

$$\omega\tau = \frac{1}{\sqrt{\alpha}} \text{ to give } \tau = 0.025, \qquad \angle G_c(j\omega) = \sin^{-1}\frac{(1-\alpha)}{(1+\alpha)},$$

to give a maximum phase lead of 46.4° and, hence, a phase margin of 46.4°.

Step 3.

$$G_c(s) = \frac{(1+0.025s)}{(1+0.004s)}.$$

The modified OLTF Bode diagram is shown as follows:

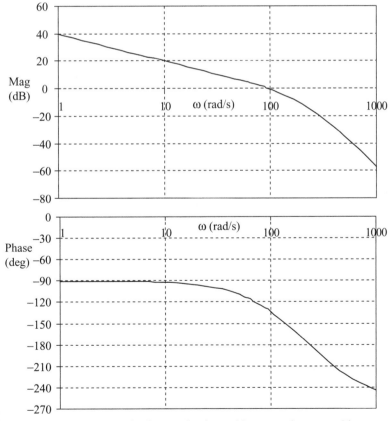

OLTF Bode diagram for the position control system with
phase lead compensation

Worked Example 6.7c.

The phase margin has clearly been increased from 22° to 46° as designed. A comparison of the step responses for a demanded position change of 10 mm is shown in the following figure:

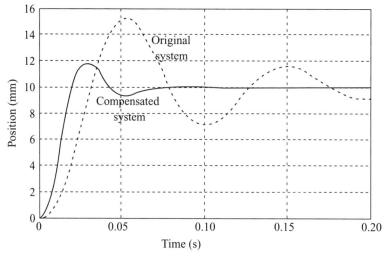

Worked Example 6.7d.

The improvement is clear, but the phase lead network has to either be constructed as an electrical circuit or created as a computer algorithm. The latter issue will be considered later in this chapter.

PID, proportional + integral + derivative, control is another technique for improving the closed-loop response, again by placing the PID network in the error loop as shown in Fig. 6.39.

The derivative term may be used to improve damping, and the integral term is used to remove steady-state error. Integral control is not required for actuator position control because of the existing and inherent integrator provided by the motor or linear actuator. It might be considered appropriate to use PID control for motor speed, although the integral term will not remove speed drop that is due to the effect of load pressure on the servovalve characteristic and leakage. The author has found it difficult to implement PI control for motor speed because of the restrictions on integrator time constant τ_i that can be selected and the error signal noise that is due to motor speed ripple and its varying frequency change with speed.

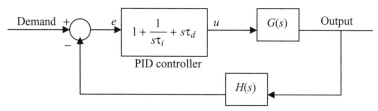

Figure 6.39. PID control for a hydraulic system.

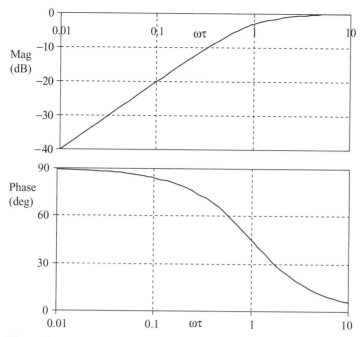

Figure 6.40. Bode diagram for a dynamic pressure transfer function.

6.4.3 Dynamic Pressure Feedback

Given that poorly designed closed-loop systems result in severe pressure oscillations, it makes sense to consider some form of pressure feedback in applications
other than those that actually desire pressure control. Using load pressure feedback
directly is of limited use for damping because this will result in a constant steady-
state signal that is due to the finite load force. One way of avoiding this is to use
dynamic pressure feedback such that only higher-frequency oscillations are used for
damping. A suitable transfer function is given by:

$$G_p(s) = \frac{s\tau}{(1 + s\tau)}. \tag{6.136}$$

The Bode diagram for this transfer function is shown in Fig. 6.40.

This transfer function provides phase lead, and it will be seen that only
the higher-frequency signals are unaffected by the transfer function, the lower-
frequency signals being highly attenuated. This type of network was originally con-
ceived as a hydromechanical device (Guillon, 1969) but now can easily be generated
by a simple computer algorithm.

Consider therefore the example that introduced Chapter 5 and reproduced here
as Fig. 6.41.

Dynamic pressure feedback is added by taking the pressure differential from the
servovalve mounting manifold, multiplying by a gain H_p, and then passing the signal
through the dynamic pressure transfer function generated by a software algorithm.
In this study:

$$G_p(s) = H_p \frac{s\tau}{(1 + s\tau)}. \tag{6.137}$$

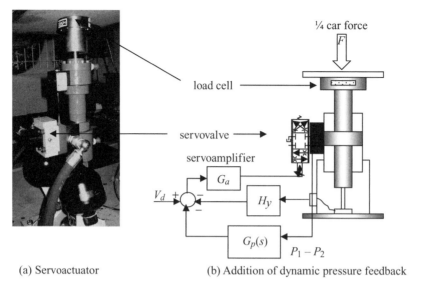

(a) Servoactuator   (b) Addition of dynamic pressure feedback

Figure 6.41. A servoactuator, one of four forming part of a four-poster vehicle test rig (Cardiff University, School of Engineering).

The use of a linearized transfer function block diagram may give some insight into dynamic pressure feedback design. Following earlier work in Section 6.2, a modified block diagram may be constructed to include dynamic pressure feedback, as shown in Fig. 6.42, neglecting servovalve dynamics.

Consider therefore the actual system as it exists with the addition of dynamic pressure feedback. From Fig. 6.42, the OLTF is given by:

$$
G(s)H(s) = \frac{\dfrac{H_y G_a k_f \sqrt{P_s/2}}{A}}{s\left(1 + \dfrac{CR_v}{2}s + \dfrac{LC}{2}s^2\right)} \frac{\left(1 + s\tau + \dfrac{A\tau H_p R_v}{H_y}s^2 + \dfrac{A\tau H_p L}{H_y}s^3\right)}{(1 + s\tau)}.
$$

(6.138)

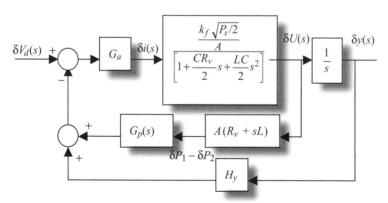

Figure 6.42. Closed-loop block diagram for a servocylinder position drive with dynamic pressure compensation.

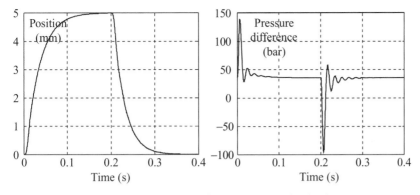

(a) Original system with no dynamic pressure feedback

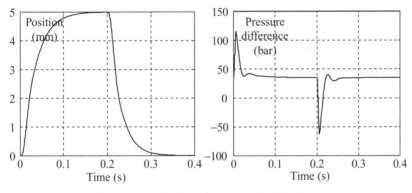

(b) System with dynamic pressure feedback

Figure 6.43. The dynamic response of the position control system with dynamic pressure feedback.

It may be seen from Eq. (6.138) that the dynamic pressure network results in the OLTF being multiplied by a modified phase lead network that includes the dynamic pressure network, given by:

$$G_c(s) = \frac{\left(1 + s\tau + \dfrac{A\tau H_p R_v}{H_y}s^2 + \dfrac{A\tau H_p L}{H_y}s^3\right)}{(1 + s\tau)}. \qquad (6.139)$$

At high frequencies, this modified network will produce a phase lead of 180°, but the magnitude will also be modified.

The most appropriate way to design the dynamic pressure network is to first consider the pressure oscillations recorded without compensation. Measurements of pressure difference oscillations, shown later in Fig. 6.43, for the original system shows a frequency of oscillation of approximately 60 Hz $\rightarrow$ 377 rad/s and close to the open-loop undamped natural frequency ω_n. A dynamic pressure filter circuit designed for this cutoff frequency would have a time constant of at most $\tau = 1/377 = 2.65$ ms. Therefore, choose an actual cutoff time constant less than this to ensure that the 60 Hz oscillations is attenuated for example $\tau = 1$ ms.

Now the compensator circuit gain H_p must be selected. The Routh array technique could be used, with limited accuracy, once the closed-loop characteristic equation has been established. From OLTF equation (6.138), and neglecting servovalve dynamics, this gives:

$$b_4 s^4 + b_3 s^3 + b_2 s^2 + b_1 s + b_0 = 0, \tag{6.140}$$

$$b_4 = \frac{LC\tau}{2},$$

$$b_3 = \frac{LC}{2} + \frac{CR_v \tau}{2} + \frac{KALH_p \tau}{H_y},$$

$$b_2 = \frac{CR_v}{2} + \tau + \frac{KAR_v H_p \tau}{H_y},$$

$$b_1 = 1 + K\tau,$$

$$b_0 = K = \frac{H_y G_a k_f}{A} \sqrt{\frac{P_s}{2}},$$

$$R_v = B_v / A^2, \quad C = V(0)/\beta, \quad L = M/A^2.$$

The condition for closed-loop stability is difficult to visualize in general terms; for example, purposes the data previously stated are used, recalling that τ has already been preselected as 0.001 s. This gives:

$$b_4 = 7 \times 10^{-9}, \qquad\qquad b_3 = 10^{-6}(7.42 + 1.19 H_p),$$
$$b_2 = 10^{-3}(1 + 0.071 H_p), \qquad b_1 = 1.033 \, b_0 = 33. \tag{6.141}$$

Note that here the pressure gain H_p is expressed in millivolts per bar. The fourth-order characteristic equation indicates a stable closed-loop system, providing:

$$\text{(i) } b_2 b_3 - b_1 b_4 > 0; \quad \text{(ii) } b_2 b_3 - b_1 b_4 > \frac{b_0 b_3^2}{b_1}. \tag{6.142}$$

Satisfying the second criterion of Eq. (6.142) gives $H_p > 1.4$ mV/bar. For this example, a wide range of gains are possible, and a selected gain of $H_p = 10$ mV/bar is used to illustrate the effect of dynamic pressure compensation, as shown in Fig. 6.43.

Comparing the compensated system pressure response with the uncompensated system pressure response shows the reduction of the undesirable pressure oscillations, particularly when lowering, together with no significant change on the position response. In addition, high-pressure peaks that occur at each position demand change have been reduced by 30 bar when lifting and 40 bar when lowering. Note that increasing H_p significantly higher, with a view to reducing peak values even more, will result in instability that is due to servovalve dynamic effects not considered in the analysis.

6.4.4 State Feedback

Now consider the use of all the output states to improve the dynamic behavior of a circuit. For example, if servoactuator position control is being considered, then the use of additional velocity and acceleration feedback would constitute state feedback. This is illustrated in Fig. 6.44.

The feedback signal is:

$$V_f = k_y y + k_u U + k_a \frac{du}{dt}. \tag{6.143}$$

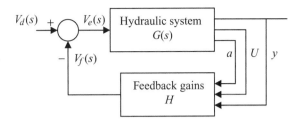

Figure 6.44. State feedback control of a servoactuator.

In Laplace transform notation, the feedback transfer function is determined from:

$$V_f(s) = (K_y + K_u s + K_a s^2)y(s),$$
$$H(s) = (K_y + K_u s + K_a s^2). \tag{6.144}$$

The problem now is to determine the appropriate state feedback gains in the $(1 \times n)$ feedback vector H, preferably those that optimize the dynamic response. In the current example, these gains are k_y, k_u, and k_a, although the position sensor gain k_y is usually known in advance.

Note that up to this point, it is assumed that actuator velocity and acceleration can be measured; that is, sensors are available. This may not be the case if angular acceleration is required for a motor drive. If sensors are not available or not feasible for the application – for example, because of difficult environmental conditions – then additional states beyond position may be computed with a suitable algorithm.

Selection of the closed-loop poles is always a difficult issue, an ideal solution being to make them match the poles of the appropriate ITAE form.

Worked Example 6.8

A servoactuator position control system with full output-state feedback has an OLTF given by:

$$G(s)H(s) = \frac{K(H_y + H_u s + H_a s^2)}{s\left(1 + \dfrac{2\zeta}{\omega_n}s + \dfrac{s^2}{\omega_n^2}\right)}.$$

Determine the feedback gains if the closed-loop poles must satisfy the appropriate ITAE criterion.

The characteristic equation is given by $1 + G(s)H(s) = 0$ and becomes:

$$\frac{s^3}{KH_y\omega_n^2} + \frac{s^2}{KH_y}\left(\frac{2\zeta}{\omega_n} + KH_a\right) + \frac{s}{KH_y}(1 + KH_u) + 1 = 0.$$

The third-order ITAE appropriate criterion is:

$$\frac{s^3}{\omega_o^3} + \frac{1.75s^2}{\omega_o^2} + \frac{2.15s}{\omega_o} + 1 = 0,$$

$$\text{Let } \overline{K} = \frac{KH_y}{\omega_n};$$

then equating coefficients gives:

$$\omega_o = \omega_n \overline{K}^{1/3},$$

$$H_a = \frac{1.75\overline{K}^{1/3} - 2\zeta}{K\omega_n}, \qquad \overline{K} > 1.49\zeta^3,$$

$$H_u = \frac{2.15\overline{K}^{2/3} - 1}{K}, \qquad \overline{K} > 0.317.$$

Hence, gain selection is possible for only a specific range of $\overline{K}$ that is set by the OLTF parameters and the position feedback transducer gain. In practice, the low damping ratio for a linear actuator in good condition, and with a significant load mass, will mean that the design requirement is probably $\overline{K} > 0.317$.

In more complex systems, the state-space techniques can be used formulate the state gains in a more generalized way. It will be recalled from Chapter 5 that linear differential equations may be written in state-space notation, which for hydraulic control systems with u as the input and x as the output states is adequately expressed as follows:

$$\text{system} \quad \dot{x} = Ax + Bu, \tag{6.145}$$

$$\text{output} \quad y = Cx. \tag{6.146}$$

Considering just the relationship between the output $y(s)$ and the input $x(s)$, from these two equations, following Laplace transformation:

$$y(s) = C[sI - A]^{-1}[x(0) + Bu(s)], \tag{6.147}$$

where $x(0)$ is the vector of initial conditions. If all the output states are available for measurement, position, velocity, and acceleration, then a closed-loop system may be designed to often produce an optimum response characteristic by using all the n states in feedback. The feedback signal is given by:

$$r_f = Hx, \quad H = [k_1 \; k_2 \ldots k_n], \tag{6.148}$$

where H is the $(1 \times n)$ feedback vector. The block diagram is shown in Fig. 6.45.

It is then an easy matter to show that the closed-loop state solution in the s domain is given by:

$$x(s) = [sI - A + BH]^{-1}[x(0) + Br(s)]. \tag{6.149}$$

Closed-loop stability therefore depends on the roots of the characteristic equation and obtained from the determinant:

$$\text{characteristic equation} = |sI - A + BH| = 0. \tag{6.150}$$

Before state feedback design is pursued, it is important at this stage to consider the controllability of a system. A system of order n is completely state controllable if the rank of the $n \times n$ controllability matrix U is equal to n; that is:

$$\text{rank of U} = [B \; AB \; A^2 B \ldots A^{n-1} B] = n. \tag{6.151}$$

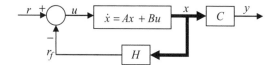

Figure 6.45. Closed-loop control using state feedback.

U is called the *controllability matrix*. The rank of a matrix is defined as follows:

A matrix A has rank m if there exists an $m \times m$ submatrix R

whose determinant $\neq 0$ and the determinant of every other $r \times r$ submatrix
is 0, where $r \geq m + 1$. (6.152)

For example, a system is defined:

$$\begin{bmatrix} \dot{x}_1 \\ \dot{x}_2 \end{bmatrix} = \begin{bmatrix} 1 & 1 \\ 0 & -1 \end{bmatrix} \begin{bmatrix} x_1 \\ x_2 \end{bmatrix} + \begin{bmatrix} 1 \\ 0 \end{bmatrix} u, \tag{6.153}$$

$$B = \begin{bmatrix} 1 \\ 0 \end{bmatrix}, \quad AB = \begin{bmatrix} 1 & 1 \\ 0 & -1 \end{bmatrix} \begin{bmatrix} 1 \\ 0 \end{bmatrix} = \begin{bmatrix} 1 \\ 0 \end{bmatrix},$$

$$U = \begin{bmatrix} B & AB \end{bmatrix} = \begin{bmatrix} 1 & 1 \\ 0 & 0 \end{bmatrix}. \tag{6.154}$$

The two determinants of U are:

$$\begin{matrix} R_1 & R_2 \\ 1 & 0 \end{matrix}. \tag{6.155}$$

Therefore, the rank of U is $1 \neq n = 2$ so the system is not state controllable. This will be obvious once the system block diagram construction is attempted; it is not possible to link the state variable x_2 into the block diagram. In hydraulic control systems, it is most unlikely that the system of interest is not state controllable, although there will inevitably be stability constraints on the feedback gains possible.

The design problem is to choose stable roots, the closed-loop poles, such that the closed loop has the desired characteristic, and this may be readily determined if the system is defined in *control canonical form* as follows:

$$A = \begin{bmatrix} 0 & 1 & 0.. & 0 \\ 0 & 0 & 1.. & 0 \\ \vdots & \vdots & \vdots & 1 \\ -a_0 & -a_1 & \cdots & -a_{n-1} \end{bmatrix}, \quad BH = \begin{bmatrix} 0 & 0 & 0.. & 0 \\ 0 & 0 & 0.. & 0 \\ \vdots & \vdots & \vdots & \vdots \\ K_1 & K_2 & \cdots & K_n \end{bmatrix}. \tag{6.156}$$

The *desired closed-loop poles* are defined in the usual way:

$$\alpha_c(s) = s^n + \alpha_{n-1} s^{n-1} + \cdots + \alpha_1 s + \alpha_0 = 0. \tag{6.157}$$

The solution for the required feedback gains is then given by:

$$K_{i+1} = \alpha_i - a_i, \quad i = 0 \to n. \tag{6.158}$$

If the *system is not defined in control canonical form*, then *Ackermann's formula may be applied* to give the solution:

$$\begin{aligned} H &= [0\, 0 \ldots 1] U^{-1} \alpha_c(A), \\ U &= [B\ AB\ A^2 B \ldots A^{n-1} B], \\ \alpha_c(A) &= A^n + \alpha_{n-1} A^{n-1} + \cdots + \alpha_1 A + \alpha_0 I, \end{aligned} \tag{6.159}$$

where $\alpha_c(A)$ is the matrix form of the desired closed-loop poles and U is the controllability matrix previously defined. Consider the simple dynamic model of a vehicle active suspension discussed in Section 5.15, and now select a control scheme as indicated in Fig. 6.46.

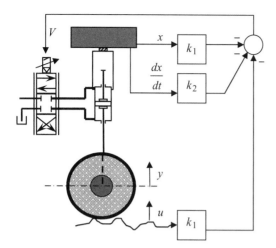

Figure 6.46. A vehicle active suspension control concept.

The control law is selected as:

$$V = -k_1 x - k_2 \frac{dx}{dt} - k_1 u. \tag{6.160}$$

Using the defined state-space notation then gives:

$$V = -k_1 x_1 - k_2 x_2 - k_1 u. \tag{6.161}$$

The dynamic model of the system was previously defined as:

$$\begin{bmatrix} \dot{x}_1 \\ \dot{x}_2 \\ \dot{x}_3 \end{bmatrix} = \begin{bmatrix} 0 & 1 & 0 \\ 0 & 0 & -k_t/M \\ 0 & 1 & -k_t/R_i A^2 \end{bmatrix} \begin{bmatrix} x_1 \\ x_2 \\ x_3 \end{bmatrix} + \begin{bmatrix} 0 \\ k_t/M \\ k_t/R_i A^2 \end{bmatrix} u + \begin{bmatrix} 0 \\ 0 \\ -G_a k_i \end{bmatrix} V, \tag{6.162}$$

$$\dot{x} = Ax + Bu + GV,$$

where the state variable $x_3 = y$. Inserting control law equation (6.161) then gives:

$$\begin{bmatrix} \dot{x}_1 \\ \dot{x}_2 \\ \dot{x}_3 \end{bmatrix} = \begin{bmatrix} 0 & 1 & 0 \\ 0 & 0 & -k_t/M \\ G_a k_i k_1 & (1 + G_a k_i k_2) & -k_t/R_i A^2 \end{bmatrix} \begin{bmatrix} x_1 \\ x_2 \\ x_3 \end{bmatrix} + \begin{bmatrix} 0 \\ k_t/M \\ k_t/R_i A^2 + G_a k_i k_1 \end{bmatrix} u,$$

$$\dot{x} = A_c x + B_c u. \tag{6.163}$$

Note that Eq. (6.163) includes feedback; that is, it is the closed-loop state-space description. The closed-loop characteristic equation then becomes:

$$|sI - A_c| = \begin{vmatrix} s & -1 & 0 \\ 0 & s & k_t/M \\ -G_a k_i k_1 & -(1 + G_a k_i k_2) & (s + k_t/R_i A^2) \end{vmatrix} = 0, \tag{6.164}$$

$$s^3 + s^2 k_t/R_i A^2 + s(1 + G_a k_i k_2) k_t/M + G_a k_i k_1 k_t/M = 0. \tag{6.165}$$

This third-order system is therefore stable, providing:

$$\frac{k_2}{R_i A^2} > \frac{k_1}{k_t} - \frac{1}{G_a k_i R_i A^2}. \tag{6.166}$$

Stability is absolutely guaranteed if:

$$k_2 > \frac{k_1 R_i A^2}{k_t}. \tag{6.167}$$

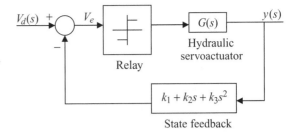

Figure 6.47. Sliding-mode control of a position control system.

This requirement relies on actuator piston seal leakage and is not desirable because leakage will change with wear during the operational lifetime. However, this is the outcome of using a simple system dynamic model, although it does give an insight into possible control problems when more detailed hydraulic characteristics are considered. This is pursued in more detail in Chapter 7 for a real active suspension on a $\frac{1}{4}$ car test rig.

Another state feedback technique that has received attention is *sliding-mode control*. A switching characteristic is used on the error signal in an attempt to drive the closed-loop system response shaped by the state feedback dynamic characteristic. The basic concept is shown in Fig. 6.47 for actuator position control using a linearized transfer function for the servoactuator.

Because the relay switches about the zero voltage point, the error can be approximated as follows:

$$V_e(s) = V_d(s) - (k_1 + k_2 s + k_3 s^2)y(s) \approx 0,$$

$$\frac{y(s)}{V_d(s)} \approx \frac{1}{(k_1 + k_2 s + k_3 s^2)}. \tag{6.168}$$

The sliding-mode transfer function dynamics defined by Eq. (6.168) should normally be slower than the actual system dynamics defined by OLTF without state feedback. Sliding-mode control can be considered as an attempt to place the transient response on a *sliding plane* defined by:

$$V_d = k_1 x_1 + k_2 x_2 + k_3 x_3, \tag{6.169}$$

where the state variables are defined as x_1, displacement; x_2, velocity; and x_3, acceleration, in the usual canonical state-space form. Hence, sliding-mode control attempts to drive the output response onto the sliding plane and then toward the steady-state condition, as shown in Fig. 6.48.

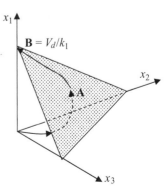

Figure 6.48. Sliding-mode control of a third-order position control system.

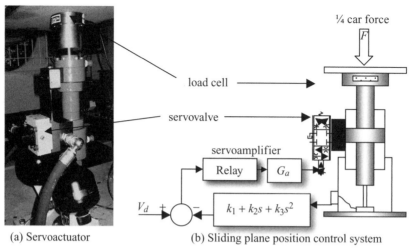

(a) Servoactuator   (b) Sliding plane position control system

Figure 6.49. A servoactuator, one of four forming part of a four-poster vehicle test rig (Cardiff University, School of Engineering).

Some main points that arise are as follows:

- The initial trajectory is defined from the origin, the initial condition in this example, until it reaches the sliding plane at point A. The control then attempts to drive the remaining trajectory along the plane to the steady-state condition at point B.
- The initial position error creates the $+1$ control voltage from the relay and the position response is initially governed by the open-loop dynamics, which will be dominated by the integral term. Consequently, if the open-loop dynamics are fast, then the sliding plane will be rapidly reached with an approximate linear variation of position with time until the relay switches to its -1 control voltage.
- The phase-plane dynamics then come into play, and the subsequent motion depends on the overall system design.
- The main problem with this type of control as it stands is that the relay is continually switching at the desired steady-state condition. This can be reduced by the addition of a dead-band in the relay, but this will result in a steady-state position error.
- An important point is therefore that the response may be much faster than the conventional closed-loop position control system because of the very first switched condition that has no connection with the sliding plane.
- A design may be readily deduced from linearized transfer function theory, but the design may be unstable when implemented into the real system that is nonlinear, may have different dynamics when extending and retracting, and has additional dynamics possibly not taken into account such as servovalve dynamics.

Consider the servoactuator system previously discussed in Sections 6.2 and 6.4, and again shown here as Fig. 6.49.

The design proceeds as follows:

- The open-loop transfer function was shown previously to have a second-order component with an undamped natural frequency of $\omega_n = 379$ rad/s and a damping ratio $\zeta = 0.079$.

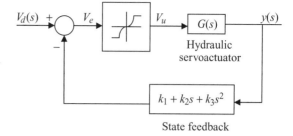

Figure 6.50. Quasi-sliding-mode control of a position control system.

- Consequently, design sliding plane transfer function equation (6.168) to have an undamped natural frequency of $\omega_s = 190$ rad/s and a damping ratio $\zeta_s = 0.7$, although many variations on these data will be acceptable.
- Therefore, given that the position transducer gain $k_1 = 100$ V/m, the velocity and accelerometer gains self-select from Eq. (6.168) to give $k_2 = 0.74$ V/ms^{-1} and $k_3 = 0.0028$ V/ms^{-2}.
- Select a dead-band, in this example ± 0.05 V, which represents a possible position error band of ± 0.5 mm. Other values will affect the transient response, and the value chosen depends on what is considered acceptable.
- The transient response is acceptable, much faster than the original closed-loop response for reasons previously described, but unacceptable switching could not be removed.
- However, when the feedback law and relay were applied to the actual system, then the response was unstable and could not be stabilized by varying parameters about the design values. The reason for this is postulated as the inability of the linearized analysis, which excluded servovalve dynamics, to provide a good design basis for the real system.

One way of overcoming this apparent failure of sliding-mode control is to replace the relay with a cubic switching function, as shown in Fig. 6.50.

This is referred to as quasi-sliding-mode control because the error is driven to zero across a nonlinear plane, the cubic switching function having a zero slope at the origin, which aids closed-loop stability. By selecting a high gain within the cubic switching function, together with a ± 1 saturation component, the effect is similar to relay switching control but without the need for a dead-band and, of course, the removal of continual switching of the relay. The function finally selected was:

$$V_u = (8V_e)^3, \quad -1 < V_u < 1. \tag{6.170}$$

Consequently, the saturation point is reached when the error voltage is ± 0.125 V and representing a position of ± 1.25 mm before the cubic function begins to operate. A good response characteristic can be seen from Fig. 6.51, although a steady-state position error does exist, typically 0.02 mm.

Worked Example 6.9

A servoactuator position control system is considered with servovalve dynamics to be significant but with a simplified actuator transfer function because of a low load moving mass. The servovalve has spool position feedback, and this feedback

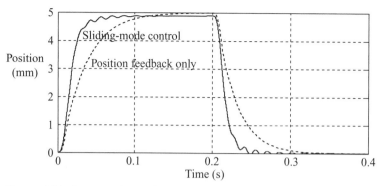

Figure 6.51. A comparison of position feedback and modified sliding-mode control with a cubic–saturation error control law.

voltage is also to be used as a control state variable in addition to load velocity and position. Determine the feedback controller gains k_1, k_2, and k_3.

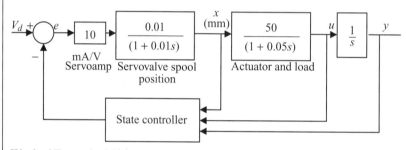

Worked Example 6.9(a).

Define the state variables: $x_1 = y$, $x_2 = u$, and $x_3 = x$. From the block diagram:

(i) $\dot{x}_1 = x_2$,

(ii) $\dfrac{u}{x} = \dfrac{50}{(1 + 0.05s)} \rightarrow \dot{x}_2 = -20x_2 + 1000x_1$,

(iii) $\dfrac{x}{e} = \dfrac{0.1}{(1 + 0.01s)} \rightarrow \dot{x}_3 = -100x_3 + 10e$.

The state equations are then placed in matrix notation:

$$\begin{bmatrix} \dot{x}_1 \\ \dot{x}_2 \\ \dot{x}_3 \end{bmatrix} = \begin{bmatrix} 0 & 1 & 0 \\ 0 & -20 & 1000 \\ 0 & 0 & -100 \end{bmatrix} \begin{bmatrix} x_1 \\ x_2 \\ x_3 \end{bmatrix} + \begin{bmatrix} 0 \\ 0 \\ 10 \end{bmatrix} e,$$

$$\dot{x} = Ax + Be.$$

This description in not in control canonical form, but Ackermann's formula may be used to determine the vector of feedback gains. The controllability matrix is given by:

$$U = [B \ AB \ A^2 B] = \begin{bmatrix} 0 & 0 & 10^4 \\ 0 & 10^4 & -12 \times 10^5 \\ 10 & -10^3 & 10^5 \end{bmatrix}.$$

The rank of U is 3 and therefore the system is state controllable. The inverse is:

$$U^{-1} = \begin{bmatrix} 0.2 & 0.01 & 1 \\ 0.012 & 0.0001 & 0 \\ 0.0001 & 0 & 0 \end{bmatrix}.$$

The matrix equivalent of the characteristic equation is given by:

$$\alpha_c(A) = A^3 + \alpha_2 A^2 + \alpha_1 A + \alpha_0 I.$$

Choosing the third-order ITAE criterion and selecting $\omega_o = 100$ rad/s gives:

$$\alpha_c(s) = s^3 + 1.75\omega_o s^2 + 2.15\omega_o^2 + \omega_o^3$$
$$= s^3 + 1.75 \times 10^2 s^2 + 2.15 \times 10^4 s + 10^6;$$

$$\alpha_c(A) = 10^4 \begin{bmatrix} \overset{\alpha_2}{100} & \overset{\alpha_1}{1.84} & \overset{\alpha_0}{5.5} \\ 0 & 63.2 & 1290 \\ 0 & 0 & -40 \end{bmatrix}.$$

The state feedback gain vector is then given by:

$$H = [001]U^{-1}\alpha_c(A) = [100 \quad 1.84 \quad 5.5].$$

load position transducer gain, $k_1 = 100$ V/m
load velocity transducer gain, $k_2 = 1.84$ V/m s^{-1}
spool position transducer gain, $k_3 = 5.5$ V/m s^{-2}

The transient response is shown with and without servovalve spool feedback.

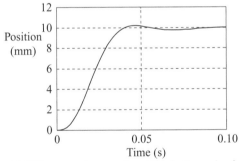

(a) With servovalve spool position in the major feedback loop

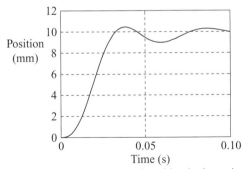

(b) Without servovalve spool position in the major feedback loop

Worked Example 6.9(b).

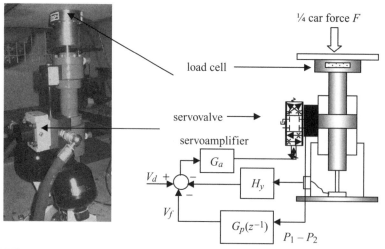

(a) Servoactuator   (b) Addition of digital dynamic pressure feedback

Figure 6.52. A servoactuator, one of four forming part of a four-poster vehicle test rig (Cardiff University, School of Engineering).

6.5 Feedback Controller Implementation

6.5.1 Analog-to-Digital Implementation

The direct use of analog signals is still common, but the trend is clearly toward digital control simply because of the relatively low cost of using microcomputers, combined with the programming and system upgrading facility. Once a control strategy is decided on, the digital implementation must be given some thought. The discrete algorithm to be used can, of course, be tested by computer simulation, but often the first step is to use linear systems theory to determine the approximate form of the control law. Consider, for example, the servoactuator system that was analyzed in some detail earlier in this chapter. Consider also the use of dynamic pressure feedback to improve the closed-loop response. The system is shown again in Fig. 6.52.

If V_f is the voltage to be generated by the pressure differential $(P_1 - P_2)$, then it was shown that the transfer function to be computed is given by:

$$\frac{V_f}{P_1 - P_2} = H_p \frac{s\tau}{(1 + s\tau)}. \tag{6.171}$$

This transfer function can be converted into its discrete equivalent; for example, using either the backward-difference or bilinear z transforms.

(i) Backward difference transformation:

$$s = \frac{(1 - z^{-1})}{T} \rightarrow \frac{V_f(z)}{(P_1 - P_2)(z)} = H_p \frac{\frac{\tau}{T}[1 - z^{-1}]}{\left[1 + \frac{\tau}{T} - \frac{\tau}{T}z^{-1}\right]}, \tag{6.172}$$

$$V_f(t) = \frac{\tau/T}{(1 + \tau/T)} V_f(t - T) + H_p \frac{\tau/T}{(1 + \tau/T)}[(P_1 - P_2)(t) - (P_1 - P_2)(t - T)].$$

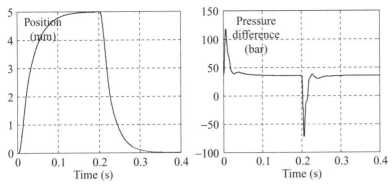

Figure 6.53. The dynamic response of the position feedback control system with a digital algorithm for dynamic pressure feedback.

Hence, the previous samples of output voltage and pressure difference must be stored at each step.

(ii) Bilinear transformation:

$$s = \frac{2}{T}\frac{(1-z^{-1})}{(1+z^{-1})} \rightarrow \frac{V_f(z)}{(P_1-P_2)(z)} = H_p\frac{\dfrac{2\tau}{T}(1-z^{-1})}{\left[1+\dfrac{2\tau}{T}+\left(1-\dfrac{2\tau}{T}\right)z^{-1}\right]},$$

$$\text{(6.173)}$$

$$V_f(t) = \frac{(2\tau/T-1)}{(1+2\tau/T)}V_f(t-T) + H_p\frac{2\tau/T}{(1+2\tau/T)}[(P_1-P_2)(t)-(P_1-P_2)(t-T)].$$

Again, just the previous samples of output voltage and pressure difference must be stored at each step. Recalling that $H_p = 10^{-7}$ V/N m^{-2} and $\tau = 1$ ms was chosen by analog design, then selecting a sampling period $T = 1$ ms gives:

backward difference transformation:

$$\frac{V_f(z)}{(P_1-P_2)(z)} = H_p\frac{(1-z^{-1})}{(2-z^{-1})},$$

bilinear transformation:

$$\frac{V_f(z)}{(P_1-P_2)(z)} = H_p\frac{(2-2z^{-1})}{(3-z^{-1})}.$$

$$\text{(6.174)}$$

Both controllers give almost identical results, with the position and pressure differential responses being little different from the case with an analog controller. The performance using the backward-difference transformation is shown in Fig. 6.53.

Comparing the results with the previous response by use of an analog dynamic pressure feedback circuit shows little difference, with just a small increase in the peak pressure differentials. The digital control performance would, of course, deteriorate if the sampling period was increased.

6.5.2 Generalized Digital Filters

Now consider the application of two digital filters placed in the forward loop and the feedback loop, as shown in Fig. 6.54, and specifically for servoactuator position control for the purpose of example.

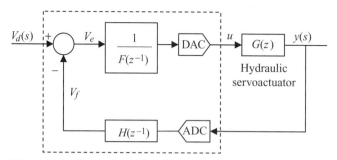

Figure 6.54. Generalized digital filters within a servoactuator position control system.

The two filters are easy to implement because they are in the following form:

$$\frac{u}{V_e} = \frac{1}{F(z^{-1})}, \quad F(z^{-1}) = f_0 + f_1 z^{-1} + f_2 z^{-2} + \cdots +,$$

$$V_e(t) = f_0 u(t) + f_1 u(t - T) + f_2 u(t - 2T) + \cdots +, \tag{6.175}$$

$$\frac{V_f}{y} = H(z^{-1}) \quad H(z^{-1}) = h_0 + h_1 z^{-1} + h_2 z^{-2} + \cdots +,$$

$$V_f(t) = h_0 y(t) + h_1 y(t - T) + h_2 y(t - 2T) + \cdots +. \tag{6.176}$$

Therefore, it is now a matter of selecting the order of each filter transfer function. Bearing in mind that position control is being considered here, the servoactuator transfer function, using ZOH theory, will usually be of the following form:

$$G(z) = \frac{K}{(z - 1)(a_0 + a_1 z + a_2 z^2 + \cdots +)}. \tag{6.177}$$

Consequently, the CLTF becomes:

$$\frac{y}{V_d} = \frac{K}{(z - 1)(a_0 + a_1 z + a_2 z^2 + \cdots +)F(z^{-1}) + K H(z^{-1})}. \tag{6.178}$$

Applying a step demand in position and then the final-value theorem requires that in the steady-state condition, $H(z^{-1})|_{z=1} = 1$; that is:

$$h_0 + h_1 + h_2 + \cdots + = 1. \tag{6.179}$$

The f and h coefficients are then determined by specifying the required closed-loop poles in the z plane.

EXAMPLE 1. Consider, for example a servoactuator with an initially estimated transfer function given by:

$$G(s) = \frac{10}{s(1 + 0.02s)}. \tag{6.180}$$

The feedback position transducer gain $H_p = 10$ V/m. Using ZOH theory and with a sampling interval of $T = 5$ ms, the z transform of Eq. (6.180) becomes:

$$G(z) = \frac{(0.0058z + 0.00526)}{(z - 1)(z - 0.779)}. \tag{6.181}$$

Consequently, the CLTF becomes:

$$\frac{10y}{V_d} = \frac{10(0.0058z + 0.00526)}{F(z^{-1})(z - 1)(z - 0.779) + 10H(z^{-1})(0.0058z + 0.00526)}. \tag{6.182}$$

Select the digital filters as follows:

$$F(z^{-1}) = f_0 + f_1 z^{-1}, \tag{6.183}$$

$$H(z^{-1}) = h_0 + h_1 z^{-1}. \tag{6.184}$$

For this particular control system, the two filters may be further considered by use of the bilinear transformation to convert each digital design to an equivalent analog design. It is then easy to show that each filter is realizable and stable, providing:

$$h_0 > 0.5, \qquad f_0 > f_1. \tag{6.185}$$

This is intended to be an approximate guide rather than a rigorous analysis, although the results do corroborate these simple design conclusions.

Closed-loop characteristic equation (6.182) then becomes:

$$F(z^{-1})(z-1)(z-0.779) + 10H(z^{-1})(0.0058z + 0.00526) = 0,$$
$$z^3 + a_2 z^2 + a_1 z + a_0 = 0, \tag{6.186}$$

$$a_0 = \frac{(0.779 f_1 + 0.0526 h_1)}{f_0},$$

$$a_1 = \frac{(0.779 f_0 - 1.779 f_1 + 0.0526 h_0 + 0.058 h_1)}{f_0},$$

$$a_2 = \frac{(-1.779 f_0 + f_1 + 0.058 h_0)}{f_0}.$$

Now select sensible closed-loop poles; for example, three equal real poles $z = z_r$. This would produce the following characteristic equations:

$$(z - z_r)(z - z_r)(z - z_r) = 0,$$
$$z^3 - 3 z_r z^2 + 3 z_r^2 z - z_r^3 = 0. \tag{6.187}$$

Comparing coefficients of Eqs. (6.187) and (6.186) then gives the solution for the filter coefficients:

$$\begin{bmatrix} z_r^3 & 0.779 & 0 & 0.0526 \\ (0.779 - 3z_r^2) & -1.779 & 0.0526 & 0.058 \\ (-1.779 + 3z_r) & 1 & 0.058 & 0 \\ 0 & 0 & 1 & 1 \end{bmatrix} \begin{bmatrix} f_0 \\ f_1 \\ h_0 \\ h_1 \end{bmatrix} = \begin{bmatrix} 0 \\ 0 \\ 0 \\ 1 \end{bmatrix}. \tag{6.188}$$

For example, selecting $z_r = 0.4$ gives:

$$f_0 = 0.512, \quad f_1 = 0.109, \qquad f_0 > f_1,$$
$$h_0 = 3.235, \quad h_1 = -2.235, \quad h_0 > 0.5. \tag{6.189}$$

The step response to a 100 mm demand change is shown in Fig. 6.55. This figure compares the original system with analog position feedback and the approach using two digital filters. The response is much improved and faster than if the gain of the original system, with analog position control, were reduced to improve the position overshoot.

EXAMPLE 2. Now consider an actual servovalve–linear actuator system similar to that of Example 1 by using a horizontally placed, low-friction-seal actuator having a bore diameter of 50.8 mm and a rod diameter of 28.58 mm. This arrangement means that

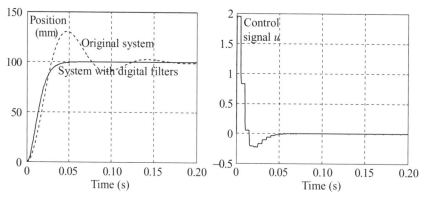

Figure 6.55. The position response of a servoactuator without and with digital filters.

load-force effects on the position extending and retracting response do not exist. However, extending and retracting responses will still be different because of the actuator's having a single rod.

- The transfer function was determined with the frequency-response method discussed in Chapter 5.
- A two-channel commercial frequency-response analyzer was used, the frequency response being automatically computed by applying a prbs signal and using the cross-correlation method.
- The system was tested under closed-loop control with a position transducer having a gain of 41.3 V/m, and the open-loop transfer function was measured.
- Results obtained with frequency response are rather limited as the frequency is increased because of the problem of obtaining reliable data.
- Results are shown in Fig. 6.56, where it will be seen that the frequency range is restricted to about 80 Hz.
- The amplitude is arbitrary because it is based on the instrument reference voltage of 1 V.

The OLTF first-estimate is given by:

$$G(s)H(s) \approx \frac{K}{s(1 + s\tau)}. \tag{6.190}$$

The time constant τ is determined by the frequency at which the phase angle has a value of $-135°$; that is, $-90°$ from the integrator and $-45°$ from the first-order lag function. This is determined as 22.7 Hz, giving a time constant $\tau \approx 7$ ms. The OLTF model is therefore valid up to a frequency of about 40 Hz for design purposes.

The servovalve has a small underlap equivalent to ± 0.27 mA, in fact, similar to that shown in Fig. 3.73. It is therefore appropriate to consider its flow gain at the closed position. Also, because the flow gains are different during motion, it is interesting to compare them with the closed-position gain. The calculations are as follows:

$$
\begin{aligned}
&\text{theoretical extending open loop,} && K = 6.27 \text{ s}^{-1}, \\
&\text{theoretical underlap steady state,} && K = 11.65 \text{ s}^{-1}, \\
&\text{theoretical retracting open loop,} && K = 5.23 \text{ s}^{-1}, \\
&\text{dynamically measured,} && K = 9.33 \text{ s}^{-1}.
\end{aligned}
\tag{6.191}
$$

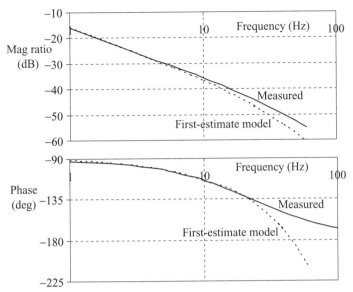

Figure 6.56. Measured open-loop frequency response and comparison with a first-estimate model (Watton, 1990).

In fact, the dynamically measured gain was found to be $K = 9.33$ s^{-1}, and this fitted the extending condition slightly better than it did the retracting condition. From exactly the same procedure as the previous example, the digital filters were designed in this example with a large sampling interval of 22 ms because of the early design of the 6502 real-time processor used at the time, with 12-bit data conversion. As in Example 1, equal closed-loop poles were selected, each having a value of $z_r = 0.4$. This gives:

$$f_0 = 0.908, \quad f_1 = -0.394, \quad f_0 > f_1,$$
$$h_0 = 1.766, \quad h_1 = 0.766, \quad h_0 > 0.5. \tag{6.192}$$

A comparison of measured and predicted results is given in Fig. 6.57.

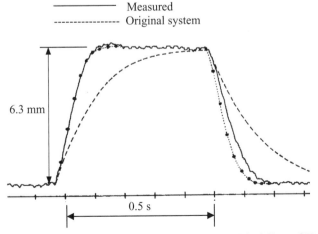

Figure 6.57. Actuator position control using digital filters (Watton, 1990).

It will be seen that the design approach has been successful, even with a large sampling period, the effect of gain change with direction being clearly seen from the measured results. It is an easy matter to change the gain K in real time, and appropriate to the direction of travel, if required.

Further analysis of Example 2 shows that the effect of using digital filters may not be significantly better than increasing the system gain in the original system. However, the approach is useful for the situation in Example 1, in which reducing the gain to avoid position overshoot may produce an unacceptably slow transient response.

6.5.3 State Estimation, Observers, and Reduced-Order Observers

The previous studies considered a variety of state feedback techniques. For full state feedback, all the states are measured with an appropriate transducer. Therefore, the question now arises concerning the control of a hydraulic system if all the states are not measured either because it is impracticable or the instrumentation is too costly. This discussion is restricted to position control systems, for which it is assumed that position only is measured. Therefore, the other states – namely, velocity and acceleration – have to be estimated with an "observer" or "state estimator."

This technique relies on having a suitable model of the system that is then capable of adequately computing the appropriate states using measured data from the system. A linearized state-space model of the system serves this purpose. Consider therefore determining the observer state vector, now defined as $\hat{x}$, using only the error signal u and the output signal $y = Cx$, where x is the position. The observer state equation may then be written:

$$\dot{\hat{x}} = A\hat{x} + Bu + L(y - \hat{y}). \tag{6.193}$$

Also recall the system state equations:

$$\dot{x} = Ax + Bu, $$
$$y = Cx. \tag{6.194}$$

The error equation relating the difference between the actual state and the observer state then becomes:

$$\dot{e} = (\dot{x} - \dot{\hat{x}}) = A(x - \hat{x}) - LC(x - \hat{x}). \tag{6.195}$$

Therefore, if the error dynamics are to rapidly decay to zero as required by the control system, this can be achieved by selecting a suitable error characteristic equation. Equation (6.195) must include only error terms, and this is achieved simply as follows:

$$\dot{e} = (A - LC)e, \tag{6.196}$$

It is then just a matter of choosing the poles of the error transfer function; that is, the zeros of the following characteristic equation:

$$\alpha_e(s) = |sI - A + LC| = 0. \tag{6.197}$$

Note that observer equation (6.193) now becomes:

$$\dot{\hat{x}} = (A - LC)\hat{x} + Ly + Bu. \tag{6.198}$$

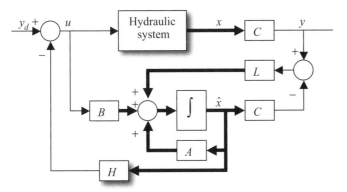

Figure 6.58. Observer-based actuator position control.

The characteristic equation of the observer is therefore the same as the characteristic equation of the error. The unknown vector L may be determined with the second Ackermann's formula:

$$L = \alpha_e(A)V^{-1}\begin{bmatrix} 0 \\ 0 \\ \vdots \\ 1 \end{bmatrix}, \quad V = \begin{bmatrix} C \\ CA \\ CA^2 \\ \vdots \\ CA^{n-1} \end{bmatrix}, \quad (6.199)$$

where V is known as the *observability matrix*. The observer poles are chosen to produce a much faster response than those of the closed-loop system and typically a factor of 4 greater than those of $\alpha_c(A)$. Considering further Eq. (6.198) then gives:

$$\dot{\hat{x}} = A\hat{x} + LC(x - \hat{x}) + Bu. \quad (6.200)$$

The state estimation and control philosophy is shown in Fig. 6.58.

The next stage is then to consider the more realistic situation in which the position is actually measured with a transducer and just the velocity and acceleration need to be estimated. This is referred to as a *reduced-order observer*; to achieve this, it is necessary to partition the appropriate state equation into two parts, one relating to the measured state x_m and the other relating to the estimated states $\hat{x}$:

$$\begin{bmatrix} \dot{x}_m \\ \cdots \\ \dot{\hat{x}} \end{bmatrix} = \begin{bmatrix} A_1 & . & R_1 \\ . & . & . \\ A_2 & . & R_2 \end{bmatrix}\begin{bmatrix} x_m \\ \cdots \\ \hat{x} \end{bmatrix} + \begin{bmatrix} B_1 \\ \cdots \\ B_2 \end{bmatrix}u. \quad (6.201)$$

Substituting into full-order estimator equation (6.198) then gives:

$$\dot{\hat{x}} = (R_2 - LR_1)\hat{x} + (A_2 - LA_1)y + (B_2 - LB_1)u + L\dot{y}. \quad (6.202)$$

It will be seen that the reduced-order equation contains the output derivative $\dot{y}$ and is not desirable for implementation because of signal noise. Therefore, a co-state equation is defined, resulting in the following usable reduced-order observer:

$$\dot{z} = (R_2 - LR_1)\hat{x} + (A_2 - LA_1)y + (B_2 - LB_1)u, \quad (6.203)$$

$$\hat{x} = \hat{z} + Ly. \quad (6.204)$$

Because the full observer is typically reduced by one state, then one integration is removed from the control algorithm. The error equation and the observer equation

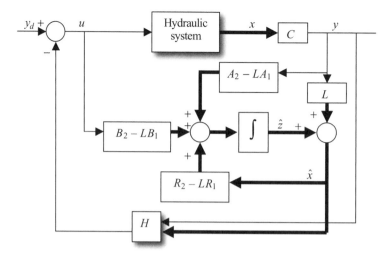

Figure 6.59. Reduced-order observer-based actuator position control.

both have the same characteristic equation, which is given by:

$$\alpha_e(s) = |sI - R_2 + LR_1| = 0. \tag{6.205}$$

The reduced-order vector is again determined by Ackermann's formula:

$$L = \alpha_e(R_2)V^{-1}\begin{bmatrix} 0 \\ 0 \\ \vdots \\ 1 \end{bmatrix}, \qquad V = \begin{bmatrix} R_1 \\ R_1 R_2 \\ R_1 R_2^2 \\ \cdot \\ R_1 R_2^{n-1} \end{bmatrix}. \tag{6.206}$$

The state estimation and control philosophy is shown in Fig. 6.59.

Consider the servoactuator position control system previously analyzed with a variety of feedback methods. In this example, the linearized state-space model of the nonlinear system will be used and with a constant gain for both extension and retraction. The system is again shown in Fig. 6.60.

The servoactuator has a linearized transfer function given by:

$$\frac{y}{u} = \frac{K}{s\left(1 + \dfrac{2\zeta s}{\omega_n} + \dfrac{s^2}{\omega_n^2}\right)}, \tag{6.207}$$

where $\zeta = 0.079$, $\omega_n = 379$ rad/s, $K = 0.33$, and the feedback position transducer has a gain $k_y = 100$ V/m. A reduced-order observer will now be developed for the general case given by Eq. (6.207) and then applied to the servoactuator. The state-space formulation of Eq. (6.207) is in control canonical form, as follows:

$$\begin{bmatrix} \dot{x}_1 \\ \dot{x}_2 \\ \dot{x}_3 \end{bmatrix} = \begin{bmatrix} 0 & 1 & 1 \\ 0 & 0 & 1 \\ 0 & -\omega_n^2 & -2\zeta\omega_n \end{bmatrix} \begin{bmatrix} x_1 \\ x_2 \\ x_3 \end{bmatrix} + \begin{bmatrix} 0 \\ 0 \\ K\omega_n^2 \end{bmatrix} u, \tag{6.208}$$

$x_1 = $ position y, $x_2 = $ velocity $\dot{x}_1$, $x_3 = $ acceleration $\dot{x}_2$.

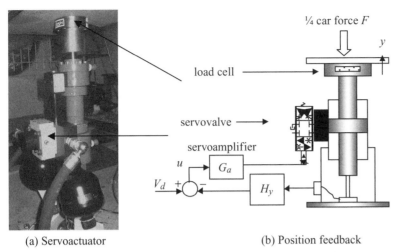

(a) Servoactuator   (b) Position feedback

Figure 6.60. A servoactuator, one of four forming part of a four-poster vehicle test rig (Cardiff University, School of Engineering).

The hydraulic system state equation is then partitioned as previously discussed:

$$\begin{bmatrix} \dot{x}_1 \\ \cdots \\ \dot{\hat{x}} \end{bmatrix} = \begin{bmatrix} A_1 & . & R_1 \\ . & . & . \\ A_2 & . & R_2 \end{bmatrix} \begin{bmatrix} x_1 \\ \cdots \\ \hat{x} \end{bmatrix} + \begin{bmatrix} B_1 \\ \cdots \\ B_2 \end{bmatrix} u,$$

$$A_1 = 0, \quad A_2 = \begin{bmatrix} 0 \\ 0 \end{bmatrix}, \quad R_1 = \begin{bmatrix} 1 & 0 \end{bmatrix}, \quad R_2 = \begin{bmatrix} 0 & 1 \\ -\omega_n^2 & -2\zeta\omega_n \end{bmatrix}, \tag{6.209}$$

$$B_1 = 0, \quad B_2 = \begin{bmatrix} 0 \\ K\omega_n^2 \end{bmatrix}.$$

The estimator characteristic equation is chosen in this example to have two equal real roots given by:

$$\alpha_e(s) = (s + \varphi)^2 = (s + 2\varphi s + \varphi^2) = 0.$$

Evaluating the estimator L from Eq. (6.206) then gives:

$$\alpha_e(R_2) = R_2^2 + 2\varphi R_2 + \varphi^2 I \quad V = \begin{bmatrix} R_1 \\ R_1 R_2 \end{bmatrix} = I,$$

$$L = \begin{bmatrix} 2\varphi \\ 4\zeta^2\omega_n^2 - 4\varphi\zeta\omega_n + \varphi^2 \end{bmatrix}. \tag{6.210}$$

The reduced-order observer equations then become:

$$\begin{aligned} \dot{\hat{z}}_2 &= -2\varphi\hat{x}_2 + \hat{x}_3, \\ \dot{\hat{z}}_3 &= \left[-\omega_n^2(1 + 4\zeta^2) + 4\varphi\zeta\omega_n - \varphi^2 \right]\hat{x}_2 - 2\zeta\omega_n\hat{x}_3 + K\omega_n^2 u, \\ \hat{x}_2 &= \hat{z}_2 + 2\varphi x_1, \\ \hat{x}_3 &= \hat{z}_3 + \left[4\zeta^2\omega_n^2 - 4\varphi\zeta\omega_n + \varphi^2 \right]x_1. \end{aligned} \tag{6.211}$$

For the particular example with $\zeta = 0.079$, $\omega_n = 379$ rad/s, $K = 0.33$, and the feedback position transducer gain $k_y = 100$ V/m, then the estimated states are compared with the actual states in Fig. 6.61.

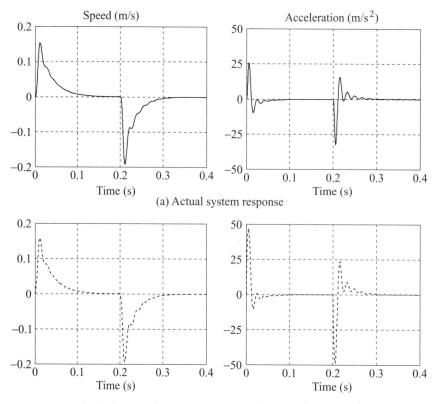

(a) Actual system response

(b) Reduced-order observer speed and acceleration estimation

Figure 6.61. Speed and acceleration estimation for a position control system and based on a reduced-order observer using a linearized system model.

The two equal roots of the characteristic equation were selected to be $\varphi = 1000$ rad/s. It can be seen that the dynamic characteristic is extremely well reproduced and particularly pleasing, given that an average gain was used and the state-space model neglects servovalve dynamic effects. The speed state is barely different from the actual state when overlayed, and the estimated acceleration differs only in that its magnitude is greater than the actual state. The estimated speed and acceleration may be used as extra feedback control signals in conjunction with selected gains. However, any predesign approach rarely produces "exact" gains, and it is common in practice to fine-tune the settings about nominal values.

6.5.4 Linear Quadratic (LQ) Optimal State Control

It is preferable that the dynamic response of a hydraulic control system be optimum in some sense. This may be just the avoidance of overshoot in position or more complex such as also minimizing the energy consumed over a specified duty cycle. Some related issues have already been used such as the use of the ITAE criterion and the IES criterion for designing closed-loop performance. The ITAE criterion has already been used on a number of design problems, and the IES criterion is now pursued further. For example, it may be desirable to minimize an index of performance

J defined by a term similar to an energy function as follows for a two-parameter problem:

$$\text{minimize } J = \int_0^\infty \left(\alpha_1 x_1^2 + \alpha_2 x_2^2\right) dt. \tag{6.212}$$

Using standard state-space notation:

$$\text{minimize } J = \int_0^\infty x^T Q x \, dt,$$

$$x^T = [\, x_1 \quad x_2 \,], \quad Q = \begin{bmatrix} \alpha_1 & 0 \\ 0 & \alpha_2 \end{bmatrix}, \quad x = \begin{bmatrix} x_1 \\ x_2 \end{bmatrix}. \tag{6.213}$$

It also may be required to restrict the control vector in a similar way; therefore, Eq. (6.213) is modified to:

$$\text{minimize } J = \int_0^\infty [x^T Q x + u^T R u] dt. \tag{6.214}$$

The structure of R will be similar to that for Q.

Quadratic forms are used because they have an engineering sense and also because it is well known that the system is asymptotically stable at the origin if Q and R are positive definite. Positive definiteness of a quadratic function is based on Lyapunov's second method. Let

$$V(x) = x^T P x. \tag{6.215}$$

$V(x)$ may be tested for positive definiteness by applying Sylvester's criterion, which states that the necessary and sufficient conditions are that all the successive principle minors of P must be positive:

$$p_{11} > 0, \quad \begin{vmatrix} p_{11} & p_{12} \\ p_{21} & p_{22} \end{vmatrix} > 0, \quad \text{etc.} \tag{6.216}$$

For example:

$$V(x) = 4x_1^2 + 2x_1 x_2 + 2x_2^2 = [\, x_1 \quad x_2 \,] \begin{bmatrix} 4 & 1 \\ 1 & 2 \end{bmatrix} \begin{bmatrix} x_1 \\ x_2 \end{bmatrix},$$

$$p_{11} = 4 > 0, \quad \begin{vmatrix} p_{11} & p_{12} \\ p_{21} & p_{22} \end{vmatrix} = \begin{vmatrix} 4 & 1 \\ 1 & 2 \end{vmatrix} = 7 > 0. \tag{6.217}$$

Therefore, $V(x)$ is positive-definite. Considering Lyapunov's second method of stability in its basic form here, it is clear that if the total energy of a system (a positive-definite function) continually decreases with time, then an equilibrium state is achieved. Hence, the derivative of the total energy is negative definite.

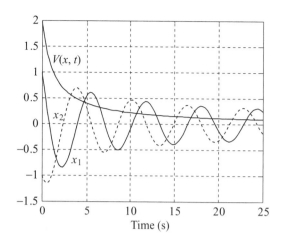

Figure 6.62. State trajectories and Lyapunov function.

Therefore, consider that if the function $V(x, t)$:

- has continuous partial derivatives,
- $V(x, t)$ is positive definite, and
- $\dot{V}(x, t)$ is negative definite [or $-\dot{V}(x, t)$ is positive definite],

then the equilibrium state at the origin is uniformly asymptotically stable.

An advantage of the method is that it can be applied to nonlinear systems. Consider the following examples:

$$\dot{x}_1 = x_2 - 0.2x_1 \left(x_1^2 + x_2^2\right),$$
$$\dot{x}_2 = -x_1 - 0.2x_2 \left(x_1^2 + x_2^2\right). \tag{6.218}$$

The transient response of this system is shown in Fig. 6.62, with initial conditions $x_1(0) = 1, x_2(0) = -1$.

It can be seen that both state trajectories decay to the equilibrium condition at the origin. Also shown is the Lyapunov function defined for this example as:

$$V(x, t) = x_1^2 + x_2^2. \tag{6.219}$$

This Lyapunov function is positive definite, which also may be deduced from Fig. 6.62 because it is continually decreasing along the time trajectory and toward a stable origin. The derivative of the Lyapunov function is:

$$\dot{V}(x, t) = 2x_1\dot{x}_1 + 2x_2\dot{x}_2. \tag{6.220}$$

Inserting Eq. (6.218) gives:

$$-\dot{V}(x, t) = 0.4 \left(x_1^2 + x_2^2\right)^2. \tag{6.221}$$

This is also positive definite, and the equilibrium state at the origin is uniformly asymptotically stable. When these concepts are collected, it is clear that there is a connection between quadratic forms and control system stability. Consider therefore a linear system defined in the usual state-space notation together with a suitable Lyapunov function:

$$\dot{x} = Ax, \qquad V(x, t) = x^T P x, \tag{6.222}$$

where P must be positive definite. The derivative of the Lyapunov function is then given by:

$$\dot{V}(x, t) = x^T[A^T P + PA]x. \tag{6.223}$$

Now this derivative must be negative definite for asymptotic stability and therefore it must be of the following form:

$$\dot{V}(x, t) = -x^T Qx. \tag{6.224}$$

Equating Eqs. (6.223) and (6.224) then gives:

$$A^T P + PA + Q = 0. \tag{6.225}$$

In practice, it is sufficient to choose $Q = I$, the identity matrix to determine whether or not P is positive definite. The solution technique is now explained with an example of a second-order system given by:

$$\begin{bmatrix} \dot{x}_1 \\ \dot{x}_2 \end{bmatrix} = \begin{bmatrix} 0 & 2 \\ -1 & -2 \end{bmatrix} \begin{bmatrix} x_1 \\ x_2 \end{bmatrix}. \tag{6.226}$$

Using Laplace transform theory, it is easy to show that the characteristic equation of this system is given by:

$$|sI - A| = s^2 + 2s + 2. \tag{6.227}$$

The system is therefore stable with an undamped natural frequency of $\omega_n = 1.414$ rad/s and a damping ratio of $\zeta = 0.707$. Equation (6.225) gives:

$$\begin{bmatrix} 0 & -1 \\ 2 & -2 \end{bmatrix} \begin{bmatrix} a & b \\ b & c \end{bmatrix} + \begin{bmatrix} a & b \\ b & c \end{bmatrix} \begin{bmatrix} 0 & 2 \\ -1 & -2 \end{bmatrix} = -\begin{bmatrix} 1 & 0 \\ 0 & 1 \end{bmatrix},$$

$$a = 7/8, \ b = 1/2, \ c = 3/4, \tag{6.228}$$

$$P = \begin{bmatrix} 7/8 & 1/2 \\ 1/2 & 3/4 \end{bmatrix}.$$

Note that P is symmetric. Also, the minors are 7/8 and 13/32; therefore $V(x, t)$ is positive definite. Now determine the Lyapunov function and its derivative:

$$V(x, t) = x^T Px = \begin{bmatrix} x_1 & x_2 \end{bmatrix} \begin{bmatrix} 7/8 & 1/2 \\ 1/2 & 3/4 \end{bmatrix} \begin{bmatrix} x_1 \\ x_2 \end{bmatrix} = \frac{7}{8}x_1^2 + x_1 x_2 + \frac{3}{4}x_2^2,$$

$$\dot{V}(x, t) = -\left(x_1^2 + x_2^2\right). \tag{6.229}$$

Both $V(x, t)$ and $-\dot{V}(x, t)$ are positive definite and validate a stable system as initially determined from Eq. (6.227).

Now consider state feedback, neglecting the system reference input, and a quadratic index of performance as follows:

$$\dot{x} = Ax + Bu, \qquad u = -Hx.$$

Minimize:

$$J = \int_0^\infty [x^T Qx + u^T Ru]dt,$$

$$J = \int_0^\infty x^T[Q + H^T RH]xdt. \tag{6.230}$$

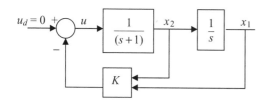

Figure 6.63. A system with optimal state feedback.

From Lyapunov it is necessary that:

$$x^T[Q + H^T RH]x = -\frac{d}{dt}[x^T Px], \qquad (6.231)$$

$$J = \int_0^\infty -\frac{d}{dt}[x^T Px]dt = -x^T(\infty)Px(\infty) + x^T(0)Px(0),$$
$$J = x^T(0)Px(0). \qquad (6.232)$$

From Eq. (6.231) this is true, providing that:

$$x^T[Q + H^T RH]x = -\dot{x}^T Px - x^T P\dot{x}$$

and because $\dot{x} = [A - BH]x$, $\qquad (6.233)$

$$[A - BH]^T P + P[A - BH] = -[Q + H^T RH].$$

Now it can be shown that the transformation $R = T^T T$ leads to the solution for the feedback gain matrix:

$$H = R^{-1}B^T P, \qquad (6.234)$$

where the positive-definite matrix P must now satisfy:

$$A^T P + PA + Q = PBR^{-1}B^T P. \qquad (6.235)$$

This is known as the reduced matrix Ricatti equation. With the right-hand side of Eq. (6.235) removed, it then reduces to that shown as Eq. (6.225). The problem in practice, assuming a state-space representation is even adequate, is determination of the P matrix. For example, the active suspension LQ control (LQC) problem discussed in more detail in Chapter 7 requires the determination of 15 constants. To get a feel for the solution approach to this optimal control technique, consider the second-order system shown in Fig. 6.63.

The state variable x_2 is actuator speed, the state variable x_1 is load position, and the state formulation is:

$$\begin{bmatrix} \dot{x}_1 \\ \dot{x}_2 \end{bmatrix} = \underbrace{\begin{bmatrix} 0 & 1 \\ 0 & -1 \end{bmatrix}}_{A} \begin{bmatrix} x_1 \\ x_2 \end{bmatrix} + \underbrace{\begin{bmatrix} 0 \\ 1 \end{bmatrix}}_{B} u. \qquad (6.236)$$

The function to be minimized is:

$$J = \int_0^\infty [x^T Qx + u^T Ru]dt, \quad Q = \begin{bmatrix} 1 & 0 \\ 0 & 2 \end{bmatrix}, R = 1. \qquad (6.237)$$

This is equivalent to minimizing $x_1^2 + 2x_2^2$, which may be considered as a measure of load energy, together with u^2, which is a measure of the control energy.

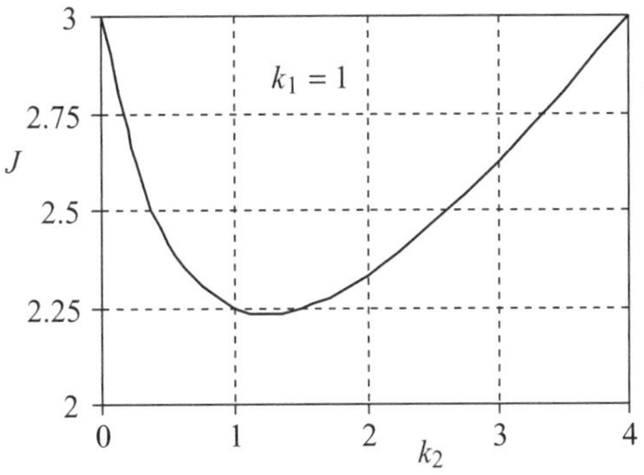

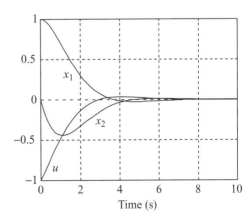

Figure 6.64. Performance index for the position control system example.

Solving Eq. (6.235) and determining Eq. (6.234) gives:

$$P = \begin{bmatrix} 1 & 1 \\ 1 & \sqrt{5}-1 \end{bmatrix}, \quad H = \begin{bmatrix} 1 & \sqrt{5}-1 \end{bmatrix}. \tag{6.238}$$

Clearly, P is positive definite. The gains calculated give an overdamped transient response, and the index of performance J is shown in Fig. 6.64 for the correct value of $k_1 = 1$ and different values of k_2. The initial conditions are $x_1(0) = 1$ and $x_2(0) = 0$.

It can be seen from Fig. 6.64 that the minimum occurs when $k_2 = 1.24$ as predicted, although the variation in J around the optimum condition does not significantly vary for typically $0.5 < k_2 < 2$. The position response is overdamped for the optimum condition. Choosing a lower value of $k_2 = 0.5$ gives a more desirable small overshoot in position response compared with the optimum condition as shown in Fig. 6.65. The performance index J has increased from 2.24 to 2.42.

6.6 On–Off Switching of Directional Valves

This approach is well established but is now having a new lease on life for high-water-content fluid applications for which directional valve technology is preferable from both a reliability and a cost point of view. There are two common approaches to using switched-valve technology:

- integral pairs of valves using pulse-width modulation (PWM)
- binary-sequenced flow valves

Figure 6.65. Transient response of the position control system with LQC.

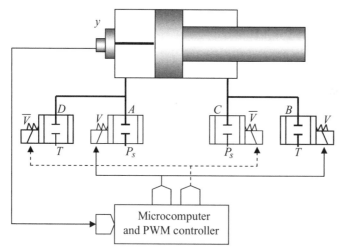

Figure 6.66. On–off switching of valve pairs using PWM for cylinder position control.

Reliability here refers to the fact that spools that are either open or closed are less prone to wear when a high-water-content fluid is used, although it must be realized that any valve that requires continual on–off switching by means of solenoids will also have reliability issues, particularly for high-flow-rate valves.

6.6.1 PWM Control

This method usually requires four single-acting valves to determine the actuator direction and a PWM drive with its frequency of operation that can be adjusted to the optimum frequency for the system. In addition, a means must be included that selects which pair of valves should be operated and also provides a dead-band such that continual switching does not occur about the reference condition. The approach for linear actuator position control is shown as Fig. 6.66.

The solenoid valves are usually switched with 12-V or 24-V dc signals, and each solenoid can draw up to typically 1–2 A of current for the larger sizes needed for larger flow rates. In the applications built by the author, two PWM drivers were used, one providing the "on" signal to the appropriate pair of valves, the other providing the complementary "off" signal to the other pair of valves. Each PWM unit must therefore be designed to supply the current needed by two valves, and the pulse width must be capable of being controlled by the position error signal. Consider Fig. 6.66:

- Valves A and B are paired to create the supply pressure and tank return to extend the actuator.
- Valves C and D are paired to create the supply pressure and tank return to retract the actuator.
- An appropriate PWM frequency is established by testing.
- The ratio between the "on" time and "off" time is proportional to the position error.
- When one pair of valves is being used, the other pair is switched off.

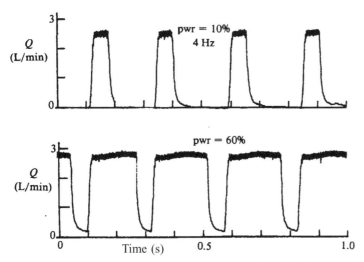

Figure 6.67. Transient flow characteristic for a solenoid valve controlled by a PWM 24-V dc signal.

Each valve has a first-order response characteristic with a time constant of about 14 ms when operated by the 24-V dc solenoid. The transient flow behavior is shown in Fig. 6.67 for a pulse frequency of 4 Hz and two different pulse-width ratios of 10% and 60% of the signal period of 0.25 s. The flow rate is measured with a flow meter that has a much faster response characteristic than the valve under test.

Figure 6.68 shows the mean flow-rate variation with pulse-width ratio and for different frequencies for one of the four valves used in this example.

This valve is restricted in the frequency range of operation, although there is certainly a minimum pulse-width ratio below which flow will not be generated. This is useful in that it can be considered as an inherent dead-band and is particularly noticeable for pulse frequencies above 4 Hz. The lack of linearity as the frequency is increased does not present a problem for position control, although for frequencies

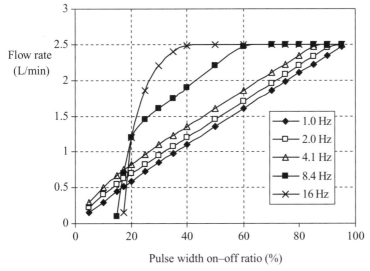

Figure 6.68. Mean flow-rate characteristic of an "off-the-shelf" commercial solenoid valve with PWM switching.

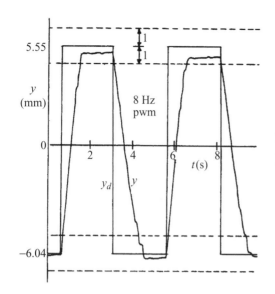

Figure 6.69. Closed-loop control of a linear actuator using PWM switching of four solenoid
valves.

above 16 Hz, the valve does tend to behave like an on–off controller with a large
dead-band.

Now consider closed-loop position control. The position error creates a proportional voltage that changes the pulse width of the appropriate pair of valves, and
within the microcomputer program. A typical transient response to a square-wave
demand is shown in Fig. 6.69. The system is set such that movement to the desired
position is at almost constant velocity, limited by the flow gain of the valve unit.
As is expected from the previous comments, the positional accuracy deteriorates as
the PWM frequency is increased beyond 8 Hz. The results of many tests also show
that positional accuracy is slightly improved if approached from the lower-velocity
retracting part of the cycle. An accuracy of better than ±0.4 mm is achieved with
the system, which may be adequate for large power control applications for which
precision control is not critical. Note, however, that the use of an industrial servovalve, of the force-feedback type, for closed-loop control did not drastically change
the performance, as shown in Fig. 6.70. This conventional analog control system is
slightly faster acting, but the positional accuracy is governed by spool underlap and
the system threshold characteristic.

Next consider replacing the linear actuator with an axial piston motor, as shown
in Fig. 6.71.

The open-loop characteristic is determined by varying the pulse-width difference between input and output valve signals and recording the motor rotation for
different pulse frequencies. The measurements are shown in Fig. 6.72.

At any desired operating frequency, there exists a maximum motor rotation
available over one cycle, and this upper limit was set arbitrarily at 327° with a
pulse-width ratio of 80% at a frequency of 2 Hz. Actually, the maximum rotation
per cycle decreases linearly with a pulse period down to a period of typically 0.1 s
(10 Hz), where there is no response to valve-switching signals.

This effective system-saturation characteristic is useful in practice because it
ensures that for large demanded position changes, the maximum rotation per pulse
is utilized. Proportional control therefore exists only when the position error falls on

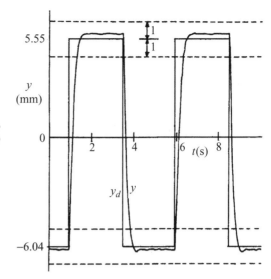

Figure 6.70. Servovalve alternative closed-loop control of the same actuator used in Fig. 6.69 (Watton, 1989).

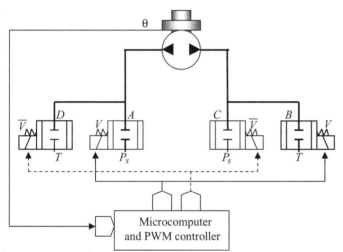

Figure 6.71. On–off switching of valve pairs using PWM for motor position control (Pierce, Coughlin, and Watton, 1985).

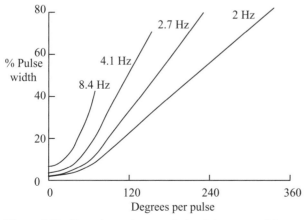

Figure 6.72. Open-loop characteristic of the motor drive.

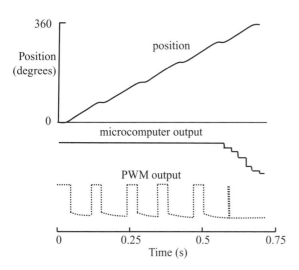

Figure 6.73. Closed-loop position response for a demand position of 360° and at a fixed pulse frequency of 4.1 Hz.

the appropriate curve of Fig. 6.72, and an equation for each curve has to be determined. Appropriate equations up to cubic form are used in the computer control algorithm, particularly for the small rotation portion of the curve at which small pulse-width changes give large changes in motor rotation. For example, considering a pulse frequency of 4.1 Hz gives:

$$\%\text{pulse width} = 0.000004\theta^3 + 0.000053\theta^2 + 0.2635\theta + 2.85, \quad 0 < \theta \le 60°,$$

$$\%\text{pulse width} = 15 + 0.583(\theta - 60°), \quad 60° < \theta \le 155°. \tag{6.239}$$

A typical performance is shown in Fig. 6.73 for a pulse frequency of 4.1 Hz.

A disadvantage in the control concept is the fact that the pulse widths are being continually changed as the system error is reduced. Once the computer begins to calculate the desired pulse width, it should be theoretically possible to make the final step in one pulse. This is an interesting point to be pursued and does suggest that it may be undesirable to have too rapid a sampling rate because one sample per pulse may be adequate. A further development to be pursued is the use of frequency control at a fixed pulse width, or perhaps a mixture of pulse width and frequency control. The open-loop and two closed-loop responses are compared in Fig. 6.74,

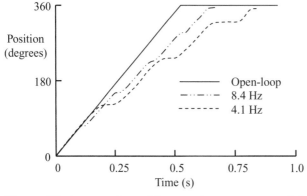

Figure 6.74. A comparison of the open-loop response and the closed-loop response at different pulse frequencies (Pierce, Coughlin, and Watton, 1985).

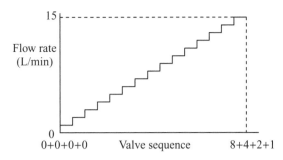

(a) Flow rate addition of binary sequenced solenoid valves

Figure 6.75. On–off switching of valve pairs using PWM for position control.

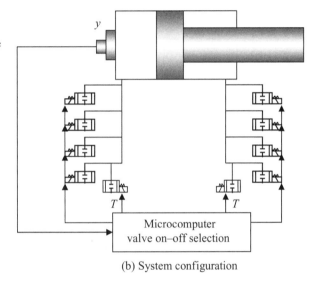

(b) System configuration

from which, as expected, it can be seen that as the pulse frequency is increased then the closed-loop response is improved.

6.6.2 Valves Sized in a Binary Flow Sequence

An alternative approach to using PWM switching is to use an array of on–off solenoid valves, the array being rated similar to a binary sequence. For example, four valves having flow rates of 1, 2, 4, and 8 L/min would suffice to provide a flow rate from 0 to 15 L/min in steps of 1 L/min, as shown in Fig. 6.75.

Figure 6.75 shows just one of several approaches possible. Because an individual valve flow rate is not high, then fast-response solenoid valves may be used. All the valves are switched off when the desired position is reached, and they can be designed to be of the poppet type and therefore more suitable for high-water-content fluids. This quasi-linear flow-gain characteristic may well be suited for many industrial applications, and some surprising levels of position accuracy have been obtained. It is not too critical that an exact binary sequence be obtained but, in practice, the chosen ratios may, reasonably designed by fine-tuning with control orifices at each valve.

Consider the approach by Linjama and Vilenius (2005) and for the system configuration shown in Fig. 6.76. Each solenoid valve has a 30-ms delay and a first-order

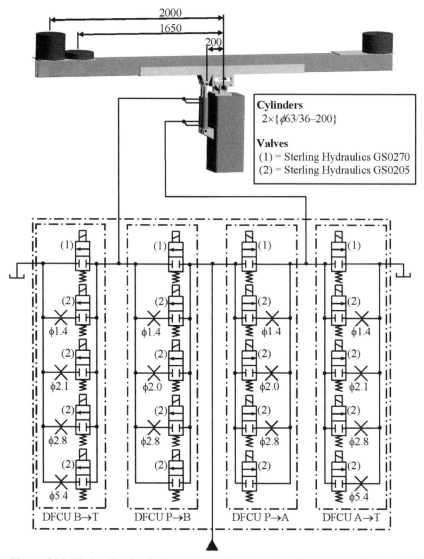

Figure 6.76. Hydraulic circuit diagram for digital control (Linjama and Vilenius, 2005).

time constant of 2 ms at a maximum rate of 200 L/min. The largest valve has a capacity of 19 L/min, and the others follow approximately a binary series. The system studied is an energy-efficient motion control of a digital joint actuator typical of a medium-sized mobile machine boom. A particular feature of the control approach is that a cost-control-function-based solution is used for on-line minimization of power losses.

Figure 6.77 shows a comparison of measured results and computer-simulated results indicating a good design simulation foundation for developing other control schemes and systems.

The future response is calculated by the selected system model, and valve controls are selected by minimizing the cost function. The steady-state velocity and pressures can be solved from a known equation if the load force and supply and tank pressures of the system are known. The system has $2^{20} \approx 10^6$ different states, which makes the real-time calculation of all combinations impossible. Therefore,

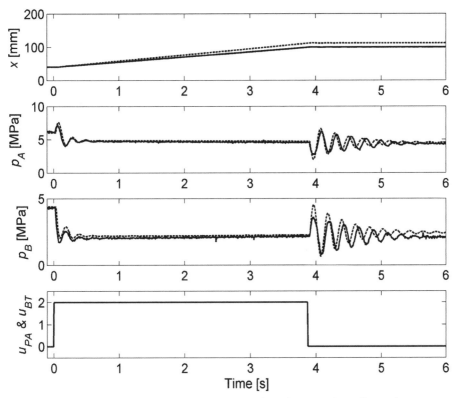

Figure 6.77. Measured and simulated responses when the second smallest valves are opened and closed (Linjama and Vilenius, 2005).

the reduced search space is defined by analyzing flow balance of both actuator chambers separately. The steady-state velocity and pressures are then solved for elements of the reduced search space by using Newton–Raphson iterations. Finally, the cost-function values are calculated for each steady-state solution, and the best valve combination is selected. This process is repeated at each sampling instant. It is important to note that the approach allows control of all four flow paths simultaneously. The inputs of the controller are target velocity and pressures, measured or estimated supply and tank pressures, and measured or estimated load force. Further details are given in references (e.g., Linjama and Vilenius, 2005), but this application does show how digitally switched valves may be used with developing on-line computational techniques to compete with servovalve systems. It is reported that the approach resulted in a 36% reduction in power losses compared with a traditional proportional valve.

6.7 An Introduction to Fuzzy Logic and Neural Network Control

This chapter is not intended to cover the more advanced aspects of control theory, but it is appropriate to make the reader aware of developing techniques that have been applied to fluid power circuits and generically labeled as *intelligent control systems*. The term is misleading in the sense that "intelligence" is preselected by the design engineer to mainly accommodate nonlinear dynamic behavior and, to some extent, handle changes in some system characteristics that reflect a change in the

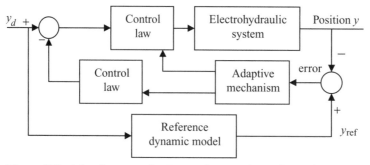

Figure 6.78. Adaptive control using a reference dynamic model.

dynamic behavior. If system parameters change, then an ideal control system will ensure that the dynamic control behavior remains ostensibly the same. In reality, this is not perfectly possible, but intelligent control aims to move toward this goal.

For example, *adaptive control systems* use controllers that are continually updated by a suitable algorithm that attempts to establish the current system dynamics and then change control laws to compensate for any change. One way of doing this is to use *model reference* control in which the actual system output is compared with the output from the reference dynamic mathematical model having the same input as the real system, as shown by the position control system in Fig. 6.78.

An alternative approach is to use a *system identification* procedure; for example, using the least-squares method, followed by control-law adaption. A continuously operating approach utilizes the recursive least-squares method, whereby the estimated coefficients of the hydraulic system model are continually updated. This approach is shown in Fig. 6.79.

Stability of these types of approach is a problem when handling large changes from the response expected because the mathematical model used in both methods may not then be valid. Other approaches therefore attempt to handle changes in system behavior in a more pragmatic way rather than a mathematical way.

Fuzzy-logic control is rule-based and utilizes the experience of practical systems design in the sense of emulating what a human controller would probably do in the presence of a particular control behavior change. For example, if a position control system had a small error, then a small correction would be made; if the error were large, then a large correction would be made. In addition, the human operator might also take into account the speed of response and make a correction in a slightly different way. However, precise values of error and speed cannot usually be

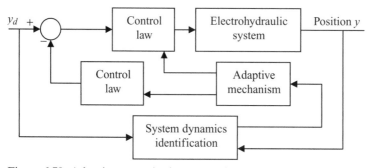

Figure 6.79. Adaptive control using system identification.

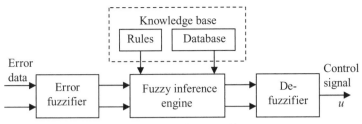

Figure 6.80. Structure of a fuzzy-logic controller.

determined and the data will have "fuzzy" boundaries that merge with one another. The elements of such a knowledge-based approach are then as shown in Fig. 6.80.

The structure shown in Fig. 6.80 is usually embodied within the error loop of the electrohydraulic control system, and there are no rigid design procedures for deciding the control logic other than the principle of common sense. If a position control system is considered, and the position error and velocity are to be used, then it must be recalled that because computer control is being implemented with sampled data, then the velocity may be easily calculated. It is not necessary to divide successive samples by the sampling interval to determine velocity because just the sample differences will suffice. Note, however, that in practice the position error signal may be noisy and any error changes computed will be subject to a degree of uncertainty.

Therefore, this fuzzy-logic control approach utilizes the error e and the change in error ce. In essence, this fuzzy logic approach is really a knowledge-based proportional + derivative $(P + D)$ approach. For example, consider just five states of c and ce as follows:

negative large, NL
negative small, NS
zero, Z
positive small, PS
positive large, PL

More states may be added to improve control performance. The fuzzy sets for both e and ce will then appear as shown in Fig. 6.81.

The fuzzy sets of control values u will also have a similar topology to that shown in Fig. 6.81 and may also be visualized as shown in Table 6.2.

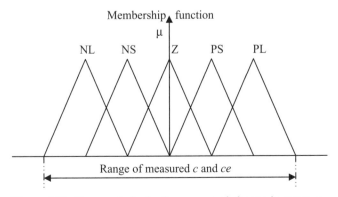

Figure 6.81. Fuzzy sets for both error e and change in error ce.

Table 6.2 *Rule base for the control action* u

		NeL	NeS	Error e eZ	PeS	PeL
	NceL			NZ	PZ	PL
	NceS					
Error change *ce*	ceZ	NL	NM	Z	PM	PL
	PceS					
	PceL		NL	NZ	PZ	

Control action *u*

To refine the output control signal, the center-of-area (COA) method may be used to determine the center of the fuzzy distribution array. Design strategies are then left to the individual; single-rod actuator area asymmetry may be accommodated as well as severe geometrical changes in load during motion. The result of applying fuzzy-logic control to a cylinder position control system is shown in Fig. 6.82 with 11 fuzzy sets for *e* and 9 fuzzy sets for *ce*. Data used are as follows:

Diameter of piston, 32 mm
Diameter of piston rod, 20 mm
Stroke, 1000 mm
Inertia load, 140–300 kg; load force, 400 N
Nominal flow of servovalve, 57 L/min
Natural frequency of servovalve, 350 rad/s
Damping factor of servovalve, 0.8

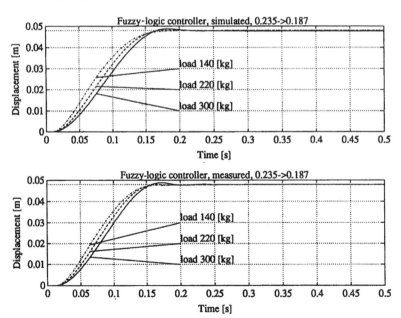

Figure 6.82. Step responses of the position control system with fuzzy-logic control, provided by T. Virvalo and also available in Virvalo and Koskinen (1992).

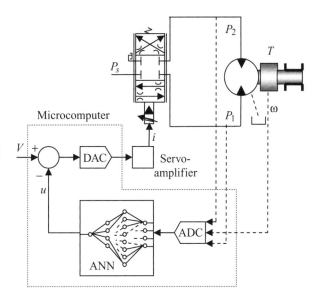

Figure 6.83. Adaptive motor speed control system incorporating an ANN (Nishiumi and Watton, 1997).

Hysteresis and threshold of servovalve, 0.5%
Resolution of feedback transducer, 0.0024 mm
DAC, 12 bits
Supply pressure, 70 bar

The application of ANNs has received some attention because they have the potential to continually update system dynamics or model nonlinear models. Updated models or model reference techniques may then also be used for control applications. In the approach adopted by Nishiumi and Watton (1997), for a servovalve–motor drive system, a combination of deterministic and heuristic techniques were used to adapt to changes in speed demand, load torque, and supply pressure, eventually attempting to maintain the reference model dynamics. The steady-state behavior and open-loop system dynamics have been discussed at some length in earlier chapters, and it was shown that a servovalve-controlled motor has a highly nonlinear speed–load pressure characteristic. The objective is to adapt the system by means of an ANN:

- using variable flow gain compensation in the forward path
- using variable accelerometer gain in the feedback path
- then also using an appropriate input–output reference model

The approach taken is shown in Fig. 6.83.
Data for this system are as follows:

Servovalve flow constant, $k_f = 1.98 \times 10^{-8}$
Motor displacement, $D_m = 5.75 \times 10^{-6}$ m^3/rad
Motor external leakage resistance, $R_e = 0.22 \times 10^{12}$ N m^{-2}/m^3 s^{-1}
Motor internal leakage resistance, $R_i = 0.37 \times 10^{12}$ N m^{-2}/m^3 s^{-1}
Motor and load inertia, $J_m = 0.021$ kg m^2
Motor Coulomb friction, $T_f = 1.55$ N m
Motor viscous friction coefficient, $B_v = 0.078$ N m/rad s^{-1}
Lines equal volumes, $V = 1.85 \times 10^{-4}$ m^3
Fluid effective bulk modulus, $\beta = 1.4 \times 10^9$ N m^{-2}

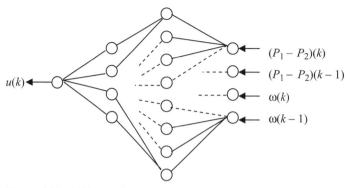

Figure 6.84. ANN topology selected for accelerometer dynamics.

A crucial aspect of the approach is the choice of an appropriate ANN topology, and foundation work has shown that acceleration feedback emulation is possible. Excellent results have been obtained for the case when the ANN is trained to track a predetermined variable gain accelerometer model (Nishiumi and Watton, 1997). The current approach is possible only by using an accurate computer simulation, the ANN being trained using appropriate simulation data. The ANN topology is then transferred to a C language program and incorporated into a real-time digital control scheme using a microcomputer with an ADC/DAC control card. Following analysis of the open-loop system, with experimental validation, the system steady-state constants and second-order transfer function were accurately determined. The accelerometer feedback transfer function is:

$$G(s) = K_v s. \tag{6.240}$$

From consideration of the system characteristics, it can be shown that the variable gain is given by:

$$K_v = a_0 - a_1 \omega. \tag{6.241}$$

This characteristic may be compensated in the ANN by utilizing the function $1/K_v$. The variable flow gain of the servovalve–motor is given by:

$$K_\omega = a_2 + a_3 P_s. \tag{6.242}$$

The supply pressure is not measured, but the pressures P_1 and P_2 are measured because they are required as inputs to the ANN. However, recalling the steady-state theory, it can be shown for most servovalve–motor systems that the sum of line pressures is slightly below supply pressure and remarkably constant for most working speeds. For this study, the supply pressure is calculated from:

$$P_s \approx 1.1(P_1 + P_2) \tag{6.243}$$

and is valid for speeds greater than 2 rad/s. To compensate for the variable flow gain, the function $1/K_\omega$ is continually computed and added to the forward path of the control loop. It was found that it was sufficiently accurate to train an ANN using pressure differential and speed. In addition, a recurrent network is needed to accommodate dynamic conditions, and just the current sampled state plus the previously sampled state of both parameters was found to be satisfactory for motor speeds up to 40 rad/s. Many topologies were considered, and the one selected is shown in Fig. 6.84.

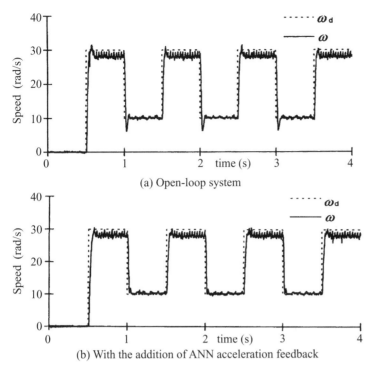

(a) Open-loop system

(b) With the addition of ANN acceleration feedback

Figure 6.85. A comparison of open-loop and ANN closed-loop control (Nishiumi and Watton, 1997).

It can be seen that the topology has eight neurons in the first hidden layer and four neurons in the second hidden layer. The ANN was trained using simulation data for various changes in demand speed, load torque, and supply pressure using motor acceleration as the desired goal u. Some results for motor speed, in response to continual step changes in demand speed, are shown in Fig. 6.85 for the open-loop system and the system with ANN feedback emulating a variable gain accelerometer.

It can be seen that the open-loop system is oscillatory, the overshoot increasing as the steady-state speed is lowered as expected. ANN acceleration feedback does not include gain compensation and therefore both responses shown in Fig. 6.85 result in a speed error from that desired. The closed-loop results were obtained using a sampling interval of 2 ms. However, the effect of ANN acceleration feedback is to significantly reduce the speed overshoot.

Now consider adapting the ANN using a model reference closed-loop desired performance. A first-order closed-loop model is chosen with a time of 15 ms. This was then implemented using the bilinear approximation to give:

$$\omega_r(k) = 0.0625\omega_d(k) + 0.0625\omega_d(k-1) + 0.875\omega_r(k-1). \qquad (6.244)$$

Adaption was achieved by on-line adjustment of all neuron weights using a simplified approach to weight change selection and given by:

$$\Delta w_j(k) = \eta[\omega(k) - \omega_r(k)]y_j(k) + m\Delta_j(k-1), \qquad (6.245)$$

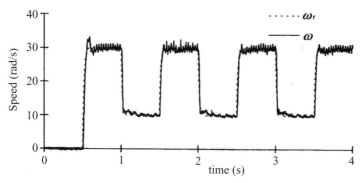

Figure 6.86. The effect of model reference adaptive control using on-line ANN weights adaption (Nishiumi and Watton, 1997).

where η is the learning rate selected as $\eta = 0.005$, m is the momentum selected as $m = 0.01$, and $y_j(k)$ is the signal from the hidden layer neuron j at iteration k. A typical result may be compared with Fig. 6.85 and is given in Fig. 6.86.

Gain compensation is included in the model reference control approach, and it can be seen that the overall approach quickly reduces the speed overshoot while virtually eliminating steady-state speed droop. Further results in Nishiumi and Watton (1997) show that the approach also adapts to large changes in supply pressure and load torque, the latter being applied by motor shaft friction. This is particularly encouraging because it indicates robustness to new conditions not met when initially training the ANN. However, longer adaption times were needed for such large changes. In addition, it was shown that the system could be successfully adapted for sinusoidal demand signals, again indicating robustness to conditions not used when training the initial ANN.

6.8 Servovalve Dither for Improving Position Accuracy

In practice, a servoactuator has a variety of nonlinear characteristics that affect the accuracy of closed-loop position control. These characteristics are dominated by:

- servovalve hysteresis due to first-stage electromagnetic effects
- servovalve underlap/overlap
- servovalve spool friction
- resolution of the position sensor
- actuator friction

Servovalve hysteresis and/or spool lap effects are usually evident from the data supplied by the manufacturer and will be significantly smaller than the rated current, as evident from Fig. 3.37. Spool underlap is usually evident by a doubling of the flow gain at the null conditions and Fig. 6.87 illustrates only the hysteresis characteristic. It may be quoted with respect to the servovalve current rating; for example, $<\pm 3\%$, with a threshold of $<0.5\%$ without dither, moving to almost 0% with dither.

Dither is no more than the addition of a high-frequency signal, usually built into the servoamplifier, with an amplitude that can be as high as 20% of the rated current and a frequency that is set to suit the servoactuator. For example, consider

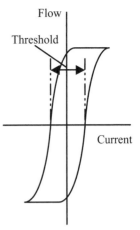

Figure 6.87. Servovalve hysteresis characteristic, not to scale.

a linear actuator closed-loop drive tested by the author and having system data as follows:

- a position transducer gain $H_p = 40$ V/m
- a servoamplifier gain $G_a = 10$ mA/V
- a demanded change in position by applying a triangular waveform having a very low frequency such that system dynamics have a negligible effect on the position measurement
- a dither signal having a frequency of 150 Hz and an amplitude that is the same as the triangular waveform; in this example, an amplitude of 0.185 mA

Three cycles are shown in Fig. 6.88 with and without the dither signal. It is clear that dither may significantly improve positional accuracy if this is required and cannot be achieved by changing the closed-loop steady-state gain.

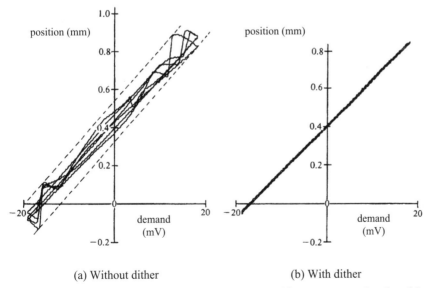

(a) Without dither   (b) With dither

Figure 6.88. The effect of a dither signal on the position accuracy of a closed-loop servo-actuator.

6.9 References and Further Reading

Alirand M, Favennec G, Lebrun M [2002]. Pressure components stability analysis: A revisited approach. *Int. J. Fluid Power* 3(1), 33–46.

Anderson T, Hansen MR, Pederson HC, Conrad F [2005]. Comparison of linear controllers for a hydraulic servo system. In *Proceedings of the 6th JHPS International Symposium on Fluid Power*, Japan Fluid Power Society, 449–454.

Anderson T, Hansen MR, Pederson HC, Conrad F [2005]. Feedback linearization applied on a hydraulic servo system. In *Proceedings of the 6th JHPS International Symposium on Fluid Power*, Japan Fluid Power Society, 167–172.

Ashley T and Mills B [1966]. Frequency response of an electrohydraulic vibrator with inertia load. *Inst. Mech. Eng. J. Mech. Eng. Sci.* 8(1), 27–35.

Barnard BW and Dransfield P [1977]. Predicting response of a proposed hydraulic control system using bond graphs. *Trans. ASME J. Dyn. Syst. Meas. Control*, 1–8.

Bruno S, Sairiala H, Koskinen KT [2006]. Analysis of the behaviour of a water hydraulic crane. *Int. J. Fluid Power* 7(2), 29–38.

Carrington JE and Martin HR [1965]. Threshold problems in electrohydraulic servomotors. *Proc. Inst. Mech. Eng. Part 1*, 180, 881–894.

Cho J, Zhang X, Manring ND, Nair SS [2002]. Dynamic modelling and parametric studies of an indexing valve plate pump. *Int. J. Fluid Power* 3(3), 37–48.

Cunha MAB, Guenther R, De Pieri ER, De Negri VJ [2002]. Design of cascade controllers for a hydraulic actuator. *Int. J. Fluid Power* 3(2), 35–46.

Davies RM and Watton J [1995]. Intelligent control of an electrohydraulic motor drive system. *J. Mechatronics* 5(5), 527–540.

Driemeyer Franco AL, De Fieri ER, Cautelan EB, Guenther R, Valdiero AC [2004]. Design and experimental evaluation of position controllers for hydraulic actuators: Backstepping and lqr-2dof controllers. *Int. J. Fluid Power* 4(3), 39–48.

Dutton K, Thompson S, Barraclough B [1997]. *Control Engineering*. Addison-Wesley Longman.

Esque S, Raneda A, Ellman A [2003]. Techniques for studying a mobile hydraulic crane in virtual reality. *Int. J. Fluid Power* 4(2), 25–34.

Ficarella A, Giuffrida A, Laforgia D [2006]. Numerical investigations on the working cycle of a hydraulic breaker: Off-design performance and influence of design parameters. *Int. J. Fluid Power* 7(3), 41–50.

Foster K and Fenney L [1989]. Characteristics and dynamic performance of electrical and hydraulic drives. In *Proceedings of the 1st JHPS International Symposium on Fluid Power*, Japan Fluid Power Society, xvii–xxiv.

Guillon M [1969]. *Hydraulic Servo Systems*. Published by Butterworths.

Healey AJ and Stringer JD [1968]. Dynamic characteristics of an oil hydraulic constant speed drive. *Proc. Inst. Mech. Eng.* 183, 682–692.

Hu H and Zhang Q [2002]. Realization of programmable control using a set of individually controlled electrohydraulic valves. *Int. J. Fluid Power* 3(2), 29–34.

Ikebe Y [1981]. Load insensitive electrohydraulic servo system. *J. Fluid Control* 13(4), 40–55.

Kazuhisa I [2005]. An adaptive pressure controller design of hydraulic servo system with dead zone. In *Proceedings of the 6th JHPS International Symposium on Fluid Power*, Japan Fluid Power Society, 161–166.

Khalil MKB, Yurkevich V, Svoboda J, Bhat RB [2002]. Implementation of single feedback control loop for constant power regulated swash plate axial piston pumps. *Int. J. Fluid Power* 3(3), 27–36.

Kontz ME and Book WJ [2007]. Flow control for coordinated motion and haptic feedback. *Int. J. Fluid Power* 8(3), 13–24.

Koskinen KT, Makinen E, Vilenius MJ, Virvalo T [1996]. Position control of a water hydraulic cylinder. In *Proceedings of the 3rd JHPS International Symposium on Fluid Power*, Japan Fluid Power Society, 43–48.

Krauss R, Book W, Brills O [2007]. Transfer matrix modelling of hydraulically actuated flexible robots. *Int. J. Fluid Power* 8(1), 51–58.

Larsson J, Krus P, Palmberg J-O [2004]. Efficient collaborative modelling and simulation with application to wheel loader design. *Int. J. Fluid Power* 4(3), 5–14.

LeVert FE [1978]. Dynamic analysis of a high speed electrohydraulic transient rod drive system. *Fluid Q.* 10, 2.

Li PY and Krishnaswamy K [2004]. Passive bilateral teleoperation of a hydraulic actuator using an electrohydraulic passive valve. *Int. J. Fluid Power* 5(2), 43–56.

Lin C-L, Chen C-H, Liu VT [2005]. A new time-delay compensating scheme for electrohydraulic systems. *Int. J. Fluid Power* 6(1), 19–28.

Linjama M and Vilenius M [2005]. Improved digital hydraulic tracking control of water hydraulic cylinder drive. *Int. J. Fluid Power* 6(1), 29–40.

Linjama M and Vilenius M [2005]. Energy-efficient motion control of a digital hydraulic joint actuator. In *Proceedings of the 6th JFPS International Symposium on Fluid Power*, Japan Fluid Power Society, 640–645.

Linjama M and Vilenius M [2007]. Digital hydraulics: Towards perfect valve technology. Presented at the 10th Scandinavian International Conference on Fluid Power, SICFP '07, Tampere, Finland.

Linjama M, Huova M, Boström P, Laamanen A, Siivonen L, Morel L, Waldén M, Vilenius M [2007]. Design and implementation of energy saving digital hydraulic control system. Presented at the 10th Scandinavian International Conference on Fluid Power, SICFP '07, Tampere, Finland.

Linjama M, Koskinen KT, Vilenius M [2003]. Accurate trajectory tracking control of water hydraulic cylinder with non-ideal on/off valves. *Int. J. Fluid Power* 4(1), 7–16.

Lisowski E and Czyzycki W [2003]. Simulation of vane pump controller by use of the DELPHI software environment. In *Proceedings of the 1st International Conference in Fluid Power Technology*, Fluid Power Net Publications, 261–270.

Martin HR and Lichtarowicz A [1966]. Theoretical investigation into the prevention of cavitation in hydraulic actuators. *Proc. Inst. Mech. Eng.* 181, 423–431.

Matsui T and Koseki H [1996]. Motion control of water hydraulic servo system. In *proceedings of the 3rd JHPS International Symposium on Fluid Power*, Japan Fluid Power Society, 61–66.

Mattila J, Siuko M, Vilenius M [2005]. On pressure/force control of a 3-DOF water hydraulic manipulator. In *Proceedings of the 6th JHPS International Symposium on Fluid Power*, 443–448.

Mikota J and Reiter H [2003]. Development of a compact and tuneable vibration compensator for hydraulic systems. *Int. J. Fluid Power* 4(1), 17–30.

Nikiforuk PN, Wilson JN, Lepp RM [1970]. Transient response of a time-optimised hydraulic servomechanism operating under cavitating conditions. *Proc. Inst. Mech. Eng.* 185, 423–431.

Nishiumi T and Watton J [1996]. Some practical considerations of real-time artificial neural network control of a servovalve/motor drive. In *Proceedings of the 9th Bath International Fluid Power Workshop*, 116–125.

Nishiumi T and Watton J [1997]. Model reference adaptive control of an electrohydraulic motor drive using an artificial neural network compensator. *Proc. Inst. Mech. Eng. Part I J. Syst. Control Eng.* 216, 357–367.

Noskievic P [2003]. Simulation and dynamic analysis of mechatronic systems with the output in the virtual reality world. In *Proceedings of the 1st International Conference in Fluid Power Technology*, Fluid Power Net Publications, 543–551.

Ogata K [1970]. *Modern Control Engineering*. Prentice-Hall.

Pai K-R and Shih M-C [1999]. Multi-speed control of a hydraulic cylinder using self-tuning fuzzy control method. In *Proceedings of the 4th JHPS International Symposium on Fluid Power*, Japan Fluid Power Society, 99–104.

Papadopoulos E and Davliakos I [2004]. A systematic methodology for optimal component selection of electrohydraulic servosystems. *Int. J. Fluid Power* 4(3), 15–24.

Papadopoulos E and Gonthier Y [2002]. On the development of a real-time simulator engine for a hydraulic forestry machine. *Int. J. Fluid Power* 3(1), 55–66.

Pierce S, Coughlin A, Watton J [1985]. Electrohydraulic position control using synchronised solenoid valves with microprocessor-controlled pulse width modulation. In *ASME Dynamic Systems Modelling and Control*, edited by M Donath, American Society of Mechanical Engineering, pp. 141–143.

Qian W, Schoenau G, Burton R [2001]. Measured performance evaluation of PID and neural net control of a hydraulically driven inertia load with nonlinear friction. *Int. J. Fluid Power* 2(1), 31–36.

Rahmfeld R and Ivantysynova M [2004]. An overview about active oscillation damping of mobile machine structure. *Int. J. Fluid Power* 5(2), 5–24.

Ramachandran S and Dransfield P [1993]. Modeling, analysis and simulation of an electrohydraulic flight control actuation system including friction. In *Proceedings of the 2nd JHPS International Symposium on Fluid Power*, Japan Fluid Power Society, 203–208.

Sampson E, Habibi S, Burton R, Chinniah [2004]. Effect of controller in reducing steady-state error due to flow and force disturbances in the electrohydraulic actuator system. *Int. J. Fluid Power* 5(2), 57–66.

Sampson E, Habibi S, Burton R, Chinniah Y [2004]. Effect of controller in reducing steady-state error due to flow and force disturbances in the electrohydraulic actuator system. *Int. J. Fluid Power* 5(2), 57–66.

Sanada K and Kitugawa A [1996]. A study on H0 control of a closed-loop pressure control system considering pipeline dynamics. In *Proceedings of the 3rd JHPS International Symposium on Fluid Power*, Japan Fluid Power Society, 177–182.

Sanada K, Kitugawa A, Pingdong W [1993]. An application of a neural network to adaptive control of a servo system. In *Proceedings of the 2nd JHPS International Symposium on Fluid Power*, Japan Fluid Power Society, 303–308.

Sato Y and Tada K [1999]. Rotational speed control of a hydraulic servomotor with a large inertia load using sliding-mode control. In *Proceedings of the 4th JHPS International Symposium on Fluid Power*, Japan Fluid Power Society, 119–124.

Scheidl R and Manhartsgruber B [1998]. On the dynamic behaviour of servo-hydraulic drives. *Nonlinear Dyn.* 17, 247–268.

Seki S and Itoh T [2005]. Application and study on the hydraulic switching control. In *Proceedings of the 6th JHPS International Symposium on Fluid Power*, Japan Fluid Power Society, 239–244.

Sepehri N, Corbet T, Lawrence PD [1995]. Fuzzy position control of hydraulic robots with valve deadbands. In *Mechatronics*, Elsevier Science, Vol. 5, 623–643.

Shih M-C and Tsai CP [1995]. Servohydraulic cylinder position control using a neuro-fuzzy controller. In *Mechatronics*, Elsevier Science, Vol. 5, 497–512.

Shukla A and Thompson DF [2005]. An investigation of the effect of feedback control on the bifurcation stability of a nonlinear servohydraulic system. *Int. J. Fluid Power* 6(2), 37–46.

Sidders JA, Tilley DG, Chapple PJ [1996]. Thermal-hydraulic performance prediction in fluid power systems. *Proc. Inst. Mech. Eng. Part I J. Sys. Control Eng.* 210, 231–242.

Siivonen L, Linjama M, Huova M, Vilenius M [2007]. Fault detection and diagnosis of digital hydraulic valve system. Presented at the 10th Scandinavian International Conference on Fluid Power, SICFP '07, Tampere, Finland.

Silberberg MY [1956]. A note on the describing function of an element with Coulomb, static and viscous friction. *Trans. Am. Inst. Electr. Eng.* Part 2, 75, 423–425.

St. Hilaire A, Ossyra J-C, Ivantysynova M [2006]. Pump-controlled active roll stabilizer. *Int. J. of Fluid Power* 7(1), 27–40.

Suzuki K, Sugi S, Ueda H [1999]. Improving the characteristics of and electrohydraulic servo system with nonsymmetrical cylinder by ZPETC and linearisation. In *Proceedings of the 4th JHPS International Symposium on Fluid Power*, Japan Fluid Power Society, 93–98.

Suzuki K and Tomioka K [1996]. Improving characteristics of electrohydraulic servo system with nonsymmetrical cylinder using DSP. In *Proceedings of the 3rd JHPS International Symposium on Fluid Power*, Japan Fluid Power Society, 201–206.

Takahashi H, Ito K, Ikeo S [2005]. Application of adaptive controller to water hydraulic servo cylinder. In *Proceedings of the 6th JHPS International Symposium on Fluid Power*, Japan Fluid Power Society, 432–436.

Ting C-S, Li T-HS, Kung F-C [1995]. Design of fuzzy controller for active suspension. In *Mechatronics*, Elsevier Science, Vol. 5, 365–383.

Tou J and Sculthesis PM [1953]. Static and sliding friction in feedback systems. *J. Appl. Phys.* 21, 1210–1217.

Ursu I and Ursu F [2004]. New results in control synthesis for electrohydraulic servos. *Int. J. Fluid Power* 4(3), 25–38.

Valdiero AC, Guenther R, De Fieri ER, De Negri VJ [2007]. Cascade control of hydraulically driven manipulators with friction compensation. *Int. J. Fluid Power* 8(1), 7–17.

Van Ham R, Verrelst B, Daerden F, Vanderborght B, Lefeber D [2005]. Fast and accurate pressure control using on–off valves. *Int. J. Fluid Power* 6(1), 53–58.

Vilenius MJ [1983]. The application of sensitivity analysis to electrohydraulic position servos. *ASME J. Dyn. Syst. Meas. Control* 106, 77–82.

Vilenius M, Koskinen KT, Lakkonen M [2002]. Water hydraulics motion control: Possibilities and challenges. In *Proceedings of the 5th JFPS International Symposium on Fluid Power*, Japan Fluid Power Society, 3–10.

Virvalo T [1999]. On the motion control of a hydraulic servo cylinder drive. In *Proceedings of the 4th JHPS International Symposium on Fluid Power*, Japan Fluid Power Society, 105–110.

Virvalo T [2002]. Comparing controllers in hydraulic motion control. In *Power Transmission and Motion Control 2002*, Professional Engineering Publications Ltd., 215–228.

Virvalo T and Koskinen [1992]. Fuzzy logic controller for hydraulic drive. In *Proceedings of the 10th Aachener Fluidtechniscches Kolloquium*, Technical University of Aachen, 225–240.

Wang PKT and Ma JTS [1963]. Cavitation in valve-controlled hydraulic actuators. *Trans. ASME J. Appl. Mech.* 537–546.

Wang Q, Luo A, Lu Y [1993]. Intelligent control for electrohydraulic proportional cylinder. In *Proceedings of the 2nd JHPS International Symposium on Fluid Power*, Japan Fluid Power Society, 309–313.

Watton J [1976]. The dynamic steering behaviour and automation of a pipe-laying machine. In *Proceedings of BHRA Fluids in Control and Automation*, British Hydrodynamics Research Association, A1-1-32.

Watton J [1986]. The stability of electrohydraulic servomotor systems with transmission lines and non-linear motor friction effects. Part A, system modelling. *J. Fluid Control* 16(2), pp. 118–136.

Watton J [1986]. The stability of electrohydraulic servomotor systems with transmission lines and non-linear motor friction effects. Part B, system stability. *J. Fluid Control* 16(2), 137–151.

Watton J [1987]. The dynamic performance of an electrohydraulic servovalve/motor system with transmission line effects. *ASME J. Dyn. Syst. Meas. Control* 109, 14–18.

Watton J [1989]. Closed-loop design of an electrohydraulic motor drive using open-loop steady-state characteristics. *J. Fluid Control* 20(1), 7–30.

Watton J [1989]. *Fluid Power Systems, Modeling, Simulation, Analog and Microcomputer Control*, Prentice-Hall.

Watton J [1990]. A digital compensator design for electrohydraulic single-rod cylinder position control systems. *ASME J. Dyn. Syst. Meas. Control* 112, 403–409.

Watton J and Al-Baldawi RA [1991]. Performance optimisation of an electrohydraulic position control system with load-dependent supply pressure. *Proc. Inst. Mech. Eng. Part I J. Syst. Control Eng.* 205, 175–189.

Wu D, Schoenau G, Burton R, Bitner D [2005]. Model and experimental validation of a load sensing system with a critically lapped regulator spool. *Int. J. Fluid Power* 6(3), 5–18.

Xiang W, Fok SC, Yap FF [2001]. A fuzzy neural network approach to model hydraulic component from input/output data. *Int. J. Fluid Power* 2(1), 37–48.

Xue Y and Watton J [1998]. Dynamics modelling of fluid power systems applying a global error descent algorithm to a self-organising radial basis function network. *Mechatronics* 8, 727–745.

Yamada H and Muto T [2003]. Development of a hydraulic tele-operated construction robot using virtual reality (new master–slave control method and an evaluation of a visual feedback system). *Int. J. Fluid Power* 4(2), 35–42.

Yamashina C, Miyakawa S, Urata E [1996]. Development of water hydraulic cylinder position control system. In *Proceedings of the 3rd JHPS International Symposium on Fluid Power*, Japan Fluid Power Society, 55–60.

Yousefi H, Handroos H, Manila JK [2007]. Application of fuzzy gain-scheduling in position control of a servo hydraulic system with a flexible load. *Int. J. Fluid Power* 8(2), 25–36.

Zhao T and Virvalo T [1993]. Fuzzy state controller and its application in hydraulic position servo. In *Proceedings of the 2nd JHPS International Symposium on Fluid Power*, Japan Fluid Power Society, 417–422.

Some Case Studies

7.1 Introduction

These studies represent a variety of mathematical and simulation solutions for a range of components and systems and also include much experimental testing with some novel measurement techniques and practical limitations. They are intended to bring together the various aspects of fluid power theory introduced in earlier chapters, but in a more comprehensive manner usually required for more complex systems studies involving the integration of components and control concepts.

7.2 Performance of an Axial Piston Pump Tilted Slipper with Grooves

7.2.1 Introduction

This study was undertaken by Bergada, Haynes, and Watton with experimental work in the author's Fluid Power Laboratory at Cardiff University as part of a comprehensive study on losses within an axial piston pump. It was concerned with a new analytical method based on the Reynolds equation of lubrication, with experimental validation, to evaluate the leakage and pressure distribution for an axial piston pump slipper, taking into account the effect of grooves.

The analytical work was developed by JM Bergada (UPC, Terrassa, Spain) with experimental work undertaken by JM Bergada and JM Haynes. Additional CFD analysis and test-rig design was undertaken by JM Haynes and J Watton. Further CFD results by R Worthing and J Watton are also presented in this overview.

The equations consider slipper spin and tilt and are extended to be used for a slipper with any number of grooves. Test rigs have been designed and used to check experimentally the applicability of the theoretical equations, and comparisons between theoretical and experimental results show a good agreement. The new theory can predict slipper leakage and pressure inside the groove with a high level of accuracy, especially at the very low slipper tilts that exist in practice.

The effect of tangential velocity on groove pressure and slipper leakage is then studied experimentally and by CFD simulation, showing that as the rotational speed increases, there is a small decrease in leakage and a small increase in the average pressure inside the groove.

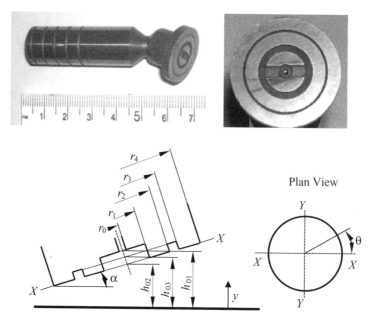

Figure 7.1. Piston–slipper assembly and slipper.

The main piston and slipper assembly that originated this study is shown in Fig. 7.1. This is one of nine pistons from a pump with a maximum volumetric displacement of 0.031 dm^3/rev. It will be seen that the slipper design uses two full lands (Bergada, Haynes, and Watton 2008).

7.2.2 Flow and Pressure Distribution, Mathematical Analysis

Considering Fig. 7.1, the following assumptions are then made:

- Flow will be considered laminar.
- The slipper-plate clearance is not uniform; the slipper is tilted.
- The fluid is hydraulic mineral oil ISO 32.
- Steady conditions are considered.
- Slipper rotation is taken into account.
- Flow is considered to be radial.
- Slipper pocket, groove, and slipper lands are flat.
- The only relative movement between slipper and swash plate is slipper rotation.

The Reynolds equation applicable to this study and its solution developed by JM Bergada are as follows (Bergada and Watton, 2002, 2005, 2008):

$$\frac{\partial}{\partial r}\left(r h^3 \frac{\partial p}{\partial r}\right) = 6\,\mu\,\omega\,r\,\frac{\partial h}{\partial \theta}. \tag{7.1}$$

The film thickness in the clearance is given by:

$$h = h_0 + \alpha\,r_m \cos\theta. \tag{7.2}$$

The average radius between land ends is used, and the film thickness is:

$$\frac{\partial h}{\partial \theta} = -\alpha\,r_m \sin\theta. \tag{7.3}$$

The first integration of differential equation (7.1) will then give:

$$\frac{\partial p}{\partial r} = \frac{-3\,\mu\,\omega\,\alpha\,r_m \sin\theta\; r}{(h_0 + \alpha\,r_m \cos\theta)^3} + \frac{k_1}{r(h_0 + \alpha\,r_m \cos\theta)^3}. \tag{7.4}$$

The second integration gives:

$$p = \frac{-3\,\mu\,\omega\,\alpha\,r_m \sin\theta\; r^2}{2(h_0 + \alpha\,r_m \cos\theta)^3} + \frac{k_1}{(h_0 + \alpha\,r_m \cos\theta)^3}\,\ln(r) + k_2. \tag{7.5}$$

The slipper leakage through a generic radius will be:

$$Q_{\text{leakage}} = \int_0^{2\pi} \int_0^h u\,r\,dy\,d\theta. \tag{7.6}$$

Assuming Poiseulle flow, the velocity distribution is given by:

$$u = \frac{1}{\mu}\frac{dp}{dr}\frac{y}{2}(y - h). \tag{7.7}$$

Then, the leakage flow is given by:

$$Q_{\text{leakage}} = \int_0^{2\pi} \int_0^h \frac{1}{\mu}\frac{dp}{dr}\frac{y}{2}(y - h)\,r\,dy\,d\theta. \tag{7.8}$$

Substituting the pressure distribution versus radius, Eq. (7.4), into Eq. (7.8) and after some integration and rearrangement gives:

$$Q_{\text{leakage}} = \int_0^{2\pi} -\frac{1}{12\,\mu}\left(-3\,\mu\,\omega\,\alpha\,r_m \sin\theta\; r^2 + k_1\right) d\theta. \tag{7.9}$$

Now a second integration cannot be normally done at this stage because the unknown constant k_1 depends on the angular position θ. However, assuming that the flow and pressure distribution in the slipper pocket and groove behave in the same way as in a conventional land, then Eqs. (7.5) and (7.9) can be applied to each slipper land to give:

Slipper pocket: $r_0 < r < r_1$,

$$p_1 = -\frac{3\mu\omega\alpha r_{m1} \sin\theta r^2}{2(h_{01} + \alpha r_{m1} \cos\theta)^3} + \frac{k_1}{(h_{01} + \alpha r_{m1} \cos\theta)^3}\ln r + k_2, \tag{7.10}$$

$$Q_{\text{leakage1}} = \int_0^{2\pi} -\frac{1}{12\mu}(-3\mu\omega\alpha r_{m1} \sin\theta r^2 + k_1)d\theta, \tag{7.11}$$

$$r_{m1} = (r_1 + r_0)/2. \tag{7.12}$$

First land: $r_1 < r < r_2$,

$$p_2 = -\frac{3\mu\omega\alpha r_{m2} \sin\theta r^2}{2(h_{02} + \alpha r_{m2} \cos\theta)^3} + \frac{k_3}{(h_{02} + \alpha r_{m2} \cos\theta)^3}\ln r + k_4, \tag{7.13}$$

$$Q_{\text{leakage2}} = \int_0^{2\pi} -\frac{1}{12\mu}(-3\mu\omega\alpha r_{m2} \sin\theta r^2 + k_3)d\theta, \tag{7.14}$$

$$r_{m2} = (r_2 + r_1)/2. \tag{7.15}$$

Slipper groove: $r_2 < r < r_3$,

$$p_3 = -\frac{3\mu\omega\alpha r_{m3} \sin\theta r^2}{2(h_{03} + \alpha r_{m3} \cos\theta)^3} + \frac{k_5}{(h_{03} + \alpha r_{m3} \cos\theta)^3}\ln r + k_6, \tag{7.16}$$

$$Q_{leakage3} = \int_0^{2\pi} -\frac{1}{12\mu}(-3\mu\omega\alpha r_{m3} \sin\theta r^2 + k_5)d\theta, \tag{7.17}$$

$$r_{m3} = (r_3 + r_2)/2. \tag{7.18}$$

Second land: $r_3 < r < r_4$,

$$p_4 = -\frac{3\mu\omega\alpha r_{m4} \sin\theta r^2}{2(h_{04} + \alpha r_{m4} \cos\theta)^3} + \frac{k_7}{(h_{04} + \alpha r_{m4} \cos\theta)^3}\ln r + k_8, \tag{7.19}$$

$$Q_{leakage4} = \int_0^{2\pi} -\frac{1}{12\mu}(-3\mu\omega\alpha r_{m4} \sin\theta r^2 + k_7)d\theta, \tag{7.20}$$

$$r_{m4} = (r_4 + r_3)/2. \tag{7.21}$$

The boundary conditions necessary to determine the constants are:

$$
\begin{aligned}
r = r_0, & \quad p_1 = p_{inlet}, \\
r = r_1, & \quad p_1 = p_2, \quad Q_{leakage1} = Q_{leakage2}, \\
r = r_2, & \quad p_2 = p_3, \quad Q_{leakage2} = Q_{leakage3}, \\
r = r_3, & \quad p_3 = p_4, \quad Q_{leakage3} = Q_{leakage4}, \\
r = r_4, & \quad p_4 = p_{outlet}.
\end{aligned} \tag{7.22}
$$

For the slipper under study with just one groove, following substantial intermediate substitutions and manipulation, the pressure distribution is given by:

$$k_1 = \frac{p_{tank} - p_{inlet} - (1.5\mu\omega\alpha \sin\theta)\delta_1 - (3\mu\omega\alpha \sin\theta)\delta_2}{H},$$

$$\delta_1 = \frac{r_{m1}(r_0^2 - r_1^2)}{(h_{01} + \alpha r_{m1} \cos\theta)^3} + \frac{r_{m2}(r_1^2 - r_2^2)}{(h_{02} + \alpha r_{m2} \cos\theta)^3}$$

$$+ \frac{r_{m3}(r_2^2 - r_3^2)}{(h_{03} + \alpha r_{m3} \cos\theta)^3} + \frac{r_{m4}(r_3^2 - r_4^2)}{(h_{04} + \alpha r_{m4} \cos\theta)^3},$$

$$\delta_2 = \frac{r_1^2(r_{m2} - r_{m1})\ln(r_2/r_1)}{(h_{02} + \alpha r_{m2} \cos\theta)^3} + \frac{[r_1^2(r_{m2} - r_{m1}) + r_2^2(r_{m3} - r_{m2})]\ln(r_3/r_2)}{(h_{03} + \alpha r_{m3} \cos\theta)^3}$$

$$+ \frac{[r_1^2(r_{m2} - r_{m1}) + r_2^2(r_{m3} - r_{m2}) + r_3^2(r_{m4} - r_{m3})]\ln(r_4/r_3)}{(h_{04} + \alpha r_{m4} \cos\theta)^3},$$

$$H = \frac{\ln(r_1/r_0)}{(h_{01} + \alpha r_{m1} \cos\theta)^3} + \frac{\ln(r_2/r_1)}{(h_{02} + \alpha r_{m2} \cos\theta)^3}$$

$$+ \frac{\ln(r_3/r_2)}{(h_{03} + \alpha r_{m3} \cos\theta)^3} + \frac{\ln(r_4/r_3)}{(h_{04} + \alpha r_{m4} \cos\theta)^3}, \tag{7.23}$$

$$p_1 - p_{inlet} = \frac{k_1\ln(r/r_0) + 3\mu\omega\alpha r_{m1} \sin\theta (r_0^2 - r^2)/2}{(h_{01} + \alpha r_{m1} \cos\theta)^3}, \tag{7.24}$$

$$p_2 - p_{\text{inlet}} = k_1 \left[\frac{\ln(r_1/r_0)}{(h_{01} + \alpha r_{m1} \cos\theta)^3} + \frac{\ln(r/r_1)}{(h_{02} + \alpha r_{m2} \cos\theta)^3} \right],$$

$$+ 1.5\mu\omega\alpha \sin\theta \left[\frac{r_{m1}(r_0^2 - r_1^2)}{(h_{01} + \alpha r_{m1} \cos\theta)^3} + \frac{r_{m2}(r_1^2 - r^2)}{(h_{02} + \alpha r_{m2} \cos\theta)^3} \right], \quad (7.25)$$

$$+ \frac{3\mu\omega\alpha \sin\theta \, r_1^2 (r_{m2} - r_{m1})}{(h_{02} + \alpha r_{m2} \cos\theta)^3} \ln(r/r_1),$$

$$p_3 - p_{\text{inlet}} = k_1 \left[\frac{\ln(r_1/r_0)}{(h_{01} + \alpha r_{m1} \cos\theta)^3} + \frac{\ln(r_2/r_1)}{(h_{02} + \alpha r_{m2} \cos\theta)^3} + \frac{\ln(r/r_2)}{(h_{03} + \alpha r_{m3} \cos\theta)^3} \right],$$

$$+ 1.5\mu\omega\alpha \sin\theta \left[\begin{array}{c} \dfrac{r_{m1}(r_0^2 - r_1^2)}{(h_{01} + \alpha r_{m1} \cos\theta)^3} + \dfrac{r_{m2}(r_1^2 - r_2^2)}{(h_{02} + \alpha r_{m2} \cos\theta)^3} \\[2mm] + \dfrac{r_{m3}(r_2^2 - r^2)}{(h_{03} + \alpha r_{m3} \cos\theta)^3} \end{array} \right], \quad (7.26)$$

$$+ 3\mu\omega\alpha \sin\theta \left[\begin{array}{c} \dfrac{r_1^2(r_{m2} - r_{m1})}{(h_{02} + \alpha r_{m2} \cos\theta)^3} \ln(r_2/r_1) \\[2mm] + \dfrac{r_1^2(r_{m2} - r_{m1}) + r_2^2(r_{m3} - r_{m2})}{(h_{03} + \alpha r_{m3} \cos\theta)^3} \ln(r/r_2) \end{array} \right],$$

$$p_4 - p_{\text{inlet}} = k_1 \left[\begin{array}{c} \dfrac{\ln(r_1/r_0)}{(h_{01} + \alpha r_{m1} \cos\theta)^3} + \dfrac{\ln(r_2/r_1)}{(h_{02} + \alpha r_{m2} \cos\theta)^3} \\[2mm] + \dfrac{\ln(r_3/r_2)}{(h_{03} + \alpha r_{m3} \cos\theta)^3} + \dfrac{\ln(r/r_3)}{(h_{04} + \alpha r_{m4} \cos\theta)^3} \end{array} \right],$$

$$+ 1.5\mu\omega\alpha \sin\theta \left[\begin{array}{c} \dfrac{r_{m1}(r_0^2 - r_1^2)}{(h_{01} + \alpha r_{m1} \cos\theta)^3} + \dfrac{r_{m2}(r_1^2 - r_2^2)}{(h_{02} + \alpha r_{m2} \cos\theta)^3} \\[2mm] + \dfrac{r_{m3}(r_2^2 - r_3^2)}{(h_{03} + \alpha r_{m3} \cos\theta)^3} + \dfrac{r_{m4}(r_3^2 - r^2)}{(h_{04} + \alpha r_{m4} \cos\theta)^3} \end{array} \right],$$

$$+ 3\mu\omega\alpha \sin\theta \left[\begin{array}{c} \dfrac{r_1^2(r_{m2} - r_{m1})}{(h_{02} + \alpha r_{m2} \cos\theta)^3} \ln(r_2/r_1) \\[2mm] + \dfrac{r_1^2(r_{m2} - r_{m1}) + r_2^2(r_{m3} - r_{m2})}{(h_{03} + \alpha r_{m3} \cos\theta)^3} \ln(r_3/r_2) \\[2mm] + \dfrac{r_1^2(r_{m2} - r_{m1}) + r_2^2(r_{m3} - r_{m2}) + r_2^2(r_{m4} - r_{m3})}{(h_{04} + \alpha r_{m4} \cos\theta)^3} \ln(r/r_3) \end{array} \right].$$

$$(7.27)$$

These equations indicate the added complexity when just one groove is added compared with the no-groove classical solution derived in Chapter 3 and with slipper

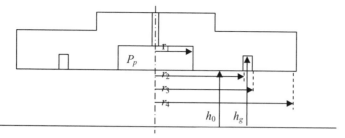

Figure 7.2. Uniform clearance condition with a single groove.

tilt. The equation for the leakage flow is found from:

$$Q_{\text{leakage}} = -\int_0^{2\pi} \frac{k_1}{12\mu} d\theta. \tag{7.28}$$

Because of the complexity of the integral, Eq. (7.28) must be integrated numerically to determine the leakage.

7.2.3 Simplification for the Nontilted Case, No-Rotation Condition

Before pursuing the effect of tilt in more detail, it is worth considering the implication of a groove addition for the slipper in its nontilted condition – a condition close to reality, as will be shown later. The previously developed equations are much simplified, particularly with no rotation, and have been discussed in Bergada and Watton (2002, 2005, 2008). For the nontilted condition, consider the simplified notation shown in Fig. 7.2.

Neglecting the inlet orifice area effect, the equations for the total force and leakage flow then become:

$$F_{\text{groove}} = P_p \pi r_1^2 \frac{\left[\dfrac{(r_2^2 - r_1^2)}{r_1^2} + \left(\dfrac{h_0}{h_g}\right)^3 \dfrac{(r_3^2 - r_2^2)}{r_1^2} + \dfrac{(r_4^2 - r_3^2)}{r_1^2}\right]}{2\left[\ln\left(\dfrac{r_2}{r_1}\right) + \left(\dfrac{h_0}{h_g}\right)^3 \ln\left(\dfrac{r_3}{r_2}\right) + \ln\left(\dfrac{r_4}{r_3}\right)\right]}, \tag{7.29}$$

$$Q_{\text{groove}} = \frac{\pi h_0^3 P_p}{6\mu} \frac{1}{\left[\ln\left(\dfrac{r_2}{r_1}\right) + \left(\dfrac{h_0}{h_g}\right)^3 \ln\left(\dfrac{r_3}{r_2}\right) + \ln\left(\dfrac{r_4}{r_3}\right)\right]}. \tag{7.30}$$

It will be recalled from Chapter 3 that the equations for the slipper without a groove are given by:

$$F_{\text{no groove}} = P_p \pi r_1^2 \frac{\left[\dfrac{(r_4^2 - r_1^2)}{r_1^2}\right]}{2\ln\left(\dfrac{r_4}{r_1}\right)}, \tag{7.31}$$

$$Q_{\text{no groove}} = \frac{\pi h_0^3 P_p}{6\mu} \frac{1}{\ln\left(\dfrac{r_4}{r_1}\right)}. \tag{7.32}$$

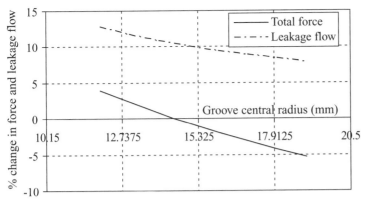

Figure 7.3. Change in force and leakage flow with a groove added (following work by Bergada JM, Watton J, Haynes JM up to 2008).

The change in force and leakage flow can then be expressed as a percentage of the no-groove case once a slipper geometry has been selected, and defined as follows:

$$\% \text{ change} = 100 \frac{(F_{\text{groove}} - F_{\text{no groove}})}{F_{\text{no groove}}} \quad \text{and} \quad 100 \frac{(Q_{\text{groove}} - Q_{\text{no groove}})}{Q_{\text{no groove}}}. \quad (7.33)$$

These percentages of changes are then independent of $P_p \pi r_1^2$ and $\pi h_0 p_p / 6\mu$. The slipper being discussed in this example has the following geometry:

Orifice radius, $r_1 = 1$ mm
Inner land inside radius, $r_2 = 10.15$ mm
Inner land outside radius, $r_3 = 14.7$ mm
Groove width $= 1$ mm, depth $= 0.8$ mm
Outside radius, $r_4 = 20.5$ mm

The result is shown in Fig. 7.3 for groove positions typically between 0.25 and 0.75 of the land width and for a constant clearance $h_0 = 21$ μm. Note from Eqs. (7.29) to (7.32) that the clearance has only a small effect on the percentage of change (7.33) because of the relatively large groove depth. ISO 32 mineral oil is assumed with $\mu = 0.032$ N s/m^2.

It can be seen that the use of a groove allows a small degree of design freedom in placing the groove to achieve the desired piston–slipper force balance and typically between +5% and −5% of the no-groove slipper force. However, this is at the expense of an increase in leakage flow, whatever the position chosen for the groove placement. The leakage flow increase is typically 10% and borne out by 3D CFD simulation, including groove-flow effects.

7.2.4 Experimental Method

To experimentally validate the equations developed, two test rigs were built: the first shown in Fig. 7.4 to hold a commercial piston–slipper unit of the type shown in Fig. 7.1. This *first test rig* was designed to measure the pressure distribution across the slipper and for a constant clearance.

Slipper data are as follows:

Orifice radius, $r_1 = 0.5$ mm
Inner land inside radius, $r_2 = 5$ mm

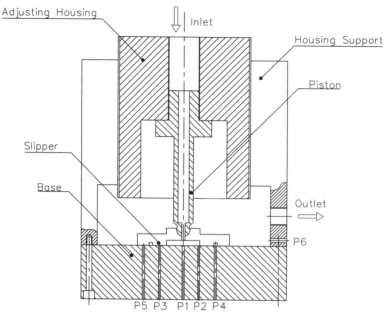

Figure 7.4. Test rig to measure the leakage and pressure distribution.

Inner land outside radius, $r_3 = 7.4$ mm
Groove width $= 0.4$ mm
Outside radius, $r_4 = 10.2$ mm
$h_{01} = h_{02} + 0.65$ mm and $h_{03} = h_{02} + 0.4$ mm

A micrometer-gauge thread was machined on the adjuster, allowing known clearances to be set; but, before the results are compared with theory, it is essential to determine the actual clearance as pressure is applied. This is due to the small yet significant compression between the adjusting housing and the housing support fine thread adjuster, the net result being that the actual clearance increases with applied pressure. The thread compression was measured with a precision position transducer mounted to the bed plate holding the test unit, and it was found that the compression increased to 4 μm as the pressure increased to 160 bar. The accuracy of the displacement transducer used to measure the relative displacement between the adjusting housing and the housing support was determined as ±0.25 μm.

Pressure tappings in the base unit then allowed the pressure distribution to be measured across one axis of the slipper, using calibrated test Bourdon gauges, and including the groove. A typical set of leakages is shown in Fig. 7.5 and a typical set of pressure distributions is shown in Fig. 7.6.

The leakage flow comparisons of Fig. 7.5 are good and support the developed theory. The comparisons for pressure, Fig. 7.6, use pressure tappings created by drilling ostensibly 0.3-mm-diameter holes in the base, and the pressure drop over this distance is 12 bar at an inlet pressure of 160 bar. The exact location of the pressure tappings with respect to the slipper cannot be precisely measured for the assembled test unit. In addition, any variation in the set clearance or any induced tilt during testing cannot be determined. It is proposed for this test rig that the experimental error for pressure measurement be ±6 bar at the highest inlet pressure used of 160 bar. Figure 7.6 shows that the comparison between theory and measurement is good for the inner land but, with experimental measurements, lower than predicted

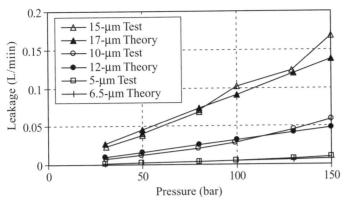

Figure 7.5. A comparison of measured and computed leakages (following work by Bergada JM, Watton J, Haynes JM up to 2008).

for the outer land, particularly at the highest pressure. A displacement error of 0.3 mm for the pressure tapping position in this region would explain the difference.

The *second test rig* is shown in Fig. 7.7 (Haynes, 2007). The test-rig slipper dimensions for the second slipper are as follows:

Orifice radius, $r_1 = 1$ mm
Inner land inside radius, $r_2 = 10.15$ mm
Inner land outside radius, $r_3 = 14.7$ mm
Groove width $= 1$ mm
Outside radius, $r_4 = 20.5$ mm
When $\alpha = 0$, then $h_{01} = h_{02} + 1.4$ mm and $h_{03} = h_{02} + 0.8$ mm

This second test rig allows rotation of the swash plate to simulate tangential velocity effects while allowing slipper tilt to be set. Three position sensors having a measurement accuracy of better than 0.25 µm are attached to the slipper at 120° intervals. These sensors require a nonferrous measuring face for optimum performance, the swash-plate assembly is manufactured from aluminum, and the slipper assembly is manufactured from stainless steel. The slipper is held in position by four screws, and the required slipper tilt orientation is achieved by adjusting four additional positioning screws. Four holes, 0.5-mm diameter and at every 90°, were drilled at the center of the slipper groove, allowing measurement of the pressure inside the groove at its

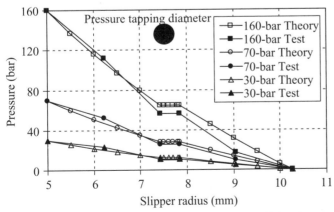

Figure 7.6. Pressure distribution across the slipper (following work by Bergada JM, Watton J, Haynes JM up to 2008).

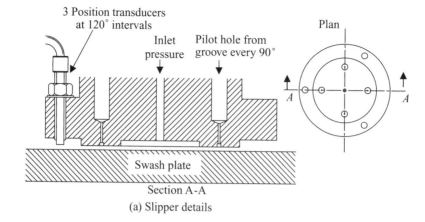

(a) Slipper details

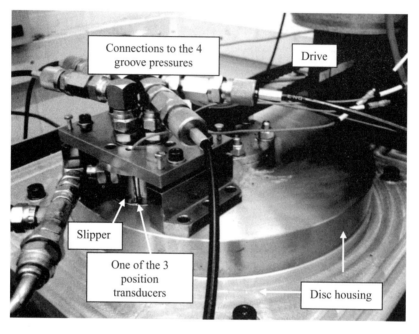

(b) Test rig assembly

Figure 7.7. Test rig with variable-speed swash-plate drive (Haynes, 2007).

four cardinal positions. In this presentation, variable-speed tests are performed with a single central clearance of 15 μm and for a single tilt of 0.035°. A set of swash-plate turning speeds are studied in the range 0–1350 rpm, the maximum turning speed corresponding to a tangential velocity on the slipper main axis of 13 m/s. It is important to realize that the exact clearance between the slipper and swash plate may need further consideration, depending on the quality of the *surface finish* on each face.

Figure 7.8 shows some typical surface measurements. The surface roughness of both the slipper and rotating disk were measured with a Talysurf machine (Taylor-Hobson Ltd., UK).

The measurements across the land of the slipper were 4 mm in length and 40 mm long across the swash plate, reflecting the diameter of the slipper contact area. The average *Ra* values from a series of tests were found to be 0.4 μm for the

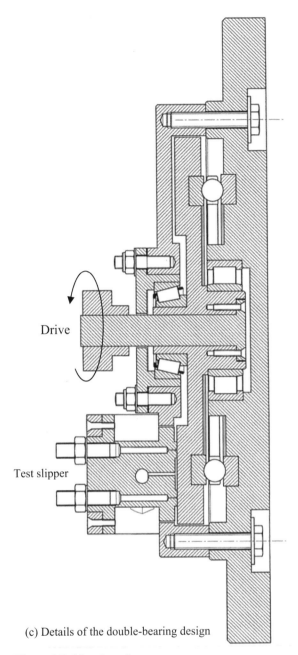

(c) Details of the double-bearing design

Figure 7.7. (*Continued*)

stainless steel slipper and 0.56 μm for the aluminum disk. This suggests that the measured clearance should be increased by typically 1 μm to reflect the combined surface finish effect.

In addition, it is impossible to manufacture such a test rig with alignment–distortion errors, part of which may develop during operation because of *disk run-out*. Although the manufacturing was to an extremely high standard, it was observed during dynamic experiments that both slipper–plate mean clearance and run-out amplitude change with operating conditions. A knowledge of the true clearance is essential for this type of testing because of the (clearance)^3 effect on leakage flow

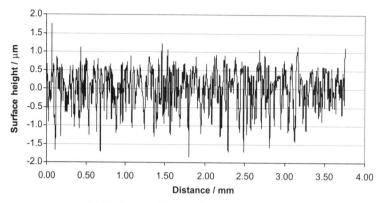

(a) Surface profile of stainless-steel slipper land

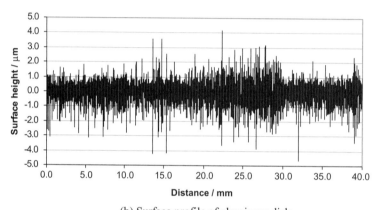

(b) Surface profile of aluminum disk

Figure 7.8. Measured surface characteristics of the slipper and swash plate.

rate, particularly at small clearances. A disk running with a sinusoidal run-out displacement superimposed on a mean clearance gives the appearance of a mean clearance increase when leakage mean flow rate is considered over one cycle.

The test conditions considered should be put into context with those that would probably exist for the test slipper when used in a real pump application – for example with the following conditions:

- a pressure distribution across the slipper that is approximated by an equivalent logarithmic decay passing through the center of the groove
- a pocket pressure marginally different from the pump pressure
- a swash-plate angle of 20°
- a maximum pump pressure of 350 bar
- an ISO 32 mineral oil with a viscosity $\mu = 0.032$ N s/m^2

The maximum hydrostatic force generated on this slipper is then 23 kN. The force balance across the slipper and piston is determined by the pump manufacturer, perfect force balance occurring for a piston diameter of 28.9 mm. If an additional hydrodynamic force is required, then this will not be greater than typically 5% of the hydrostatic force and is based on well-established design knowledge. There is no explicit theory for determining the hydrodynamic lift for a circular slipper, but a

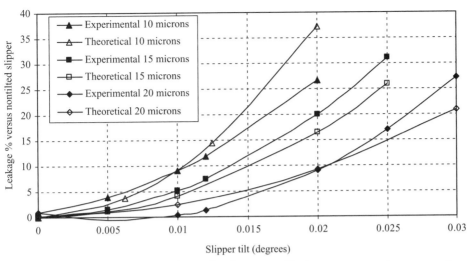

Figure 7.9. A comparison of experimental and computed leakages, expressed as a percentage of the nontilted slipper case (Bergada, Haynes, Watton, 2008).

good approximation can be made by using "equivalent" square plain bearing theory with side leakage effects taken into account, as discussed in Chapter 3. It is further assumed that:

- a square plain bearing of equivalent area, 36.3 mm × 36.3 mm, applies,
- the bearing central clearance $h_0 = 10$ μm,
- a side leakage compensating factor of 0.44 applies, and
- a tangential velocity of 13 m/s still applies.

The bearing tilt is then calculated to be equivalent to a 0.26-μm increase from the trailing edge to the leading edge. This gives a square bearing tilt angle of 0.00041° and therefore smaller by a factor of 12.2 compared with the minimum nonzero value of 0.005° that was set in the tests. Even if all the slipper lift force was created by hydrodynamic effects, a condition that would not normally occur with a nonblocked slipper orifice, the tilt angle would still be only 0.0079°.

7.2.5 Some Results with Slipper Tilt Included and for No Rotation

When leakage is represented as a percentage of the nontilted slipper, it is found that for a given central clearance, all the different curves can be brought together. Figure 7.9 presents the trend curve for all the central clearances studied, which are compared with the theoretical predictions.

It can be seen that a good agreement is found, especially at the very low tilts that exist in practice. From these results, it can be stated that leakage percentage increase versus a nontilted slipper is mostly independent of the inlet pressure. Nevertheless, it has been found experimentally that as the inlet pressure increases, the percentage of increase in the trend line curve tends to slightly increase beyond the theoretical predictions.

It must be stated at this point that it is extremely difficult to obtain consistent leakage measurements in fluid power components such as a slipper because

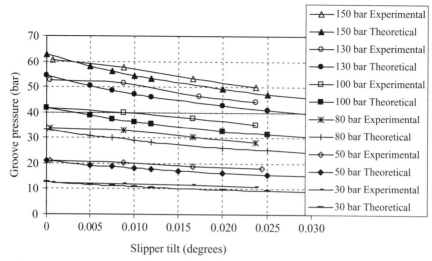

Figure 7.10. Groove pressure variation with tilt for different piston pressures, 10-μm central clearance (Bergada, Haynes, Watton, 2008).

of unknown local temperature changes and the sensitivity that is due to the well-known (clearance)[3] effect. Results presented here represent the average of many hours of repeated testing and are good indicators of a practical trend.

The equations presented are capable of predicting the pressure at all points below the slipper, and Fig. 7.10 shows just one set of results for a central clearance of 10 μm.

Because of the consideration of radial flow, the theoretical pressure differential inside the slipper groove is slightly higher than what has been found experimentally. In fact, the experiments have revealed that the pressure inside the groove is mostly constant for the set of tilts and central clearances studied. Theoretically, the pressure inside the groove decreases for a tilted slipper as the slipper clearance increases, and it can be said that the minimum theoretical pressure is the most likely to appear in reality. A well-designed groove geometry allows flow from the theoretical groove high-pressure points to move almost instantaneously toward the groove theoretical low-pressure points, thus equalizing the pressure within the groove. It can then be concluded that for the groove studied, a rate of momentum exchange exists between fluid particles at the top of the groove.

7.2.6 The Effect of Tilt and Rotation, Measurement, and CFD Simulation

Finally, consider the effect of swash-plate rotation on slipper behavior. The swash plate was able to be turned at different rotational speeds between 0 and 1350 rpm. This is opposite to the real pump method of operation using a fixed swash plate, but the effect is the same. A tilt of 0.035° is studied here, and the clearance, when corrected for surface finish and disk run-out, is 21 μm. Recalling previous work in this section, it should be noted that the tilt being considered to highlight changes in performance is much greater than will probably exist in practice. It is not possible to obtain an explicit solution for the flow through the slipper with tilt and rotation, and the use of a CFD simulation is invaluable to gain a further insight into the flow and pressure distribution. Figure 7.11 shows the grid at the centerline defined along the *XX* axis of Fig. 7.1 and for only a small part of the groove section.

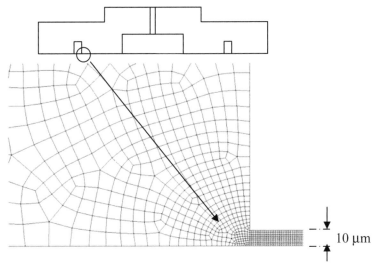

Figure 7.11. CFD grid at the centerline of the slipper and at part of the groove (analyzed by Worthing and Watton at Cardiff University in 2008).

The theory presented earlier, and for no slipper tilt, is compared with the 3D CFD prediction as shown in Fig. 7.12, the two sets of results being indistinguishable from each other.

The CFD result for the complete pressure distribution across the same XX axis with tilt only is shown in Fig. 7.13. The tilt of 0.035° is equivalent to a gap of 33.5 μm at the leading edge and 8.5 μm at the trailing edge. It can be seen from Fig. 7.13 that at zero speed, there is a slight decrease in the groove pressure with a slight distortion of the pressure profile across the two lands in the XX plane.

As the turning speed is increased, the groove pressure significantly increases, and Fig. 7.14 shows the pressure profile for a turning speed of 125 rad/s (1195 rpm). It can be seen that there is now a more noticeable pressure drop around the groove,

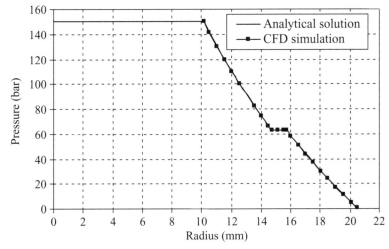

Figure 7.12. A comparison of 3D CFD simulation and theory, no tilt, central clearance 21 μm, 150 bar (analyzed by Worthing and Watton at Cardiff University in 2008).

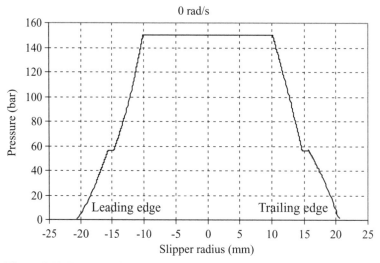

Figure 7.13. Pressure distribution across the *XX* axis using CFD simulation, central clearance 21 μm, 150 bar, tilt 0.035°, zero speed. (analyzed by Worthing and Watton at Cardiff University in 2008).

the leading-edge pressure distribution shape has barely changed, and the trailing-edge pressure distribution shape is significantly changed because of the hydro-dynamic effect of tilt. The 2D slice through the fluid shows that the pressure peaks on the two faces toward the trailing edge are actually focused around a very small area; but, clearly, a net moment at the slipper is created with an increase in total force generated.

This first study shows what might be intuitively expected:

- An increased turning speed increases the groove mean pressure in the presence of a significant slipper tilt.
- A small moment exists across the slipper in a direction to reduce tilt in the presence of a significant slipper tilt and for practical tilts.
- The leakage flow reduction and the groove mean pressure increase will probably be negligible for practical slipper tilts.

Determining the leakage flow rate experimentally is straightforward, as deduced from earlier work on the slipper with no tilt, but the mean clearance and run-out amplitude vary with speed and pressure. CFD results suggest that leakage flow barely changes with increasing speed, assuming a fixed clearance. Experimental data, following many repeatable tests and with an unadjusted clearance over the range of the tests, are shown in Fig. 7.15 for a clearance set at 15 μm. Results for a tilted slipper and a nontilted slipper are compared.

The tilt angle is 0.03°, much larger than will exist in practice to emphasize the effect, and results in a slipper leading-edge height of 25.7 μm and a trailing-edge height of 4.3 μm. The mean clearance does vary slightly with speed and pressure because of run-out amplitude effects. These variations, of course, become more significant at lower nominal clearance settings. From Fig. 7.15, it can be seen that leakages slightly increase with tilt for a given pressure. The effect of turning speed is

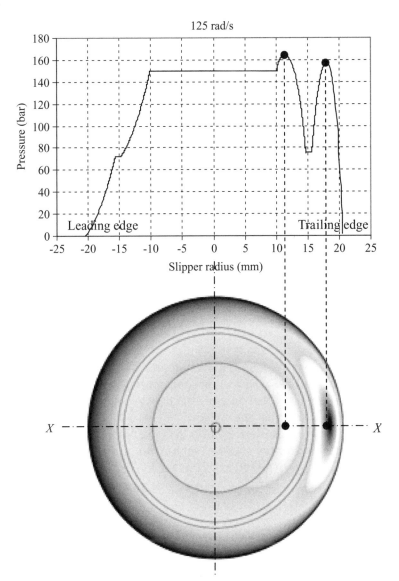

Figure 7.14. Pressure distribution across the XX axis using CFD simulation, central clearance 21 μm, 150 bar, tilt 0.035° 125 rad/s (analyzed by Worthing and Watton at Cardiff University in 2008).

negligible for low pressures, but leakage flow slightly increases at the highest pressure when tilt is present.

These results show the importance of accurate measurements of localized temperature and clearance when interpreting experimental data. Considering the inverse (clearance)^3 effect on leakage flow, combined with the entrained flow that is due to tangential velocity, the very small changes expected from theoretical considerations might be masked by small geometrical distortions in practice because of both pressure effects and run-out. The general conclusion, supported by further work not reported here, is that speed and tilt have only a small effect on slipper leakage for the particular slipper configuration considered with a single groove.

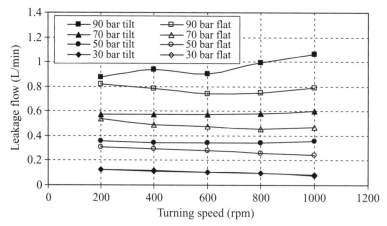

Figure 7.15. The effect of turning speed and pressure on leakage flow for a nominal clearance of 15 μm, tilt 0.03° (Bergada, Haynes, and Watton, 2008; Kumar, Bergada, and Watton, 2009).

7.3 Modeling a Forge Valve and Its Application to Press Cylinder Control

7.3.1 Introduction

Further details of this work may be found in Watton and Nelson (1993). Figure 7.16 shows the control system of the type used in some forging-press applications, whereby main cylinder motion is created by using the flow bypass valve to control flow out of the main line between pump and cylinder.

With the flow bypass valve (also termed a *forge valve*) closed, the cylinder extends with the return relief setting the back pressure. At the same time, the accumulator is charged to assist retracting motion of the main cylinder when the flow

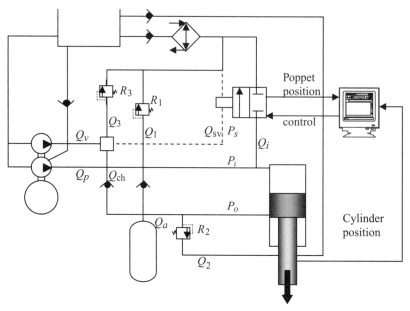

Figure 7.16. Press control circuit.

(a)

(b)

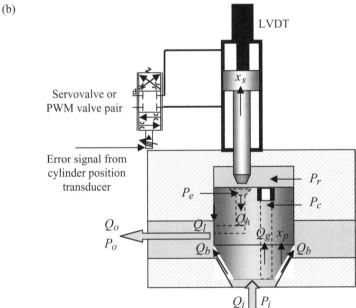

Figure 7.17. Flow bypass valve test system and a schematic of the flow bypass valve with follow-up positioning of the main poppet.

bypass valve is opened. In this way, the cylinder may be "bounced" during the forging cycle. To achieve better control of the press cylinder, closed-loop position control is desirable, and this study considers the influence of flow bypass valve design on the closed-loop dynamic performance. In addition, the servovalve used is compared with a pair of digitally switched PWM solenoid valves. The flow bypass valve used is shown in Fig. 7.17 with the PWM-controlled solenoid valves in position. Positioning of the main poppet is achieved by closed-loop position control of the control cylinder spindle using an integral transducer coupled to the spindle. Movement

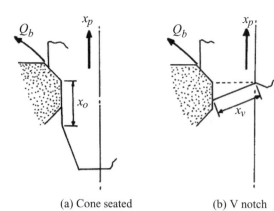

Figure 7.18. Cone-seated and V-notch poppet designs.

(a) Cone seated   (b) V notch

of the control spindle from its normally closed position allows the back pressure P_r to collapse to atmospheric pressure conditions, resulting in rapid motion of the main poppet. Precise flow across the main poppet therefore can be achieved by this follower-type characteristic. Position control of the small-size control cylinder spindle is achieved normally with an electrohydraulic servovalve; but, in this study, PWM switching of a pair of fast-acting valves is also compared.

7.3.2 Developing the Component Equations

Flow Bypass Valve Poppet Design

Two poppet designs are considered, the original cone-seated design and an alternative design proposed and referred to as a V-notch design. The flow-rate paths are quite different for each poppet design, as may be deduced from the poppet schematics shown in Fig. 7.18.

Detailed flow measurements are required for each poppet design to establish equations required for computer analysis. During operation of the flow bypass valve, a pilot control flow is needed; dynamic simulation results presented later show that a small constant displacement occurs between the control spindle and poppet because of the appropriate force balance requirement. Hence, any flow measurements made with either poppet in situ will include a combination of leakage flow, true port flow, and pilot control flow. A series of measurements is therefore required with and without pilot control flow.

For the original cone-seat design, the poppet must be displaced by an amount x_0 before the valve is finally open. During this preopening phase, the small radial clearance between the poppet and the bush provides a laminar-flow-type leakage path initially of constant axial length followed by a decreasing length until the port is finally open. For the V-notch design, the poppet is machined such that there is no region of overlap. Hence, any poppet movement will result in port opening, thus creating the appropriate flow-control area. The design for the V notch is such that any initial laminar-flow-type leakage should be much less than the original cone-seated design and negligible when compared with the flow rate across the main control port.

A comparison of the steady-state flow characteristics is shown in Fig. 7.19 for the flow bypass valve in its operating mode.

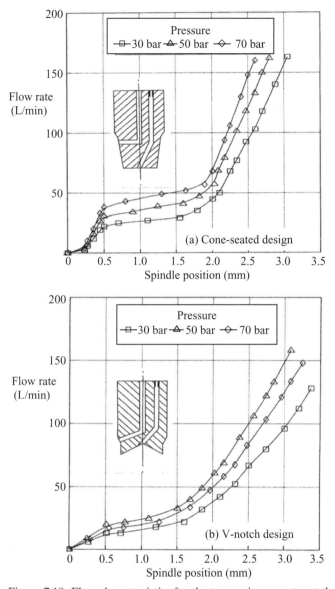

Figure 7.19. Flow characteristics for the two main poppet seat designs.

The displacement of the control spindle is obtained from a position transducer fitted to the top of the valve, with the flow rate being obtained from a positive-displacement flow meter. The results for the cone-seated poppet at small displacements show a combination of leakage and pilot control flow in which the laminar leakage contribution increases as the poppet lifts and theoretically falls to zero as the region of overlap is passed. As the poppet lifts farther, the port flow becomes dominant, the flow characteristic eventually having the expected Bernoulli form. The results for the V-notch poppet appear to be superior to the original cone-seated design because the nonlinearity at poppet displacements up to typically 2 mm is significantly improved because of the anticipated drop in leakage flow. From a feed-back control point of view, the minimization of nonlinearities is desirable, and the comparisons at first sight suggest that the V-notch design would be the preferred choice.

Detailed dynamic simulation results show that the steady-state constant displacement between control spindle and poppet occur so rapidly in relation to typical closed-loop response times that the steady-state flow characteristics presented in Fig. 7.19 may be used for simulation purposes.

Main Poppet Flow Characteristic, Cone-Seated Design:

$$Q_b = \left(k_0 + k_1 x_p^2\right) P_i^{0.775}, \quad 0.25\,\text{mm} \le x_p \le 1.66\,\text{mm},$$

$$Q_b = C_q A_b \sqrt{\frac{2P_i}{\rho}}, \quad x_p > 1.66\,\text{mm}, \tag{7.34}$$

$$A_b = 2\pi \left[r_i - \frac{\tan\theta}{2}(x_p - x_o) \right](x_p - x_o)\sin\theta, \quad x_p > x_o.$$

Main Poppet Flow Characteristic, V-Notch Design:

$$Q_b = C_q A_b \sqrt{\frac{2P_i}{\rho}},$$

$$A_b = 4r_i \left\{ x_p \sin^{-1}\left(\frac{x_p}{x_v}\right) + x_r \left[\sqrt{1 - \left(\frac{x_p}{x_v}\right)^2} - 1 \right] \right\}. \tag{7.35}$$

Determination of the pilot flow rate requires detailed modeling of the poppet design, including the pressure drops down each small-diameter line together with the poppet side leakage and the choke flow characteristic. This may be done with reference to the notation shown in Fig. 7.17, and the various equations used are listed as follows.

Pilot Flow Rates:

$$Q_g = \frac{(P_i - P_c)}{R_{fg}}, \quad Q_h = \frac{P_e}{R_{fh}}, \quad Q_\ell = \frac{p_r}{R_{f\ell}},$$

$$Q_g = C_q \pi r_c^2 \sqrt{\frac{2(P_c - P_r)}{\rho}}, \quad Q_h = C_q 2\pi r_s (x_s - x_p) \sqrt{\frac{2(P_r - P_c)}{\rho}}. \tag{7.36}$$

The solution for P_e is then readily determined to give:

$$P_e = \frac{D_k^2 R_{fh}^2}{2}\left(\sqrt{1 + \frac{4P_r}{D_k^2 R_{fh}^2}} - 1 \right),$$

$$D_k = C_q A_h \sqrt{\frac{2}{\rho}}, \quad A_h = 2\pi r_s(x_s - x_p). \tag{7.37}$$

The Flow Back to Tank:

$$Q_o = Q_\ell + Q_h + Q_b. \tag{7.38}$$

Because the flow bypass valve operates in a dynamic mode in practice, the steady-state flow characteristics previously derived must be combined with the appropriate dynamic characteristics. It is therefore necessary to consider the relevant flow-continuity and force equations.

Force Balance for the Main Poppet:

$$P_i A_{bn} - P_r A_r = F_b - F_s + B_v \frac{dx_p}{dt} + M_p \frac{d^2 x_p}{dt^2},$$

$$\text{lower face, } F_b = 2C_q \cos\theta A_b(P_i - P_o), \tag{7.39}$$

$$\text{upper face, } F_s = 2C_q \cos 69 A_h(P_r - P_e),$$

$$A_{bn} = \pi[r_i - (x_p - x_o)\tan\theta]^2.$$

Flow Continuity at the Main Poppet Upper Face:

$$Q_g - Q_h - Q_\ell = -A_r \frac{dx_p}{dt} + A_s \frac{dx_s}{dt} + \frac{V_r}{\beta_e}\frac{dP_r}{dt}, \tag{7.40}$$

Control Spindle Motion Approximation

The motion of the control spindle is determined by making a number of realistic assumptions regarding the dominant dynamic components. Because of the small cylinder volumes and mass in comparison with other components, it is realistic to neglect compressibility and inertia effects. The industrial servovalve used has a well-defined flow characteristic, the manufacturer's data being confirmed by experimental testing:

$$\frac{dx_s}{dt} \approx \frac{k_f i}{A_2}\sqrt{\frac{P_s - \alpha x_s - P_k - P_f}{(1+\gamma)^3}}, \quad \frac{dx_s}{dt} > 0,$$

$$\frac{dx_s}{dt} \approx \frac{k_f i}{A_2}\sqrt{\frac{P_s + \alpha x_s + P_k + P_f}{(1+\gamma)^3}}, \quad \frac{dx_s}{dt} < 0, \tag{7.41}$$

$$\gamma = \frac{A_1}{A_2}, \quad \alpha = \frac{k_s}{A_2}, \quad P_k = \frac{F_k}{A_2}, \quad P_f = \frac{F_s}{A_2},$$

where k_s is the weak retaining spring stiffness, F_k is the precompression force exerted by the spring, and F_s is the flow-reaction force. Equation (7.41) does assume that servovalve dynamics also may be neglected such that the application of the servocurrent i causes instantaneous and proportional positioning of the spool. This has been found again to be sufficiently accurate in terms of significant modeling trends. The manufacturer's data suggest a second-order transfer function approximation for this servovalve, having a damping ratio of typically unity and an undamped natural frequency of typically 140 Hz. This characteristic may be easily added to the theory presented, if required.

7.3.3 Developing the System Equations

Additional components now to be considered are:

- pumps
- relief valves
- check valves
- press cylinder and accumulator

The pressure–flow characteristics of the pumps, relief valve, and check valves were obtained from the test rig by use of positive displacement and turbine flow meters positioned in the appropriate lines. All the components tested had linear pressure–flow characteristics apart from the pilot relief valve attached to the servovalve supply line. Consideration of the complete set of flow data then leads to the following equations:

$$\text{vane pump, } Q_v = Q_{vo} - \frac{P_s}{R_v},$$

$$\text{axial piston pump, } Q_p = Q_{po} - \frac{P_i}{R_p},$$

$$\text{two-stage PRV, } Q_1 = \frac{(P_i - P_{vr})}{R_1}, \quad P_i > p_{vr},$$

$$\text{single-stage PRV, } Q_2 = \frac{(P_o - P_{vr})}{R_2}, \quad P_o > P_{vr}, \qquad (7.42)$$

$$\text{check valves, } Q_{ch} = \frac{P_i - (P_o + P_{vc})}{R_c}, \quad P_i > P_o + P_{vc},$$

$$\text{pilot relief valve, } Q_3 = \frac{(P_s - 0.53 P_{vr})}{k_3 \sqrt{P_{vr}}}, \quad P_s < 0.53 P_{vr},$$

$$Q_3 = \frac{(P_s - P_{vr})}{R_3}, \quad P_s \geq 0.53 P_{vr}.$$

The unusual flow characteristic of the pilot relief valve is due to the poppet seat design, and it should be noted that the valve does not normally operate at low flow rates in practice. Measured steady-state characteristics for the relief valves and check valve are given in Watton and Nelson (1993). Dynamics of each valve may be added as a second-order effect refinement and taken from either the manufacturer or in-house tests (Watton, 1988, 1989).

Flow Continuity:

$$Q_p - Q_1 - Q_i = A_c \frac{dx_c}{dt} + \frac{V_i}{\beta} \frac{dP_i}{dt},$$

$$\qquad (7.43)$$

$$Q_2 + Q_a - Q_{ch} = A_o \frac{dx_c}{dt} - \frac{V_i}{\beta} \frac{dP_o}{dt},$$

where transmission line dynamics have been neglected together with actuator leakage. Also, the check-valve flow rate Q_{ch} will exist only if transient pressure differential conditions allow the check valve to open. In normal operation, the check valve should always remain closed. The existence of check-valve flow rate will require knowledge of the vane pump circuit dynamics, the appropriate equation being:

$$Q_v - Q_{ch} - Q_3 - Q_{sv} = \frac{V_r}{\beta} \frac{dP_s}{dt}. \qquad (7.44)$$

Load Cylinder Force Equation:

$$P_i A_i - P_o A_o = F_\ell + B_a \frac{dx_c}{dt} + F_e + F_f \, \text{sign} \left(\frac{dx_c}{dt} \right) + M_\ell \frac{d^2 x_c}{dt^2}. \qquad (7.45)$$

In this model, F_ℓ is the vertically acting load force including the piston and rod mass effect, B_a is the viscous damping coefficient, F_e is an end-stop spring stiffness contribution, F_f is the sign-of-velocity-dependent stiction–friction characteristic, and M_ℓ is the total load mass including fluid volume effects. The viscous damping coefficient is varied as the cylinder enters the cushioned end zone of its stroke and the end-stop force F_e is assumed to have a linear spring characteristic with a very high stiffness when the rod becomes fully extended or retracted.

These variable characteristics are simply adjusted in the simulation to give a realistic performance at the fully extended or retracted position. Under normal position control, these conditions should not occur. A variety of stiction–Coloumb friction characteristics were introduced into the simulation using typical manufacturer's data. However, the effect on closed-loop position control was found to be insignificant, and this aspect of modeling was not considered further.

Accumulator
The accumulator used had a 4-UK-gal capacity and was of the precharged rubber-bag type with a spring-loaded poppet valve to prevent bag distortion when discharging. Such an accumulator tends to be modeled under the assumption that the compressibility of the fluid used is negligible in comparison with that for the gas. Also, a polytropic exponent in the expansion equation is assumed to vary between 1.0 and 1.4, depending on the assumption of an isothermal process or an adiabatic process. However, as discussed earlier, in real dynamic systems, the exponent may be higher than 1.4, and for this study the following equations were used:

$$PV^n = A_c, \quad n = 1.55. \tag{7.46}$$

The constant A_c is determined from the precharge pressure P_{ao} and the initial gas volume V_a. It is assumed that the flexible bag may be considered as a zero-mass and frictionless piston and that the oil and gas pressures are dynamically equal. Hence, also assuming a rigid flask and neck gives:

$$V_g = \left(\frac{A_c}{P_a}\right)^{1/n}, \quad V_f = V_a - V_g,$$

$$Q_a = C_q A_a \sqrt{\frac{2|P_o - P_a|}{\rho}} \, \text{sign}(P_o - P_a). \tag{7.47}$$

This analysis ignores the effect of the spring-loaded poppet valve at the accumulator inlet because its inclusion was found to be insignificant. The weak spring used allows the valve to rapidly open, thus allowing the fixed orifice to dominate the flow characteristic throughout the compression and decompression cycle. For pressures greater than the precharge pressure P_{ao}, the accumulator pressure is determined from the flow-continuity equation as follows:

$$Q_a = \frac{V_f}{\beta}\frac{dP_a}{dt},$$

$$\frac{V_a}{\beta} = \frac{V_f}{\beta_o} + \frac{V_g}{\beta_g}, \quad \beta_g = \frac{(P_a + 10^5)}{\gamma_g}. \tag{7.48}$$

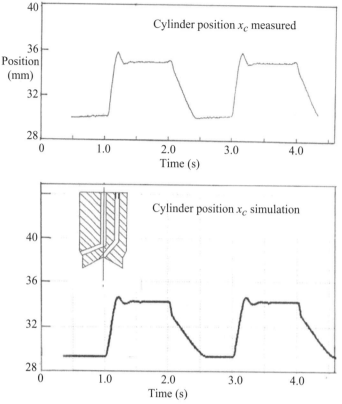

Figure 7.20. A comparison of transient responses for the main actuator and using a V-notch main poppet design.

7.3.4 System Dynamics for Closed-Loop Control

For closed-loop press control, position feedback from the main cylinder is obviously used together with closed-loop control of the control spindle position. The set of equations were modeled with Xanalog simulation software, a powerful icon-based tool at the time that has been further developed. Step demand changes of 5 mm were used at a frequency of 2 Hz, and results are shown in Figs. 7.20 and 7.21. It can be seen that the main actuator displacements are similar within typically ± 0.25 mm.

From the results, there is a slightly greater overshoot in practice than predicted by theory. This is probably caused by fine-detail modeling effects of system dynamics in the valves and accumulator. Actuator oscillation settles down more rapidly in practice than the theory predicts; this is caused by higher damping effects in the real system than are allowed for in the model. This is a common problem in fluid power system modeling and is likely to be due to assumptions made about the poppet damping characteristics, coupled with secondary resistance effects. On the actuator return stroke, the model shows two distinct regions, where actuator velocity is initially fast and then where it decreases after a short time. Although the experimental work shows a similar characteristic, it is much less pronounced. The initial movement is thought to be largely attributable to the effect of the accumulator discharging; after it has fully discharged, the actuator is moved solely by the flow rate

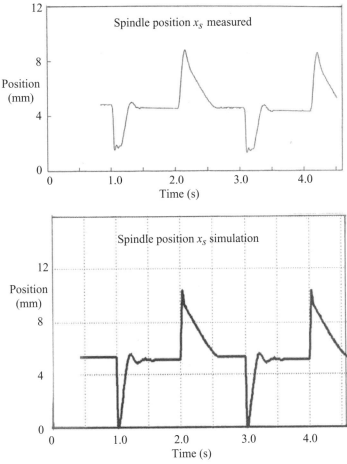

Figure 7.21. A comparison of control spindle transient responses using a V-notch main pop-pet design.

produced by the pilot system vane pump. It is therefore likely that the accumulator discharge characteristic is slightly different from that modeled.

What is most interesting, although not shown, is that changing the main-stage poppet from cone-seated to V-notch resulted in indistinguishable differences between both the measured and simulation results. This is pursued in more detail in Watton and Nelson (1993). Because the two main-stage poppet designs were developed from radically different concepts, the similarity in performance between the two designs is surprising, particularly when the steady-state flow characteristics are considered. Also noticeable is that the poppet seems to be slightly open at all times, unlike the simulation prediction. Secondary fluid inertia effects may cause the poppet to open and close more slowly than predicted. The fact that the valve does not close is important because it is clear from both experimental and modeling results that the flow-bypass-valve flow characteristic, around the closed position, is not particularly important when operating under dynamic conditions. It does explain why the cylinder closed-loop position response is not significantly affected by the change in bypass-valve poppet design.

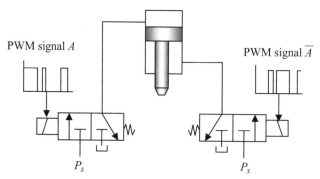

Figure 7.22. PWM switching concept for flow-bypass spindle position control.

7.3.5 The Use of PWM Control of a Pair of Fast-Acting Solenoid Valves

The servovalve controlling the bypass-valve control spindle is now replaced with a pair of fast-acting three-port, two-position solenoid valves, as shown in Fig. 7.22 and on the test equipment in Fig. 7.17.

The solenoid valves have switch-on and switch-off times of typically 5 ms and 2 ms, as reported by the manufacturer. They have a rated flow rate of 3 L/min at a pressure of 70 bar and are adequate for this particular application. A drive frequency of 50 Hz was used, and a series of tests was necessary to optimize the performance. Figure 7.23 shows some results.

The cylinder transient response is optimized by using the original servovalve and the result obtained using digital valves with 0.5-mm diameter chokes in each line. The latter is not optimized in the sense that choke diameter could be refined to improve the response. The PWM control approach used the same drive electronics used in Chapter 6.6.1 with pulse width set by the position error. It is worth noting that:

• Without chokes, the position response had noticeable yet acceptable noise contribution; there was no overshoot but there was undershoot.
• With a choke on the return line only, the noise effect was still present, and the overshoot and undershoot were the same as shown in Fig. 7.23, which has chokes in both lines.

7.4 The Modeling and Control of a Vehicle Active Suspension

7.4.1 Introduction

A car suspension incorporating a previously used Lotus UK active actuator and a TVR UK suspension–wheel unit is studied both experimentally and analytically. An emphasis is placed on hydraulic modeling using a series of transfer functions linking the hydraulic and suspension components. This is significantly aided by use of a Moog programmable Servocontroller (PSC) for nonlinear control implementation. The system equations are developed using linear state-space theory, and a suitable form is proposed for further design studies. It is shown that the hydraulic components significantly contribute to the system dynamics and, hence, cannot be neglected when control schemes are formulated. In particular, the significance of hydraulic bulk modulus on dynamic performance is evaluated, and the importance of accurately determining all components of velocity-type damping is highlighted.

Position (cm)

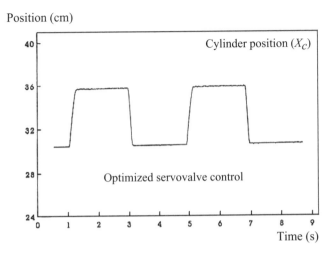

Position (cm)

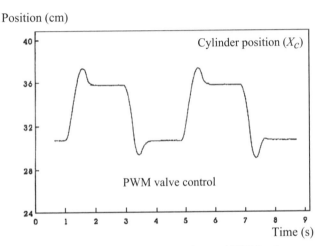

Figure 7.23. A comparison of servovalve and PWM valve control.

Fully active electrohydraulic control of the suspension is then considered using both pole assignment (PA) and LQC techniques to design the state feedback gains with a view to achieving an optimum body acceleration characteristic, based on a validated linearized mathematical model. Computer simulation of the complete system suggests that the LQC design approach gives the better performance characteristic. The PSC is implemented to include features such as gain scheduling and state gain switching to achieve improved control. It is shown that although body displacement compensation is naturally achieved for road input changes, the global optimum design for acceleration transmissibility could not be achieved because of practical limitations caused by the predicted low transducer gain between wheel and body. A further feature of the programmable controller approach was the ability to change state feedback gains during operation. This was found to be necessary to move the suspension from its initial rest position to its operating position. However, an improved performance in body acceleration amplitude control was still possible compared with the optimum passive suspension theoretical predictions.

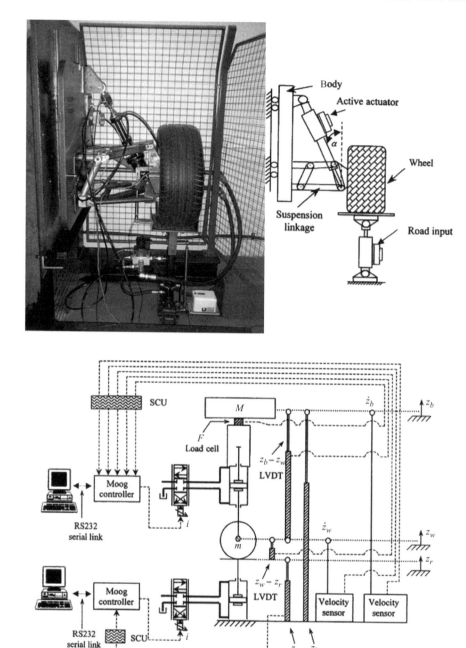

Figure 7.24. Details of the ¼ car test rig.

Consider the ¼ car active suspension system shown in Fig. 7.24. Both road and active actuator are controlled using Moog electrohydraulic servovalves and Moog M2000 PSCs operating with 2-ms sampling intervals. The suspension and wheel unit is linked at an angle α to the plane of movement of the body mass, which is constrained to move vertically by means of linear bearings. A load cell is positioned between the actuator body and car-body pivot point, and LVDT position transducers together with velocity transducers are appropriately placed to measure road,

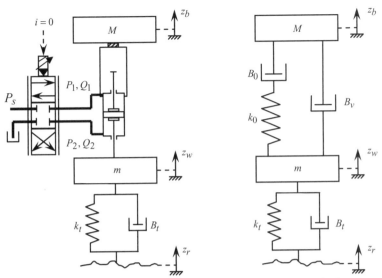

(a) Schematic of the locked actuator   (b) Model of the locked actuator

Figure 7.25. Dynamic model with no active control.

wheel, and body motion. Further details regarding this work by the principal investigators may be found in Watton, Holford, and Surawattanawan (2001, 2004).

Data were collected by a high-speed parallel-channel data-acquisition card system to enable further analysis to be undertaken using the MATLAB simulation and signal-processing environment. A vast amount of experimental testing was carried out using both frequency responses, by means of a transfer function analyzer with pseudobinary random signal excitation, and transient analysis, by means of the step response method. General trends in frequency response were clearly identified, but the Bode diagrams were not sufficiently accurate for comparison purposes beyond a frequency of typically 15 Hz and inadequate to use for parameter identification. Modeling concepts are presented by means of the transfer function approach, in keeping with most of the background literature, and experimental parameter identification and validation were achieved using step response testing.

7.4.2 Determining the Open-Loop Fluid Power Model

To obtain a feel for the dynamic behavior, Fig. 7.25 illustrates the dynamic linear model for the hypothetical situation with the actuator fixed in its central position and with no active control. This is not possible to achieve in practice because of cross-line leakage, which results in the car-body mass moving to its rest position, but serves to illustrate the frequency domain of interest. Therefore, the servovalve spool is assumed to be centered in the absence of an input signal.

The vehicle body mass $M = 240$ kg and the wheel–tire mass $m = 40$ kg. Actuator and lines create an oil stiffness contribution k_o together with velocity damping generated by oil viscosity effects and also a suspension mechanical friction component to a lesser extent. Oil stiffness is defined with the actuator in its central position and equal length lines and, hence, with equal oil volumes on either side. Note that low-cost synthetic flexible hoses were used to couple the actuator and the servovalve; hence, it was known that a low effective bulk modulus would exist. The viscous damping coefficient is B_v, and the cross-line leakage also may be considered to be equivalent to velocity damping B_o in series with oil stiffness k_o. The wheel tire has a stiffness k_t and an assumed velocity damping B_t. Related work resulted in the following data:

- Effective bulk modulus, $\beta_e = 0.22 \times 10^9$ N/m^2
- Cross-line leakage resistance, $R_i = 9.8 \times 10^{10}$ N m^{-2}/m^3 s^{-1}
- Oil stiffness, $k_o = (2\beta_e A^2/V) = 3.73 \times 10^5$ N/m
- Tire stiffness, $k_t = 2.8 \times 10^5$ N/m
- Actuator viscous damping coefficient, $B_v = 300$ N/ms^{-1}
- Tire damping coefficient, $B_t = 4000$ N/ms^{-1}
- Cross-line damping coefficient, $B_o = R_i A^2 = 5930$ N/ms^{-1}

Cross-line leakage damping is the most significant, followed by tire damping and the much lower pure viscous damping. Oil stiffness, even when corrected for actuator angle, is as significant as the tire stiffness. To obtain the equivalent linearized dynamic open-loop transfer functions for the system, it is first necessary to consider the servovalve flow equations. It will be recalled from previous chapters that at zero steady-state current, the servovalve flow gains are:

$$\text{extending, } k_f\sqrt{\frac{P_s - P_L}{2}}, \quad \text{retracting, } k_f\sqrt{\frac{P_s + P_L}{2}}, \quad P_L = \frac{Mg}{A}. \tag{7.49}$$

For this study, the gain ratio is then given by:

$$\frac{\text{gain retracting}}{\text{gain extending}} = \sqrt{\frac{P_s + P_L}{P_s - P_L}} = 1.69. \tag{7.50}$$

A particular advantage of using a PSC is that the control loop forward gain may be modified by a real-time algorithm, depending on whether the actuator is extending or retracting. In this study, the forward gain was increased by a factor of 1.3 when extending and decreased by a factor of 1.3 when retracting.

Considering the equations using state-space notation for derivatives, then the actuator flow-rate equations including compressibility and cross-line leakage may be written.

$$Q_1 = Q_2 = k_i i, \tag{7.51}$$

$$Q_1 = A(\dot{z}_b - \dot{z}_w) + \frac{V}{\beta_e}\dot{P}_1 + \frac{(P_1 - P_2)}{R_i}, \tag{7.52}$$

$$Q_2 = A(\dot{z}_b - \dot{z}_w) - \frac{V}{\beta_e}\dot{P}_2 + \frac{(P_1 - P_2)}{R_i}. \tag{7.53}$$

The actuator hydraulic force is given by:

$$F = A(P_1 - P_2). \tag{7.54}$$

The suspension equations of motion are:

$$M\ddot{z}_b = F\cos\alpha - B_v(\dot{z}_b - \dot{z}_w), \tag{7.55}$$

$$m\ddot{z}_w = -F\cos\alpha + B_v(\dot{z}_b - \dot{z}_w) + k_t(z_r - z_w) + B_t(\dot{z}_r - \dot{z}_w). \tag{7.56}$$

Therefore, taking Laplace transforms and neglecting initial conditions allows the force transfer function to be written:

$$F = \frac{k_i A}{\left(\dfrac{sV}{2\beta_e} + \dfrac{1}{R_i}\right)} i - \frac{s A^2}{\left(\dfrac{sV}{2\beta_e} + \dfrac{1}{R_i}\right)} (z_b - z_w). \qquad (7.57)$$

For the locked actuator condition with no active feedback, $i = 0$:

$$F = -\frac{s A^2}{\left(\dfrac{sV}{2\beta_e} + \dfrac{1}{R_i}\right)} (z_b - z_w.) \qquad (7.58)$$

The transfer function relating road input z_r to car-body displacement z_b is:

$$\frac{z_b}{z_r} = \frac{b_0 + b_1 s + b_2 s^2 + b_3 s^3}{a_0 + a_1 s + a_2 s^2 + a_3 s^3 + a_4 s^4 + a_5 s^5}. \qquad (7.59)$$

These coefficients are obtained from:

$$a_0 = \frac{4k_t}{R_i}\left(A^2\cos\alpha + \frac{B_v}{R_i}\right),$$

$$a_1 = 2A^2\cos\alpha\left(\frac{k_t V}{\beta_e} + \frac{2B_t}{R_i}\right) + \frac{4B_v}{R_i}\left(\frac{k_t V}{\beta_e} + \frac{B_t}{R_i}\right) + \frac{4k_t M}{R_i^2},$$

$$a_2 = \frac{4m}{R_i}\left(A^2\cos\alpha + \frac{B_v}{R_i}\right) + \frac{B_v V}{\beta_e}\left(\frac{4B_t}{R_i} + \frac{k_t V}{\beta_e}\right),$$

$$+ \frac{4M}{R_i}\left(A^2\cos\alpha + \frac{B_v}{R_i} + \frac{k_t V}{\beta_e} + \frac{B_t}{R_i}\right) + \frac{2A^2 B_t V\cos\alpha}{\beta_e},$$

$$a_3 = \frac{4m}{R_i}\left(\frac{B_v V}{\beta_e} + \frac{M}{R_i}\right) + \frac{MV}{\beta_e}\left(\frac{4B_v}{R_i} + \frac{2B_t}{R_i} + 2A^2\cos\alpha + \frac{k_t V}{\beta_e} + \frac{2B_t}{R_i}\right),$$

$$+ \frac{B_t B_v V^2}{\beta_e^2} + \frac{2A^2 mV\cos\alpha}{\beta_e},$$

$$a_4 = \frac{V}{\beta_e}\left(\frac{4mM}{R_i} + \frac{B_v mV}{\beta_e} + \frac{B_v MV}{\beta_e} + \frac{B_t MV}{\beta_e}\right),$$

$$a_5 = \frac{mMV^2}{\beta_e^2},$$

$$b_0 = \frac{4k_t}{R_i}\left(\frac{B_v}{R_i} + A^2\cos\alpha\right),$$

$$b_1 = \frac{4B_v}{R_i}\left(\frac{B_t}{R_i} + \frac{k_t V}{\beta_e}\right) + 2A^2\cos\alpha\left(\frac{k_t V}{\beta_e} + \frac{2B_t}{R_i}\right),$$

$$b_2 = \frac{4B_t B_v V}{R_i \beta_e} + \frac{2A^2 B_t V\cos\alpha}{\beta_e} + \frac{B_v k_t V^2}{\beta_e^2}, \quad b_3 = \frac{B_t B_v V^2}{\beta_e^2}, \qquad (7.60)$$

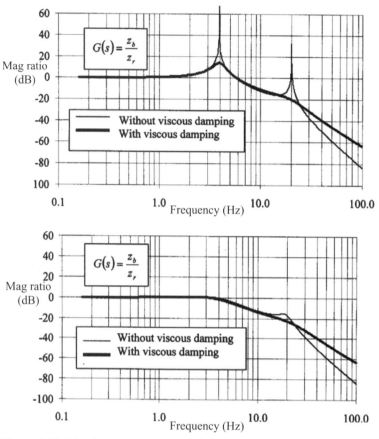

Figure 7.26. Magnitude response of the vehicle body with no active control.

where for this study the coefficients are:

$$a_0 = 6.51 \times 10^{-13}, \quad a_1 = 4.82 \times 10^{-14}, \quad a_2 = 2.10 \times 10^{-15},$$
$$a_3 = 3.48 \times 10^{-17}, \quad a_4 = 2.37 \times 10^{-19}, \quad a_5 = 1.01 \times 10^{-21},$$
$$b_0 = 6.51 \times 10^{-13}, \quad b_1 = 2.02 \times 10^{-14}, \tag{7.61}$$
$$b_2 = 1.64 \times 10^{-16}, \quad b_3 = 1.26 \times 10^{-19}.$$

Figure 7.26 shows the response magnitude of only the body, for a road input disturbance, and illustrates the two fundamental resonant frequencies of 3.9 and 20.2 Hz
in the absence of cross-line leakage at the actuator.

These frequencies may be easily verified from the previous transfer function,
which produces the following undamped characteristic equation:

$$1 + \left(\frac{M+m}{k_t} + \frac{M}{k_o \cos \alpha} \right) s^2 + \frac{Mm}{k_t k_o \cos \alpha} s^4 = 0. \tag{7.62}$$

It can be seen from Eq. (7.62) that the oil stiffness must be corrected for the actuator
angle.

For an infinitely stiff actuator, the single natural frequency would be 5 Hz; for an
infinitely stiff tire, the single natural frequency would be 5.9 Hz. Hence, for this system, it is not a simple matter to couple stiffnesses and masses to define convenient
"body" or "wheel" modes of oscillation. However, it is clear that the two natural
frequencies are isolated by distinct body and wheel mass effects. Figure 7.26 shows

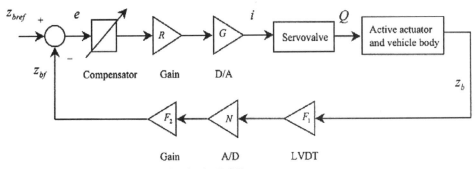

Figure 7.27. Block diagram for the single-DOF test.

that, in reality, the experimentally determined effects of actuator damping B_v and tire damping B_t are evident, illustrating amplitude attenuation throughout the frequency range of interest. When cross-line leakage is introduced, and using the value experimentally identified from dynamic tests, the amplitude response is further and significantly changed.

7.4.3 Actuator Dynamic Stiffness

Consider again Eq. (7.58), which is reproduced here:

$$F = -\frac{s\,A^2}{\left(\dfrac{sV}{2\beta_e} + \dfrac{1}{R_i}\right)}(z_b - z_w).$$ (7.63)

It can be seen that there are two frequency regimes, as follows:

low frequency, $F \to B_0(\dot{z}_b - \dot{z}_w)$, a cross-line viscous damper;   (7.64)

high frequency, $F \to k_o(z_b - z_w)$, a hydraulic spring.   (7.65)

This high-pass, phase-advance filter characteristic that is due to cross-line leakage is useful for damping lower-frequency components, and inserting data into Eq. (7.63) gives a break frequency of 10 Hz. Therefore, its damping effect will be of value for frequencies below this break frequency, which in this example has a fortuitous influence on the body mode of vibration.

7.4.4 The Introduction of Feedback, the One-Degree-of-Freedom (1 DOF) Test to Identify Actuator Viscous Damping B_v and Leakage Resistance R_i

To determine the actuator viscous damping B_v and leakage resistance R_i, it is necessary to introduce feedback control of the servovalve, as shown in Fig. 7.27.

The simplest approach is to use position control of the actuator relative to the wheel with no ground motion. The programmable controller was used with an LVDT position transducer linked between the wheel axle and the body. For the expected frequency of oscillation of the vehicle body, the dynamics of the servovalve may be neglected. Assuming the position transducer gain $F1$, an additional data-acquisition gain N, the feedback gain $F2$, the forward gain R, a D/A and servoamplifier gain G, then the system equations become:

$$2k_i\,Ai = 2A^2\dot{z}_b + \frac{V}{\beta_e}\dot{F} + \frac{2}{R_i}F,$$ (7.66)

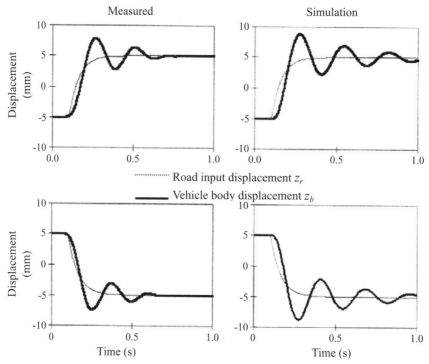

Figure 7.28. Typical single-DOF transient responses.

$$M\ddot{z}_b = F\cos\alpha - B_v\dot{z}_b, \tag{7.67}$$

$$i = GR(z_{bref} - z_{bf}), \tag{7.68}$$

$$z_{bf} = F_1 F_2 N z_b. \tag{7.69}$$

Therefore, the closed-loop transfer function relating z_{bf} to z_{bref} (the voltage equivalent of z_{bf}) is given by:

$$\frac{z_{bf}}{z_{bref}} = \frac{a_0}{a_0 + a_1 s + a_2 s^2 + a_3 s^3}. \tag{7.70}$$

The coefficients are given by:

$$a_0 = 2AF_1F_2\,Gk_i\,NR\cos\alpha, \qquad a_1 = \left(\frac{2B_v}{R_i} + 2A^2\cos\alpha\right),$$

$$a_2 = \left(\frac{2M}{R_i} + \frac{B_v V}{\beta_e}\right), \qquad a_3 = \frac{MV}{\beta_e},$$

$$a_0 = 2.30 \times 10^{-6}, \qquad a_1 = 1.14 \times 10^{-7},$$

$$a_2 = 5.00 \times 10^{-9}, \qquad a_3 = 7.78 \times 10^{-11}. \tag{7.71}$$

A comparison between simulation and experiment is shown in Fig. 7.28 with gain scheduling being implemented in both cases. Demand changes of ± 5 mm are illustrated and similar comparisons have been validated for a range of inputs.

The coefficients of this third-order transfer function were identified using standard software tools within the MATLAB environment, the prediction error method producing the most stable approach from those available in the library. Experimental data, using step changes in demand for both extending and retracting, were obtained at a sampling frequency of 500 Hz, and the resulting discrete transfer

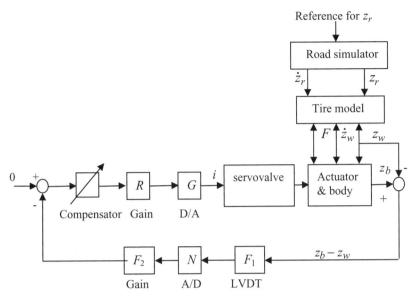

Figure 7.29. Block diagram of the 2-DOF test.

function was transformed to continuous time using the bilinear transformation. This then allowed explicit determination of the unknown parameters, calculated to be:

actuator viscous damping coefficient, $B_v = 300 \, \text{N/m s}^{-1}$;

actuator leakage coefficient, $R_i = 9.8 \times 10^{10} \, \text{Nm}^{-2}/\text{m}^3 \, \text{s}^{-1}$. $\qquad$ (7.72)

In addition, parameter identification from the 1-DOF test may be used to validate the servovalve flow gain. The identified value was found to be the mean of the extending and retracting gains and within 1% of the value obtained from the servovalve no-load steady-state flow test.

7.4.5 The Introduction of Feedback, the Two-Degree-of-Freedom (2 DOF) Test to Identify Tire Viscous Damping B_t and Validate Tire Stiffness k_t

To determine the tire damping coefficient B_t and validate tire stiffness k_t, the road input z_r is used with servovalve control having a zero reference input but with negative feedback from the measured position $(z_b - z_w)$ for identification purposes only. Figure 7.29 shows the general block diagram, and for this system the equations are:

$$2k_i \, Ai = 2A^2 \left(\dot{z}_b - \dot{z}_w \right) + \frac{V}{\beta_e} \dot{F} + \frac{2}{R_i} F, \qquad (7.73)$$

$$M\ddot{z}_b = F \cos \alpha - B_v \left(\dot{z}_b - \dot{z}_w \right), \qquad (7.74)$$

$$m\ddot{z}_w = -F \cos \alpha + B_v \left(\dot{z}_b - \dot{z}_w \right) + k_t (z_r - z_w) + B_t \left(\dot{z}_r - \dot{z}_w \right), \qquad (7.75)$$

$$i = GP \left[0 - NF_1 F_2 \left(z_b - z_w \right) \right]. \qquad (7.76)$$

The transfer function relating road input z_r to car-body displacement z_b is:

$$\frac{z_b}{z_r} = \frac{b_0 + b_1 s + b_2 s^2 + b_3 s^3 + b_4 s^4}{a_0 + a_1 s + a_2 s^2 + a_3 s^3 + a_4 s^4 + a_5 s^5 + a_6 s^6}. \qquad (7.77)$$

The transfer function coefficients are given by:

$$a_0 = \frac{4\,AGk_i\,NPF_1F_2k_t\cos\alpha}{R_i},$$

$$a_1 = 2\,AGk_i\,NP\left(\frac{F_1F_2k_tV\cos\alpha}{\beta_e} + \frac{2B_tF_1F_2\cos\alpha}{R_i}\right) + \frac{4k_t}{R_i}\left(\mathrm{A}^2\cos\alpha + \frac{B_v}{R_i}\right),$$

$$a_2 = 2\,AGk_i\,NP\left(\frac{2F_1F_2m\cos\alpha}{R_i} + \frac{B_tF_1F_2V\cos\alpha}{\beta_e} + \frac{2F_1F_2M\cos\alpha}{R_i}\right),$$

$$+ 2A^2\cos\alpha\left(\frac{k_tV}{\beta_e} + \frac{2B_t}{R_i}\right) + \frac{4B_v}{R_i}\left(\frac{k_tV}{\beta_e} + \frac{B_t}{R_i}\right) + \frac{4k_tM}{R_i^2},$$

$$a_3 = 2\,AGk_i\,NP\left(\frac{F_1F_2mV\cos\alpha}{\beta_e} + \frac{F_1F_2MV\cos\alpha}{\beta_e}\right) + \frac{4m}{R_i}\left(A^2\cos\alpha + \frac{B_v}{R_i}\right),$$

$$+ \frac{B_vV}{\beta_e}\left(\frac{4B_t}{R_i} + \frac{k_tV}{\beta_e}\right) + \frac{4M}{R_i}\left(A^2\cos\alpha + \frac{B_v}{R_i} + \frac{k_tV}{\beta_e} + \frac{B_t}{R_i}\right) + \frac{2A^2B_tV\cos\alpha}{\beta_e},$$

$$a_4 = \frac{4m}{R_i}\left(\frac{B_vV}{\beta_e} + \frac{M}{R_i}\right) + \frac{MV}{\beta_e}\left(\frac{4B_v}{R_i} + \frac{2B_t}{R_i} + 2A^2\cos\alpha + \frac{k_tV}{\beta_e} + \frac{2B_t}{R_i}\right),$$

$$+ \frac{B_tB_vV^2}{\beta_e^2} + \frac{2A^2mV\cos\alpha}{\beta_e},$$

$$a_5 = \frac{V}{\beta_e}\left(\frac{4mM}{R_i} + \frac{B_vmV}{\beta_e} + \frac{B_vMV}{\beta_e} + \frac{B_tMV}{\beta_e}\right),$$

$$a_6 = \frac{mMV^2}{\beta_e^2},$$

$$b_0 = \frac{4\,AGk_i\,NPF_1F_2k_t\cos\alpha}{R_i},$$

$$b_1 = 2\,AGk_i\,NP\left(\frac{F_1F_2k_tV\cos\alpha}{\beta_e} + \frac{2B_tF_1F_2\cos\alpha}{R_i}\right) + \frac{4k_t}{R_i}\left(\frac{B_v}{R_i} + A^2\cos\alpha\right),$$

$$b_2 = 2\,AGk_i\,NP\left(\frac{B_tF_1F_2V\cos\alpha}{\beta_e}\right) + \frac{4B_v}{R_i}\left(\frac{B_t}{R_i} + \frac{k_tV}{\beta_e}\right) + 2A^2\cos\alpha\left(\frac{k_tV}{\beta_e} + \frac{2B_t}{R_i}\right),$$

$$b_3 = \frac{4B_tB_vV}{R_i\beta_e} + \frac{2A^2B_tV\cos\alpha}{\beta_e} + \frac{B_vk_tV^2}{\beta_e^2},$$

$$b_4 = \frac{B_tB_vV^2}{\beta_e^2}. \tag{7.78}$$

Comparisons between experiment and model prediction for the 2-DOF test, for step demand changes in road position, are shown in Fig. 7.30. Only two parameters need to be identified from this test: the tire stiffness k_t and the tire damping B_v. These parameters were then varied to give the best fit between experimental step response data and transfer function equation (7.77), which resulted in the following:

$$a_0 = 2.80 \times 10^{-11}, \quad a_1 = 1.50 \times 10^{-12}, \quad a_2 = 8.25 \times 10^{-14},$$
$$a_3 = 2.55 \times 10^{-15}, \quad a_4 = 3.48 \times 10^{-17}, \quad a_5 = 2.37 \times 10^{-19},$$
$$a_6 = 1.01 \times 10^{-21}, \quad b_0 = 2.80 \times 10^{-11}, \quad b_1 = 1.50 \times 10^{-12},$$
$$b_2 = 2.65 \times 10^{-14}, \quad b_3 = 1.64 \times 10^{-16}, \quad b_4 = 1.26 \times 10^{-19},$$

tire stiffness, $B_t = 2.8 \times 10^5$ N/m,

tire velocity damping coefficient, $B_t = 4000\,\text{N/m s}^{-1}$. (7.79)

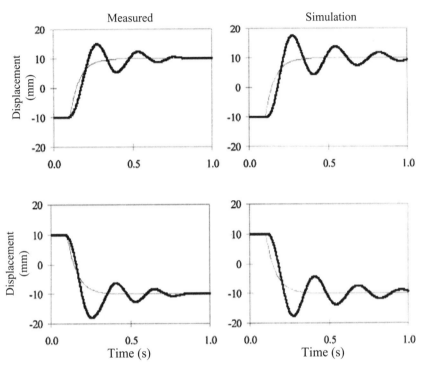

Figure 7.30. 2-DOF model validation for a road displacement of ± 10 mm.

The tire stiffness identified from dynamic testing is a very accurate validation of the value that may easily be calculated from a compressive test using a laboratory materials testing machine. This test was done with the tire compressed with diametral displacements up to 40 mm. This produced a linear force–displacement characteristic with a small amount of hysteresis. The stiffness measured from this test is doubled to validate the dynamic test in which the tire is in its normal mode of compression between the hub and the road. This static test gave a value of $k_t = 2.8 \times 10^5$ N/m, the same as the dynamic test and both for a tire preload pressure of 2.04 bar.

It is important that the road input hydraulic response is correctly modeled to allow comparisons to be made between experiment and theory, although a detailed mathematical knowledge of this separate position control loop is not necessary. A second programmable controller was used to close the position control loop, and it was deduced that a simple first-order transfer function, with a time constant of 0.06s, was a sufficiently accurate representation of the dynamic behavior over all the test conditions used. Again, additional tests were used to validate the model for other input conditions. Generally, the results are good; usually, the simulation predictions are a little more lightly damped than the experimental data.

7.4.6 A State-Space Model for the Active Suspension

The state variables are defined as follows:

$$x_1 = \dot{z}_b, \quad x_2 = \dot{z}_w, \quad x_3 = F,$$
$$x_4 = z_b - z_w, \quad x_5 = z_w - z_r, \tag{7.80}$$

$$x = \text{state vector} = \begin{bmatrix} \dot{z}_b \\ \dot{z}_w \\ F \\ z_b - z_w \\ z_w - z_r \end{bmatrix}. \tag{7.81}$$

The open-loop equations may then be written in the following state-space format:

$$\dot{x} = Ax + Be + G_d \dot{z}_r, \tag{7.82}$$

$$A = \begin{bmatrix} -\dfrac{B_v}{M} & \dfrac{B_v}{M} & \dfrac{\cos\alpha}{M} & 0 & 0 \\[2mm] \dfrac{B_v}{m} & -\dfrac{(B_v + B_t)}{m} & -\dfrac{\cos\alpha}{m} & 0 & -\dfrac{k_t}{m} \\[2mm] -\dfrac{2\beta_e A^2}{V} & \dfrac{2\beta_e A^2}{V} & -\dfrac{2\beta_e}{V R_i} & 0 & 0 \\[2mm] 1 & -1 & 0 & 0 & 0 \\[1mm] 0 & 1 & 0 & 0 & 0 \end{bmatrix}, \tag{7.83}$$

$$B = \begin{bmatrix} 0 \\ 0 \\ \dfrac{2k_i AGP\beta_e}{V} \\ 0 \\ 0 \end{bmatrix}, \tag{7.84}$$

$$G_d = \begin{bmatrix} 0 \\ \dfrac{B_t}{m} \\ 0 \\ 0 \\ -1 \end{bmatrix}. \tag{7.85}$$

The control signal is e and $\dot{z}_r$ is the disturbance signal (i.e., velocity of road disturbance). For full state feedback, and considering the implementation of the Moog PSC, the control signal is given by:

$$e = -Kx,$$
$$e = -NK_1 K_2 x, \tag{7.86}$$
$$e = -N[I_1 I_2 \dot{z}_b + J_1 J_2 \dot{z}_w + H_1 H_2 F + F_1 F_2 (z_b - z_w) + L_1 L_2 (z_w - z_r)],$$

where:

$K = 1 \times 5$ state feedback gain vector,

$N = $ A/D gain,

$K_1 = $ state feedback gain vector for the controller,

$$K_1 = \begin{bmatrix} I_1 & J_1 & H_1 & F_1 & L_1 \end{bmatrix} \tag{7.87}$$

$$K_2 = \text{transducer gain matrix} \begin{bmatrix} I_2 & 0 & 0 & 0 & 0 \\ 0 & J_2 & 0 & 0 & 0 \\ 0 & 0 & H_2 & 0 & 0 \\ 0 & 0 & 0 & F_2 & 0 \\ 0 & 0 & 0 & 0 & L_2 \end{bmatrix}. \tag{7.88}$$

Validation of the state-space model is indicated by Figs. 7.31 and 7.32, which show measured and computed state variable responses for a step input to the servovalve controlling the road input system. The results represent nonoptimized gains that were selected to give a stable behavior, with only feedback $z_b - z_w$ being used.

A range of input conditions from ± 5 to ± 15 mm were validated and indicate comparisons similar to those selected. The results show excellent directional symmetry because of gain scheduling for both measured and simulated data, the simulation results indicating a small but detectable lower damping characteristic than that of the measured performance. The measured wheel-hub velocity shows excessive damping compared with that of the predicted response, although the unusual characteristic around the peak velocity is validated by the simulation. The wire potentiometer–tachogenerator used to measure velocity also contributes to the measured overdamping, and it would appear that the measured results are not reliable at speeds below 0.05 m/s.

7.4.7 Closed-Loop Control Design by Computer Simulation

Closed-loop control is based around the Moog M2000 PSC. Two separate units were used, with each unit capable of controlling two servovalves. Key features of the PSC are as follows:

- The engineering user interface is a text-based programming language specifically designed for the configuration and programming of the PSC. The software is installed in each of the supervisory PCs and runs automatically when the computers are switched on.
- Programs are created off-line using ASCII code called a "log file." When a program is loaded, it is automatically compiled (alerting the user of any errors) and transferred to the PSC by the RS232 serial link.
- The program continues to run until either the computer is switched off or another program is loaded.
- While the PSC is running a program, the operator is presented with a screen displaying key system parameters. The screen is split into two with a maximum of 16 parameters displayed on the left-hand side, which can be modified, and up to a further 16 parameters on the right-hand side for real-time monitoring.

Note that since this work was undertaken, the PSC has been updated to the M3000 PSC, with a faster sampling time. Figure 7.33 shows the control concept.

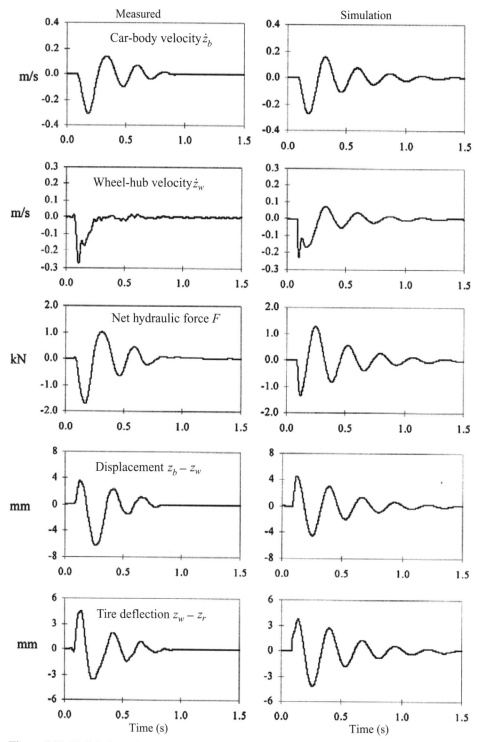

Figure 7.31. Validation of the state variables for a road step change of $+10$ mm $\rightarrow -10$ mm.

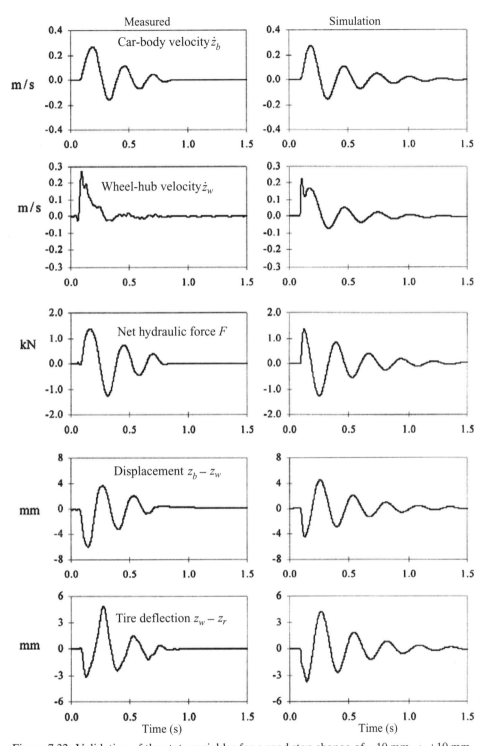

Figure 7.32. Validation of the state variables for a road step change of -10 mm $\rightarrow$ $+10$ mm.

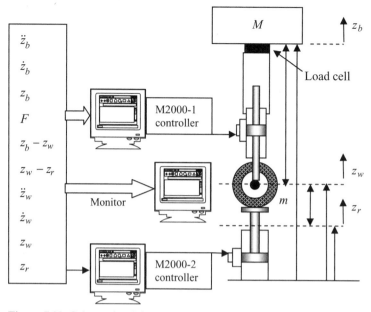

Figure 7.33. Schematic of the control concept.

Two design methods are considered, PA and LQC, and initially through computer simulation. This was done to minimize the possibility of control-loop instability of the test rig by random selection of individual transducer gains. A diagram of the modeling and state control approach is shown in Fig. 7.34.

One novel feature of the PSC approach is the ability to gain schedule the servovalve control signal in real time using existing algorithm features. This effectively

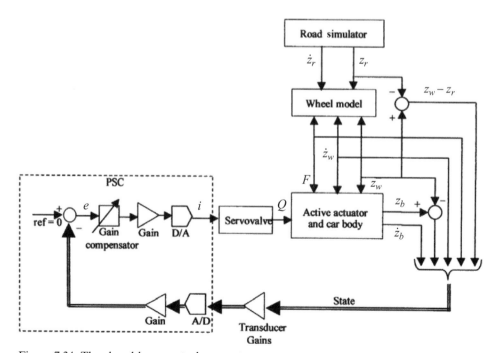

Figure 7.34. The closed-loop control concept.

compensates for the effect of servovalve flow gain change, depending on whether the active actuator is extending or retracting, as discussed earlier in this section. This effect is significant and the implementation of gain scheduling has two advantages:

- The transient response in either direction may be matched within the constraints of the practical limitation, principally because of dynamic pressure differential change effects.
- The linearized dynamic model may be used with more confidence because it assumes constant servovalve flow gains in each direction.

Considering the state-space equations derived in the previous sections combined with the measured data then gives:

$$
A = \begin{bmatrix} -1.25 & 1.25 & 3.7 \times 10^{-3} & 0 & 0 \\ 7.5 & -1.08 \times 10^2 & -2.22 \times 10^{-2} & 0 & -7 \times 10^3 \\ -3.73 \times 10^5 & 3.73 \times 10^5 & -63 & 0 & 0 \\ 1 & -1 & 0 & 0 & 0 \\ 0 & 1 & 0 & 0 & 0, \end{bmatrix}, \quad B = \begin{bmatrix} 0 \\ 0 \\ 186 \\ 0 \\ 0 \end{bmatrix}.
$$

$$(7.89)$$

The controllability matrix U is given by:

$$
U = [B \ AB \ A^2B \ A^3B \ A^4B].
$$

$$(7.90)$$

It will be deduced that because the rank of U is 5, then the system is state controllable.

Closed-Loop Design Using Pole Assignment

The characteristic equation for the open-loop model is given by:

$$
|sI - A| = a_0 + a_1 s + a_2 s^2 + a_3 s^3 + a_4 s^4 + a_5 s^5.
$$

$$(7.91)$$

The desired closed-loop characteristic equation is given by:

$$
\alpha_0 + \alpha_1 s + \alpha_2 s^2 + \alpha_3 s^3 + \alpha_4 s^4 + \alpha_5 s^5.
$$

$$(7.92)$$

Therefore, the solution for the feedback signal is given by:

$$
\begin{aligned}
& e = -Kx, \\
& K = [(\alpha_0 - a_0) \ (\alpha_1 - a_1) \ (\alpha_2 - a_2) \ (\alpha_3 - a_3) \ (\alpha_4 - a_4)] T^{-1}, \\
& T = UW, \\
& W = \begin{bmatrix} a_1 & a_2 & a_3 & a_4 & 1 \\ a_2 & a_3 & a_4 & 1 & 0 \\ a_3 & a_4 & 1 & 0 & 0 \\ a_4 & 1 & 0 & 0 & 0 \\ 1 & 0 & 0 & 0 & 0 \end{bmatrix}.
\end{aligned}
$$

$$(7.93)$$

For the present study:

$$W = \begin{bmatrix} 1.02 \times 10^7 & 5.96 \times 10^5 & 3.27 \times 10^4 & 1.72 \times 10^2 & 1 \\ 5.96 \times 10^5 & 2.37 \times 10^4 & 1.72 \times 10^2 & 1 & 0 \\ 3.27 \times 10^4 & 1.72 \times 10^2 & 1 & 0 & 0 \\ 1.72 \times 10^2 & 1 & 0 & 0 & 0 \\ 1 & 0 & 0 & 0 & 0 \end{bmatrix},$$

$$T = \begin{bmatrix} 7.55 \times 10^{-10} & 4.82 \times 10^3 & 68.8 & 0.688 & 0 \\ -9.22 \times 10^{-9} & 1.58 \times 10^{-11} & 4.69 \times 10^{-13} & -4.13 & 0 \\ -1.24 \times 10^{-5} & 1.62 \times 10^6 & 1.32 \times 10^6 & 2.02 \times 10^4 & 186 \\ 4.82 \times 10^3 & 68.8 & 4.82 & 0 & 0 \\ 1.58 \times 10^{-11} & 4.69 \times 10^{-13} & -4.13 & 0 & 0 \end{bmatrix}. \tag{7.94}$$

To start the design process, two dominant poles were selected to have values of $-7.45 \pm j6.37$, giving an approximated second-order transfer function an undamped natural frequency of 1.56 Hz and a damping ratio of 0.76. The three remaining poles were placed at equal values of -75. This condition actually minimizes body acceleration, as will be shown later. The solution is:

$$K = \begin{bmatrix} -756 & 2500 & 0.359 & 8250 & 91{,}000 \end{bmatrix}. \tag{7.95}$$

From Eqs. (7.87) and (7.88):

$$K_1 = N^{-1} K K_2,$$

$$K_2 = \begin{bmatrix} 5 & 0 & 0 & 0 & 0 \\ 0 & 5 & 0 & 0 & 0 \\ 0 & 0 & 66.7 \times 10^{-6} & 0 & 0 \\ 0 & 0 & 0 & 57.2 & 0 \\ 0 & 0 & 0 & 0 & 18.2 \end{bmatrix}, \tag{7.96}$$

$$K_1 = \begin{bmatrix} -0.0944 & 0.313 & 3.37 & 0.09 & 3.13 \end{bmatrix}.$$

Consider Now Closed-Loop Design Using LQC

Considering earlier work in Chapter 6, a performance index is selected as follows:

$$J = \int_0^\infty (y^T Q y + e^T \mathrm{R} e)\, dt,$$

$$Q = \begin{bmatrix} q_1 & 0 & 0 \\ 0 & q_1 & 0 \\ 0 & 0 & q_1 \end{bmatrix}. \tag{7.97}$$

The output is given by:

$$y = Cx,$$

$$C = \begin{bmatrix} 0 & 0 & 1 & 0 & 0 \\ 0 & 0 & 0 & 1 & 0 \\ 0 & 0 & 0 & 0 & 1 \end{bmatrix}. \tag{7.98}$$

Recalling that $e = -Kx$, the performance index then becomes:

$$J = \int_0^\infty (x^T C^T Q C x + x^T K^T R K x)\, dt. \tag{7.99}$$

Minimizing the performance index then results in the solution of the matrix Ricatti equation given by:

$$A^T P + PA + C^T QC - P^T BR^{-1} B^T P = 0. \tag{7.100}$$

For this application, P contains 15 constants and the feedback gains are then given by:

$$K = R^{-1} B^T P,$$
$$K_1 = N^{-1} K K_2^{-1}. \tag{7.101}$$

The difficult aspect of this approach is the selection of the four weighting parameters, R, q_1, q_2, q_3. Recalling previously published work (Surawattanawan, 2000; Thomson, 1976), the global minimum point determination is not affected by the choice of R, which is then selected to have a value $R = 1$. Because q_2 and q_3 vary with the square of the measured variables $(z_b - z_w)$ and $(z_w - z_r)$, it is proposed that the ratio q_1/q_2 is also specified as a square law evaluated with each maximum displacement – in this study, 25 and 6.6 mm. This gives $q_1/q_2 = 13.5$, and the problem is reduced to selecting just q_1 and q_2. If the design is initiated by selecting $q_1 = 43.7$ and $q_2 = 2 \times 10^6$, then it follows that:

$$K = [-420 \quad 1690 \quad 6.25 \quad 1410 \quad -8100],$$
$$K_1 = [-0.053 \quad 0.211 \quad 58.5 \quad 0.015 \quad -0.278]. \tag{7.102}$$

Following the establishment of ostensibly workable control laws using both PA and LQC methods, the active control performance may be further evaluated by means of computer modeling; in this study, within the MATLAB Simulink environment. In practice, this means a large number of simulations, selected as 150 for this study, as the following parameters are varied:

- the undamped natural frequency and damping ratio for PA control
- the weighting factors q_1 and q_2 for LQC

A random road input model was used, and the rms values of the appropriate system variable were used to compare results. An integrator with a gain of 0.223 was used to represent the transfer function relating the road-input displacement to Gaussian white noise representing the road-input velocity. The gain represents the situation in which a vehicle runs at a relatively high speed of 150 km/h on a relatively rough road surface, and a simulation time of 40 s was used for all simulations. Therefore, the average level of the road input displacement changes with time and is considered to be representative of a real road surface. For completeness, it is useful to also compare the purely passive suspension, and this is done by assuming typical, yet hypothetical for this study, suspension stiffnesses in the range $k_s = 20–118$ kN/m. A suspension damping rate is implied by assuming the equivalent damping ratio of the resulting second-order transfer function for the suspension mode of vibration. Hence, a complete plot of body rms acceleration may be obtained, and the results for the three control schemes are shown in Fig. 7.35. For all the simulation results shown, the maximum rms values are $(z_w - z_b)_{max} = 25$ mm and $(z_r - z_w)_{max} = 6.8$ mm.

It can be seen that LQC gives the lowest value of rms acceleration, but this will be at the expense of an increased suspension displacement.

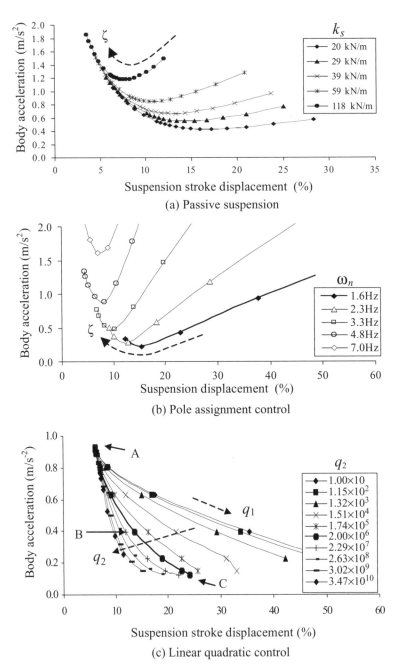

Figure 7.35. A comparison of strategies for a road random input profile.

7.4.8 Experimental Validation of the Preferred LQC System

Both optimum PA and LQ control designs result in extremely low feedback gains for the body-to-wheel ($z_w - z_b$ in practice) displacement transducer, as evident from the calculations shown in Eqs. (7.96) and (7.102). The design is also limited to the midposition of the active actuator because this is the condition on which the open-loop hydraulic linearized model is based.

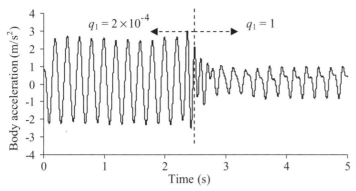

Figure 7.36. The effect of state feedback gain changes for no-optimal conditions developed from LQC theory.

However, when the system is shut down, the active actuator naturally moves to its fully retracted position because of the body mass combined with hydraulic leakage. These combined aspects mean that for the preferred closed-loop operating condition, the initial servovalve current is insufficient to move the body from its rest position to the operating midposition following subsequent start-up of the system. Fortunately, a further novel feature of the Moog PSC is the ability to switch gains in real time, thus allowing different settings to be used for both rest and operating conditions. Before this gain switching is done, it is preferable to consider its implication from a theoretical point of view.

Consider Fig. 7.35(c) and the trajectory containing the optimum condition as indicated by the path A–B–C. None of these conditions allowed initial positioning of the suspension for the reason previously stated, and it was concluded that the theoretical optimum given by $q_1 = 44$ and $q_2 = 2 \times 10^6$ could not be implemented. Moving to the left-hand side of the mapping reduces the suspension stroke displacement; for example, selecting $q_2 = 3 \times 10^9$ reduces the maximum stroke displacement by 1/3. Moving down the trajectory via similar points A–B–C as before means that the dominant closed-loop poles move off the negative real axis in the s plane such that the natural frequencies vary between 2.9 and 0.88 Hz with a damping ratio change from 0.65 to 0.72. Thus, it can be seen that increasing the penalty function for $z_w - z_b$ causes a degradation in the acceleration isolation performance. However, under this new trajectory, it was found possible to move the active actuator to its midposition after control is initiated, although points A $\rightarrow$ B were only experimentally feasible.

Comparisons between theory and experiment were then made by selecting $q_2 = 3 \times 10^9$ with an initial value of $q_1 = 2 \times 10^{-4}$ to determine the gain settings. With the system now operating with the active actuator in its midposition, and for a sinusoidal road input, the new gains were calculated using $q_1 = 1$, the maximum feasible, and then switched. An accelerometer was attached to the body, and the effect of the gain change at a frequency of 5 Hz is shown in Fig. 7.36.

It is clear from Fig. 7.36 that the gain switch is stable and significantly reduces the vibration amplitude. A comparison between simulation and measurement is shown as Fig. 7.37 for $q_1 = 2 \times 10^{-4}$ and $q_1 = 1$.

The LQ design approach was found to be better than the PA approach because it effectively shifts the dominant natural frequency to a lower value. This leads to a "softer" control approach and allows the damping rate to be increased to near

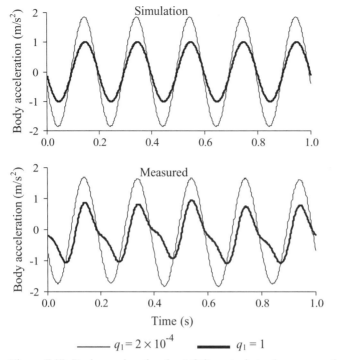

Figure 7.37. Body acceleration for LQC control at a frequency of 5 Hz.

its critical value. It was found possible to reduce the acceleration amplitude by about 30% of the passive suspension case. The LQ approach has a greater flexibility because it allows different weights to be assigned to each measured state. It was found that a great deal of experience was needed to select the LQ design weights and, following many analytical solutions to the matrix Ricatti equation, a global minimum point could be theoretically established. This resulted in a very low suspension position gain to the extent that the suspension could not be moved from its rest position to its operating midposition. It is concluded that:

• The complex system mathematical model, because of the inclusion of practical hydraulic characteristics, means that only suboptimal solutions are possible;
• the theoretical design approaches considered were considerably aided by the ability to gain schedule the servovalve gain in practice, allowing a linearized theory to be more applicable to the open-loop nonlinear equations; and
• the programming facility of the PSC was also particularly useful in selecting different state feedback gains to allow the suspension to be moved to its operating position.

7.5 The Performance of a Car Hydraulic Power-Steering System

7.5.1 Introduction

This study was undertaken principally by colleague Y Xue with support from TRW Steering Systems UK, which provided the commercially available power-steering unit and its design details, some of which are confidential (Xue and Watton, 2005).

Power-assisted steering (PAS) systems are force–torque amplification equipment installed in vehicles with the obvious aim to reduce drivers' steering maneuvering efforts. PAS has now become a necessary requirement on most domestic vehicles and particularly on large vehicles. Good power steering reduces driver fatigue and enables the car to be rapidly maneuvered out of difficulties. There are mainly three types of PAS systems in land and water vehicle applications; that is, hydraulic, electric–hydraulic, and electric power-steering systems. Hydraulic PAS systems use an engine accessory belt to drive a pump, which delivers fluid to operate a piston in the power-steering rack cylinder. If the steering wheel is being turned, a pressure difference is built up between the two chambers of the rack cylinder, assisting the driver's maneuvering. This is achieved by the mechanism of a pinion–control-valve assembly in which a small steering torque is applied at the steering wheel by the driver. The pump pressure increases when steering is operated, and internal mechanical feedback ensures that the pump pressure returns to normal, which is typically 1–2 bar. The advantage of a hydraulic PAS is its large power amplification of steering power, whereas energy and space consumption together with maintenance are some disadvantages.

New types of power steering of electrohydraulic and direct electric systems are becoming attractive for car makers and some customers. In electric–hydraulic PAS systems, the hydraulic pump is driven by an electric motor, and pump speed is regulated by an electric controller to vary pump pressure and flow, providing steering effort tailored for different driving situations. The pump can be run at low speeds or shut off to provide energy savings during straight-ahead driving. Direct electrical PAS uses an electric motor to drive the steering rack directly by a gear mechanism, no pump or fluid is needed, and the power-steering function is therefore independent of engine speed, resulting in energy savings. Microprocessors can be incorporated to control steering dynamics and driver effort by considering the vehicle speed and turning rate. Electrohydraulic and direct electrical power steering are developing trends, and conventional hydraulic power steering is not losing its requirement in many applications because of its advantage of large power amplification. Modern steering units use a faceted spool approach whereby flats, or facets, are placed at strategic parts of the spool to improve the steering characteristic. Commercial units require component design to be executed to extremely fine tolerances that significantly influence the dynamic behavior. This project makes a unique contribution because the few research publications available do not discuss detailed models of hydraulic PAS nonlinear steady-state characteristics together with dynamics, including the effect of nonlinear friction.

7.5.2 Experimental Setup and Operation

The laboratory setup is shown in Fig. 7.38. The road load against steering action is achieved using two pneumatic cylinders at each end of the rack, with an adjustable airflow restrictor in the link tube between the two chambers of the cylinders. Although the pneumatic loading built is not exactly the same as the road load, it has little effect on the model identification being presented here. In the test rig, a torque transducer and an angular displacement transducer were installed at the steering column. The angular displacement transducer was designed using two pulleys, one fitted with the steering column and the other with a potentiometer. Pressures at the hydraulic delivery–return lines at the steering rack are measured by two

(a)

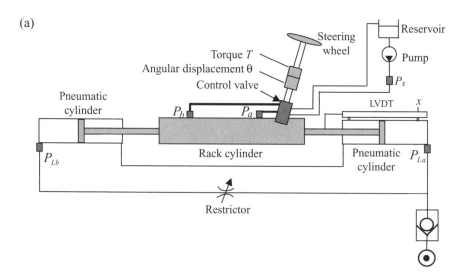

(b)

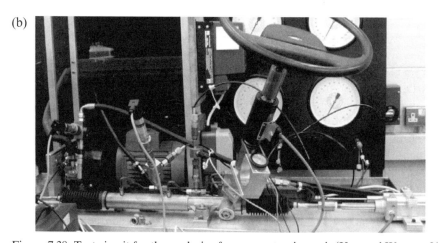

Figure 7.38. Test circuit for the analysis of a power-steering unit (Xue and Watton, 2005).

pressure transducers. The steering system output, the rack linear displacement, was measured by a LVDT transducer. The oil was supplied by a constant-displacement vane pump, which was driven by an electric motor with variable speed control. The outlet pressure of the pump was also measured. All data measured by the transducers were captured by a digital signal processor (DSP) installed in a computer.

7.5.3 Steady-State Characteristics of the Steering Valve

The system operation can be observed from the mechanism of power assistance, Fig. 7.39, and shown *for a nonfaceted control edge valve.*

The control valve consists of three main components: a torsion bar, an inner element, and a sleeve. Slots are made on the inner surface of the sleeve to act as control edges, which operate in conjunction with the control edges made on the outer surface of the inner element. This directs the flow from the pump to the rack cylinder chambers and to the reservoir. The top of the torsion bar is directly fitted with the steering column and the top of the inner element, the bottom of which is free. Thus, the top of the torsion bar and the inner element obey the steering-wheel

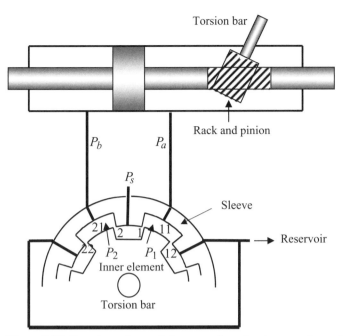

Figure 7.39. Steering valve and the control mechanism, nonfaceted control edge valve.

movements. The lower end of the torsion bar is fitted with the pinion rod, and the valve sleeve is tightly keyed to the pinion as well; therefore, the torsion bar lower end and the valve sleeve faithfully obey any rotary motion made by the pinion. The control edges on the inner element and on the sleeve are in neutral position until the torsion bar is twisted.

Oil is supplied to the control valve from the pump, and when there is no steering, the oil is allowed to pass through paths in the valve and return to the reservoir with very low pressure generated. The shaft and sleeve are designed symmetrically with six oil paths to and from the cylinder, only two paths being shown in Fig. 7.39. If a torque is applied to the steering column, the torsion bar will twist; a relative movement will take place between the inner element and the sleeve. The oil's normal return path will be restricted to a smaller orifice in the valve; therefore, the oil's delivery path pressure will increase and return line pressure decrease. Thus, a pressure difference will be built up between the steering rack cylinder chambers. This pressure difference will drive the rack to move in the desired direction, realizing power-assisted steering. The angular displacement difference between the inner element and sleeve decreases with the rack moving, and the valve will return to the neutral condition and pressure will drop back to the initially very small value. The power-steering unit shown in Fig. 7.39 may be represented by the flow circuit path shown in Fig. 7.40.

To obtain a feel for the steady-state characteristic, it is assumed that the cylinder piston is fixed in position. It then follows that:

$$p_1 = \frac{p_s \left(\dfrac{t_1}{t_{12}}\right)^2}{1 + \left(\dfrac{t_1}{t_{12}}\right)^2}, \qquad p_2 = \frac{p_s \left(\dfrac{t_2}{t_{22}}\right)^2}{1 + \left(\dfrac{t_2}{t_{22}}\right)^2}, \qquad (7.103)$$

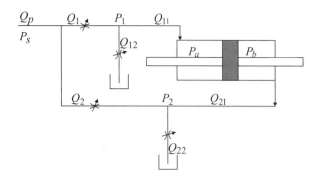

Figure 7.40. Hydraulic circuit schematic.

where t is the orifice width and subscripts are the orifices shown in Figs. 7.39 and 7.40. The pump outlet pressure is P_s and can be determined from the valve-flow equations and the pump characteristic. According to the nonfacet control edge design in Fig. 7.39, changes of t_1, t_2, t_{11}, and t_{22} against angular relative movement of the shaft and sleeve can be determined, and the steady-state characteristic is shown in Fig. 7.41.

With relative twist increasing, the orifices t_1 and t_{22} will increase and t_{12} and t_2 decrease; thus, the pressure P_1 will increase and P_2 will decrease. These changes result in the pressure difference $P_1 - P_2$ increasing with steering-wheel relative rotation, and the pump pressure P_s always increases with steering in either direction.

For a good steering feeling, the pressure difference curve is expected to have a smaller gradient around zero-degree displacement for safe maneuvering, and a larger gradient elsewhere for effective torque assistance. However, too large a gradient may also result in losing control, and the gradient needs to be carefully designed to accommodate the driver's demand. *The control edge of a two-facet valve* used in this study is shown in Fig. 7.42 together with its pressure characteristic. The facets are designed to achieve the desired steering feeling.

Prior to the facet engaging, the orifice area is simply the distance from the slot edge in the sleeve to the outer facet edge of the inner element, multiplied by the respective width. After the outer facet is engaged, the orifice area can be found by calculating the gap between the valve sleeve and the inner element, normal to the facet, and multiplying this distance by the facet width. The effects of facets on

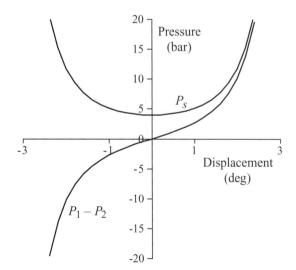

Figure 7.41. Steady-state characteristic of a nonfaceted steering valve.

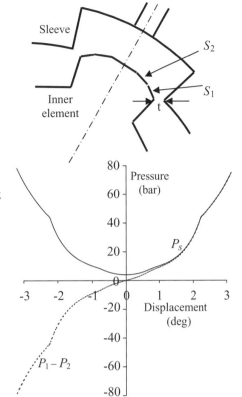

Figure 7.42. Characteristics of the two-faceted steering valve.

the pressure difference and pump pressure are shown in the consequent steady-state characteristic. The small–large–small gradient profile in practice enhances the steering feeling and the driver's confidence in maneuvering for the required power assistance and reduced risk in operation.

7.5.4 Dynamic Behavior of the Power-Steering Unit

The dynamics modeling carried out is concerned with oil compressibility, combined friction forces at the cylinder piston and between the shaft and sleeve, inertia of the moving parts, and the pneumatic loading dynamics. The flows to and from the cylinder considering rack displacement and fluid compressibility are as follows:

$$Q_{11} = A\frac{dx}{dt} + \frac{V_1}{\beta}\frac{dP_1}{dt},$$
$$Q_{21} = A\frac{dx}{dt} - \frac{V_2}{\beta}\frac{dP_2}{dt},$$

(7.104)

where A is the piston working area, x is the rack displacement, β is the fluid bulk modulus, and V_1 and V_2 are the volumes in the cylinder working chambers, which are varying during steering. Considering the torque applied to the pinion and valve sleeve gives:

$$\frac{J}{r_p}\frac{d^2x}{dt^2} = K_b\Delta\theta - f_m r_p - T_f - T_r,$$

(7.105)

If θ_i is the input steering angle at the steering wheel given by the driver, then $\Delta\theta = \theta_i - \theta$, where θ is the angle of twist at the pinion that is due to the displacement of the rack. K_b is the torsion bar stiffness, f_m is the meshing force between the pinion and rack, and r_p is the pinion radius. T_f is the torque induced by the friction at the sleeve, and T_r is the torque applied to the pinion by means of the sleeve and inner element, which is induced by the combined flow-reaction forces at the orifices. J is the moment of inertia of the pinion and valve sleeve with the fittings. The flow-reaction force on a moving part caused by velocity vector changes due to an orifice can be approximated as:

$$f_r = 2C_q a\rho \Delta P \cos\alpha, \tag{7.106}$$

where ΔP is the pressure drop across the orifice, a is the orifice area, and α is the jet angle at the vena contracta. Considering the control valve, the flow-reaction forces applied to the sleeve are at orifices "12" and "22," as follows:

$$f_{r12} = 2C_q K_a t_{12}\rho\, P_{12} \cos\alpha,$$
$$f_{r22} = 2C_q K_a t_{22}\rho\, P_{22} \cos\alpha. \tag{7.107}$$

The flow-reaction forces applied to the inner element are as follows:

$$f_{r1} = 2C_q K_a t_1\rho (P_s - P_1) \cos\alpha,$$
$$f_{r2} = 2C_q K_a t_2\rho (P_s - P_1) \cos\alpha. \tag{7.108}$$

The overall effect of these forces in Eqs. (7.107) and (7.108) is against the sleeve to follow the steering movement. These forces contribute to the flow-reaction torque T_r in Eq. (7.105) as follows:

$$T_r = (f_{r2} - f_{r1} + f_{22} - f_{11})r_i, \tag{7.109}$$

where r_i is the radius of the inner element. The dynamics at the rack cylinder is:

$$M\frac{d^2x}{dt^2} = (P_a - P_b)A + f_m - F_L - F_r, \tag{7.110}$$

where P_a and P_b are gauge pressures in the rack cylinder; $F_L = (P_{Lb}-P_{La})A_L$ is the load; P_{Lb} and P_{La} are the loading cylinder absolute pressures; A_L is the loading cylinder bore; and F_f is the combined friction force in the control valve, rack cylinder, and the load cylinders. F_f is nonlinear because of the inherent existence of stiction, Coulomb friction, and viscous effect. The friction model used in simulation is then:

$$F_f \begin{cases} C\left(\dfrac{dx}{dt} - e\right)^2 + d, & \dfrac{dx}{dt} \le e, \\[2ex] B_v\left(\dfrac{dx}{dt} + e\right), & \dfrac{dx}{dt} > e \end{cases}, \tag{7.111}$$

where B_v is the viscous coefficient and C, d, and e are coefficients assessed from experimental measurements on the actual equipment. The manufacturer was able to estimate only the static friction component. However, as indicted by Eq. (7.111), the friction was found to be highly nonlinear. In the modeling of a nonfacet power-steering valve, it was found that the flow coefficient varies with the valve rotation while around a mean value. The flow coefficient C_q at the valve orifice affects the simulation, and its value at the valve neutral position can be determined using the pump model and the valve flow equations. Consider the initial condition in which the valve and the sleeve are neutral to each other; then, the following equations apply:

$$t_1 = t_2 = t_{12} = t_{22},$$

$$Q_p = Q_0 - \frac{P_s}{R_p} = 2C_q K_a t_1 \sqrt{\frac{P_s}{\rho}}, \tag{7.112}$$

where R_p is the pump leakage resistance, which can be determined by experiment. Thus, the flow coefficient at the valve neutral position can be determined using measured data such as:

$$C_q = \frac{Q_0 - \dfrac{P_s}{R_p}}{2K_a t_1 \sqrt{\dfrac{P_s}{\rho}}}. \tag{7.113}$$

The linear movement of the pneumatic loading system follows the steering-rack responses. The pressure changes in the load cylinder are assumed to obey the gas law, and a model is given as follows:

$$\frac{dm_a}{dt} = \frac{1}{RT_p}(P_{La}\dot{V}_{La} + \dot{P}_{La}V_{La}),$$

$$\frac{dm_b}{dt} = \frac{1}{RT_p}(P_{Lb}\dot{V}_{Lb} + \dot{P}_{Lb}V_{Lb}),$$

$$\frac{dm_b}{dt} = C_d C_m a_p p_{La}/\sqrt{T_p},$$

$$\frac{dm_b}{dt} = -\frac{dm_a}{dt}, \tag{7.114}$$

$$\frac{dV_{La}}{dt} = -A_L \frac{dx}{dt},$$

$$\frac{dV_{Lb}}{dt} = A_L \frac{dx}{dt},$$

where R is the universal gas constant, T_p is the absolute temperature, a_p is the restrictor orifice area, $\dot{m}_a$ and $\dot{m}_b$ are the mass flow rates, C_d is the gas discharge coefficient, C_m is the mass flow parameter, V_{La} and V_{Lb} are the cylinder chamber volumes, and A_L is the net cross-sectional area of the piston.

The numerical simulation can be described by a flow chart, Fig. 7.43, in which the core point is to find the actual twist angle of the torsion bar at each iteration. Thus, the orifice sizes can be determined that are needed in the next iteration in processing the dynamic equations of the pinion, rack, and loading system. States of the system are needed in the models of flow-reaction forces, friction force, and pump characteristics. The solution routine was programmed in C code.

7.5.5 Results

The steering-wheel input was applied over a rotation of $360°$, and the actual measured input was also used in the numerical simulation. Results are shown in Fig. 7.44, and it can be seen that the angular steering response follows the input displacement given at the steering column.

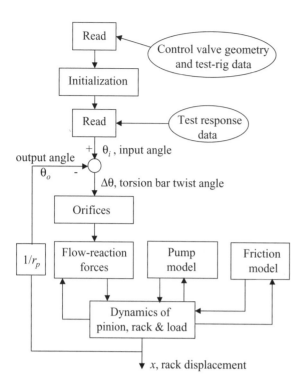

Figure 7.43. Computer numerical simulation flow chart.

The predicted variation in control orifice length t_{12} is shown in Fig. 7.45. The trends of pressures at the pump outlet and in the rack cylinder are in line with measured data and errors can be clearly observed. The effect of nonlinear combined friction is evident regarding a peak pressure at the beginning of the response at each pressure. This peak pressure happens because the rack does not move until the pressure differential across the cylinder can produce a sufficient force to overcome the combined inherent static friction force, which exists between the inner element and valve sleeve and at the hydraulic and pneumatic cylinders. When the rack is starting to move, the friction reduces and the pressure drops. After the initial pressure peak, the pressure keeps increasing until the demanded steering angle profile is reflecting negative acceleration. The changes of pressure at the pump outlet and the driving chamber are of the same sign, whereas the exhaust chamber pressure is in the opposite direction.

The profiles of changes of pump pressure P_s and driving line pressure P_1 compare well between simulation and experiment. However, little changes can be seen in the return line pressure P_2, although its changes in simulation can be observed. This may be due to the effect of line pressure drop, which was not considered in the simulation. Either the valve is in the neutral or the nonneutral condition, and all the oil returns to the reservoir via the same line. If the pump flow rate changes little at different working conditions and return line resistance is not negligible, the return line pressure P_2 at the valve will not be zero but will have little change in the steering process.

7.6 Onboard Electronics (OBE) for Measurement and Intelligent Monitoring

The electrohydraulic control valve is clearly one application on which OBE has made an impact; for example, integral sensors for spool position measurement. In

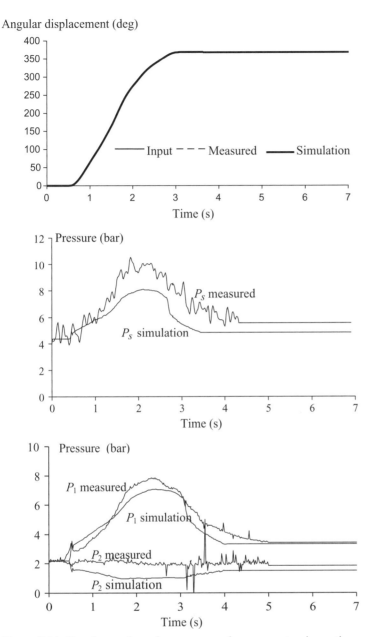

Figure 7.44. Steering angle and pressures under power-steering action.

addition, the use of integrated pressure sensors and communication bus systems allows valves to be used in various control modes and to communicate with other components in the system to optimize performance. The addition of OBE also has the bonus of being able to provide condition monitoring and fault diagnostic information. For example, the awareness of pump deterioration within a manufacturing system can have a significant impact on:

- pump performance awareness and assessment,
- reduced repair time and costs,
- plant operational knowledge,

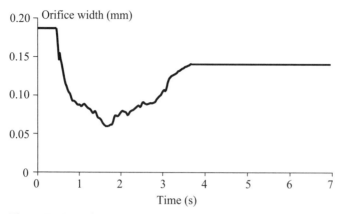

Figure 7.45. Variation in control orifice length t_{12} during power steering.

- maintaining product quality,
- minimizing revenue loss,
- a maintenance cost saving,
- improved plant life, efficiency, and safety assurance,
- reduced personnel risk, and
- maintaining customer relationship

As sensor technology costs are reduced, then fluid power components will inevitably integrate both sensing and wireless signal transmission techniques within a monitoring and control environment. This means that onboard information can be transmitted to external devices and would be significantly enhanced if electrical power could also be onboard-generated. The concept, applied to a pump, is illustrated in Fig. 7.46 and will add significant design and diagnostic power for this application to piston pumps and motors, as shown in Fig. 7.47.

The onboard processor calls for the pressure transient to be measured for one piston, and diagnostics are then performed on the data recorded. In a nine-piston pump, for example, the remaining eight pistons may be used to charge the processor

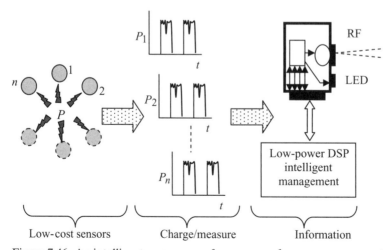

Figure 7.46. An intelligent sensor array for pump performance assessment.

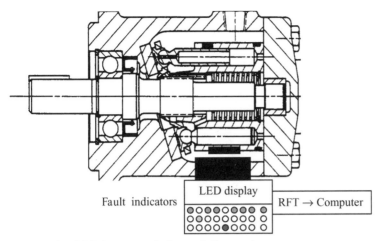

Figure 7.47. OBE for pump design and diagnostics.

card by reversing the role of the pressure transducer from sensing to charging. This is sometimes called *power harvesting*. Alternatively, all sensors may be used to charge the electronics until sufficient power is available to obtain the necessary pressure transient from the selected piston. Each piston condition could be simply displayed on a three-level indicator attached to the pump, and detailed transient pressure information may be stored for postprocessing and so on. Clearly, this approach is too advanced for probably most industrial applications at the time of writing this book, notwithstanding the high instrumentation cost compared with the pump cost. More advanced OBE systems therefore will need to develop new low-cost approaches.

This preliminary study is concerned with developing a technique to measure all piston pressures within an axial piston pump and using OBE.For the first level of appraisal, just three of the nine piston chambers were considered. Measuring individual kidney-port pressures is relatively easy; for example, by using well-established slip-ring technology as indicated briefly in Section 3.1. The actual system used in the author's laboratory is shown in Fig. 7.48.

Miniature piezoelectric pressure transducers were used and were capable of recording the actual ripple content of the signal by selection of the appropriate time constant of the high-impedance electronic circuit. The new design developed in the author's laboratory moves toward the intelligent array concept previously discussed. Specific feature are as follows:

- Each pressure is monitored by an integrated and flexible electronics card that is mounted around the pump barrel.
- The controlling microprocessor is able to select the piston to be monitored.
- Power is fed directly to the electronic card through the pump body by a single bus embedded on the barrel perimeter.
- A wireless RF transmission unit on the board is used to transmit the pressure transient to an outside receiver–PC system.

The approach is shown in Fig. 7.49, again for only three pressure-measuring channels and now using much lower-cost strain gauge pressure transducers.

Pressure transducers with a constant current source were used with associated amplifiers. A 10-bit ADC was selected with a microcontroller with external RAM

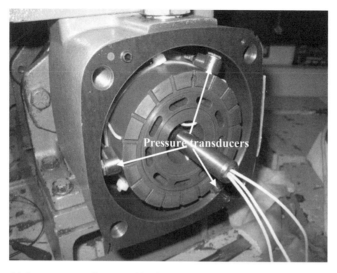

(a) Pressure transducer positioning and connection via the drive shaft

(b) Front plate in place with slip ring

Figure 7.48. Slip-ring assembly for measurement of cylinder pressures.

to control the whole operation of the circuit while having enough storage space for sampled data and the radio link. The receiver module consisted of a radio receiver, microcontroller, and a TTL-to-RS232 converter. To aid the development, process-task-orientated modules were designed and built. These modules, when connected together, allow each individual module to be tested in isolation from the others. The microcontroller module is based around a high-performance RISC CPU, incorporating 22 input–output lines (each capable of sinking or sourcing 25 mA), three timer modules (two 16-bit and one 8-bit), synchronous serial port, and a universal synchronous asynchronous receiver transmitter, all in a small 28-pin SDIP package. The microcontroller was set up to operate at 20 MHz to ensure a good response for the task in hand. Two microcontroller circuits are used within the system as a whole,

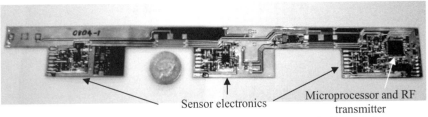

(a) Electronic 3-channel unit with data acquisition and RF transmitter

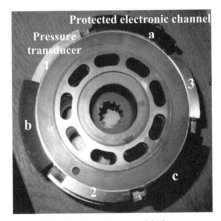

(b) Barrel with assembled electronics and sensors

(c) Assembling the barrel and intelligent sensor array

Figure 7.49. Intelligent sensor array prototype for three pressure-sensing channels.

one in the transmitter unit and one in the receiver unit. Both use the same module configuration just described but with different software control.

The ADC and RAM module is a single module incorporating two circuits. The first circuit, the ADC circuit, uses a 12-bit serial ADC. This circuit is capable of operating with a clock speed of 200 kHz, giving a sampling rate of 12.5 kHz. The unit operates from a 5-V supply and utilizes a 2.7-V reference, giving a 2.7-V voltage range. Only 10 bits of data are used in this application because the microcontroller ignores the first two least-significant bits. When operating with the microcontroller module, a maximum sampling rate of approximately 4 kHz was achieved. The second circuit on this module is the static RAM chip. This device is a serial device offering 2048 bits of memory, which in our case offers the ability to store more than 200 10-bit samples. The serial communications protocol used by this device is the I^2C bus, which when used connected to the microcontroller gives a read–write rate of more than 6 kHz.

The TTL-to-RS232 converter module converts the 5-V logic system used by the microcontroller to the RS232 standard 12-V rail system. The limiting factor on the baud rate used with this module is the oscillator crystal used with the microcontroller. The crystal used was a 20-MHz crystal and this placed a baud rate limit on the microcontroller circuit of 38,400 bps. The chip set used in this module was a transceiver chip. The radio transmitter and receiver pair used in this design operates on the 433.92-MHz band and is capable of digital data rates of 128 kbps but significantly reduced in practice because of the design and coding techniques used. A signal transmitting aerial is placed around the inside of the casing but, in practice,

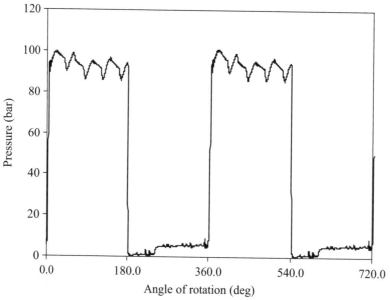

Figure 7.50. A pressure measurement for a pump pressure of 100 bar.

it was found unnecessary to connect the receiver to it up to typically 2 m away from the pump. A measurement is shown in Fig. 7.50.

All three measured pressure transients had the same waveform for this relatively new pump, and some issues that arise are as follows:

- Problems with noise at high pressures need to be overcome, with alternative ways required for powering the electronics.
- Notice that the ripple frequency within the kidney port is 9 × pump speed for this nine-piston pump; this pressure amplitude is greater than expected. Each half-cycle of the pump barrel illustrates 4.5 cycles of pressure fluctuation.
- Also notice the slight increase in inlet pressure, which requires further investigation.

Next-generation pressure arrays will allow pressure sensors, particularly piezoelectric type and not interrogated, to reverse their role and to be used as charging elements. The OBE system will then be completely autonomous, self-powering, and have the ability to be externally controlled. It also may be possible to use internally generated information to optimize the operating characteristic of the pump in real time by means of the control valve adjusting circuits for pump displacement and/or operating pressure; the concept of intelligent control will eventually arrive.

7.7 References and Further Reading

Acarman T and Redmill KA [2002]. A robust controller design for drive by wire hydraulic power steering system. In *Proceedings of the American Control Conference*, American Automatic Control Council, 2522–2527.

Adams FJ [1983]. Power steering "road feel." Presented at the SAE Passenger Car Meeting, Paper 830998, Dearborn, MI.

Appleyard M and Wellstead PE [1995]. Active suspensions: Some background. *IEE Proc. Control Theory Applications*, 142, 123–132.

Bergada JM, Haynes JM, Watton J [2008]. Leakage and groove pressure of an axial piston pump slipper with multiple lands. *Tribol. Trans.* 51, 469–482.

Bergada JM and Watton J [2002]. A direct leakage flow rate calculation method for axial pump grooved pistons and slippers, and its evaluation for a 5/95 fluid application. In *Proceedings of the 5th JFPS International Symposium on Fluid Power*, Japan Fluid Power Society.

Bergada JM and Watton J [2002]. Axial piston pump slipper balance with multiple lands. In *Proceedings of the ASME International Mechanical Engineering Congress and Exposition, IMECE 2002*, American Society of Mechanical Engineering, Vol. 2, Paper 39338.

Bergada JM and Watton J [2005]. Force and flow through hydrostatic slippers with grooves. In *Proceedings of the 8th International Symposium on Fluid Control, Measurement and Visualization FLUCOME 2005*, China Aerodynamics Research Society, Paper 240.

Birsching, JE [1999]. Two-dimensional modelling of a rotary power-steering valve. SAE Technical Paper 1999–01-0396, International Congress and Exposition, Society of Automotive Engineers, 1–4.

Darling J, Tilley DG, Hickson LR [1999]. A centralized hydraulic system for passenger cars. *Proc. Inst. Mech. Eng.* D 213, 425–434.

Elbeheiry ED, Karnopp DC, Elaraby ME, Abdelraaout AM [1995]. Advanced ground vehicle suspension systems: A classified bibliography. *Journal of Vehicle Syst. Dyn.* 24(3), 231–258.

Engelmann GH and Rizzoni G [1993]. Including the force generation process in active suspension control formulation. In *Proceedings of the American Control Conference*, American Automatic Control Council, 701–705.

Fisher MJ [1962]. A theoretical determination of some characteristics of a tilted hydrostatic slipper bearing. BHRA Rep. RR 728, April 1962.

Freeman P [1962]. *Lubrication and friction*. Pitman.

Harris RM, Edge KA, Tilley DG [1993]. Predicting the behaviour of slipper pads in swashplate-type axial piston pumps. In *Proceedings of the ASME Winter Annual Meeting*, American Society of Mechanical Engineering, 1–9.

Harris RM, Edge KA, Tilley DG [1996]. Slipper pads in swashplate-type axial piston pumps. *ASME J. Dyn. Syst. Meas. Control* 41–47.

Haynes JM [2007]. Axial piston pump leakage modelling and measurement. Ph.D. thesis, Cardiff University, School of Engineering.

Heinreichson N and Santos IF [2008]. Reducing friction in tilting-pad bearings by use of enclosed recesses. *ASME J. Tribol.* 130, 011009/1–9.

Hooke CJ and Kakoullis YP [1978]. The lubrication of slippers on axial piston pumps. In *Proceedings of the 5th International Fluid Power Symposium*, British Hydrodynamics Research Association, B2–13–26.

Hooke CJ and Kakoullis YP [1981]. The effects of centrifugal load and ball friction on the lubrication of slippers in axial piston pumps. In *Proceedings of the 6th International Fluid Power Symposium*, British Hydrodynamics Research Association, 179–191.

Hooke CJ and Kakoullis YP [1983]. The effects of nonflatness on the performance of slippers in axial piston pumps. *Proc. Inst. Mech. Eng.* 197 C, 239–247.

Hooke CJ and Li KY [1988]. The lubrication of overclamped slippers in axial piston pumps centrally loaded behaviour. *Proc. Inst. Mech. Eng.* 202 C, 287–293.

Hooke CJ and Li KY [1989]. The lubrication of slippers in axial piston pumps and motors. The effect of tilting couples. *Proc. Inst. Mech. Eng.* 203 C, 343–350.

Howe JG, Rapp MY, Jang B-C, Woodburn CM, Guenther DA, Heydinger GJ [1997]. Improving steering feel for the national advanced driving simulator. SAE Special Publications, Paper Number 970567, Vol. 1228, 135–145.

Hrovat D [1997]. Survey of advanced suspension developments and related optimal control applications. *Automatica* 33, 1781–1817.

Iboshi N and Yamaguchi A [1982]. Characteristics of a slipper bearing for swash-plate-type axial piston pumps and motors, theoretical analysis. *Bull. JSME* 25, 1921–1930.

Iboshi N and Yamaguchi A [1983]. Characteristics of a slipper bearing for swash-plate-type axial piston pumps and motors, experimental. *Bull. JSME* 26, 1583–1589.

Kazama T [2005]. Numerical simulation of a slipper model for water hydraulic pumps/motors in mixed lubrication. In *Proceedings of the 6th JFPS International Symposium on Fluid Power,* Japan Fluid Power Society, 509–514.

Kobayashi S, Hirose M, Hatsue J, Ikeya M [1988]. Friction characteristics of a ball joint in the swashplate-type axial piston motor. In *Proceedings of the 8th International Symposium on Fluid Power*, J2-British Hydromachanics Research Association, 565–592.

Koc E and Hooke CJ [1996]. Investigation into the effects of orifice size, offset, and overclamp ratio on the lubrication of slipper bearings. *Tribol. Int.* 29, 299–305.

Koc E and Hooke CJ [1997]. Considerations in the design of partially hydrostatic slipper bearings. *Tribol. Int.* 30, 815–823.

Koc E, Hooke CJ, Li KY [1992]. Slipper balance in axial piston pumps and motors. *Trans. ASME J. Tribol.* 114, 766–772.

Kumar S, Bergada JM, Watton J [2009]. Axial piston pump grooved slipper analysis by CFD simulation of three-dimensional NVS equation in cylindrical coordinates. *Comput. Fluids,* 38(3), 648–663.

Li KY and Hooke CJ [1991]. A note on the lubrication of composite slippers in water-based axial piston pumps and motors. *Wear* 147, 431–437.

Long, MR [1999]. Isolating hydraulic noise from mechanical noise in power rack & pinion steering systems. SAE Technical Paper 1999–01-0397, International Congress and Exposition, Society of Automotive Engineers.

Mrad RB et al. [1991]. A nonlinear model of an automobile hydraulic active suspension. *ASME Adv. Automation Technol. DE* 40, 347–359.

Nelson RM [1990]. The microcomputer modelling and control of a flow bypass valve operated hydraulic press rig. Ph.D. thesis, Cardiff University, School of Engineering.

Nishimura, S and Matsunaga [2000]. Analysis of response lag in hydraulic power-steering system. *J. Soc. Automotive Eng. Rev.* 21, 41–46.

Rajamani R and Hedrick JK [1994]. Performance of active automotive suspensions with hydraulic actuators: Theory and experiment. In *Proceedings of the American Control Conference*, American Automatic Control Council, 1214–1218.

Rajamani R and Hedrick JK [1995]. Adaptive observers for active automative suspensions: Theory and experiment. *IEEE Trans. Control Syst. Technol.* 3, 86–93.

Sharp RS and Crolla DA [1987]. Road vehicle suspension system design: A review. *J. Veh. Syst. Dyn.* 16, 167–192.

Surawattanawan P [2000]. The influence of hydraulic system dynamics on the behaviour of a vehicle active suspension. Ph.D. thesis, Cardiff University, School of Engineering.

Synflex Hydraulic Hose Catalogue. Eaton Corporation Engineered Polymer Products Division, Aurora, OH 44202.

Takahashi K and Ishizawa S [1989]. Viscous flow between parallel disks with time varying gap width and central fluid source. In *Proceedings of the JHPS International Symposium on Fluid Power*, Japan Fluid Power Society, 407–414.

Thompson AG and Chaplin PM [1996]. Force control in electrohydraulic active suspensions. *Journal of Vehicle Syst. Dyn.* 25, 185–202.

Thompson AG and Davis BR [1992]. A technical note on the Lotus suspension patents. *Vehicle Syst. Dyn.* 20, 381–383.

Thomson, AG [1976]. An active suspension with optimal state feedback. Vehicle System Dynamics, 5, 187–203.

Tsuta T, Iwamoto T, Umeda T [1999]. Combined dynamic response analysis of a piston-slipper system and lubricants in hydraulic piston pump. Emerging technologies in fluids, structures, and fluid/structure interactions. *ASME* 396, 187–194.

Watton, J [1988]. The design of a single-stage relief valve with directional damping. *J. Fluid Control* 18(2), 22–35.

Watton, J [1989]. Transient analysis of a back-pressure controlled relief valve for controlling the supply pressure rate of rise. In *Proceedings of the 1st JHPS International Symposium on Fluid Power*, Japan Fluid Power Society, 373–378.

Watton J [1990]. *Fluid Power Systems: Modeling, Simulation, Analog and Microcomputer Control*, Prentice-Hall.

Watton J, Holford KM, Surawattanawan P [2001]. Electrohydraulic effects on the modelling of a vehicle active suspension. *Proc. Inst. Mech. Eng. J. Syst. Control Eng.* 215 D, 1077–1092.

Watton J, Holford KM, Surawattanawan P [2004]. The application of a programmable servo-controller to state control of an electrohydraulic active suspension. *Proc. Inst. Mech. Eng. J. Automobile Eng.* 218 D, 1367–1377.

Watton J and Nelson RJ [1993]. Evaluation of an electrohydraulic forge valve behaviour using a CAD package. *Appl. Math. Model.* 17, 355–368.

Wieczoreck U and Ivantysynova M [2000]. CASPAR-A computer-aided design tool for axial piston machines. In *Proceedings of the Power Transmission Motion and Control International Workshop*, Professional Engineering Publishing Ltd., 113–126.

Wieczoreck U and Ivantysynova M [2002]. Computer-aided optimization of bearing and sealing gaps in hydrostatic machines: The simulation tool CASPAR. *Int. J. Fluid Power* 3(1), 7–20.

Williams DA and Wright PG [1986]. U.S. Patent 4,625,993. Vehicle suspension system.

Williams RA [1997]. Automotive active suspensions part 1: Basic principles. *Proc. Inst. Mech. Eng.* D 211, 415–426.

Williams RA [1997]. Automotive active suspensions part 2: Practical considerations. *Proc. Inst. Mech. Eng.* D 211, 427–444.

Wong, T [2001]. Hydraulic power-steering system design and optimization simulation. Technical Paper 2001–01-0479, SAE 2001 World Congress.

Wright PG and Williams DA [1984]. The application of active suspension to high performance road vehicles. *Inst. Mech. Eng.* C 239, 23–28.

Xanalog Corporation, 799 Middlesex Turnpike, Billerica, MA 01821; web@xanalog.com.

Xue Y and Watton J [2005]. Modelling of a hydraulic power-steering system. *Int. J. Vehicle Des.* 38, 162–178.

Index